VANADIUM COMPOUNDS: BIOCHEMICAL AND THERAPEUTIC APPLICATIONS

Developments in Molecular and Cellular Biochemistry

Series Editor: Naranjan S. Dhalla, Ph.D., FACC

1. V.A. Najjar (ed.): *Biological Effects of Glutamic Acid and Its Derivatives.* 1981 ISBN 90-6193-841-4
2. V.A. Najjar (ed.): *Immunologically Active Peptides.* 1981 ISBN 90-6193-842-2
3. V.A. Najjar (ed.): *Enzyme Induction and Modulation.* 1983 ISBN 0-89838-583-0
4. V.A. Najjar and L. Lorand (eds.): *Transglutaminase.* 1984 ISBN 0-89838-593-8
5. G.J. van der Vusse (ed.): *Lipid Metabolism in Normoxic and Ischemic Heart.* 1989 ISBN 0-7923-0479-9
6. J.F.C. Glatz and G.J. van der Vusse (eds.): *Cellular Fatty Acid-Binding Proteins.* 1990 ISBN 0-7923-0896-4
7. H.E. Morgan (ed.): *Molecular Mechanisms of Cellular Growth.* 1991 ISBN 0-7923-1183-3
8. G.J. van der Vusse and H. Stam (eds.): *Lipid Metabolism in the Healthy and Diseased Heart.* 1992 ISBN 0-7923-1850-1
9. Y. Yazaki and S. Mochizuki (eds.): *Cellular Function and Metabolism.* 1993 ISBN 0-7923-2158-8
10. J.F.C. Glatz and G.J. van der Vusse (eds.): *Cellular Fatty-Acid-Binding Proteins, II.* 1993 ISBN 0-7923-2395-5
11. R.L. Khandelwal and J.H. Wang (eds.): *Reversible Protein Phosphorylation in Cell Regulation.* 1993 ISBN 0-7923-2637-7
12. J. Moss and P. Zahradka (eds.): *ADP-Ribosylation: Metabolic Effects and Regulatory Functions.* 1994 ISBN 0-7923-2951-1
13. V.A. Saks and R. Ventura-Clapier (eds.): *Cellular Bioenergetics: Role of Coupled Creatine Kinases.* 1994 ISBN 0-7923-2952-X
14. J. Slezák and A. Ziegelhöffer (eds.): *Cellular Interactions in Cardiac Pathophysiology.* 1995 ISBN 0-7923-3573-2
15. J.A. Barnes, H.G. Coore, A.H. Mohammed and R.K. Sharma (eds.): *Signal Transduction Mechanisms.* 1995 ISBN 0-7923-3663-1
16. A.K. Srivastava and J.-L. Chiasson (eds.): *Vanadium Compounds: Biochemical and Therapeutic Applications.* 1995 ISBN 0-7923-3763-8

KLUWER ACADEMIC PUBLISHERS – DORDRECHT / BOSTON / LONDON

Vanadium Compounds: Biochemical and Therapeutic Applications

Edited by

ASHOK K. SRIVASTAVA

Centre de Recherche,
Hotel-Dieu de Montreal,
3850, rue Saint Urbain,
Pavillon Marie de la Ferre,
Montreal, Quebec H2W 1T8, Canada

and

JEAN-LOUIS CHIASSON

Hotel-Dieu de Montreal,
Centre de Recherche,
3850, rue Saint Urbain,
Pavillon Marie de la Ferre,
Montreal, Quebec H2W 1T8, Canada

Kluwer Academic Publishers
Dordrecht / Boston / London

Library of Congress Cataloging-in-Publication Data

Vanadium compounds: biochemical and therapeutic applications / edited by Ashok K. Srivastava, Jean-Louis Chiasson.
p. cm. – (Developments in molecular and cellular biochemistry; v. 16)
ISBN 0-7923-3762-X
1. Vanadium–Physiological effect–Congresses. 2. Vanadium-Therapeutic use–Testing–Congresses.
I. Srivastava, Ashok, K. II. Chiasson, Jean-Louis. III. Series.
QP535.V2V35 1996
612'.01524–dc20 95–38780
CIP

ISBN 0-7923-3762-X

Published by Kluwer Academic Publishers,
P.O. Box 17, 3300 AA Dordrecht, The Netherlands.

Kluwer Academic Publishers incorporates
the publishing programmes of
D. Reidel, Martinus Nijhoff, Dr W. Junk and MTP Press.

Sold and distributed in the U.S.A. and Canada
by Kluwer Academic Publishers,
101 Philip Drive, Norwell, MA 02061, U.S.A.

In all other countries, sold and distributed
by Kluwer Academic Publishers Group,
P.O. Box 322, 3300 AH Dordrecht, The Netherlands.

Printed on acid-free paper

Printed in the Netherlands

Molecular and Cellular Biochemistry:

An International Journal for Chemical Biology in Health and Disease

CONTENTS VOLUME 153, Nos. 1-2, 1995

VANADIUM COMPOUNDS: BIOCHEMICAL AND THERAPEUTIC APPLICATIONS
A.K. Srivastava and J.-L. Chiasson, guest editors

Molecular and Cellular Biochemistry **153:** 1, 1995.

Preface

This focussed issue contains invited reviews as well as original research papers presented at the Vanadium Symposium 1994, held on July 29–31, 1994 in Montreal.

Vanadium is a trace element and its compounds have been shown to exert a wide variety of insulin like effects including the ability to lower hyperglycemia in several experimental models of diabetes mellitus. Because of the possibility that vanadium compounds may be able to serve as potential therapeutic agents for the treatment of diabetes and possibly other diseases, this trace element has attracted the attention of biomedical researchers from a variety of fields. Therefore the Vanadium Symposium 1994 was organized to facilitate exchange of ideas and increase interaction among researchers of different disciplines actively engaged in studying the biological actions of vanadium compounds. The Organizing Committee for this symposium in addition to us included Drs. John H. McNeill (University of British Columbia) and Barry I. Posner (McGill University). The major sponsors of this symposium were: Canadian Diabetes Association, Fonds de la recherche en sante du Quebec, Juvenile Diabetes Foundation of Canada, Marion Merrell Dow Canada, McGill University, Medical Research Council of Canada, Medisense Canada Inc., Merek Frosst, Canada Inc. and Miles Canada Inc. We thank these institutions and organizations for financial assistance.

The papers in this issue are written by leading vanadium researchers and are grouped in 3 main sections. The first section deals with the chemistry, the second section contains studies on the biochemical and physiological aspects, and finally, the last section addresses the potential therapeutic use and toxic effects of vanadium compounds. We hope that this special issue will serve as a good source of information on vanadium chemistry and biology.

We wish to thank all the contributors for their help and cooperation. We also wish to thank Mrs Susanne Bordeleau-Chenier for secreterial help. We are grateful to Dr. N.S. Dhalla, Editor-in-chief of Molecular and Cellular Biochemistry for his interest in the Vanadium Symposium 1994 and for agreeing to publish the proceedings of this symposium in a special issue of the journal.

ASHOK K. SRIVASTAVA
AND JEAN-LOUIS CHIASSON
Centre de Recherche, Hôtel Dieu de Montréal and
Department of Medicine, University of Montréal,
Montréal (Québec) H2W 1T8, Canada

PART I

CHEMISTRY

Molecular and Cellular Biochemistry **153:** 5–15, 1995.

The chemistry of peroxovanadium compounds relevant to insulin mimesis

Alan Shaver, Jesse B. Ng, David A. Hall and Barry I. Posner
Departments of Chemistry and Medicine, McGill University, Montréal, Québec, H3A 2K6, Canada

Abstract

The inorganic coordination chemistry of peroxovanadium compounds relevant to insulin mimesis is reviewed. The structure and kinetic reactivity of solutions of vanadate anion, vanadyl complexes and peroxovanadate complexes are briefly compared. Peroxovanadium compounds contain an oxo group, one or two peroxo ligands (O_2^{2-}) and an ancillary ligand which is usually bidentate. These compounds approximate a trigonal bipyramidal structure which can be divided conceptually into a polar 'oxo' half and a relatively non-polar organic half. This presents a number of interesting design variations which are discussed with respect to the development of a rudimentary structure-activity correlation of insulin mimetic ability. (Mol Cell Biochem **153:** 5–15, 1995)

Key words: peroxovanadium compounds, structure-activity relation of vanadium, insulin mimetic activity

Abbreviations: phen – 1,10-phenanthroline; 4,7-Me_2phen – 4,7-dimethyl-1,10-phenanthroline; 3,4,7,8-Me_4phen – 3,4,7,8-tetramethyl-1,10-phenanthroline; 5-CH_3phen – 5-methyl-1,10-phenanthroline; 5-NO_2phen – 5-nitro-1,10-phenanthroline; 5-NH_2phen – 5-amino-1,10-phenanthroline; bipy – 2,2′-bipyridine; bipyH – 2,2′-bipyridinium; 4,4′-Me_2bipy – 4,4′-dimethyl-2,2′-bipyridine; bipy-4,4′-(COO)2 – 2,2′-bipyridine-4,4′-dicarboxylato; pic – pyridine-2-carboxylato; 3-OHpic – 3-hydroxypyridine-2-carboxylato; 3-acetpic – 3-acetatoxypyridine-2-carboxylato; m,n-pdc – pyridine-m,n-dicarboxylato; ox – oxalato; pzc – pyrazine-2-carboxylato; 3-NH_2pzc – 3-aminopyrazine-2-carboxylato; 4OH-2,6pdc – 4-hydroxy-2,6-pyridinedicarboxylato; IDA –iminodiacetato; cit – citrato; EDTA – ethylenediaminetetraacetato; HEDTA – ethylenediaminetetraacetic acid; isoquin – isoquinoline-2-carboxylato; quin – quinolato; NTA – nitrilotriacetato; glyH – glycine; cystH – cysteine; nicH – nicotinic acid; Hheida – *N*-(2-hydroxyethyl)iminodiacetato

Introduction

Swedish chemist Nils Sefstrom was so impressed by the rainbow of colours possessed by compounds of the new element he discovered in 1830 that he named the element after Vanadis, the Norse goddess of beauty. Actually Sefstrom had rediscovered erythronium which had been discovered by del Rio in 1801, but del Rio had, mistakenly, retracted his claim! Vanadium is a common element but usually occurs in relatively low concentration. The average abundance in the earth's crust is approx. 0.02% and, while there are over 65 vanadium-containing minerals, few deposits are concentrated enough to mine. As a result vanadium is obtained commercially as a by-product from the processing of the ores of other metals [1].

Vanadium compounds are also found in the biosphere. Vanadium is found in the specialized blood cells of sea squirts in concentrations as high as 0.15 M in the form of tunichrome which is thought to play an important role in cell metabolism [3]. Vanadoenzymes (haloperoxidases) are found in marine algae [4–7] and certain nitrogenases of nitrogen-fixing bacteria (Azotobacter) [8, 9] contain vanadium. In mammals it is an ultratrace element which is widely distributed in tissues [10]. The total body pool is estimated to be about 100 μg based on a daily intake of 10–60 μg [10]. The bioinorganic chemistry of vanadium has been extensively reviewed [11, 12].

In 1979 vanadate solutions (V+5) were shown to exert insulin-like effects on glucose metabolism in rat diaphragms *in vitro* [13]. This was followed by reports of the insulin

Address for offprints: A. Shaver, Departments of Chemistry and Medicine, McGill University, Montréal, Québec, H3A 2K6, Canada mimetic actions of vanadate on isolated adipocytes [14, 15]. A major breakthrough was the demonstration in 1985 that oral administration of vanadate solutions lowered blood sugar in diabetic rats [16]. The relatively high toxicity of vanadate combined with the postulate that it is reduced intracellularly to vanadyl (V+4) which may be responsible for insulin-like effect [15], led to the development of new vanadyl reagents in the late 1980's and early 1990's [17–20]. At about the same time aqueous solutions containing peroxovanadium complexes [21] pV, and later discrete peroxovanadium complexes themselves [22] were shown to have strong *in vitro* and *in vivo* insulin mimetic properties. This review sketches the inorganic coordination chemistry of peroxovanadate complexes which we hope will prove useful to the study of their insulin mimetic properties.

Coordination theory

Every metal ion has an outer coordination sphere and an inner coordination sphere (Fig. 1). Species such as ammonia, chloride ion, water and peroxide ion may occupy either sphere. In the inner sphere ligands donate a free lone pair of electrons to the metal and form a bond with a high degree of covalent character. The inner sphere usually has one of a handful of well defined geometries (Fig. 2). The outer sphere is a less well-defined environment where molecules interact relatively weakly with the metal-coordinated ligands via electrostatic forces or hydrogen bonding, the latter being most important in biochemistry. Coordination compounds are well suited to X-ray crystallographic structure determinations and many have been characterized to a high level of resolution. The geometry of the ligands in the inner sphere is usually readily recognizable while the counter ions and any other included molecules (ie solvent) are distributed regularly throughout the lattice but often with no recognizable regular structure with respect to the metal atom. In solution the ligands of 'inert' complexes do not enter or leave the inner coordination sphere rapidly. On the other hand the ligands in 'labile' complexes enter and leave the inner sphere easily. The molecules and ions in the outer sphere are only weakly associated and they come and go easily.

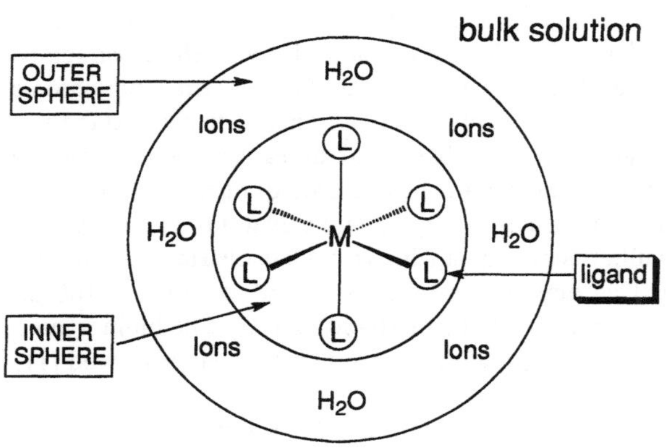

Fig. 1. A diagrammatic representation of the essential components of a transition metal complex salt.

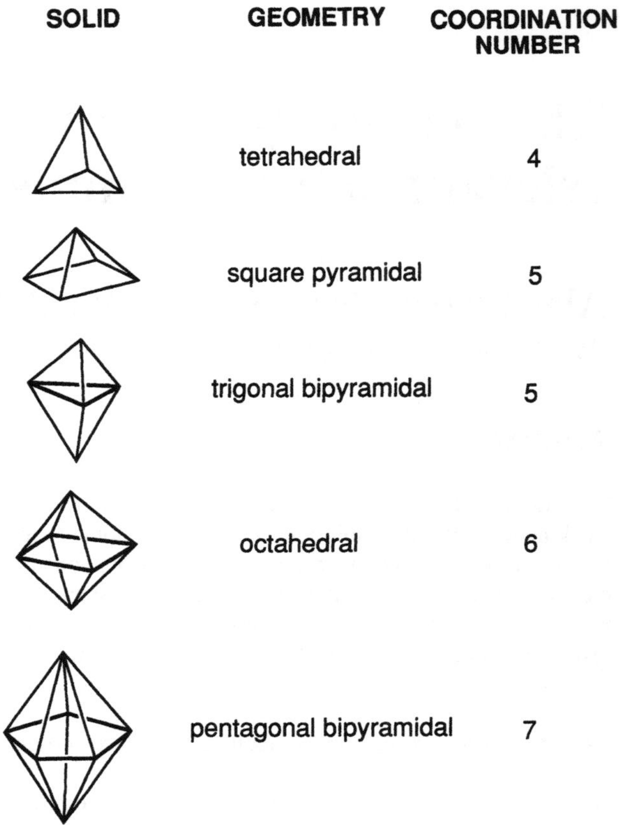

Fig. 2. Geometrical shapes associated with coordination numbers 4–7.

Vanadate solutions

The ^{51}V NMR spectrum of a solution vanadate at neutral pH will normally reveal at least four different peaks [23]. These correspond to $H_2VO_4^-$ or HVO_4^{2-}, $V_2O_7^{4-}$, $V_4O_{12}^{4-}$ and $V_5O_{15}^{5-}$ which result from a series of complex, rapid hydrolysis and polymerization reactions which are concentration and pH dependent. Thus vanadate is a very labile system which also rapidly interacts with other potential ligands such as nitrogen bases and especially OH groups to give mixtures of complexes that are labile and difficult to isolate. The complex aqueous chemistry of vanadate solutions [23] and its relevance to biochemical studies has been reviewed by Crans *et al.* [24]. It is usually very difficult to know for certain the nature of the vanadate complexes present in biological tests and applications. Vanadate can form complexes with many compounds used in buffers (e.g. EDTA, tricine, citrate, phosphate) and thus Hepes is recommended as the buffer of choice [23, 24]. This reactivity has been exploited by using desferrioxamine-B, a powerful natural chelating agent, to reverse vanadate inhibition of an enzyme [25].

Vanadium (IV)

Reduction of vanadate by mild reagents in aqueous media give vanadyl species ($VO(H_2O)_5^{2+}$) which are said to be less toxic but poorly absorbed from the intestine [17]. Discrete complexes of VO^{2+} (vanadyl ion) are very common so it is not surprising that inert and well characterized compounds have been prepared possessing ligands designed to increase both solubility and absorption. The complex bis(maltolato)oxovanadium (IV) [18] 1, is typical, having a square pyramidal geometry with the oxo ligand (O^{2-}) in the apical site (Fig. 3). The oxophilicity of the element is apparent in this and in other vanadyl complexes recently shown to have insulin mimetic properties. The maltol ligand in **1** is naturally occurring while other insulin mimetic vanadyl complexes can contain oxygen, sulfur and nitrogen donors. Vanadyl complexes are paramagnetic and thus not suitable for NMR studies; however, they usually give excellent ESR spectra which are very useful for characterization [26].

Peroxovanadium complexes [27]

Synthesis

The combination of vanadate and H_2O_2 under physiological conditions generates several different peroxovanadium species in equilibrium with one another depending upon the pH of the solution, and the concentrations of both vanadate and H_2O_2. The equilibria are less complex than those for solutions of vanadate and are generally well understood [28]. At low vanadate and low H_2O_2 concentrations, mono and diperoxo products are favoured but an increase in both vanadate and H_2O_2 concentrations leads to triperoxo and tetraperoxovanadates. Most species have pH-dependent ^{51}V NMR chemical shifts especially $V(O_2)^{3+}$ and $V(O_2)_2^+$, arising from protonation and deprotonation reactions. In aqueous solution, the number of peroxo ligands per vanadium increases with alkalinity; on the other hand, a low pH favours polymerization and decreases the number of peroxo groups per vanadium atom. Increasing the concentration of hydrogen peroxide decreases the degree of polymerization. Thus, addition of H_2O_2 to V_2O_5 results in an acidic red solution, because of the formation of $VO(O_2)^+$. A mixture of V_2O_5 and KOH yields an almost colourless solution which, upon addition of H_2O_2, gives a yellow-orange solution believed to be $[VO(O_2)_2(H_2O)_2]^-$, and at high pH the latter is converted to $VO(O_2)_3^{3-}$ and tetraperoxovanadates.

The study of solutions containing vanadate and H_2O_2 can be complicated by the vanadate-catalyzed disproportionation of H_2O_2 to O_2 and H_2O; however, the stability of the complexes is enhanced by the use of suitable ancillary ligands. Thus, in moderately basic medium (pH ~ 7–8), addition of a ligand to $[VO(O_2)_2(H_2O)_2]^-$ results in substitution of H_2O and in the formation of complexes which are represented by the formulae: $M_n[VO(O_2)_2(L)]$, $M_n[VO(O_2)_2(LL')]$ or $M_n[VO(O_2)(H_2O)(LL'L'')]$ where M = NH_4 or K; n = 1–3; and L, LL′ and LL′L″ are mono-, bi- and tridentate ancillary ligands. Ligands that form stable complexes include oxalic acid, 1,10-phenanthroline, 2,2′-bipyridine, picolinic acid and their derivatives which give diperoxovanadates; and 2,6-pyridinedicarboxylic acid which forms a monoperoxo complex. The trisperoxo compounds $K[V(O_2)_3phen]\cdot H_2O$ and $K[V(O_2)_3bipy]\cdot H_2O$ have been reported [29] but these appear to be misidentifications [30] of the appropriate oxobisperoxo compounds.

The formation of bisperoxovanadates is favoured at pH 7–8, while monoperoxovanadates are synthesized in an acidic environment (pH 3–5). Most peroxovanadates are stabilized with a bidentate ligand which forms a 5-membered ring with the vanadium atom. The relatively high stability observed for complexes with the 2,6-pyridinedicarboxylato ligand is due to a tridentate coordination mode of the ligand in the pentagonal plane of the complex. Peroxovanadium compounds containing amino acid or peptide ligands are reported but they are unstable [31].

Table 1 lists almost 60 peroxovanadium complexes reported in the literature or recently prepared in our laboratory. The complexes are usually yellow or red free-flowing crystalline powders. They are stable indefinitely when stored cold in the dark. Those compounds isolated as salts are freely soluble in H_2O but insoluble in most organic solvents. Aqueous solutions (0.1 M) are generally tasteless. Those few non ionic pV compounds whose water solubility is small are often soluble in organic solvents. The preparation, isolation and storage of pV compounds seems to present few hazards; however, it is important to note the toxicity of V_2O_5 dust, the strong oxidizing power of H_2O_2 and the fact that these com-

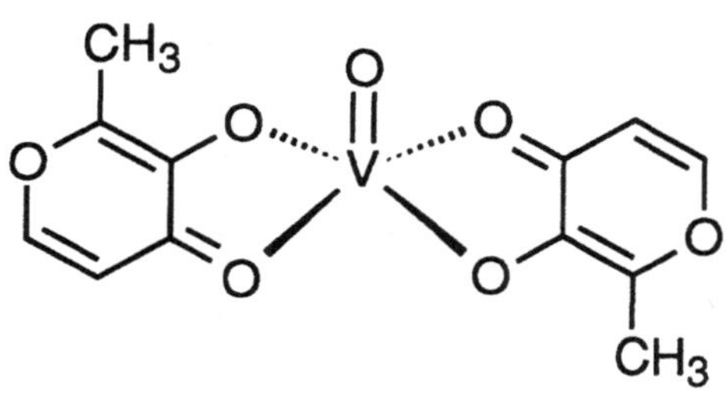

1

Fig. 3. The structure of bis(maltolato)oxovanadium(IV).

Table 1. A list of peroxovanadium compounds with their abbreviations, selected ^{51}V NMR data and references to their preparation and X-ray structural determination, where available

Abbrevn.	Complex	$(^{51}V)^{a}$	References (X-Ray)
bpV(phen)	$K[VO(O_2)_2(phen)]\cdot 3H_2O$	–746	32
bpV(bipy)	$K[VO(O_2)_2(bipy)]\cdot 5H_2O$	–747	32, (33)[b]
bpV(ox)	$K_3[VO(O_2)_2(ox)]\cdot 2H_2O$	–739	32, 34, (35)
$bpV(NH_3)$	$NH_4[VO(O_2)_2(NH_3)]$	–711, –748, –753	36, (37)
mpV(2,6-pdc)	$NH_4[VO(O_2)(2,6\text{-}pdc)(H_2O)]\cdot H_2O$	–598	38, (39)
mpV(pic)	$VO(O_2)(pic)(H_2O)_2$	–600	(40)
bpV(pic)	$K_2[VO(O_2)_2(pic)]\cdot 2H_2O$	–744	41, (42)
$bpV(Me_2phen)$	$K[VO(O_2)_2(4,7\text{-}Me_2phen)]\cdot 0.5H_2O$	–748	
mp(but)V(2,6pdc)	$VO(OO\text{-}Bu^t)(H_2O)(2,6\text{-}pdc)$	–597	(43)
mpV(ox)	$K_3[VO(O_2)(ox)_2]\cdot 0.5H_2O$	–592	(44)
mpV(picbipy)	$VO(O_2)(pic)(bipy)\cdot H_2O$	–579	(45)
mpV(picphen)	$VO(O_2)(pic)(phen)\cdot H_2O$	–581	
bpV(OHpic)	$K_2[VO(O_2)_2(3\text{-}OHpic)]\cdot H_2O$	–741	(42)
$bpV(Me_4phen)$	$K[VO(O_2)_2(3,4,7,8\text{-}Me_4phen)]\cdot 5H_2O$	–745	41
$bpV[bipy(CO_2^-)]$	$K_3[VO(O_2)_2\{bipy\text{-}4,4'\text{-}(COO)_2\}]$	–747	
$bpV(CO_3)$	$K_3[VO(O_2)_2(CO_3)]\cdot 3H_2O$	–762	34, 46, (47)
bpV(2,3-pdc)	$K_3[VO(O_2)_2(2,3\text{-}pdc)]\cdot 2H_2O$	–743	48
bpV(2,4-pdc)	$K_3[VO(O_2)_2(2,4\text{-}pdc)]\cdot 3.25H_2O$	–743	(48)
bpV(2,5-pdc)	$K_3[VO(O_2)_2(2,5\text{-}pdc)]\cdot 2H_2O$	–742	48
bpV(Mephen)	$K[VO(O_2)_2(5\text{-}Mephen)]\cdot 2H_2O$	–743	
bpV(NH2phen)	$K[VO(O_2)_2(5\text{-}NH_2phen)]\cdot 0.5H_2O$	–744	
$bpV(NO_2phen)$	$K[VO(O_2)_2(5\text{-}NO_2phen)]\cdot 2H_2O$	–739	(49)
mpV(OHpdc)	$NH_4[VO(O_2)(H_2O)(4\text{-}OH\text{-}2,6\text{-}pdc)]\cdot H_2O$	–599	
bpV(acetpic)	$K_3[VO(O_2)_2(3\text{-}acetpic)]\cdot 2H_2O$	–745	(48)
mpV(quin)	$NH_4[VO(O_2)(isoquin)]\cdot 5H_2O$	–746	41
bpV(pzc)	$K_2[VO(O_2)_2(pzc)]\cdot xH_2O$	–732	40
$bpV(3\text{-}NH_2pzc)$	$K_2[VO(O_2)_2(3\text{-}NH_2pzc)]\cdot xH_2O$	–739	
mpV(ida)	$M[VO(O_2)(IDA)]\cdot 0.5H_2O$; M: K, NH_4		
mpV(cit)	$K_2[VO(O_2)(cit)]_2\cdot 2H_2O$	–541, –548	51
bpV(F)	$M_2[VO(O_2)_2F]$ (M: Na, K, Rb, Cs, NH_4)		52, (53)
bpV(F2)	$(NH_4)_3[VO(O_2)_2F_2]$		54
mpV(EDTA)	$M_3[VO(O_2)(EDTA)]\cdot 2H_2O$; M: Na, NH_4		55
	$H[VO(O_2)(H_2O)(2,6\text{-}pdc)]\cdot H_2O$		56
mpV(HEDTA)	$M_2[VO(O_2)(HEDTA)]\cdot 4H_2O$		57
bpV(acetate)	$M_2[VO(O_2)_2(CH_3COO)]$	–705	58
	$NH_4[VO(O_2)_2(bipy)]\cdot 4H_2O$		59
	$(bipyH)[\{VO(O_2)_2(bipy)\}_2]\cdot xH_2O_2\cdot(6\text{-}x)H_2O$ (x:0.5)		(60)
	$(bipyH)[VO(O_2)_2(bipy)]\cdot(3+x)H_2O_2\cdot(2\text{-}x)H_2O$ (x:0.4)		(61)
	$PPh_4[VO(O2)(pic)2]\cdot H2O$	-596^{c}	
	$VO(O_2)(quin)L$ (L: EtOH, DMF, DMSO, THF)		62
	$M_3[VO(O_2)_2(HPO_4)]\cdot 2H_2O$ (M: K, NH_4)		62
	$K_2[V_2O_2(O_2)_3(cystH)_2]\cdot H_2O$		63
	$M_2[V(O_2)_3F]$ (M: Na, K, NH_4)		64
	$M[V(O_2)_3]$ (M: Na, K)		65
	$NH_4[VO(O_2)(malato)]\cdot H_2O$		55
mpV(NTA)	$Na_2[VO(O_2(NTA)]\cdot 5H_2O$	–549	66
	$K_2[VO(O_2)(NTA)]$		55, (67)
	$Ba[VO(O_2)(NTA)]\cdot 3H_2O$		68
	$M_2[VO(O_2)_2Cl]$ (M: Na, K, NH_4)		69
	$M_2[V(O_2)_3Cl]$ (M: Na, K, NH_4)		69
	$M[VO(O_2)_2(glyH)]\cdot H_2O$ (M: K, NH_4)		70
$bpV(H_2O)$	$K[VO(O_2)_2(H_2O)x]$ (x: 1 or 2)	–688 to –695	34, 71, 72
	$(NH4)4[\{VO(O2)_2\}_2O]$		73
	$(NH_4)_3[\{VO(O_2)_2\}_2OH]\cdot H_2O$		74
	$K_2[\{VO(O_2)_2\}_2(\mu\text{-}nicH)]\cdot H_2O$		75
	$V_2O_2(O_2)_3(glyH)2(H_2O)_2$		70
	$K[VO(O_2)_2(Hheida)]$		76

[a]All ^{51}V NMR spectra were obtained on aqueous solutions on a Varian XL-300 NMR spectrometer operating at 78.891 MHz. Chemical shifts are in parts per million (ppm) with respect to $VOCl_3$ as external reference at 0.00 ppm, with negative shifts being upfield.
[b]X-ray of NH_4^+ salt.
[c]In CD_3OD.

pounds combine oxidants and reductants at the molecular level which can lead to explosives. While they should always be handled with respect, no unusual or violent decompositions have been observed in our laboratory. Vials containing the compounds display no build-up of pressure even after prolonged storage.

Stability

Aqueous solutions of peroxovanadates are usually stable for days at neutral or slightly basic pH but at low pH bisperoxovanadates can decompose to vanadate species within hours (possibly with some reduction). Light seems to accelerate decomposition. The oxophilicity and lability of vanadate species leads to strong interactions with buffers containing potential oxygen donors such as triethanolamine and phosphate [23, 34]. Vanadates react with reductants such as dithiothreitol (DTT) but not with Hepes the recommended buffer. Peroxovanadium compounds on the other hand are fairly inert and do not interact with non-reducing buffers such as phosphate. However the complexes do react with Hepes, DTT, β-mercaptoethanol, catechol, ascorbic acid, citrate and catechol derivatives such as norepinephrine and dopa. The use of near neutral phosphate buffers has proven most satisfactory in our laboratory.

In the preparation of peroxovanadates, the strongly oxidizing medium can lead to oxidative cleavage of certain ligands [77] with the concomitant formation of unexpected products. In the attempted preparation of $K_2[VO(O_2)_2(4\text{-Mepic})]$, where 4-Mepic = 4-methylpyridine-2-carboxylato, oxidation of the ligand occurred to give $K_3[VO(O_2)_2(ox)]\cdot H_2O_2$.

Spectroscopy

Many of the peroxovanadium complexes listed in Table 1 have been characterized in solution by techniques such as infrared, nuclear magnetic resonance and ultraviolet spectroscopy and in the solid state by X-ray crystallography. The infrared spectra of peroxovanadium(V) compounds show a ν(V=O) stretching vibration at ~940 cm^{-1}, ν(O-O) bands between 800 and 900 cm^{-1}, and ν(V-O_2) bands between 500 and 650 cm^{-1} [32]. A modestly intense peroxo ligand-to-metal charge transfer (LMCT) band in the range 300–460 nm is observed in the ultraviolet spectra. In the +5 state, no *d-d* transitions are possible and the yellow colour of the complexes is due to strong tailing of the LMCT band into the visible region [78].

The majority of peroxovanadium compounds contain ligands bound to vanadium through O and N. They show ^{51}V NMR shifts far upfield of $VOC1_3$ (Fig. 4) with monoperoxo species giving rise to resonances near -600 ppm whereas diperoxo species show resonances between –700 and –800 ppm [72, 79] (Table 1). A relationship between the ^{51}V chemical shift and the sum of ligand electronegativities has been

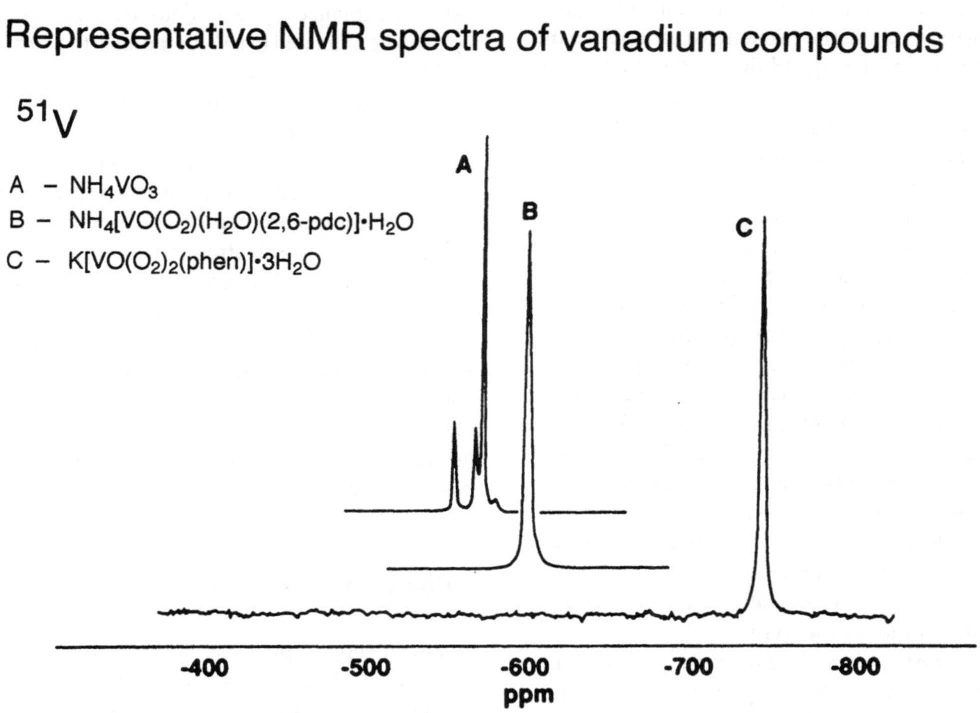

Fig. 4. Representative ^{51}V NMR spectra of vanadium compounds: A, vanadate; B, a monoperoxovanadate; C, a bisperoxovanadate.

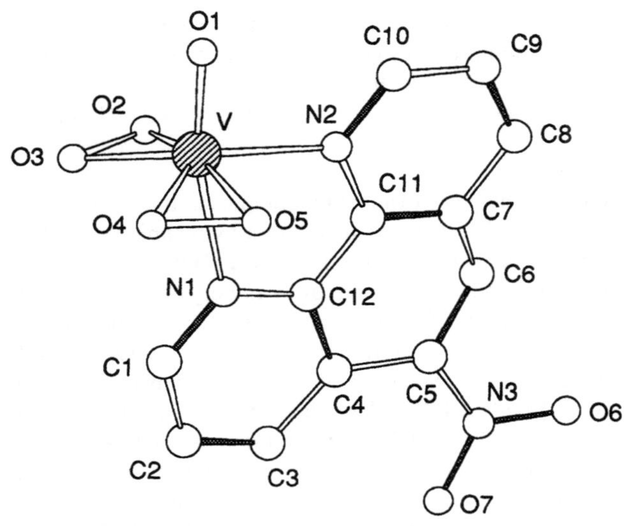

Fig. 5. Solid state structure of a typical bisperoxovanadate anion, $[VO(O_2)_2(5\text{-}NO_2phen)]^-$.

proposed for a large set of complexes [80].

Structural features

The structure of a typical peroxo complex (Fig. 5), displays the distorted pentagonal bipyramidal geometry of the inner sphere commonly observed for these compounds. The oxo group and one nitrogen atom of the phenanthroline ligand occupy apical positions while the two peroxo groups and the second nitrogen atom of the ligand define the pentagonal plane. In most cases, the O-O bond distance is between 1.4–1.5 Å. The trans effect of the V = O moiety results in the bond trans to it being lengthened, a feature commonly observed in most structures.

The general structure of a pV complex can be divided conceptually into a polar 'oxo' half and a non-polar organic half. This presents a number of interesting design variations the influence of which on the insulin mimetic properties of this class have been tested [22]. For example, compounds 2–4 in (Fig. 6) are structurally very similar, however bpV(ox), 2, a bisperoxo compound, is about twice as effective a phosphotyrosine phosphatase (PTPase) inhibitor as mpV(ox), 3, a monoperoxo compound. Compound 4 is a bisoxo compound wherein the angle between the two oxo ligands is 90°. This is a reasonable model of a peroxo complex but it displays much less activity than 3 and is about as active as vanadate. Similarly 5 is much more active than 6. Even a seemingly minor change such as alkylating the peroxo ligand in 5 to give 7 results in loss of activity.

Compounds 8 possess identical structures except that the central metal atom is vanadium 8a, molybdenum 8b, or tungsten 8c. All three are strong inhibitors of PTPase [22] but only the vanadium complex is a strong kinase activator and a strong *in vivo* insulin mimic. Thus the presence of vanadium(V) with at least one peroxo group results in significant insulin mimetic activity.

The organic half of the molecule permits extensive modifications and it has been shown that some of the biochemical effects of pV compounds are dependent on the nature of relatively small changes in the ancillary ligand (Fig. 7) [22]. For example, complex 9a is an active PTPase inhibitor but placing methyl groups on the ends of the phenanthroline ligand as in 9b and 9c dramatically reduces the activity of the complexes. On the other hand substitution at the 5-position of the phenanthroline ligand as in 9d–f does not reduce the activity even though the substituents vary considerably in character. Similarly the complexes 10a–f all possess about the same degree of PTPase inhibitory activity. This suggests that substituents that do not increase the steric bulk of the ancillary ligand around the vanadium atom have little effect on activity. This opens the door to extensive ligand design.

PTPase inhibition

There is a good correlation between the PTPase inhibitory abilities of pV complexes, their abilities to promote activation of the insulin receptor [22], and their *in vivo* insulin mimetic activities [81]. The X-ray crystal structure of the first PTPase enzyme was recently reported [82], and the active site contains an essential cysteine residue. Gresser and Tracey developed an analogy [83] between phosphate and vanadate esters to explain enzyme inhibition by the latter and this analogy can be extended to pV compounds to a certain extent (Fig. 8). We suggest that pV complexes are able to bind to the active site of a PTPase because they are quite similar to a phosphate ester. Since pV complexes are good oxidants [27], it is tempting to speculate that they inhibit PTPase by irreversibly oxidizing the cysteine residue in the active site of the enzyme.

pV complexes are capable of oxidizing a variety of organic substrates by electron transfer mechanisms or by oxygen atom transfer. Of interest here is the oxidation of sulfides to sulfoxides [84] and the oxidative coupling of cysteine to cystine (Fig. 9) [42]. The water soluble sodium tris(3-sulfonatophenyl)-phosphine is oxidized in minutes by pV complexes [48]. Thus it seems quite possible for the complexes to oxidize one or more critical functional groups in the active sites of PTPase enzymes. Further studies in this direction are being actively pursued.

Acknowledgements

Financial support provided by the Medical Research Council (MRC) of Canada, the Department of Education of Quebec, and Nordic Merrell Dow Research Inc. of Montreal is gratefully acknowledged.

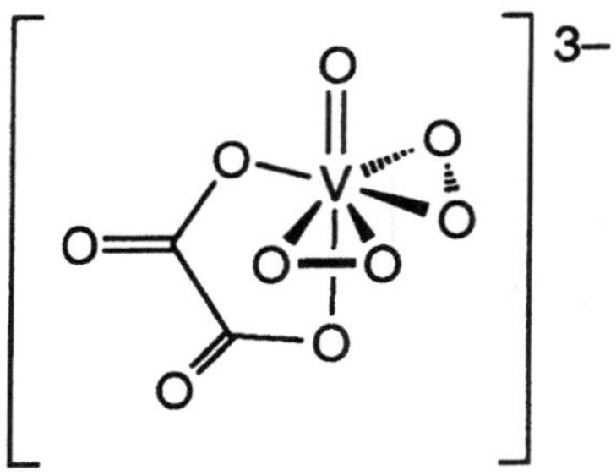

bpV(ox), **2**

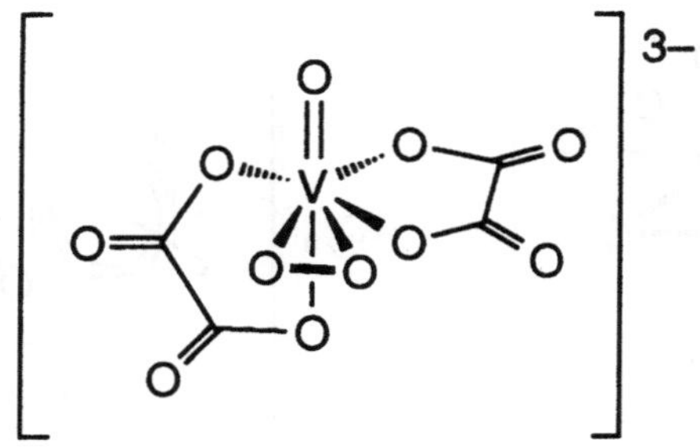

mpV(ox), **3**

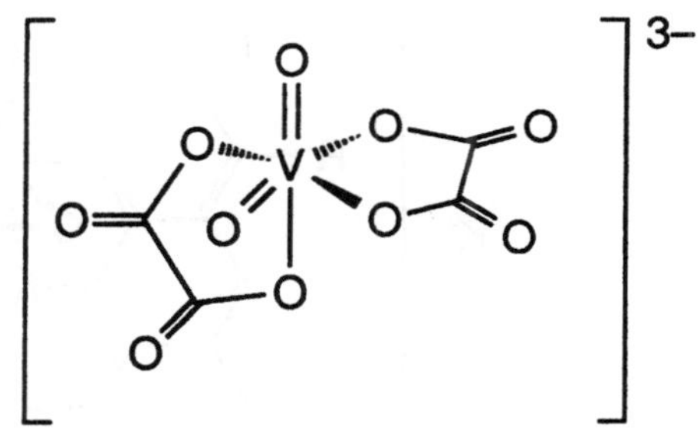

$V(O)_2(ox)_2$, **4**

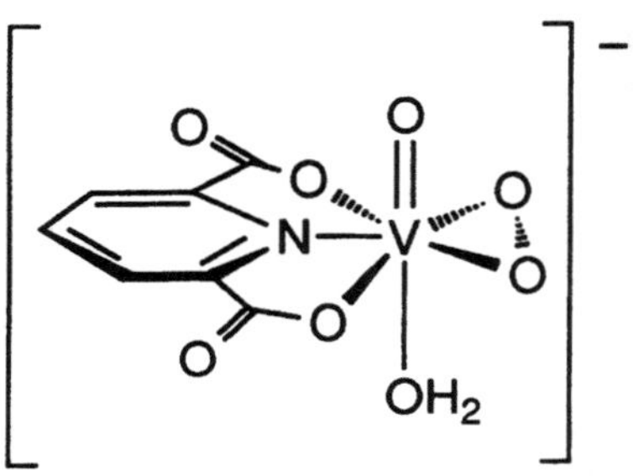

mpV(2,6-pdc), **5**

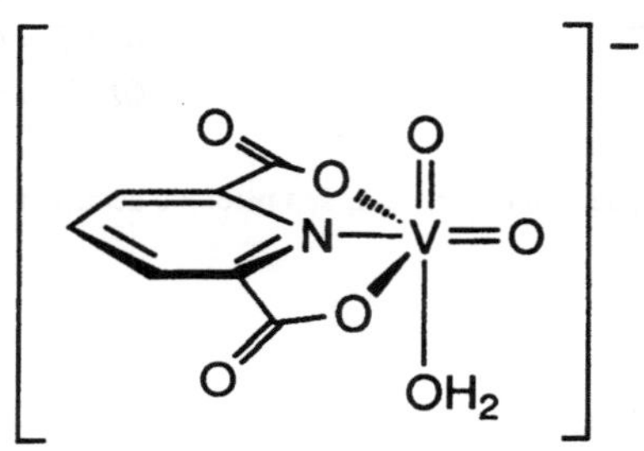

6

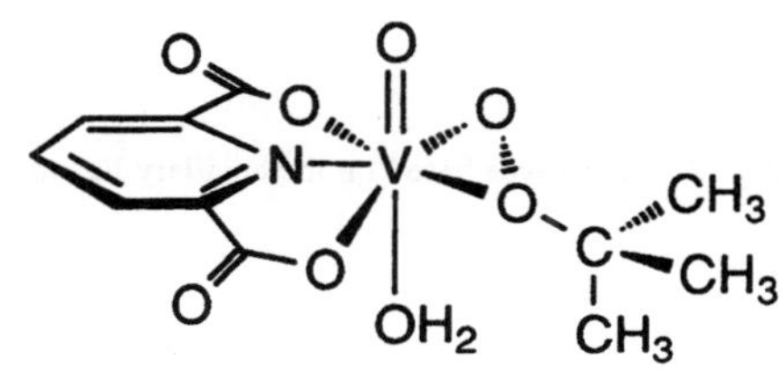

7

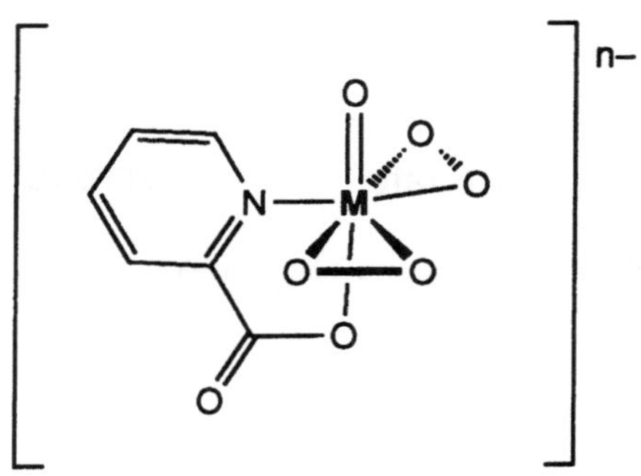

8a M = V; n = 2
8b M = Mo; n = 1
8c M = W; n = 1

Fig. 6. Structures of the peroxovanadium complexes 2, 3, 5 and 8a, the cis-dioxovanadium analogues 4 and 6, a t-butylperoxo analogue 7, and the molybdenum and tungsten analogues 8b and 8c.

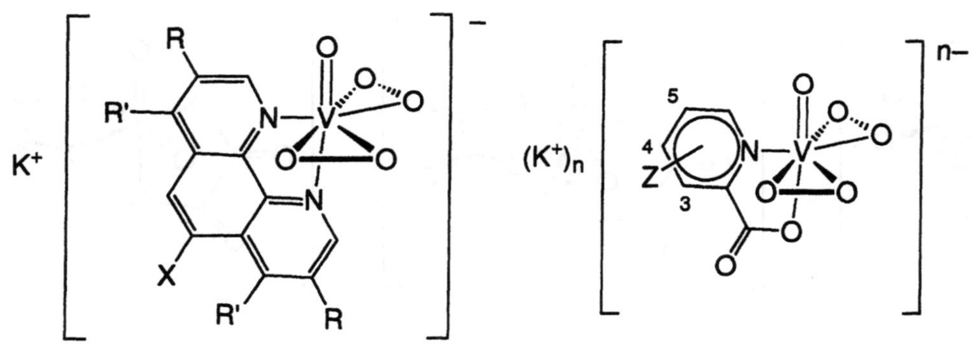

9a: R = R' = H; X = H
9b: R = H; R' = CH_3; X = H
9c: R = R' = CH_3; X = H
9d: R = R' = H; X = CH_3
9e: R = R' = H; X = NH_2
9f: R = R' = H; X = NO_2

10a: Z = 3-H
10b: Z = 3-OH
10c: Z = 3-$OCH_2CO_2^-$
10d: Z = 3-CO_2^-
10e: Z = 4-CO_2^-
10f: Z = 5-CO_2^-

Fig. 7. Derivatives based on the ancillary ligand 1,10-phenanthroline, 9a–f, and picolinic acid, 10a–f.

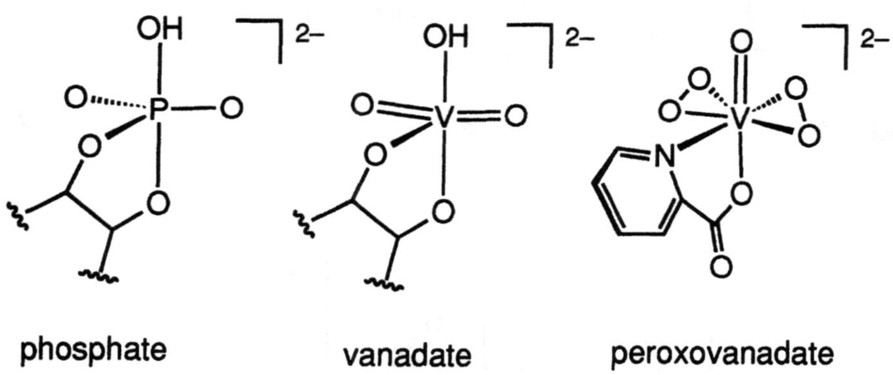

Fig. 8. Structural analogy linking phosphate and vanadate esters to a peroxovanadium complex.

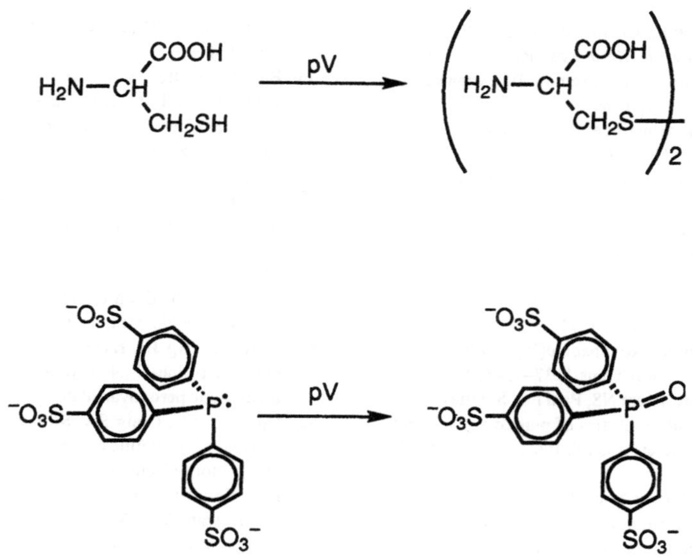

Fig. 9. Oxidative coupling of cysteine to cystine, and oxidation of a phosphine to a phosphine oxide.

References

1. Clark RJH: Vanadium. In: J.C. Bailar Jr, H.J. Emeléus, R. Nyholm (eds). A.F. Trotman-Dickenson (exec ed.). Comprehensive Inorganic Chemistry. Pergamon Press, Toronto, 1973, vol 3, pp 451–551.
2. Oltz EM, Bruening RC, Smith MJ, Kustin K, Nakanishi K: The tunichromes. A class of reducing blood pigments from sea squirts: Isolation, structures and vanadium chemistry. J Am Chem Soc 110: 6162–6172, 1988
3. Smith MJ, Kim D, Horenstein B, Nakanishi K, Kustin K: Unraveling the chemistry of tunichrome. Acc Chem Res 24: 117–124, 1991
4. Vilter H: Vanadium-an element involved in the biosynthesis of halogenated compounds and nitrogen fixation. Phytochemistry 23: 1387–1390, 1984
5. Weber R, de Boer E, Plat H, Krenn BE: Peroxidases from phaeophyceae: a vanadium(V)-dependent peroxidase from *Ascophyllum nodosum*. FEBS Lett 216: 1–3, 1987
6. Plat H, Krenn BE, Wever R: The bromoperoxidase from the lichen *Xanthoria parietina* is a novel vanadium enzyme. Biochem J 248: 277–279, 1987
7. Krenn BE, Plat H, Wever R: The bromoperoxidase from the red alga *Ceranium rubrum* also contains vanadium as a prosthetic group. Biochim Biophys Acta 912: 287–291, 1987
8. Hales BJ, Case EE, Morningstar JE, Dzeda MF, Mauterer LA: Isolation of a new vanadium containing nitrogenase from *Azotobacter vinelandii*. Biochemistry 25: 7251–7255, 1986
9. Robson RL, Eady RR, Richardson TH, Miller RW, Hawkins M, Postgate JR: The alternative nitrogenase of *Azotobacter chroococcum* is a vanadium enzyme. Nature 322: 388–390, 1986
10. Nechay BR, Nanninga LB, Nechay PSE, Post RL, Grantham JJ, Macara IG, Kubena LF, Phillips TD, Nielsen FH: Role of vanadium in biology. Fed Proc 45: 123–132, 1986
11. Rehder D: The bioinorganic chemistry of vanadium. Angew Chem Int Ed Engl 30: 148–167, 1990
12. Chasteen ND (ed.): Vanadium in biological systems. Kluwer Academic Publishers, Dordrecht, 1990
13. Tolman EL, Barris E, Burns M, Pansini A, Partridge R: Effects of vanadium on glucose metabolism *in vitro*. Life Sci 25: 1159–1164, 1979
14. Dubyak GR, Kleinzeller AD: The insulin-mimetic effects of vanadate in isolated rat adipocytes. Dissociation from the effects of vanadate as a (Na^+/K^+) ATPase inhibitor. J Biol Chem 255: 5306–5312, 1980
15. Shechter Y, Karlish SJD: Insulin-like stimulation of glucose oxidation in rat adipocytes by vanadyl (IV) ions. Nature 284: 556–558, 1980
16. Heyliger CE, Tahiliani AG, McNeill JH: Effect of vanadate on elevated blood glucose and depressed cardiac performance of diabetic rats. Science 227: 1474–1477, 1985
17. Ramanadham S, Mongold JJ, Brownsey RW, Cros GH, McNeill JH: Oral vanadyl sulfate in treatment of diabetes mellitus in rats. Am J Physiol 257: H904–H911, 1989
18. McNeill JH, Yuen VG, Hoveyda HR, Orvig C: Bis(maltolato)-oxovanadium(IV) is a potent insulin mimic. J Med Chem 35: 1489–1491, 1992
19. Shechter Y, Shisheva A, Lazar R, Libman J, Shanzer A: Hydrophobic carriers of vanadyl ions augment the insulinomimetic actions of vanadyl ions in rat adipocytes. Biochemistry 31: 2063–2068, 1992
20. Sakurai H: A new orally active insulin-mimetic vanadyl complex: bis(pyrrolidine-N-carbodithiotato)oxovanadium(IV). J Med Chem 37: 876–877, 1994
21. Kadota S, Fantus IG, Deragon G, Guyda HJ, Hersh B, Posner BI: Peroxide(s) of vanadium: a novel and potent insulin mimetic-agent which activates the insulin receptor kinase. Biochem Biophys Res Commun 147: 259–266, 1987
22. Posner BI, Burgess JW, Faure R, Bevan AP, Lachance D, Zhang-Sun G, Fantus IG, Ng JB, Hall DA, Soo Lum B, Shaver A: Peroxovanadium compounds: a new class of potent phosphotyrosine phosphatase inhibitors which are insulin mimetics. J Biol Chem 269: 4596–4604, 1994
23. Crans DC: Aqueous chemistry of labile oxovanadates: relevance to biological studies. Comments Inorg Chem 16: 1–33, 1994
24. Crans DC: Enzyme interactions with labile oxovanadates and other

polyoxometalates. Comments Inorg Chem 16: 35–76, 1994

25. Butler A, Parsons SM, Yamagata SK, De la Rosa RI: Reactivation of vanadate-inhibited enzymes with desferrioxamine B, a vanadium(V) chelator. Inorg Chim Acta 163: 1–3, 1989
26. Chasteen ND: Vanadyl(IV) EPR spin probes. Inorganic and biochemical aspects. Biol Magn Reson 3: 53–119, 1981
27. Butler A, Clague MJ, Meister GE: Vanadium peroxide complexes. Chem Rev 94: 625–638, 1994
28. Campbell NJ, Dengel AC, Griffith WP: Studies on transition metal peroxo complexes-X. The nature of peroxovanadates in aqueous solution. Polyhedron 8: 1379–1386, 1989
29. Sala-Pala J, Guerchais JE: Peroxo-complexes: triperoxovanadates(V) complexes. J Chem Soc (A) 1132–1136, 1971
30. Vuletic N, Djordjevic C: Oxodiperoxovanadate(V) complexes with bidentate ligands. J Chem Soc Dalton Trans 1137–1141, 1973
31. Bhattacharjee M, Chaudhuri MK, Islam NS, Paul PC: Synthesis, characterisation and physicochemical properties of peroxo-vanadium(V) complexes with glycine as the hetero-ligand. Inorg Chim Acta 169: 97–100, 1990
32. Vuletic N, Djordjevic C: Oxodiperoxovanadate(V) complexes with bidentate ligands. J Chem Soc Dalton Trans: 1137–114, 1973
33. Szentivanyi H, Stomberg R: The crystal structure of ammonium (2,2′-bipyridine) oxodiperoxovanadate (V) tetrahydrate, $NH_4[VO(O_2)_2(C_{10}H_8N_2)]\cdot 4H_2O$, at –100°C. Acta Chem Scand A37: 553–559, 1983
34. Schwendt P, Petrovič P, Uskert D: Oxo-peroxokomplexe des vanadiums(V). Z Anorg Allg Chem 466: 232–236, 1980
35. Begin D, Einstein FWB, Field J: An asymmetrically coordinated diperoxo compound. Crystal structure of $K_3[VO(O_2)_2(C_2O_4)]\cdot H_2O$. Inorg Chem 14: 1785–1790, 1975
36. Schwendt P, Pisárcik M: Vibrational spectra of ammonium amine-oxo-diperoxovanadate. Coll Czech Chem Commun 47: 1549–1555, 1982
37. Drew RE, Einstein FWB: The crystal structure of ammonium oxodiperoxoamminevanadate (V). Inorg Chem 11: 1079–1083, 1972
38. Hartkamp H: Darstellung von ammonium-[(pyridin-2,6-dicarboxylato)-oxo-peroxo-vanadat(V)]. Angew Chem 71: 553, 1959
39. Drew RE, Einstein FWB: Crystal structure at –100°C of ammonium oxoperoxo(pyridine-2,6-dicarboxylato)vanadate (V) hydrate, $NH_4[VO(O_2)(H_2O)(C_5H_3N(COO)_2)]\cdot xH_2O$ (x=1.3). Inorg Chem 12: 829–835, 1973
40. Mimoun H, Saussine L, Daire E, Postel M, Fischer J, Weiss R: Vanadium(V) peroxo complexes. New versatile biomimetic reagents for epoxidation of olefins and hydroxylation of alkanes and aromatic hydrocarbons. J Am Chem Soc 105: 3101–3110, 1983
41. Quilitzsch U, Wieghardt K: Kinetics of the diperoxovanadate(V)-monoperoxovanadate(V) conversion in perchloric acid media. Inorg Chem 18: 869–871, 1979
42. Shaver A, Ng JB, Hall DA, Soo Lum B, Posner BI: Insulin-mimetic peroxovanadium complexes: preparation and structure of potassium oxodiperoxo(pyridine-2-carboxylato)vanadate(V), $K_2[VO(O_2)_2(C_5H_4NCOO)]\cdot 2H_2O$, and potassium oxodiperoxo(3-hydroxypyridine-2-carboxylato)vanadate(V), $K_2[VO(O_2)2(OHC_5H_3NCOO)]\cdot 3H_2O$, and their reactions with cysteine. Inorg Chem 32: 3109–3113, 1993
43. Mimoun H, Chaumette P, Mignard M, Saussine L: First d^0 metal alkylperoxidic complexes. Synthesis, X-ray structure and hydroxylating properties of vanadium(V) dipicolinato alkylperoxides. Nouv J Chim 7: 467–475, 1983
44. Stomberg R: The crystal structures of potassium bis(oxalato)oxoperoxovanadate(V) hemihydrate, $K_3[VO(O_2)(C_2O_4)_2]\cdot 1/2H_2O$, and potassium bis(oxalato)dioxovanadate(V) trihydrate, $K_3[VO(O_2)(C_2O_4)_2]\cdot 3H_2O$. Acta Chem Scand A40: 168–176, 1986
45. Szentivanyi H, Stomberg R: The crystal structure of (2,2′-bipyridine)-oxoperoxo(pyridine-2-carboxylato)vanadium(V) hydrate, $[VO(O_2)(C_5H_4NCOO)(C_{10}H_8N_2)]\cdot H_2O$ at –100°C. Acta Chem Scand A37: 709–714, 1983
46. Basumatary JK, Chaudhuri MK, Dutta Purkayastha RN, Hiese Z: First synthesis and structural assessment of alkali-metal carbonato-dioxoperoxouranate(VI) monohydrates, $A_2[UO_2(O_2)CO_3)]\cdot H_2O$, and carbonatooxodiperoxovanadate(V) trihydrates, $A_3[VO(O_2)_2(CO_3)]\cdot 3H_2O$. J Chem Soc Dalton Trans: 709–711, 1986
47. Stomberg R: The crystal structure of potassium carbonatooxo-diperoxovanadate(V), $K_3[VO(O_2)2(CO_3)]$. Acta Chem Scand A39: 725–731, 1985
48. Shaver A, Hall DA, Ng JB, Lebuis AM, Hynes RC, Posner BI: Bisperoxovanadium compounds: Synthesis and reactivity of some insulin-mimetic complexes. Inorg Chim Acta 229: 253–260, 1995
49. Shaver A, Ng JB, Hynes RC, Posner BI: Potassium (5-nitro-1,10-phenanthroline-N^1,N^{10})oxodiperoxovanadate(V) dihydrate: an insulin mimetic peroxovanadate. Acta Cryst Sect C C50: 1044–1046, 1994
50. Djordjevic C, Craig SA, Sinn E: A polymeric peroxo heteroligand vanadate(V). Synthesis, spectra, and structure of $M^I[VO(O_2)(C_4H_5O_4N)$. Inorg Chem 24: 1281–1283, 1985
51. Djordjevic C, Lee M, Sinn E: Oxoperoxo(citrato)- and dioxo-(citrato)vanadates(V): synthesis, spectra and structure of a hydroxyl oxygen bridged dimer $K_2[VO(O_2)(C_6H_6O_7).]_2\cdot 2H_2O$ Inorg Chem 28: 719–723, 1989
52. Chaudhuri MK, Ghosh SK: Synthesis of alkali oxydiperoxy-fluorovanadates(V). Polyhedron 1: 553–555, 1982
53. Stomberg R: The crystal structure of potassium fluorooxodiperoxo-vanadate(V), $K_2[VFO(O_2)_2]$. Acta Chem Scand A38: 223–228, 1984
54. Stomberg R: The crystal structure of ammonium difluorooxo-diperoxovanadate(V), $(NH_4)_3[VF_2O(O_2)_2]$. Acta Chem Scand A38: 541–545, 1984
55. Djordjevic C, Wampler GL: Antitumour activity and toxicity of peroxo heteroligand vanadates(V) in relation to biochemistry of vanadium. J Inorg Biochem 25: 51–55, 1985
56. Wieghardt K: Preparation and characterization of dipicolinato-vanadium(V) complexes. Kinetics and mechanism of their reaction with hydrogen peroxide in acidic media. Inorg Chem 17: 57–64, 1978
57. Schwendt P, Sivak M, Lapshin AE, Smolin YI, Shepelev,YF, Gyepesova D: Ethylenediaminetetraacetato-monoperoxo complexes of vanadium(V). Trans Metal Chem 19: 34–36, 1994
58. Vilter H, Rehder D: 51V NMR investigation of a vanadium(V)-dependant peroxidase from *Ascophyllum nodosum (L) le jol.* Inorg Chim Acta 136: L7-L10, 1987
59. Campbell NJ, Capparelli MV, Griffith WP, Skapski AC: On the existence of triperoxo vanadium complexes. X-Ray crystal structures of $K_3[VO(O_2)_2(C_2O_4)]\cdot H_2O_2$ and of $(NH_4)[VO(O_2)_2(bipy)]\cdot 4H_2O$. Inorg Chim Acta 77: L215–L216, 1983
60. Szentivanyi H, Stomberg R: The crystal structure of 2,2′-bipyridinium(1+)μ-hydrogen-bis[(2,2′-bipyridine)oxodiperoxovanadate](1-)-x-hydrogen peroxide-(6-x)-water, $(Hbipy)[H\{VO(O_2)_2bipy\}_2]\cdot xH_2O_2\text{-}(6\text{-}x)H_2O$, x ≈ 0.5, at –100°C. Acta Chem Scand A38: 101–107, 1984
61. Stomberg R, Szentivanyi H: The crystal structure of 2,2′-bipyridinium(1+) (2,2′-bipyridine)oxodiperoxovanadate(1-)-(3+x)-hydrogen peroxide-(2-x)-water, $(C_{10}H_9N_2)[VO(O_2)_2(C_{10}H_9N_2)]\cdot(3+x)H_2O_2\text{-}(2\text{-}x)H_2O$, x ≈ 0.4, at –100°C. Acta Chem Scand A38: 121–128, 1984
62. Bhattacharjee M, Chaudhuri MK, Paul PC: Synthesis and physico-chemical studies of newer mono and diperoxo heteroligand complexes of vanadium. Can J Chem 70: 2245–2248, 1992
63. Chaudhuri, M. K., Paul PC: Complex peroxovanadates(V) containing the chelated cysteinate as the co-ligand: synthesis and spectroscopic studies. Ind J Chem 31A: 466–468, 1992
64. Chaudhuri MK, Ghosh SK:- Alkali-metal and ammonium triperoxy-fluorovanadates(V), $A_2[V(O_2)_3F]$. Inorg Chem 21: 4020–4022, 1982

65. Chaudhuri MK, Ghosh SK, Islam NS: First synthesis and structural assessment of alkali-metal triperoxovanadates(V), $A[V(O_2)_3]$. Inorg Chem 24: 2706–2707, 1985
66. Wu D, Lei X, Cao R, Hong M: Synthesis and structure of the peroxovanadium(V) complex containing the nitriloacetate $Na_2[VO(O_2)N(CH_2COO)_3]\cdot 5H_2O$. Jiegou Huaxue 11: 65–67, 1992
67. Wei YG, Zhang SW, Huang GQ, Shao MC: Synthesis, spectrum and single crystal structure of a peroxovanadium complex of nitrilotriacetate ligand without water of crystallization, $K_2[VO(O_2)NTA]$. Polyhedron 13: 1587–1591, 1994
68. Kuchta L, Sivak M, Pavelcik F: Synthesis, characterization and crystal structure of barium (nitrilotriacetato)oxoperoxovanadate(V) trihydrate. J Chem Res S 393, 1993
69. Chaudhuri MK, Ghosh SK: The $[VO(O_2)_2Cl]^{2-}$-$[V(O_2)_3Cl]^{2-}$ pair in peroxovanadium(V) chemistry: Synthesis of the first chloroperoxyvanadate(V) compounds and evidence for diperoxyvanadate(V)-triperoxyvanadate(V)interconversion. Inorg Chem 23: 534–537, 1984
70. Bhattacharjee M, Chaudhuri MK, Islam NS, Paul PC: Synthesis characterization and physicochemical properties of peroxo-vanadium(V) complexes with glycine as the heteroligand. Inorg Chim Acta 169: 97–100, 1990
71. Bhattacharjee MN, Chaudhuri MK, Islam NS: Reactivity of coordinated peroxide at a highly peroxygenated vanadium(V) center in an aqueous medium. Inorg Chem 28: 2420–2423, 1989
72. Jaswal JS, Tracey AS: Formation and decomposition of peroxovanadium complexes in aqueous solution. Inorg Chem 30: 3718–3722, 1991
73. Stomberg R, Olson S, Svensson IB: The crystal structure of ammonium-μ-oxobis(oxodiperoxovanadate) (4-), $(NH_4)_4[O\{VO(O_2)_2\}_2]$. A refinement. Acta Chem Scand. A38: 653–656, 1984
74. Campbell NJ, Flanagan J, Griffith WP, Skapski, AC: X-ray structure of the peroxovanadate $[HV_2O_{11}]^{3-}$ ion in the ammonium peroxovanadate salt $(NH_4)_3[HO\{VO(O_2)_2\}_2]\cdot H_20$. Trans Met Chem 10: 353–354, 1985
75. Djordjevic C, Puryear BC, Vuletic N, Abelt CJ, Sheffield SJ: Preparation, spectroscopic properties, and characterization of novel peroxo complexes of vanadium(V) and molybdenum(VI) with nicotinic acid and nicotinic acid N-oxide. Inorg Chem 27: 2926–2932, 1988
76. Colpas GJ, Hamstra BJ, Kampf JW, Pecoraro VL: A functional model for vanadium haloperoxidase. J Am Chem Soc 116: 3627–3628, 1994
77. Szentivanyi H, Stomberg R: The crystal structure of (2,2′-bipyridine)oxoperoxo(pyridine-2-carboxylato)vanadium(V) hydrate. Acta Chem Scand A37: 709–714, 1983
78. Lever ABP, Gray HB: Electronic spectra of metal-dioxygen complexes. Acc Chem Res 11: 348–355, 1978
79. Rehder D, Wieghardt K: The ^{51}V NMR spectra of some oxo-, peroxo-, nitrosyl- and hydroxylamidovanadium complexes. Z Naturforsch 36B: 1251–1254, 1981
80. Rehder D, Weidemann C, Duch A, Priebsch W: ^{51}V shielding in vanadium(V) complexes: A reference scale for vanadium binding sites in biomolecules. Inorg Chem 27: 586–587, 1988
81. Yale J-F, Lachance D, Bevan AP, Vigeant C, Shaver A, Posner, BI: Hypoglycemic effects of peroxovanadium compounds in Sprague-Dawley and diabetic BB rats. Diabetes in press
82. Barford D, Flint AJ, Tonks NK: Crystal structure of human protein tyrosine phosphatase 1B. Science 263: 1397–1404, 1994
83. Gresser MJ, Tracey AS: Vanadates as phosphate analogs in biochemistry. In: N.D. Chasteen (ed.). Vanadium in Biological systems. Kluwer Academic Publishers, Dordrecht, 1990, pp 63–79
84. Ballistreri FP, Tomaselli GA, Toscano RM, Conte V, DiFuria F: Application of the thianthrene 5-oxide mechanistic probe to peroxometal complexes. J Am Chem Soc 113: 6209–6212, 1991

Molecular and Cellular Biochemistry **153**: 17–24, 1995.

Vanadium chemistry and biochemistry of relevance for use of vanadium compounds as antidiabetic agents

Debbie C. Crans, Mohammed Mahroof-Tahir and Anastasios D. Keramidas
Department of Chemistry and Cell and Molecular Biology Program, Colorado State University, Colorado 80523, USA

Abstract

The stability of 11 vanadium compounds is tested under physiological conditions and in administration fluids. Several compounds including those currently used as insulin-mimetic agents in animal and human studies are stable upon dissolution in distilled water but lack such stability in distilled water at pH 7. Complex lability may result in decomposition at neutral pH and thus may compromise the effectiveness of these compounds as therapeutic agents; Even well characterized vanadium compounds are surprisingly labile. Sufficiently stable complexes such as the VEDTA complex will only slowly reduce, however, none of the vanadium compounds currently used as insulin-mimetic agents show the high stability of the VEDTA complex. Both the bis(maltolato)oxovanadium(IV) and peroxovanadium complexes extend the insulin-mimetic action of vanadate in reducing cellular environments probably by increased lifetimes under physiological conditions and/or by decomposing to other insulin mimetic compounds. For example, treatment with two equivalents of glutathione or other thiols the (dipicolinato)peroxovanadate(V) forms (dipicolinato)oxovanadate(V) and vanadate, which are both insulin-mimetic vanadium(V) compounds and can continue to act. The reactivity of vanadate under physiological conditions effects a multitude of biological responses. Other vanadium complexes may mimic insulin but not induce similar responses if the vanadate formation is blocked or reduced. We conclude that three properties, stability, lability and redox chemistry are critical to prolong the half-life of the insulin-mimetic form of vanadium compounds under physiological conditions and should all be considered in development of vanadium-based oral insulin-mimetic agents. (Mol Cell Biochem **153**: 17–24, 1995)

Key words: vanadium chemistry, vanadium biochemistry, compound stability, compound lability, insulin- mimetic, metabolic involvement

Abbreviations: ADP – adenosine 5′-diphosphate; ATP - adenosine 5′-triphosphate, ADP-V – adenosine 5′-diphosphate-vanadate; bpV – bis(peroxo)oxovanadium(V); (bpV)2 – bis(peroxo)oxovanadium(V) dimer; bpVpic – bis(peroxo)picolinatooxovanadate(V); ^{13}C – carbon-13; EDTA – ethylenediaminetetraacetic acid; EPR – electron paramagnetic resonance; EXSY – exchange spectroscopy; ^{1}H – proton; HSG – glutathione; NAD – β-nicotinamide adenine dinucleotide; NADP – β-nicotinamide adenine dinucleotide phosphate; NADV – β-nicotinamide adenine dinucleotide vanadate; NMR – nuclear magnetic resonance (also referred to as magnetic resonance imaging); pVdipic – (dipicolinato)peroxovanadate(V); Vcit – (citrato)dioxovanadate(V); VEDTA – (ethylenediaminetetraacetato)dioxovanadate(V); Vmalto – bis(maltolato)-oxovanadium(IV); Voxal – bis(oxalato)dioxovanadate(V); ^{51}V – vanadium-51; V_1 – vanadate monomer; V_2 - vanadate dimer; V_4 – vanadate tetramer; V_5 – vanadate pentamer; UV-vis spectroscopy – ultraviolet-visible spectroscopy

Address for offprints: D.C. Crans, Department of Chemistry, Colorado State University, Fort Collins, Colorado 80523, USA

Introduction

In vitro and animal studies show that vanadate and other vanadium compounds increase glucose transport activity and normalize glucose metabolism [1–5]. Furthermore, these insulin-mimetic compounds can be administered orally. Vanadate enhances the phosphoprotein formation which is attributed to either the activation of protein kinases or inhibition of protein phosphatases. Despite the interest in documenting the effects of vanadate on protein kinases, most reports have used indirect methods and studies with purified kinases show weak, if any, interaction of vanadate with kinases as a group of enzymes (reviewed in Refs. [6–8]). Vanadate interacts potently with phosphatases and the inhibition is attributed to a five-coordinate vanadate complex which mimics the transition state of the phosphate ester hydrolysis reaction (reviewed in Refs. [7, 9]). Given the multitude of alternative mechanisms by which vanadate can act at the cellular level consideration of these effects should assist in identification of the undesirable modes of action [6–9]. Since mechanistic information will enable us to optimize insulin-mimetic effects and minimize toxicity, it is appropriate to briefly discuss some alternative modes of action of vanadate in cells.

Vanadate as a phosphate analog is a potent inhibitor for ATPases; the lowest K_i value for the Na^+-K^+-ATPase has been reported to 9 nM thus justifying the considerations of this enzyme system as a major contributor to the biological effects of vanadate [10]. However, the effects of vanadate extend way beyond the ATPases and vanadate as a phosphate analog. Vanadate is believed to enter cells through phosphate transport systems (path a, Fig. 1) [11]. Based on the reactivity of vanadate and the cellular components and metabolites, vanadate is likely to undergo several reactions inside the cell (illustrated in Fig. 1). Since the *in vivo* responses to vanadate represent the cumulative responses to all the vanadium compounds present in the cell, the intracellular form vanadate becomes essential to the overall responses to vanadate. First, vanadate reacts with alcohols to form esters (eq. 1) [12]. Second, vanadate reacts with phosphates to form vanadate-phosphate anhydrides (eq. 2) [13]. In the presence of glutathione (HSG) [14, 15], or ascorbate [16], vanadate is reduced to vanadium(IV). Once administered vanadate eventually reduces to vanadium(IV) depending on intracellular location and cell type [11, 16–20]. Whether the slow formation of vanadium(IV) is related to a slow reaction or a slow penetration of cells by vanadate is not clear [11]. Since free vanadyl cation is stable to only 10^{-7} M at neutral pH, generated vanadium(IV) will combine with available ligands (path f, Fig. 1) [21].

$$H_2VO_4^- + HOR \rightleftarrows ROVO_3H^- + H_2O \qquad (1)$$

$$H_2VO_4^- + H_2PO_4^- \rightleftarrows HOVO_2OPO_3H^{2-} + H_2O \qquad (2)$$

Because most cellular components contain hydroxyl and/or phosphate groups, vanadate reacts as shown in eq. 1, and 2 with a variety of metabolites. For example, the reaction of vanadate with the 2′-hydroxyl group of the cofactor NAD generates an NADP analog, NADV (path b) [22]. NADV is an excellent cofactor for enzymes such as glucose-6-phosphate dehydrogenase, 6-phosphogluconate dehydrogenase, and alcohol dehydrogenase [22]. The presence of NADV could affect the levels of reducing equivalents in the cell, important in maintaining a normal glucose metabolism. This type of mechanism is distinct from the vanadate-induced NADH oxidation by plasma membranes [23]. Organic vanadates have been shown to substitute for organic phosphates in many of the enzymes related to glucose metabolism and could effect these enzymes as well [24, 25]. Other organic vanadates are potent inhibitors of phosphatases [9], ribonucleases [26] and phosphoglucomutase [27, 28]. Formation of a vanadate-protein tyrosine phosphatase complex (path d) is particularly relevant to the vanadate-induced insulin-mimetic effects [9].

Nucleotides are other metabolites that could react with vanadate as shown in eq. 2 and Fig. 2 to form metabolite-conjugates. Specifically, ADP-V is accepted by myosin in place of ATP (path c) [29]. The corresponding GDP-V is accepted by adenylate cyclase in place of GTP and activates this enzyme for synthesis of cAMP (path c) [30]. Analogous processes could involve other metabolites and enzyme systems [7]. Although most of this discussion focuses on vanadate and vanadate derived compounds, vanadium(IV) compounds sometimes cause similar responses (see for example Refs. [22, 25, 31]).

The reactions shown in Fig. 1 underscore the importance of determining the type of intracellular vanadium compounds that form. Vanadate and vanadium compounds have commonly been used in animal studies with little consideration of their chemistry or information on the active species: Do all these compounds convert to a common form? Despite the few lability studies of vanadium complexes, we will here show that three physicochemical properties of vanadium compounds are essential to drug administration; stability, lability and redox properties. In this paper we will describe the aqueous chemistry of relevance to administration of vanadium compounds. Second, studies probing the lability and stability of a series of vanadium compounds are discussed and finally, the redox stability of these complexes are considered.

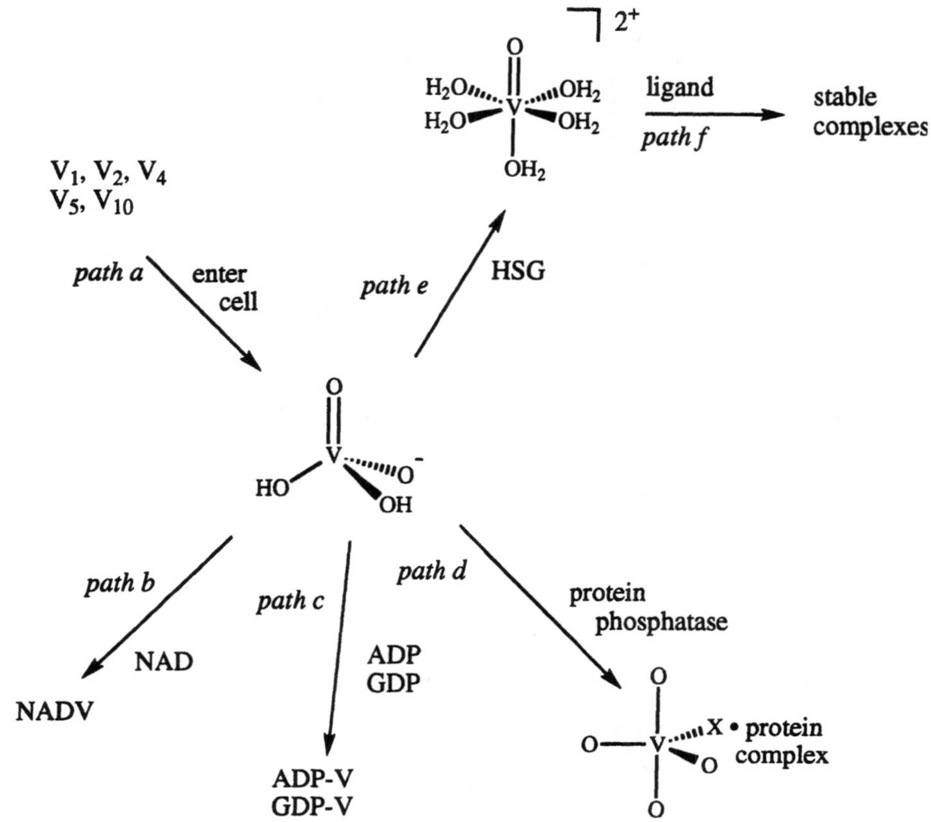

Fig. 1. Intracellular vanadium(V) compounds that can form when vanadate has been administered. Each compound has been found to act as substrate, cofactor or inhibitor for enzymes related to glucose metabolism.

Materials and methods

Materials

The vanadium complexes (see list of abbreviations) were prepared as described previously for bpV [32], $(bpv)_2$ [33], bpVpic [34], Vdipic [35], pVdipic [35], Vmalto [36], VEDTA [37], Vcit [38], and Voxal [39]. The reagents used in these synthesis reactions were purchased from Aldrich and used without further purification unless indicated otherwise.

Stock solutions

Vanadate stock solutions were prepared from sodium orthovanadate (Na_3VO_4) or sodium metavanadate ($NaVO_3$). When pH was adjusted by the addition of HC1, the stock solution immediately turned yellow-orange indicating the presence of vanadate decamer; such solutions were heated or stored until the yellow-orange color disappeared [40]. The pH often changed until the solution was colorless so pH adjustments were repeated until the desired pH was obtained.

Sample preparation of vanadium compounds

All NMR samples were prepared at ambient temperatures and contained 20% D_20. Solutions were prepared by dissolution of the crystalline compound into double-distilled water, and pH was adjusted with NaOH or HC1. No solution changed color during pH adjustment except for sodium orthovanadate. Some samples were prepared in buffer (HCO_3^-/CO_3^{2-} or Hepes) to maintain constant pH. The Vmalto solution was prepared under nitrogen and the stability studied when the nitrogen was removed. Identical samples for UV-vis spectroscopic studies were prepared in the absence of D_2O. 2-Mercaptoethanol was added to some samples immediately before UV-vis spectroscopy.

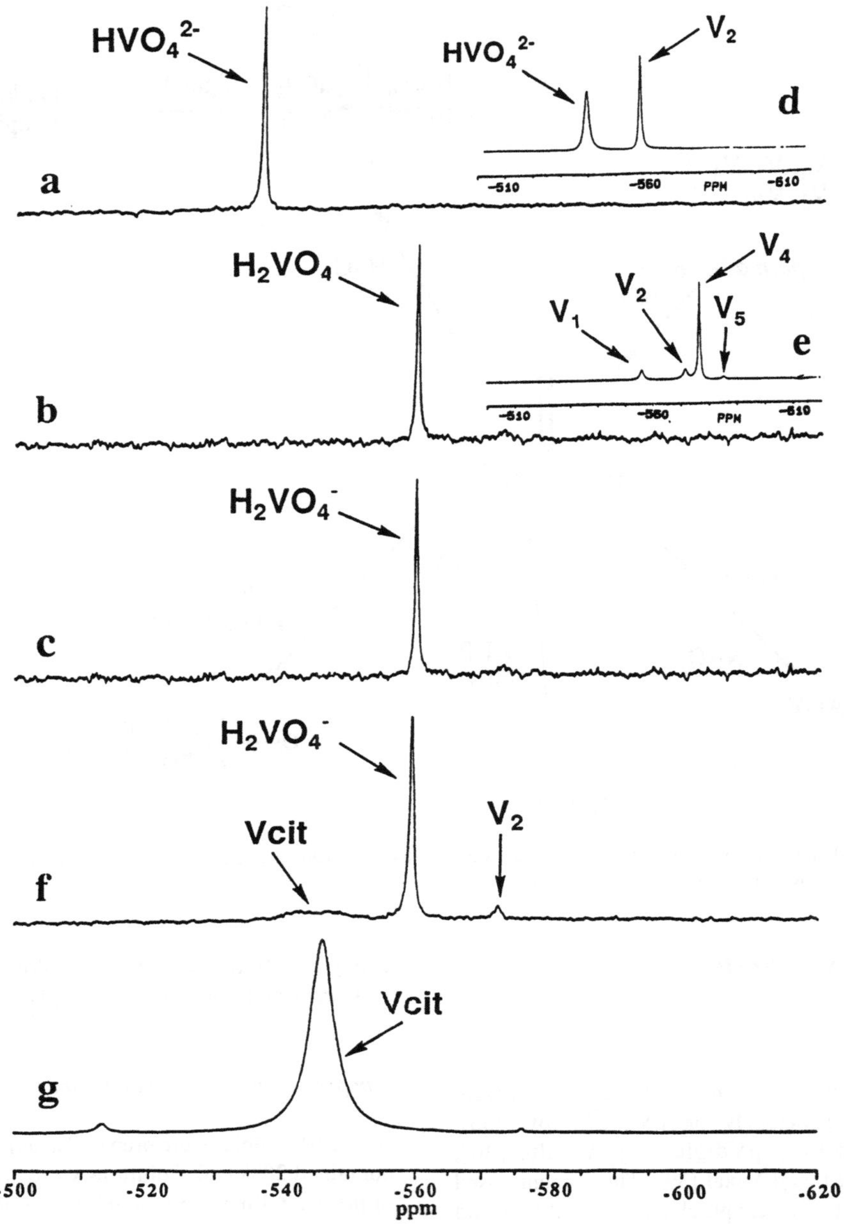

Fig. 2. ^{51}V NMR spectra of vanadate solutions prepared from various sources of vanadate: a) 0.38 mM vanadate prepared from sodium orthovanadate (pH 10.50), b) 0.38 mM vanadate prepared from sodium metavanadate (pH 7.04), c) 0.4 mM colorless vanadate prepared from sodium orthovanadate and adjusted to pH 7.04, d) 10.0 mM vanadate prepared from sodium orthovanadate (pH 12.5), e) 10.0 mM vanadate prepared from sodium metavanadate (pH 7.04), f) 0.38 mM sodium orthovanadate adjusted to pH 7.04 with citric acid, and g) 10.0 mM sodium metavanadate and 40.0 mM citrate at pH 7.04. The data presented were confirmed in duplicate or triplicate experiments.

Spectroscopy

^{51}V NMR, UV-vis and EPR spectroscopy were used to monitor the decomposition of vanadium compounds. Each ^{51}V NMR spectrum took 4.5 min to record using routine parameters [40]. ^{51}V NMR spectroscopy is limited to measuring the vanadium(V) in solution, so to confirm the NMR studies, we used UV-vis spectroscopy (Perkin Elmer kamda 4B) which gives different absorption spectra for the vanadium(V) and vanadium(IV) compounds in solution. Extinction coefficients and/or absorbances for each vanadium derivative were measured first, and then the UV absorbance was used to measure the disappearance and/or appearance at various wavelengths of a specific complex and other components in solution. In some cases EPR spectroscopy was used to measure the vanadium(IV) compounds.

Results and discussion

Vanadate salts

Several vanadate salts are commercially available, but all form the same vanadate anions on dissolution (reviewed in Ref. [40]) (Fig. 2). A solution of sodium orthovanadate in distilled water at 0.38 mM vanadate has a pH of 10.50 and contains mostly HVO_4^- (Fig. 2A). ^{51}V NMR spectroscopy is a convenient method to monitor the form of vanadate in solution since the chemical shift is very sensitive to the form of the vanadium atom (see Ref. [40] and references therein). A solution of 0.38 mM vanadate from sodium metavanadate has pH 7.04 and contains mainly $H_2VO_4^-$ with some HVO_4^{2-} (Fig. 2B). Adjusting the pH of the solution in Fig. 2A will produce solutions with the same anionic form of vanadate as shown in Fig. 2B (Fig. 2C). The spectrum recorded of a stock solution of 10.0 mM sodium orthovanadate (pH 12.5) is shown in Fig. 2D. The additional resonance observed is vanadate dimer (V_2) which hydrolyzes upon dilution or administration [41]. The spectrum recorded of a stock solution of 10.0 mM sodium metavanadate (pH 7.04) contains three additional signals: V_2, V_4 and V_5 which all rapidly hydrolyze to form vanadate upon dilution or administration [41]. The presence of these other forms in stock solutions is therefore of no concern. However, the addition of for example citrate to drinking water for rats (see, for example, Ref. [42]) can be problematic; see ^{51}V NMR spectra in Fig. 2F and 2G of vanadate solutions containing various concentrations of citrate. If possible, additives should be avoided because they often generate new complexes with vanadate and with vanadyl cation.

Stability of vanadium compounds in aqueous solution

The stability of 11 vanadium compounds was measured using ^{51}V NMR, UV and EPR spectroscopy in aqueous solution including compounds previously used as insulin-mimetic agents (vanadate, $VOSO_4$, bpV, bpVpic, pVdipic, Vmalto), and related compounds (VEDTA, Vcit, Voxal, $(bpV)_2$ and Vdipic) (Fig. 3). As shown in Fig. 3A, vanadate, $VOSO_4$, $(bpV)_2$, pVdipic, Vdipic, VEDTA, Vcit and Voxal showed little evidence for decomposition after 4 days when dissolved in distilled water (see Fig. 3A caption for pH information). The results obtained by ^{51}V NMR, EPR and UV-vis spectroscopy agreed in all cases when both carried out. Vmalto decomposed 50% in 1 day, at which level it remained for 4 days. Both bpV and bpVpic completely decomposed within 1 day, however, some (if not all) peroxo compounds are sensitive to light [43], and since peroxovanadium species decompose through radical mechanisms [44] significant variation in decomposition times can be expected depending on purity of bpVpic, water and other solution additives. Some variation was observed even in our hands in repeat decomposition runs particularly in the case of bpVpic; the results shown represent an average of three experiments.

Adjusting the solutions to pH 7 and maintaining this pH by 15 or 20 mM Hepes decreased (or remained the same) the stability of all but one complex (Fig. 3B), bpVpic, which had an increased half-life. The reactions were monitored both in distilled water and in Hepes, and unless indicated, no differences were observed. At pH 7 Voxal and $VOSO_4$ hydrolyzed completely within 5 min, and 30–50% of the vanadyl cation oxidized to vanadate within 4 days. Three compounds bpV, pVdipic and Vmalto hydrolyzed completely within 2 days. The decomposition rate of bpVpic in Hepes at pH 7 decreased, and only after approximately 2 days, the compound began to decompose rapidly as observed for radical decomposition mechanisms [44]. Within 5 min 70% of Vdipic decomposed and 15% of Vcit decomposed; both remained at these levels after 4 days. In contrast, vanadate, $(bpV)_2$, and VEDTA complexes showed no change for 4 days at neutral pH.

Most of the vanadium complexes examined are stable in solution for days when dissolved in distilled water. However, if these compounds are administered orally as insulin-mimetic agents, the stability of the compounds at acidic pH (not shown here) and neutral pH is of concern. We have shown that the stability of most compounds examined here decrease at neutral pH, suggesting that intracellular life-times of these particular compounds are limited.

Lability of vanadium compounds in aqueous solution

Little information is available on the lability of vanadium complexes including insulin-mimics under physiological conditions. Given space limitations, we will not provide experimental evidence here for lability of even well-known vanadium compounds. Suffice to say, that despite the documentation by ^{51}V NMR spectroscopy of the stability of a particular vanadium compound, such complexes can be exceedingly labile [45, 46]. This point is important, because labile complexes will convert to vanadate under physiological conditions which accesses the multitude of reactions shown in Fig. 1 for vanadate. Lability is an important property for vanadium compounds used in diabetic drug therapy and should be considered.

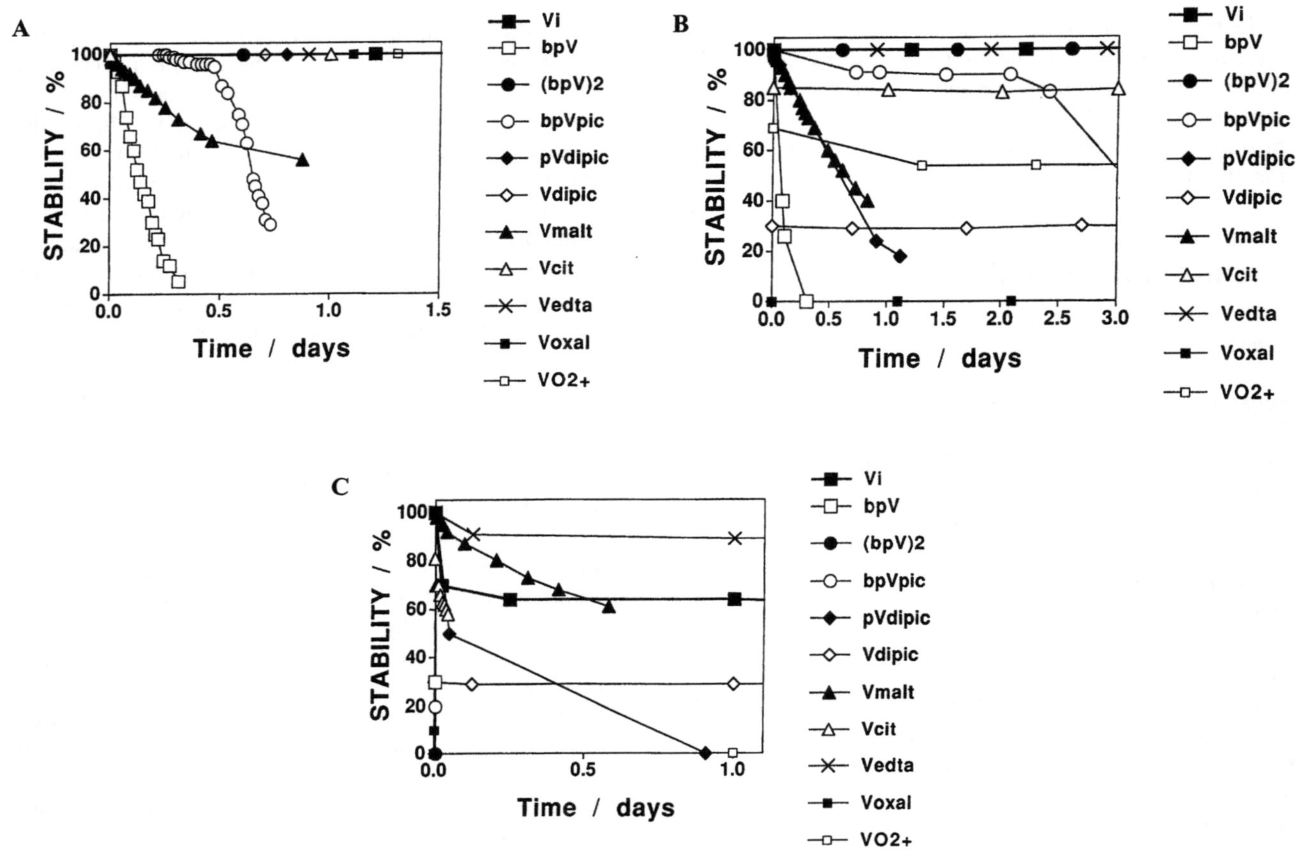

Fig. 3. The stability of a series of vanadium compounds dissolved in: A) distilled water, B) distilled water containing 15 or 20 mM Hepes at 7.0 (± 0.2). Hepes was added to avoid changes in pH during decomposition. C) distilled water containing 15–20 mM Hepes added 2 equivalents of 2-mercaptoethanol at pH 7.0 (± 0.2). The measurement was carried out using both ^{51}V NMR spectroscopy (2 and 10 mM solutions), UV spectroscopy (10 mM solutions) and in some/other cases confirmed by EPR spectroscopy. Specifically in the thiol studies, the thiol absorbence spectra often overlap with the original vanadium compound necessitating us to monitor the formation of new presumably vanadium(IV) species at higher wavelengths. The wavelength and the absorbence for the species examined by UV spectroscopy are as follows: $NaVO_3$ (203 nm, 0.48 mM, at 100% A = 2.09), $VOSO_4$ (204 nm, 0.48 mM, at 100% A = 0.331), Vcit (569 nm, 10 mM, at 100% A = 0.0, at 0% A = 0.187), Vmalto (877 nm, 4 mM, at 100% A = 0.116), Voxal (800 nm, 10 mM, at 100% A = 0.0, at 0% A = 0.180), Vdipic (860 nm, 10 mM, at 100% A = 0.0, at 0% A = 0.349), VEDTA (779 nm, 10 mM, at 100% A = 0.0, at 0% A = 0.244). The symbols for each compound and the pH of the solution represented in Fig. 3A are indicated: vanadate, 7.1 (■), bpV, 6.8 (□), $(bpv)_2$, 6.6 (●), bpVpic, 7.3 (O), pVdipic, 6.2 (◆), Vdipic, 5.4 (◊), V-malto, 5.3 (▲), V cit, 3.3 (Δ), VEDTA, 5.9 (x), V-oxal, 5.1 (■) and vanadyl cation (VO^{2+}), 3.5 (□).

Stability of vanadium compounds in reducing intracellular environments

In Fig. 3C we follow the decomposition of 11 vanadium complexes in the presence of 2 equivalents of 2-mercaptoethanol (see below for glutathione). Vanadate reduces in the presence of a large excess of thiols and these findings were confirmed in this study [14, 15]. However, 2 mM vanadate in the presence of 15 mM Hepes and 4 mM 2-mercaptoethanol formed 23% of a new vanadium(V) complex (–356 ppm) whereas 65% remained as vanadate. After 4 days only 50% of the vanadate remained. VOSO4 immediately formed EPR silent complexes of which 30–40% were vanadium(V) compounds and included the new complex at -356 ppm. This solution remained stable for 4 days.

Four other compounds were immediately reduced by the thiol: bpV, $(bpV)_2$, bpVpic, and Voxal. Vcit completely decomposed within 12 h and the Vmalto decomposed to 50% after 1 day. Upon dissolution at pH 7 only 30% of Vdipic remained, and slow reduction decreased this concentration to 25% after 4 days. The pVdipic converted to Vdipic and vanadate upon treatment with 2 equivalents of 2-mercaptoethanol within 12 h, consistent with reaction of the hydrogen peroxide with the thiol. VEDTA, alone, persisted for 4 days at 90%.

As anticipated, most vanadium(V) compounds were not stable in the presence of glutathione and other thiols. The reduction was, in some cases, very sensitive to the ratio of thiol to vanadium compound; low ratios would prolong the lifetime of the vanadium compound. Studies were carried out with glutathione, but with this thiol the reduction occurred faster than with 2-mercaptocthanol. However, the fact VEDTA is stable in the presence of 2-mercaptoethanol was very encouraging and probably in part due to the very low equilibrium concentration of vanadate as well as the complex's redox potential. The pVdipic, on the other hand, formed Vdipic when reacting with thiol. Although Vdipic may not have the same insulin-mimetic activities as pVdipic [5], this reactivity pattern prolongs the presence of insulin-mimetic agents under physiological conditions. Further examination of vanadium compounds with desirable insulin mimetic properties could lead to identification of vanadium compounds with enhanced insulin-mimetic responses.

Conclusion

Three key properties, stability, lability and redox chemistry, are critical to the insulin mimetic action of vanadium compounds. These insulin-mimetic effects presumably result from interactions of vanadium(V) (and perhaps vanadium(IV)) compounds with metabolites and other cellular components. Observed stability of a vanadium compound in stock solutions does not imply stability after administration and uptake by cells. The stability under the latter conditions should specifically be examined. The reactivity of vanadate and the other vanadium compounds will affect the compounds insulin-mimetic properties. Exploring compounds with increased stability and modified reactivity pattern could play an important role in development of insulin mimetic agents. For example, despite the lability of the VEDTA complex, its rate of reduction by thiols is slow, which should prevent many of the intracellular reactions observed with vanadate. Alternatively, thiols react with compounds such as bpVdipic to form Vdipic and vanadate thus generating vanadium compounds with continued insulin mimetic action. By both these strategies compounds can extend their life-times and effects under physiological conditions. Recognizing the complex aqueous vanadium chemistry and information on stable compounds and compounds with altered reactivity pattern will facilitate the development of vanadium compounds for therapeutic use as oral insulin substitutes.

Acknowledgment

We thank Drs. D. A. Roess and G. R. Willsky for stimulating discussions and reading early versions of this manuscript. We thank NIH, American Heart Association and the Sloan Foundation for partially funding this work.

References

1. Shechter Y: Insulin-mimetic effects of vanadate. Possible implications for future treatment of diabetes. Diabetes 39: 1–5, 1990
2. Posner BI; Shaver A, Fantus IG: Insulin mimetic agents: Vanadium and peroxovanadium compounds. In: C.J. Bailey, P.R. Flatt (ed.). New Antidiabetic Drugs. Smith, Gordon, 1990, pp. 107–118
3. Shechter Y, Shisheva A, Lazar R, Libman J, Shanzer A: Hydrophobic carriers of vanadyl ions augment the insulinomimetic actions of vanadyl ions in rat adipocytes. Biochem 31: 2063–2068, 1992
4. Orvig C, Thompson KH, Battell M, McNeill JH: Vanadium compounds as insulin mimics. In: H. Sigel, A. Sigel (ed.). Metal lons in Biological Systems. Marcel Dekker Inc., New York, 1995, 31: 575–594
5. Posner BI, Faure R, Burgess JW, Bevan AP, Lachance D, Zhang-Sun G, Fantus I G, Ng JB, Hall DA, Soo Lum B, Shaver A: Peroxovanadium compounds. A new class of potent phosphotyrosine phosphatase inhibitors which are insulin mimetics. J Biol Chem 269: 4596–4604, 1994
6. Crans DC: Enzyme interactions with labile oxovanadates and other oxometalates. Comments on Inorganic Chemistry 16: 35–76, 1994
7. Stankiewicz PJ, Tracey AS, Crans DC: Inhibition of phosphate-metabolizing enzymes by oxovanadium complexes. In: H. Sigel, A. Sigel (ed.). Metal Ions in Biological Systems. Marcel Dekker Inc., New York, 1995, 31: 287–324
8. Stankiewicz PJ, Tracey AS: Stimulation of enzyme activity by oxovanadium complexes. In: H. Sigel, A. Sigel (ed.). Metal lons in Biological Systems. Marcel Dekker Inc., New York, 1995, 31: 249–286
9. Gresser MJ, Tracey AS, Stankiewicz PJ: The interaction of vanadate with tyrosine kinases and phosphatases. Adv Prot Phosphatases 4: 35–57, 1987
10. Nechay BR, Nanninga, LB, Nechay PSE, Post, RL, Grantham JJ, Macara IG, Kubena LF, Phillips TD, Nielsen FH: Role of vanadium in biology. FASEB 45: 123–132, 1986
11. Willsky GR: Vanadium in the biosphere. In: N.D. Chasteen (ed.). Vanadium in Biological Systems: Physiology and Biochemistry. Kluwer Academic Publishers: Boston, 1990, p. 1–24
12. Gresser MJ, Tracey AS: Vanadium(V) oxyanions: The esterification of ethanol with vanadate. J Am Chem Soc 107: 4215–4220, 1985
13. Gresser MJ, Tracey AS, Parkinson KM: Vanadium(V) oxyanions: The interaction of vanadate with pyrophosphate, phosphate, and arsenate. J Am Chem Soc 108: 6229–6234, 1986
14. Macara IG, Kustin K, Cantley LC, Jr.: Glutathione reduces cytoplasmic vanadate; mechanism and physiological implications. Biochim Biophys Acta 629: 95–106, 1980
15. Sakurai H, Shimomura S, Ishizu K: Reduction of vanadate(V) to oxovanadium(IV) by cysteine and mechanism and structure of the oxovanadium(IV)-cysteine complex subsequently formed. Inorg Chim Acta 55: L67–L69, 1981
16. Kustin K, McLeod GC, Gilbert TR, Briggs LBR 4th: Vanadium and other metal ions in the physiological ecology of marine organisms. Structure and Bonding 53: 139–161, 1983
17. Chasteen ND, Grady JK, Holloway CE: Characterization of the binding, kinetics, and redox stability of vanadium(IV) and vanadium(V) protein complexes in serum. Inorg Chem 25: 2754–2760, 1986
18. Willsky GR, White DA, McCabe BC: Metabolism of added orthovanadate to vanadyl and high-molecular-weight vanadates by Saccharomyces cerevisiae. J Biol Chem 259: 13273–13281, 1984
19. Cantley LC, Jr., Aisen P: The fate of cytoplasmic vanadium. J Biol Chem 254: 1781–1784, 1979
20. Crans DC, Cortizo AM, Etcheverry SB, Mahroof-Tahir M: Vanadate

proliferation in osteoblast: Studies probing the active species, submitted, 1994

21. Chasteen ND: Vanadyl(IV) EPR spin probes inorganic and biochemical aspects. In: J. Reuben (ed.). Biological Magnetic Resonance. Plenum Press: New York, 1981, p. 53–119
22. Crans DC, Simone CM, Blanchard JS: Chemically induced modification of cofactor specificity of glucose-6-phosphate dehydrogenase. J Am Chem Soc 114: 4926–4928, 1992
23. Liochev SI, Fridovich I: Vanadate-stimulated oxidation of NAD(P)H in the presence of biological membranes and other sources of O_2^-. Arch Biochem Biophys 279: 1–7, 1990
24. Nour-Eldeen AF, Craig MM, Gresser MJ: Interaction of inorganic vanadate with glucose-6–phosphate dehydrogenase. Nonenzymic formation of glucose–6-vanadate. J Biol Chem 260: 6836–6842, 1985
25. Drueckhammer DG, Durrwachter JR, Pederson RL, Crans DC, Daniels L, Wong C-H: Reversible and in situ formation of organic arsenates and vanadates as organic phosphate mimics in enzymatic reactions: Mechanistic investigation of aldol reactions and synthetic applications. J Org Chem 54: 70–77, 1989
26. Lindquist RN, Lynn JL, Jr., Lienhard GE: Possible transition-state analogs for ribonuclease. The complexes of uridine with oxo-vanadium(IV) ion and vanadium(V) ion. J Am Chem Soc 95: 8762–8768, 1973
27. Ray WJ, Jr., Puvathingal JM: Characterization of a vanadate-based transition-state-analogue complex of phosphoglucomutase by kinetic and equilibrium binding studies. Mechanistic implications. Biochem 29: 2790–2801, 1990
28. Percival MD, Doherty K, Gresser MJ: Inhibition of phospho-glucomutase by vanadate. Biochem 29: 2764–2769, 1990
29. Goodno CC: Inhibition of myosin ATPase by vanadate ion. Proc Natl Acad Sci USA 76: 2620–2624, 1979
30. Combest WL, Johnson RA: Detergent-induced distinctions between fluoride- and vanadate-stimulated adenylate cyclases and their responses to guanine nucleotides. Arch Biochem Biophys 225: 916–927, 1983
31. Lopez V, Stevens T, Lindquist RN: Vanadium ion inhibition of alkaline phosphatase-catalyzed phosphate ester hydrolysis. Arch Biochem Biophys 175: 31–38, 1976
32. Kadota S, Fantus IG, Deragon G, Guyda HJ, Hersh B, Posner BI: Peroxide(s) of vanadium: A novel and potent insulin-mimetic agent which activates the insulin receptor kinase. Biochem Biophys Res Commun 147: 259–266, 1987
33. Svensson IB, Stomberg R: Studies on peroxovanadates. I. The crystal structure of ammonium μoxo-bis(oxodiperoxovanadate(V)), $(NH_4)_4[O(VO(O_2)_2)_2]$. Acta Chem Scand 25: 898–910, 1971
34. Shaver A, Ng JB, Hall DA, Lum BS, Posner BI: Insulin-mimetic peroxovanadium complexes: Preparation and structure of potassium oxodiperoxo(pyridine-2-carboxylato)vanadate(V), $K_2[VO(O_2)_2(C_5H_4NCOO)]\cdot 2H_2O$, and potassium oxodiperoxo(3-hydroxypyridine-2-carboxylato)vanadate(V), $K_2[VO(O_2)_2(OHC_5H_3NCOO)]\cdot 3H_2O$, and their reactions with cysteine. Inorg Chem 32: 3109–3113, 1993
35. Wieghardt K: Preparation and characterization of dipicolinato-vanadium(V) complexes: Kinetics and mechanism of their reaction with hydrogen peroxide in acidic media. Inorg Chem 17: 57–64, 1978
36. McNeill JH, Yuen VG, Hoveyda HR, Orvig C: Bis(maltolato)-oxovanadium(IV) is a potent insulin mimic. J Med Chem 35: 1489–1491, 1992
37. Przyborowski L, Schwarzenbach G, Zimmerman T: Komplexe XXXVII. Die EDTA-komplexe des vanadiums(V). Helvectia Chimica Acta 48: 1556–1565, 1965
38. Djordjevic C, Lee M, Sinn E: Oxoperoxo(citrato)- and dioxo-(citrato)vanadates(V): Synthesis, spectra, and structure of a hydroxyl oxygen bridged dimer, $K_2[VO(O_2)(C_6H_6O_7)]_2\cdot 2H_2O$. Inorg Chem 28: 719-723, 1989
39. Rieskamp H, Gietz P, Mattes R: Tetrameric dioxo(oxalato)-vanadates(V). The crystal structure of $K[VO_2(C_2O_4)_2]\cdot{}_2H_2O$. Chem Ber 109: 2090–2096, 1976
40. Crans DC: Aqueous chemistry of labile oxovanadates: Of relevance to biological studies. Comments on Inorganic Chemistry 16: 1–33, 1994
41. Crans DC, Rithner CD, Theisen LA: Applicadon of dme-resolved ^{51}V 2D NMR for quantitation of kinedc exchange pathways between vanadate monomer, dimer, tetramer, and pentamer. J Am Chem Soc 112: 2901–2908, 1990
42. Brichard SM, Bailey CJ, Henquin J-C: Marked improvement of glucose homeostasis in diabetic ob/ob mice given oral vanadate. Diabetes 39: 1326–1332, 1990
43. Arransio D, Suber L, Shul-pin GB: Photochemical oxidation of hydrocarbons by vanadium(v)peroxo complex. Izv Akad Nauk Ser Khim 8: 1918–1921, 1992
44. Bonchio M, Conte V, Di Furia F, Modena G, Moro S: Nature of the radical intermediates in the decomposition of peroxovanadium species in protic and aprotic media. Inorg Chem 33: 1631–1637, 1994
45. Crans DC, Ehde PM, Shin PK, Pettersson L: Structural and kinetic characterization of simple complexes as models for vanadate-protein interactions. J Am Chem Soc 113: 3728–3736, 1991
46. Crans DC, Shin PK, Armstrong KB: Application of NMR spectroscopy to studies of aqueous coordination chemistry of vanadium(V) complexes. In: H. Thorp, V. Pecoraro (ed.). Mechanistic Bioinorganic Chemistry. American Chemical Society: Washington, DC, 1995, in press

Molecular and Cellular Biochemistry **153**: 25–29, 1995.

Peroxo heteroligand vanadates(V): Synthesis, spectra-structure relationships, and stability toward decomposition

Cirila Djordjevic[1], Nikola Vuletic[2], Myunghi Lee Renslo[3], Bruce C. Puryear[4] and Ramin Alimard[5]

[1]*Department of Chemistry, The College of William and Mary, Williamsburg, VA 23187–8795, USA;* [2] *Institute of Tobacco, 41000 Zagreb, Croatia;* [3]*Kaiser M. D. Associated, Los Angeles, CA 90274;* [4]*Department of Anesthesiology, College of Medicine, University of Florida, Gainesville, FL 32610 and* [5]*Medical College of Virginia, VCU, Richmond, VA 23298–0462, USA*

Abstract

Effect of heteroligands (L) on the properties of vanadium peroxides was investigated by preparing a number of peroxovanadium complexes, which were characterized by analysis, IR, UV/V and NMR spectra. X-ray structures for some were obtained. The vanadates(V) contain the cation M(I) = Na, K, NH_4, Rb or Cs. Diperoxo complexes include $M(I)[VO(O_2)_2 L]$, where L = dipyridyl, o-phenanthroline; $M(I)_3[VO(O_2)_2(C_2O_4)]$; $K_2[(\text{nicotinic acid})\{VO(O_2)_2\}_2]H_2O$;$M(I)_4[O\{VO(O_2)_2\}_2\text{cystine}]2H_2O$; $H_4[O\{VO(O_2)_2(\text{adenine})_2\}_2]2H_2O$; and $K_2H_2[O\{VO(O_2)_2(\text{adenosine})\}_2]2H_2O$. Monoperoxo vanadates(V) correspond to the formula $M(I)_2[VO(O_2)L]_2$ for L = citrate and malate; $M(I)_2[VO(O_2)L]$ for L = nitrilotriacetate; $M(I)[VO(O_2)L]$ for L = iminodiacetate, tartrate and EDTA; and $[HVO_2(O_2)(\text{adenosine})]2H_2O$. Syntheses of these heteroligand peroxovanadium compounds are sensitive to pH, temperature and the concentration of the components. The stability towards decomposition in solid state, mother-liquid and pure water solutions depends upon the heteroligand. Characteristic (V=O) and (O-O) stretching frequency bands in IR can be correlated with the corresponding bond lengths and the [peroxo~V(V)] charge transfer bands in UV/V spectra. Intramolecular one-electron transfer in peroxo vanadates(V) can trigger the generation of radicals, and its dependency upon the nature of the heteroligand is discussed. (Mol Cell Biochem **153:** 25–29, 1995)

Key words: vanadium peroxo complexes, peroxo aminopolycarboxylato vanadates, peroxo cystine vanadate, peroxo adenine vanadate, peroxo adenosine vanadate, peroxo-vanadium charge transfer bands

Introduction

Insulin mimetic properties of peroxovanadium compounds have revived and intensified the interest in these complexes [1, 2]. Over the years we have synthesized and characterized about 45 novel peroxoheteroligand vanadates [3–11]. An overview of these complexes is presented here, including some new peroxovanadium compounds. Chemical properties of all these vanadium peroxides vary significantly with the nature of the heteroligand present.

In moderately acid or alkaline aqueous solutions the coordination sphere of peroxo vanadates includes one or two peroxo groups, in addition to one or two oxo groups. The composition of vanadium peroxides depends upon pH, the peroxide and the vanadium concentration. Generally, several complex species are present, governed by complex equilibria [12, 13]. In the presence of molecules or ions which contain suitable donor atoms, the oxo-peroxo ligand spheres tend to incorporate molecule(s) or ion(s) present, which are then referred to as "heteroligands". Heteroligands (L) can be monodentate, i.e. occupying one coordination site, or polydentate, where commonly oxygen or nitrogen donor atoms occupy two or more sites in the V(V) coordination polyhedron, forming chelate rings. The latter tend to stabilize the

Address for offprints: C. Djordjevic, Department of Chemistry, The College of William and Mary, Williamsburg, VA 23187–8795, USA coordinated peroxo group(s), as well as the whole complex V(V) entity. The presence of a peroxo group always stabilizes V(V) of a d^0 electron configuration [12].

Heteroligand peroxo vanadates are of interest to biochemistry because they enclose two biologically relevant components: vanadium, recognized as important in a number of bio-processes [2, 12–18], and peroxides, widely respected dioxygen species in biomatter. A third bio-interesting component can be the heteroligand, made of a biomolecule or an ion participating in the metabolism. Vanadium is an interesting metal because of its V(V)/V(IV) interplay of the electron configuration d^0/d^1. The thermodynamic potential of this one-electron transition changes in the presence of heteroligands, and it can overlap with the peroxide oxidation, as well as with the Fe(III)/Fe(II) redox process, involving d^5/d^6 electron configuration. Consequently, a proper tuning of V(V)/V(IV) potential by heteroligands can generate oxygen radicals, such as the superoxide ion, O_2^-, OH·, or singlet oxygen. Combined with the stereochemical nonrigidity of vanadium ion, and the lability of coordination sites, the presence of peroxo vanadates can therefore perturb a variety of bioprocesses [12].

Materials and methods

Crystalline peroxoheteroligand vanadates(V) are generally obtained from concentrated aqueous solutions [3–11]. Syntheses are not always reproducible, and are sensitive to the temperature, pH, the concentration of components, and sometimes to the sequence and the rate of reagents' addition. Coprecipitation of a yellow vanadium peroxide, or of an excess of the heteroligand is often a problem. Occasionally minor explosions occur, and impure products are more sensitive toward decomposition in the solid state. An unusual preparation of the novel cystine-peroxo vanadium compound is described below.

Synthesis of $K_4[O\{VO(O_2)_2\}_2$ cystine$]2H_2O$. V_2O_5(0.91 g, 10.0 mmole V) was dissolved in KOH (1.15 g, 20.5 mmoles in 10 ml H_2O) by stirring at 50–60°C. H_2O_2 (30%, about 0.5 ml) was then added to insure complete dissolution. Cysteine (1.21 g, 10.0 mmoles) was added to the clear solution at room temperature, producing a clear purple-black solution of pH ~ 10. The reaction mixture was put in an ice bath, and H_2O_2 (30%, 6.0 ml, 58 mmoles) was added dropwise under stirring. The solution was then stirred gently at room temperature for about 2 h, and a yellow solid precipitated. Reaction mixture was then left to settle for about 15–30', and was afterwards filtered (pH about 6). Crystalline precipitate was washed once with ice-cold water and then ethanol (three times). Yield: 2.1–2.8 g, 60–80%. The microcrystalline compound is readily soluble in water. A violent reaction sometimes redissolves the precipitate, if left standing in mother liquid. The same compound can be obtained with cystine, but the reaction is more violent and usually does not yield a precipitate of uniform composition. Anal. calcd. for $C_6H_{16}K_4N_2O_{17}S_2V_2$: C, 10.1; H, 2.3; N, 3.9; O, 9.0; V, 14.3; O_2^{2-}, 18.0. Found: C, 10.1; H, 2.4; N, 4.0; O, 9.1; V, 14.4; O_2^{2-}, 18.1.

Analyses and instrumental measurements were carried out as described previously [3, 5–7].

Results

Peroxo heteroligand vanadates always contain V(V), which is stabilized by the presence of peroxo group. We have prepared crystalline diperoxo and monoperoxo compounds, briefly summarized below. M(I) usually represents K or NH_4 cation. Diperoxo complexes are of the type $M(I)_x[VO(O_2)_2 L]$, containing neutral bidentate heteroligands L = dipyridyl and o-phenanthroline, or the oxalate ion, $C_2O_4^{2-}$ [3]. Diperoxo compounds with bio-heteroligands correspond to the formulae:

K_2 [(nicotinic acid)$\{VO(O_2)_2\}_2]H_2O$ [6], $M(I)_4[O\{VO(O_2)_2\}_2$-cystine$]2H_2O$ [11], $H_4[O\{VO(O_2)_2$(adenine$)_2\}_2]2H_2O$, and $K_2H_2[O\{VO(O_2)_2$(adenosine$))_2]2H_2O$ [9], and the syntheses of the last three were not yet published. All diperoxo complexes are yellow, displaying a broad [peroxo → vanadium] charge transfer band in the vicinity of 320 nm. Monoperoxo complexes with aminopolycarboxylato and hydroxypolycarboxylato heteroligands are orange to red: $M(I)_2[VO(O_2)L]$ for L = NTA = nitrilotriacetate [10], $M(I)[VO(O_2)L]$ for L = IDA = iminodiacetate [5], tartrate, and EDTA. Salts of M(I) = Na, K, NH_4, Rb and Cs were synthesized. Stability of these complexes towards decomposition in solid state and pure water solution decreases in the order NTA > IDA > EDTA. The NTA complex is unusually stable in the solid state and aqueous solutions. NMR spectrum in D_2O did not change for 24 h, and the visible spectrum in aqueous solutions does not change in the pH range 1–6. Crystals have remained unchanged for years, and they can grow in the mother liquid for months. Citrato [7] and malato [11] complexes are interesting dimers bridged by two alkoxy oxygens, as shown by the X-ray structure analysis, and correspond to the formula $M(I)_2[VO(O_2L]_2$. Tartrate compounds are not easy to prepare and they readily decompose. Stability in hydroxypolycarboxylato series in solid state and pure water solutions follows the sequence citrato > malato > tartrato. In general, stability towards decomposition in the solid state or pure water solutions depends upon the heteroligand. Some compounds, kept in a dry, dark atmosphere remained unchanged for years, and others decomposed in a few days or after a few hours. Loss of peroxides in the solid state can occur without much change in color, and can be followed by peroxide analyses. Decomposition in aqueous solution can be followed by changes in

NMR. Loss of peroxides finally leads to reduction to V(IV). The yellow or orange-red solutions usually turn blue, a typical color for V(IV).

IR spectra of heteroligand peroxo vanadates are very informative and useful. In the first place, the specific shifts of the heteroligand bands occur upon the coordination on comparing with the spectra of the free ligand. They reveal the coordination sites in chelating molecules or ions. Additional very strong bands appear due to the stretching frequency of the (V=O) and (O-O) bonds. A selected set of compounds is shown in Table 1, where the corresponding IR bands are listed along with the observed bond lengths and the charge transfer bands in UV/V spectra. A distinct difference is found between the spectra of mono- and diperoxo complexes in UV/V. The [peroxo →vanadium] charge transfer band appears in diperoxo complexes around λ_{max} ~ 320 nm. Monoperoxo complexes absorb at a much lower energy with the λ_{max} ~ 420 nm. Related to it is a shorter (O-O) bond length in monoperoxo complexes (1.43–1.44 Å), as compared to a longer bond (1.45 - 1.47 ~) in the diperoxo species. The (O-O) bond length difference is further reflected in the (O-O) stretching frequencies in IR, where monoperoxo complexes display a strong band in vicinity of 930 cm^{-1}, at a higher energy than the diperoxo compounds, which show a strong doublet at about 890, 860 cm^{-1} [3–11].

X-ray structure analyses of a number of peroxoheteroligand vanadates have shown that in most of mono- and diperoxo complexes the V(V) achieves the coordination number seven, best viewed as a distorted pentagonal bipyramid [1, 5, 7, 10, 11, 19–21]. The vanadyl V=O group is located at the apical position, cis to the equatorially placed peroxo groups, and trans to an usually elongated vanadium-ligand bond. The (O-O) bond lengths observed in monoperoxo complexes shown in Table 1 are definitely shorter than in diperoxo vanadates. The citrato and malato peroxovanadates are interesting dimers, bridged by two alkoxy oxygens in an almost regular rhomboidal V_2O_2 plane [11]. The new compounds described here, cystine, adenine and adenosine peroxovanadates were microcrystalline and too small for an X-ray structure analysis. Cystine complex was obtained by oxidation of cysteine in the process of the synthesis. The existence of these complexes demonstrates that unlike cysteine [1, 8], cystine, adenine and adenosine [9] can coexist with peroxo group in a vanadate complex.

Discussion

Crystalline peroxoheteroligand vanadates are separated from aqueous solutions, in which several peroxo species are usually present, governed by complex equilibria [12–15]. The crystals contain the complex ion which is probably more abundant in the solution, and is favored, in addition, by the lattice energy of the solid. Extrapolating the properties of the complexes from solid state to aqueous solution is just as risky as it is to make extrapolations from the behavior in pure aqueous solutions to aqueous systems in the biomatter. However, in the absence of more sophisticated techniques hopefully developed in the future for examining reactions in living matter, we have to use with proper reservation data obtained from instrumentation and methods available today.

The presence of peroxides stabilizes V(V) of a d° electron configuration. In the absence of peroxides, V(V) and V(IV)

Table 1. (V=O) and (O-O) bond lengths [Å], IR stretching frequencies [cm^{-1}], and charge transfer bands in UV/V [nm] for selected peroxo heteroligand vanadates(V)

Compound/ion	$V_{v=o}$ [cm^{-1}]	V=O [Å]	$V_{(O-O)}$ cm^{-1}	O–O[Å]	λ_{max} [nm]	Reference
$(NH_4)_4[O\{VO(O_2)_2\}_2]$	975	1.613(3)	890	1.474(4)	320	20
	940	1.601(3)	870	1.463(4)		
$K_3IVO(O_2)_2(C_2O_4)]2H_2O$	930	1.622(4)	877, 852	1.460(6)	325	3, 19
				1.451(6)		
$K[VO(O_2)_2bipy]5H_2O$	930		875, 856		320	3
$(NH_4)[VO(O_2)_2(phen)]2H_2O$	930		870, 850		315	3
$K_2[VO(O_2)NTA]2H_2O$	960	1.617(2)	926	1.438(4)	425	10
$NH_4[VO(O_2)IDA]$	970	1.587(1)	930	1.435(5)	420	5
$K_2[VO(O_2)citrato]_22H_2O$	988	1.601(1)	933	1.427(2)	415	7
$K_2[VO(O_2)malato]_22H_2O$	977	1.603(1)	925	1.441(2)	425	11
$K_2[nicotinic\ acid\{VO(O_2)_2\}_2]H_2O$	941		879		329	6
			863			
$K_4[O\{VO(O_2)_2\}_2cystine]2H_2O$	971		886		315	8
	943		870			
$H_4[O\{VO(O_2)_2(adenine)_2\}_2]2H_2O$	970		890			9
	940		860			
$K_2H_2[O\{VO(O_2)_2(adenosine)\}_2]2H_2O$	950		885			9
	940		875			
$[HVO_2(O_2)\ (adenosine)]2H_2O$	975		930			9
	930					

species coexist in any solution containing water, including therefore the biomatter at physiological pH [12,15,22]. Thermodynamic potentials indicate that intramolecular one-electron transfer can occur in peroxoheteroligand vanadates [12]. The V(V)/V(IV) potential is tuned by the nature of heteroligand present in the coordination sphere, and under certain condition V(V) can be reduced to V(IV) by peroxide, which is oxidized to superoxide, O_2^-. This process then triggers generation of radicals, including OH· and singlet oxygen. Biological activity of peroxovanadates may well be related (among other factors) to this radical generation ability, which can also function in the opposite direction, operating as a radical scavenging activity. The mechanism of action of peroxovanadium compounds has previously been attributed to radical mechanisms [12, 18]. Recently, radical intermediates have been observed in the decomposition of peroxovanadium species in protic and aprotic media [23].

Biochemical interactions of peroxo heteroligand vanadates in the biomatter are very complex essentially because of the following three characteristic properties of V(V) and V(IV) ions: 1) the redox interplay of V(V)/V(IV), which can overlap with peroxide and Fe(III)/Fe(II) redox processes, including the generation of radicals; 2) the stereochemical nonrigidity of V(V) and V(IV), enabling low energy rearrangements of the coordination numbers 4, 5, 6 and 7; and 3) the lability of ligands at coordination sites, which enables a rapid exchange of donor atoms in vanadium coordination sphere [12]. Many vanadium compounds are known to hydrolyze fast in water and rapidly exchange ligands. Interactions of peroxo vanadium compounds in biomatter are expected to be of similar nature. Rearrangements and ligand exchanges must occur easily, when necessary combined with the electron transfer, in order to obey equilibria between complex species of V(V) and V(IV) called for at physiological pH.

Biological activity and toxicity of vanadium have been found to depend upon the nature of the compounds and the route of administering it [2, 12, 15, 17]. The complexity of biological interaction of peroxovanadium compounds inevitably brings about the possibility of interesting novel medicinal applications. Among the most fascinating recent discoveries are the insulin mimetic properties [1, 2] of peroxovanadium complexes, including the three compounds [2] given in Table 1. One of these, $K[VO(O_2)_2bipy]5H_2O$ proved to be among the most potent vanadium compounds tested. All three are diperoxovanadates, and the monoperoxo tested derivatives have shown a lesser activity [2]. It would be very interesting to test in this respect the monoperoxo $K_2[VO(O_2)$-NTA]$2H_2O$, which is exceptionally stable toward decomposition in the solid state, and aqueous solutions over a wider pH range [10]. In addition, the series of dimeric peroxovanadium complexes such as $(NH_4)_4[O\{VO(O_2)_2\}_2]$ and $M(I)_4[O\{VO(O_2)_2L\}_2]$, L = cystine, adenine, adenosine (shown in the Table I) look attractive. These compounds represent a set of water soluble diperoxo derivatives, which should behave differently than the monomers, and which contain molecules familiar to bioenvironment as heteroligands. These changes in vanadium ligand sphere are expected to affect the insulin mimetic activity. These dimers differ from the peroxides tested [2] in solubility, stability towards decomposition, and most likely also toxicity and related properties of importance for medicinal use. The probability of finding a peroxovanadium compound with outstanding insulin mimetic potency is very real, and investigations in this direction are expected to bring about impressive and consequential results.

Acknowledgement

Partial support of this work by the NSF (Grant CHE-8521376) is gratefully acknowledged.

References

1. Shaver A, Ng TB, Hall DA, Lum BS, Posner I: Insulin-mimetic peroxovanadium complexes: Preparation and structure of potassium oxodiperoxo (pyridine-2-carboxylato) vandate(V), $K_2[VO(O_2)_2$-$(C_5H_4NCOO)]\cdot 2H_2O$, and potassium oxodiperoxo (3-hydroxy-pyridine-2-carboxylato) vanadate(V), $K_2[VO(O_2)_2(OHC_5H_3N$-$COO)]\cdot 3H_2O$, and their reactions with cysteine. Inorg Chem 32: 3109–3113, 1993
2. Posner BI, Faure R, Burgess JW, Bevan AP, Lachance D, Zhang-Sun G, Fantus IG, Ng JB, Hall DA, Soo Lum B, Shaver A: Peroxovanadium Compounds. A new class of potent phosphotyrosine phosphatase inhibitors which are insulin mimetics. J Biol Chem 269: 4596–4604, 1994
3. Vuletic N, Djordjevic C: Oxoperoxovanadate(V) complexes with bidentate ligands. J Chem Soc Dalton: 1137–1141, 1973
4. Djordjevic C: Heteroligand peroxo transition metal complexes, Chem Br 18: 554–557, 1982
5. Djordjevic C, Craig SA: A polymeric peroxoheteroligandvanadate(V). Synthesis, spectra and structure of $M(I)[VO(O_2)(C_4H_5O_4N)]$. Inorg Chem 24: 1281–1283, 1985
6. Djordjevic C, Puryear BC, Vuletic N, Abelt CJ, Sheffield SJ: Preparation, spectroscopic properties and characterization of novel peroxo complexes of vanadium(V) and molybdenum(VI) with nicotinic acid and nicotinic acid N-oxide. Inorg Chem 27: 2926–2932, 1988
7. Djordjevic C, Lee M, Sinn E: Oxoperoxo(citrato)- and dioxo(citrato) vanadates(V): synthesis, spectra and structure of a hydroxyl oxygen bridged dimer, $K_2[VO(O_2)(C_6H_6O_7)_2]_2 2H_2O$. Inorg Chem 28: 719–723, 1989
8. Djordjevic C: Peroxo heteroligand complexes of vanadium(V) and molybdenum(VI): Analogy and differences. 29th ICCC, Lausanne, Switzerland, Abstract, p. 108, 1992
9. Djordjevic C, Vuletic N, Puryear B.C: Vanadium(V) and molybdenum(VI) peroxo complexes with adenine and adenosine, and the choice of donor atoms in peroxo heteroligand spheres. J Inorg Biochem 51: 308, 1993
10. Djordjevic C, Wilkins P.L, Sinn E, Butcher R.J: Peroxoaminopoly carboxylatovanadate(V) of an unusual low toxicity: synthesis and

structure of the very stable $K_2[VO(O_2)(C_6H_6NO_6)]2H_2O$. Inorg Chim Acta 230: 241–244, 1995

11. Djordjevic C, Lee-Renslo M, Sinn E: Peroxo malato vanadates(V): synthesis, spectra, and structure of $(NH_4)_2[VO(O_2)(C_4H_4O_5)]_2 2H_2O$ dimer with a rhomboidal V_2O_2(hydroxyl) bridging core. Inorg Chim Acta 233: 97–102, 1995
12. Djordjevic C: Antitumor Activity of vanadium compounds. In: H. Sigel, A. Sigel (eds.). Metal ions in biological systems, Vol. 31. Vanadium and its role for life. Marcel Dekker, New York, 1995
13. Conte V, Di Furia F, Moro S: ^{51}V-NMR investigationon the formation of peroxo vanadium complexes in aqueous solution: some novel obeservations. J Molec Catal, 94: 323–333, 1994
14. Jaswal SJ, Tracey AS: Reactions of mono- and diperoxovanadates with peptides containing functionalized side chains. J Am Chem Soc 115: 5600–5607, 1993
15. Chasteen N.D. (ed.): Vanadium in biological systems. Kluwer Academic Publishers, Dordrecht, 1990
16. Wever R, Kustin K: Vanadium: a biologically relevant element. Adv. Inorg. Chem. 35: 81–115, 1990
17. Nechay BR, Nanninga LB, Nechey PSE, Post RL, Grantham JJ, Macara IG, Kubena LF, Phillips TD, Nielsen FH: Role of vanadium in biology, Feder Proc 45: 123–132, 1966
18. Djordjevic C, Wampler G.L: Antitumor activity and toxicity of peroxo heteroligand vanadates(V) in relation to biochemistry of vanadium, J. Inorg. Biochem. 25: 51–55, 1985
19. Begin D, Einstein FWB, Field J: An assymmetrically coordinated diperoxo compound. Crystal structure of $K_3[VO(O_2)_2(C_2O_4)]H_2O$. Inorg Chem 14: 1785–1790, 1975
20. Stomberg R, Olson S, Svenson I.B: The crystal structure of ammonium μ-oxo bis(oxodiperoxovanadate)(4-),$(NH_4)_4[O\{VO(O_2)_2\}_2]$. A Refinement. Acta Chem.Scand. A 38: 653–656, 1984
21. Stomberg R, Szentivanyi H: The crystal structure of 2,2′-Bipyridinium(1+) (2,2′-Bipyridine)-oxodiperoxovanadate(1-)(3+x)-hydrogen peroxide–(2-x)–water, $(C_{10}H_9N_2)[VO(O_2)_2(C_{10}H_8N_2)]\cdot(3+x)H_2O_2$–$(2\text{-}x)H_2O$, x = 0.4, at –100°C. Acta Chem Scand A38: 121–128, 1984
22. Chasteen ND, Grady JK, Holloway CE: Characterization of the binding, kinetics and redox stability of vanadium(IV) and vanadium(V) protein complexes in serum. Inorg Chem 25: 2754–2760, 1986
23. Bonchio M, Conte V, Di Furia F, Modena G, Moro S: Nature of the radical intermediates in the decomposition of peroxovanadium species in protic and aprotic media. Inorg Chem 33: 1631–1637, 1994

Molecular and Cellular Biochemistry **153:** 31–36, 1995.

Chemically and photochemically initiated DNA cleavage by an insulin-mimetic *bis*peroxovanadium complex

Catharina Hiort, Jerry Goodisman and James C. Dabrowiak
Department of Chemistry, Syracuse University, Syracuse, NY 13244–4100, USA

Abstract

Chemically and photochemically induced cleavage of DNA by the insulin-mimetic compound $NH_4[VO(O_2)_2$-(1,10-phenanthroline)], bpV(phen), have been studied. ^{51}V NMR and absorption indicate that photoirradiation with low energy UV light of aqueous solutions containing bpV(phen) leads to the conversion of the compound to simple vanadates. Photoillumination of the compound in the presence of supercoiled pBR322 DNA results in cutting of the plasmid to produce nicked circular and linear DNA. Quantitative analysis of agarose gel data shows that bpV(phen) is a single strand nicking agent exhibiting sequence and/or base specificity. (Mol Cell Biochem **153:** 31–36, 1995)

Key words: DNA cleavage, insulin-mimetic, *bis*peroxovanadium, photolysis, photocleavage

Introduction

Studies involving vanadium in oxidation states IV and V have been numerous since the finding of several potent biological effects of this element. For example, vanadate functions as an inhibitor of many enzymes [1]. Vanadate has also been reported to photocleave a variety of proteins, among them tubulin [2]. Vanadium has considerable therapeutic potential, and vanadium salts have long been known to decrease glycosuria in diabetic patients [3]. Furthermore, vanadium(+4) and vanadium(+5) peroxo-complexes show antitumor activity against several types of tumor cells [4–6]. The basis for the antitumor properties of these compounds is unknown, but it has been suggested that the V(+4) complexes, upon reaction with cellular hydrogen peroxide, introduce strand breaks in DNA, whereas the peroxo compounds are thought to produce superoxide ions which kill the cell [7]. Recently, compounds of V(+4) and V(+5) containing organic ligands, have proved to be highly potent insulin mimetic agents [8–12].

Here we investigate the cleavage of DNA by the bisperoxovanadium(+5) complex, $NH_4[VO(O_2)_2$-(1,10-phenanthroline)] (bpV(phen)). We find that the compound can be photochemically and chemically activated to produce breaks in closed circular DNA.

Materials and methods

NH_4VO_3 was purchased from Johnson Matthey Materials Technology, UK, and 1,10 phenanthroline was obtained from Aldrich Chemical Company. $(NH_4)[VO(O_2)_2$(1,10-phenanthroline)] was synthesized according to Vuletic and Djordjevic [13].

Spectroscopy

The samples, contained in 1 cm quartz cuvettes or 1 cm NMR tubes, were irradiated at a distance of 30 cm from a Blak Ray 100 watt mercury lamp (emission maximum at 365 nm). The absorption was measured in phosphate buffer solutions (0.1 M, pH 6) using a Varian Cary 1 spectrophotometer and 1 cm quartz cuvettes. Anaerobic conditions were obtained by purging the solution of bpV(phen) for >15 min with a low, constant flow of nitrogen. The ^{51}V NMR spectra were recorded

Address for offprints: J.C. Dabrowiak, Department of Chemistry, Syracuse University, Syracuse, NY 13244-4100, USA at 65.81 MHz on a 250 MHz Bruker FTNMR spectrometer using a tunable 10 mm probe. The NMR measurements were carried out in D_2O solutions containing 25 mM bpV(phen). The spectral width was 50,000 Hz, the pulse angle 90°C, the accumulation time 0.082 sec, and no relaxation delay between pulses was used. The chemical shifts are reported relative to the external reference standard $VOC1_3$ (0 ppm).

DNA cleavage

pBR322 DNA was purchased from Boehringer Mannheim Biochemicals, USA, and used as obtained. The photocleavage reactions were performed in a volume of 25 µl (or 75 µl for the oxygen-depleted samples) in 0.1 M phosphate buffer containing 0.33 mM TRIS-HC1, pH 6. The samples were irradiated in Pyrex test tubes (Ø = 5 mm). Oxygen depletion was obtained by purging the DNA/bpV(phen) mixtures with nitrogen in the reaction tubes for up to 3 min. Prior to irradiation, the volumes of the solutions were readjusted to 75 µl by adding nitrogen-purged distilled water. The pBR322 DNA was separated via electrophoresis on a 1% agarose gel and DNA was visualized through ethidium bromide staining. The transilluminated gel was photographed with a Polaroid camera using type 55 film, and the negative was scanned with a Molecular Dynamics model 300A microdensitometer.

Results

Figure 1 shows the absorption of bpV(phen) in water solution. The absorption envelope of the complex consists of two

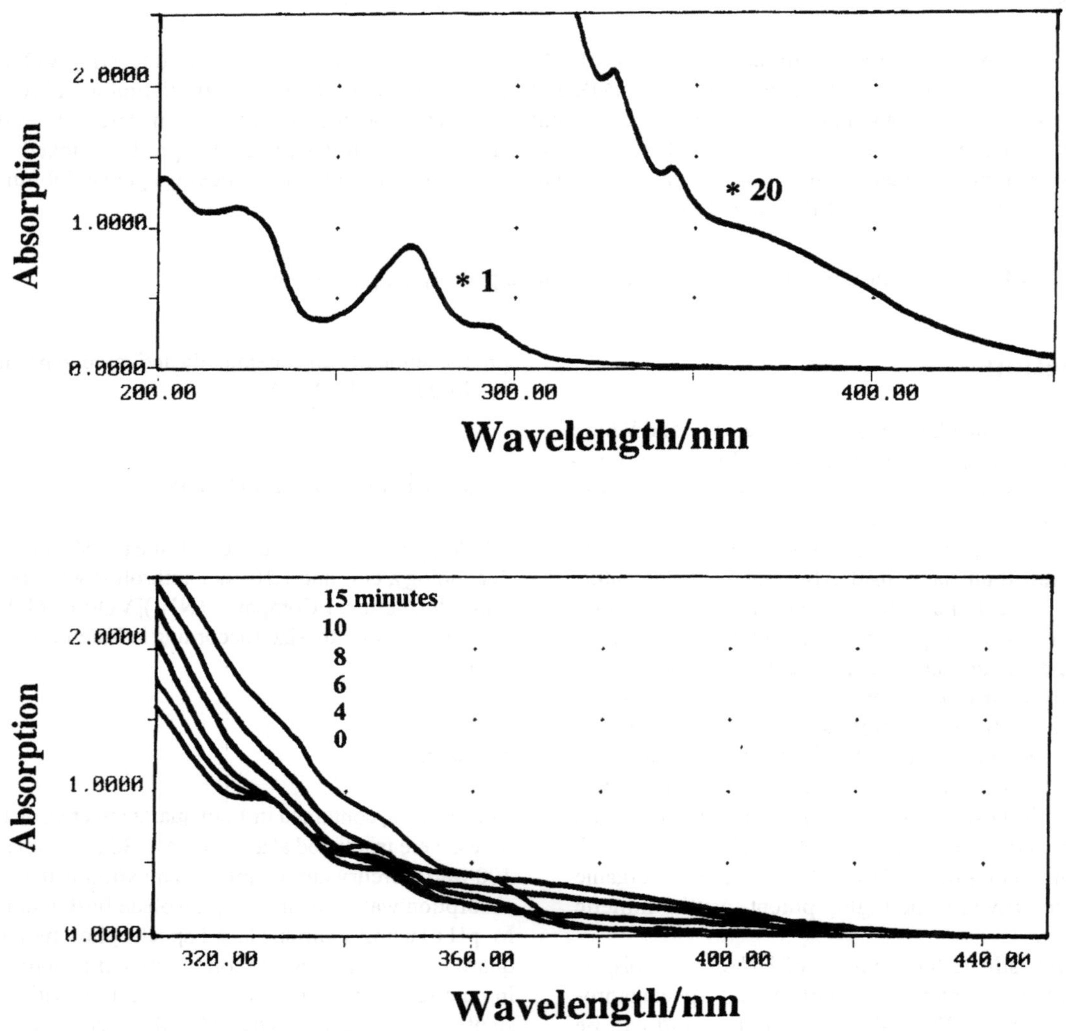

Fig. 1. Top panel: Absorption spectrum of 35 µM bpV(phen) in 0.1 phosphate buffer, pH 6. The signal intensity for the spectrum labeled *20 has been multiplied by a factor of 20. Bottom panel: Spectral changes induced by 365 nm photolysis of bpV(phen). Photolysis times for the curves are, from bottom to top at 320 nm; 0, 4, 6, 8, 10 and 15 min.

major features. (i) A series of weak bands (e ≈ 900 $cm^{-1}M^{-1}$) centered around 345 nm, arising from ligand-to-metal charge transfer (LMCT) transitions between the peroxo groups and the vanadium ion [14]. (ii) Strong p-p* intra ligand phenanthroline bands at 270 and 223 nm (found at 263 and 229 nm for the free ligand dissolved in ethanol).

Upon irradiation with UV light (λ_{max}=365 nm), into the ligand-to-metal charge transfer band of the metal complex, the absorption and the ^{51}V NMR spectrum of bpV(phen) change considerably and non-reversibly. No isosbestic point is obtained in the absorption when the irradiation time is increased. The ^{51}V NMR results (Fig. 2) show that bpV(phen) (–744 ppm) is converted into three new NMR-active species, the monomer ($H_2VO_4^-$, –558 ppm), dimer ($H_2V_2O_7^{-2}$, –571 ppm) and tetramer (cyclic $V_4O_{12}^{-4}$, –576 ppm) of

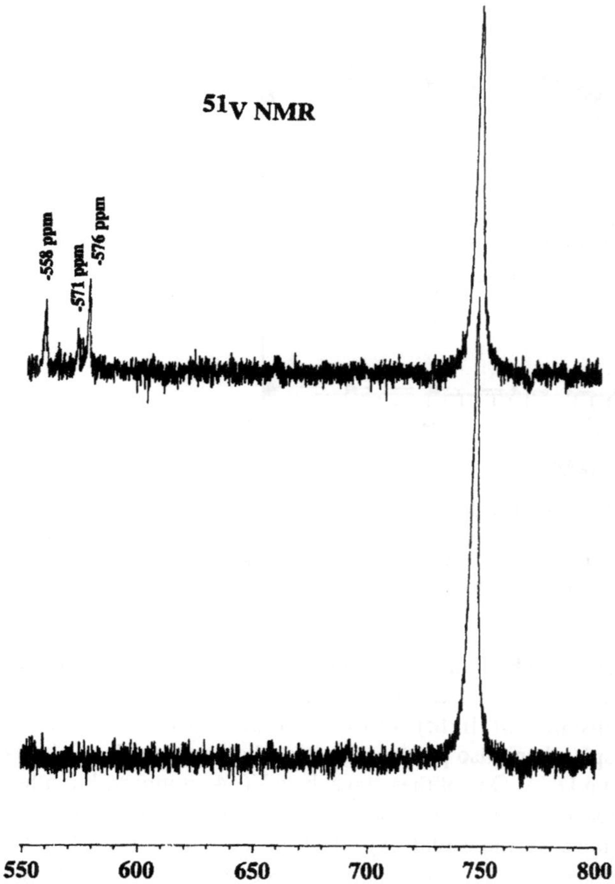

Fig. 2. Photodecomposition of bpV(phen) into simple vanadates compounds observed by ^{51}V NMR spectroscopy. Bottom panel: 25 mM bpV(phen) in D_2O, before irradiation. Top panel: Same sample after irradiation for 30 min. The resonance at –744 ppm originates from bpV(phen). The signals at –558, –571 and –576 ppm arise from the monomer, dimer and tetramer vanadate complexes, respectively.

vanadate. Under anaerobic conditions, the bpV(phen) decomposition rate is considerably higher as judged by earlier appearing changes in absorption and ^{51}V NMR spectra as a result of photoirradiation.

Photoirradiation of pBR322 plasmid DNA in the presence of bpV(phen) results in the conversion of the closed circular form (form I) to the nicked circular (form II) and the linear (form III) forms. The amounts of the products formed can be determined due to their different electrophoretic mobilities. Figure 3 shows plots of the spot intensities for the three different forms of DNA as a function of bpV(phen) concentration. At low concentrations, the cutting produces exclusively nicked circular DNA, indicating that the cleavage is single-strand. At higher concentrations of the metal complex (> 5 mM) linear form III DNA is formed.

Analysis of the kinetics was performed similar to previously [15]. The fit between experimental and calculated data is also shown in Fig. 3. The ratio k_2/k_1 was found to be 0.042, where k_2 and k_1 are the rate constants for converting form I to form II DNA and form II to form III DNA, respectively. If the nicking process is random, the ratio k_2/k_1 is expected to be about 10^{-3} [15], and since k_2/k_1 was found to be circa 40 times higher than the random value we suggest that the cleavage is sequence and/or base specific.

The cleavage efficiency increases with increasing irradiation times as shown in Fig. 4. After circa 10 minutes of photoillumination, at a bpV(phen) concentration of 15 μM, 50% of the supercoiled plasmids are cut to produce form II DNA. No cutting takes place if the DNA is added directly after irradiation of the bpV(phen) solution, indicating that the active agent is a short-lived species.

DNA photocleavage by simple vanadates (a mixture of the monomer, dimer and tetramer of vanadate) was studied under the same conditions as for bpV(phen), and the cleavage efficiency was compared with that obtained for bpV(phen) by comparing the respective vanadium concentration at which 50% of the form I DNA was cut upon photoirradiation for 30 min. The concentrations were found to be approximately 4 and 25 μM for bpV(phen) and vanadate, respectively.

In the absence of light, the cleavage reaction can be initiated by adding the biologically abundant reducing agent NADPH to the reaction medium (results not shown), which may suggest that the active vanadium species is a (+4) complex.

Discussion

The changes in absorption indicate that bpV(phen) undergoes major chemical modifications upon irradiation with UV light, and an overall non-isosbestic behavior with increasing illu-

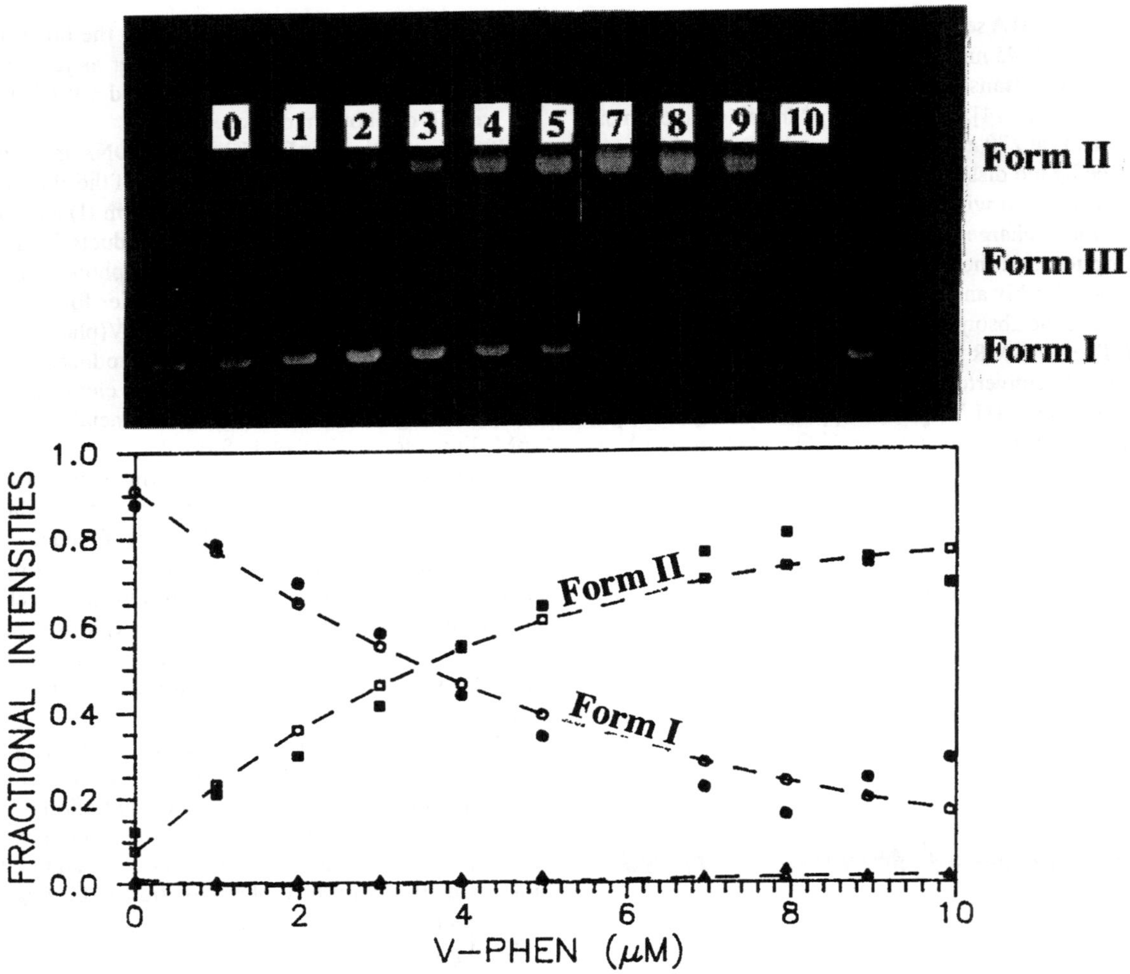

Fig. 3. Photocleavage efficiency as a function of bpV(phen) concentration. Top: Ethidium bromide stained form I, II and III DNA in an agarose gel. Lane 1, left: control, native DNA; lane 2: DNA + light (30 min); lanes 3–11: DNA + bpV(phen) + light (30 min); lane 12: DNA + bpV(phen) + dark. The respective bpV(phen) concentrations, in μM, are given in the figure. [DNA phosphate] = 23 μM. Bottom: Normalized spot intensities (divided by the total intensity in the lane) of forms I, II and III DNA (obtained by densitometric scanning of the gel shown above) versus concentration of bpV(phen). The curves are fit to a model for calculated intensities (open symbols) [15].

mination time shows that more than one new species is obtained. In agreement with this, the NMR results show that bpV(phen) is converted into the monomer, dimer and tetramer of vanadate. The simple mono and polynuclear vanadates formed upon irradiation of bpV(phen) can, however, not be solely responsible for the DNA cleavage exhibited by the complex, since the DNA photocleavage of bpV(phen) is at least 5 times more efficient than that of the simple vanadates alone. More likely, either an intermediary vanadium complex, possibly a V(+4) species, or an oxoradical formed during the photoreduction and subsequent decomposition of bpV(phen), is responsible for the DNA strand breakage. This suggestion is supported by the following observations. Firstly, the cleavage agent is a relatively short-lived species. Exposing DNA to preirradiated solutions of bpV(phen) does not result in cutting. Secondly, the fact that the DNA cleavage can be initiated by NADPH (in the absence of light) shows that chemical reduction of bpV(phen) also produces species that are capable of cleaving DNA. One of these must be a V(+4) complex, but other reactive species are probably released as well. Thirdly, removal of molecular oxygen from the sample prior to irradiation leads to increased formation of DNA cleavage products. Since the presence of dissolved oxygen gas retards photoreduction of bpV(phen) (as is evident from the spectroscopic changes in the absence of DNA) the dependence of DNA degradation on oxygen suggests that the cleavage reaction may involve a V(+4) species. However, further work is necessary to unravel the details of the mechanism.

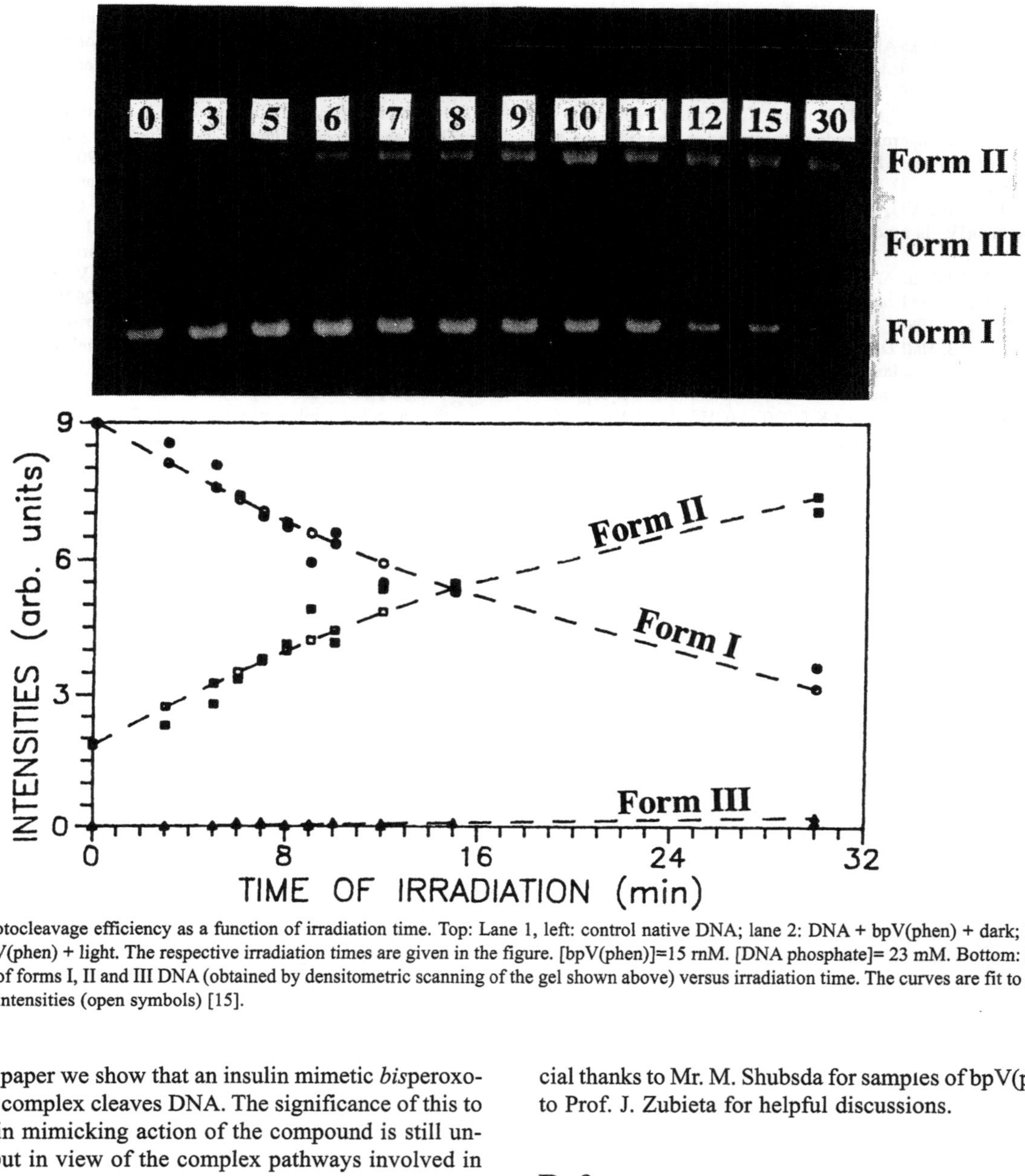

Fig. 4. Photocleavage efficiency as a function of irradiation time. Top: Lane 1, left: control native DNA; lane 2: DNA + bpV(phen) + dark; lanes 3–12: DNA + bpV(phen) + light. The respective irradiation times are given in the figure. [bpV(phen)]=15 mM. [DNA phosphate]= 23 mM. Bottom: Plot of spot intensities of forms I, II and III DNA (obtained by densitometric scanning of the gel shown above) versus irradiation time. The curves are fit to a model for calculated intensities (open symbols) [15].

In this paper we show that an insulin mimetic *bis*peroxovanadate complex cleaves DNA. The significance of this to the insulin mimicking action of the compound is still unknown, but in view of the complex pathways involved in glucose regulation together with the observation that vanadium compounds can induce oncogene production [16], direct interaction of the compounds with cellular DNA cannot yet be disregarded.

Acknowledgments

This work was in part supported by a Shannon Award from the NIH (lROlGM45924-01A1). Catharina Hiort is grateful to the Swedish Institute for a post-doctoral fellowship. Special thanks to Mr. M. Shubsda for samples of bpV(phen) and to Prof. J. Zubieta for helpful discussions.

References

1. Crans DC: Enzyme interactions with labile oxovanadates and other oxometalates. Comments Inorg Chem 16: 35–76, 1994
2. Correia JJ, Lipscomb LD, Dabrowiak JC, Isern N, Zubieta J: Cleavage of tubulin by vanadate ion. Arch Biochem Biophys 309: 94–104, 1994
3. Lyonnet B, Martz ME, Martin E: L'emploi therapeutique des derives du vanadium. La Presse Med 1: 191–192, 1899
4. Djordjevic C, Wampler GL: Antitumor activity and toxicity of peroxo heteroligand vanadates(V) in relation to biochemistry of vanadium. J Inorg Biochem 25: 51–55, 1985
5. Thompson HJ, Chasteen ND, Meeker LD: Dietary vanadyl(IV)sulfate inhibits chemically induced mammary carcinogenesis. Carcinogenesis

5: 849–851, 1984
6. Hanauske V, Hanauska AR, Marshall MH, Muggia VA, von Hoff DD: Biphasic effect of vanadium salts on *in vitro* tumor colony growth. Int J Cell Cloning 189: 1090–1095, 1987
7. Sakurai H, Nakai M, Mika T, Tsuchiya K, Takada J, Matsushita R: DNA cleavage by hydroxyl radicals generated in a vanadyl ion-hydrogen peroxide system. Biochem. Biophys Res Commun 189: 1090–1095, 1992
8. McNeill JH, Yuen VG, Hoveyda HR, Orvig CJ: Bis(maltolato)-oxovanadium(IV) is a potent insulin mimic. Med Chem 35: 1489, 1992
9. Heyliger CE, Tahiliani A, McNeill JH: Effect of vanadate on blood glucose and depressed cardiac performance of diabetic rats. Science 227: 1474–1477, 1985
10. Shaver A, Ng JB, Hall DA, Soo Lum B, Posner BI: Insulin-mimetic peroxovanadium complexes: Preparation and structure of potassium oxodiperoxo(pyrididine-2-carboxylate)vanadate(V), $K_2[VO(O_2)_2(C_5H_4N\text{-}COO)]\cdot 2H_2O$, and $K_2[VO(O_2)_2(OHC_5H_3NCOO)]\cdot 3H_2O$, and their reactions with cysteine. Inorg Chem 32: 3109–3113, 1993
11. Fantus JG, Kadota S, Deragon G, Foster B, Posner BI: Pervanadate [peroxide(s) of vanadate] mimics insulin action in rat adipocytes via activation of the insulin receptor tyrosine kinase. Biochemistry 28: 8864–8871, 1989
12. Heffetz D, Bushkin I, Dror R, Zick Y: The insulin mimetic agents H_2O_2 and vanadate stimulate protein tyrosine phosphorylation in intact cells. J Biol Chem: 265: 2896–2902, 1990
13. Vuletic N, Djordjevic C: Oxodiperoxovanadate(V) complexes with bidentate ligands. J Chem Soc Dalton Trans: 1137–1143, 1973
14. Djordjevic C, Puryear BC, Vuletic N, Abelt CJ, Sheffields SJ: Preparation, spectroscopic properties and characterization of novel peroxo complexes of vanadium(V) and molybdenum(VI) with nicotinic acid and nicotinic acid N-oxide. Inorg Chem 27: 2926–2932, 1988
15. Kishikawa H, Jiang YP, Goodisman J, Dabrowiak JC: Coupled kinetic analysis of cleavage of DNA by esperamicin and calicheamicin. J Am Chem Soc 113: 5434–5440, 1991
16. Wang H, Scott RE: Adipocyte differentiation selectively represses the serum inducibility of *c-jun* and *jun B* by reversible transcription-dependent mechanism. Proc Natl Acad Sci 91: 4649–4653, 1994

PART II

BIOCHEMICAL AND PHYSIOLOGICAL STUDIES

Molecular and Cellular Biochemistry **153:** 39–47, 1995.

Insulin-like actions of vanadate are mediated in an insulin-receptor-independent manner via non-receptor protein tyrosine kinases and protein phosphotyrosine phosphatases

Yoram Shechter[1], Jingping Li[1], Joseph Meyerovitch[2], Dov Gefel[3], Rafael Bruck[4], Gerard Elberg[1], David S. Miller[5] and Assia Shisheva[1]

[1]*Departments of Hormone Research, The Weizmann Institute of Science, Rehovot 76100;* [2]*Department of Pediatrics, Sheba Medical Center, Tel-Hashomer, Tel Aviv;* [3]*Department of Internal Medicine B, Kaplan Hospital, Rehovot;* [4]*Department of Gastroenterology, Wolfson Hospital, Holon, 58100, Israel;* [5]*National Institute of Health, NIEHS, Research Triangle Park, North Carolina 27709, USA*

Abstract

Most or all mammalian cells contain vanadium at a concentration of 0.1–1.0 μM. The bulk of the vanadium in cells is probably in the reduced vanadyl (IV) form. Although this element is essential and should be present in the diet in minute quantities, no known physiological role for vanadium has been found thus far. In the years 1975–1980 the vanadate ion was shown to act as an efficient inhibitor of Na^+,K^+-ATPase and of other related phosphohydrolyzes as well. In 1980 it was observed that vanadate vanadyl, when added to intact rat adipocytes, mimics the biological actions of insulin in stimulating hexose uptake and glucose oxidation. This initiated a long, currently active, field of research among basic scientists and diabetologists. Several of the aspects studied are reviewed here. (Mol Cell Biochem **153:** 39–49, 1995)

Key words: vanadium salts, insulin action, cytosolic protein tyrosine kinase, protein phosphotyrosine phosphatase, molybdate – permolybdate, tungstate – pertungstate

In vitro actions of vanadium

Isolated, intact cells are the least complex experimental systems for analyzing the actions of insulin. The observations made in cells can be studied later in more complex systems such as perfused liver or muscle and finally be applied to intact experimental animal models (i.e. to mice and rats). The insulin like effects of vanadium were initially observed in isolated rat adipocytes. Exogenously added vanadate (V) and vanadyl (IV) mimicked insulin in stimulating hexose uptake [1] glucose oxidation [2], lipogenesis [3] and in inhibiting lipolysis [4]. Further studies have demonstrated that vanadate (VO_3^-) mimics many of the documented actions of insulin in other insulin-responsive tissues as well. The current list also includes activation of glycogen synthetase in muscle and fat, and even extends to other 'less physiological' actions of the hormone such as stimulation of potassium uptake and calcium influx and increase of intracellular pH (Table 1 and reviewed in [5]). Indeed, vanadate even mimics insulin at the level of the genome by increasing, for example, the L-type pyruvate kinase gene expression in primary culture of adult rat hepatocytes. Therefore vanadate can be considered as a general mimicker of the bioactions of the hormone.

Initially it was thought that the insulin like effects of exogenously added VO_3^-, related to the anion's potency to inhibit Na^+,K^+-ATPase [6–8]. This, however, does not seem to be the case in intact cellular systems. Vanadate was shown to mimic insulin in K^+-depleted medium [2]. Also, direct measurement of rubidium uptake, a measure for Na^+,K^+-ATPase activity, indicated that this parameter is not blocked even in the presence of high (1 mM) concentrations of $V0_3^-$ [1]. Further studies aimed at elucidating the mode of action

Address for offprints: Y. Shechter, Department of Hormone Research, The Weizmann Institute of Science, Rehovot 76100, Israel

Table 1. Current list of the insulin-like actions mediated also by vanadate in various insulin responsive tissues (from Shechter, 1990 ref. [5])

Activity	Direction of effect	Target tissue
Hexose transport	Stimulated	Skeletal muscle, Rat adipocytes
Lipogenesis	Stimulated	Rat adipocytes
Glucose oxidation	Stimulated	Rat adipocytes
LIpolysis	Inhibited	Rat adipocytes
Glycogensynthase	Stimulated	Skeletal muscle, rat adipocytes Rat hepatocytes
Mitogenic activity	Augmented	Various cultured cells
Translocation of IGF-II	Stimulated or augmented	Rat adipocytes
K^+ uptake	Stimulated	Cardiac muscle cells
Ca^{2+}-Mg^{2+}-ATPase	Inhibited	Plasma membranes, from rat adipocytes
Ca^{2+} influx	Stimulated	Adipose tissue
Intracellular pH	Elevated	A-431 cells

of vanadate yielded the following conclusions: (a) Vanadate, which is a structural analog of phosphate [7], permeates cell interiors via the phosphate or the anion carrier system. (b) Inside the cell vanadate is reduced to vanadyl ions (oxidation-state +4). (c) Vanadyl ions are complexed to reduced glutathione (GSH). Several such types of VO^{2+}·GSH complexes were observed by using electron-spin resonance spectroscopy [4]. This intracellular reduction of vanadate to vanadyl explains the lack of inhibition of Na^+,K^+-ATPase in vanadate treated adipocytes. Vanadyl is an ineffectual inhibitor of Na^+,K^+-ATPase [9]. It is also suggested at this stage that vanadyl ion (+4 oxidation state) must be the form that mimics the actions of insulin.

Effect of vanadate in isolated perfused rat liver

Perfused, isolated liver is a suitable system that connects between the relatively simpler *in vitro* (i.e. isolated cells) systems and the more sophisticated and complex intact animal models. *In vivo*, insulin reaches first the liver and suppresses the release of glucose, potassium and amino acids [10]. Inhibition of glucose output is primarily due to arresting the biochemical pathways that facilitate glycogen breakdown [11].

Vanadate, at very low concentrations (IC_{50} = 0.5–1 μM), effectively inhibited release of glucose from isolated perfused rat liver [12]. The magnitude of the effect was larger than that obtained by insulin. Several other parameters such as rate of initiation and the rate of termination of the effect resembled those of the hormone. Thus, this physiologically important hepatic function of insulin is also mimicked by vanadate.

Studies in intact experimental animals

Effects in experimental animals representing insulin-dependent diabetes mellitus

A new turning point occurred in 1985, when Heyliger, McNeil and coworkers demonstrated that oral administration of vanadate to streptozocin-treated diabetic rats (STZ rats), a representative model of Type I diabetes, lowered their high levels of blood glucose to normal values [13]. Unlike insulin, which is not absorbed orally [14], vanadate has the physical attributes of being a low molecular weight substance and a phosphate analog, enabling it to permeate plasma membranes and the intestinal wall with relative ease. Sodium metavanadate ($NaVO_3$) in drinking water was found to be optimal at a concentration of 0.2 mg/ml [15]. Under these conditions, blood glucose levels of STZ-diabetic rats dropped to near normal values within 4 days of treatment, and stable normoglycemia was retained for 2–3 weeks, as long as $NaVO_3$ was included in the drinking water (Fig. 1). Further studies from several laboratories have demonstrated that vanadate therapy ameliorated tissue aberrations that occurred predominantly due to chronic hyperglycemia. For example, in liver, insulin binding capacity was fully normalized. Key enzymes involved in carbohydrate metabolism, as well as the level of glycogen which is greatly diminished in these diabetic rats, were dramatically restored following two weeks of oral treatment. The level of glycogen was restored from 40 to 109% of that of healthy rats, that of glucokinase from 0 to 65%. The level of 6-phosphofructo-2-kinase, another key enzyme of carbohydrate metabolism, was restored from 20 to 122% in vanadate-treated STZ-rats [16].

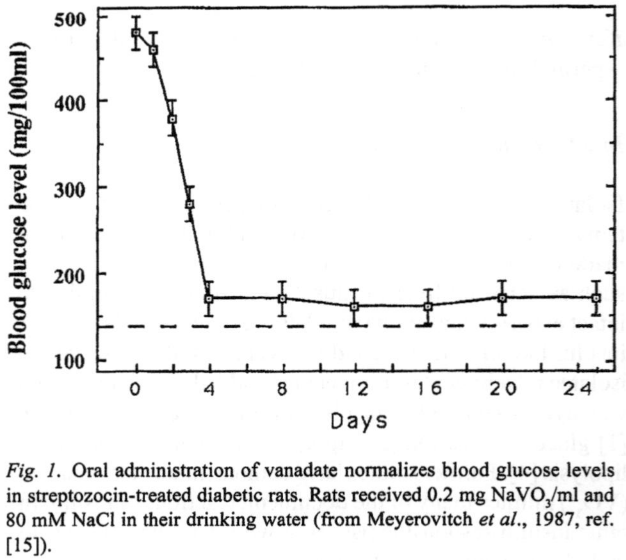

Fig. 1. Oral administration of vanadate normalizes blood glucose levels in streptozocin-treated diabetic rats. Rats received 0.2 mg $NaVO_3$/ml and 80 mM NaCl in their drinking water (from Meyerovitch *et al.*, 1987, ref. [15]).

It should, however, be mentioned that in spite of these improvements, insulin therapy is superior to vanadium therapy in the insulin-dependent diabetic experimental model. For example, daily gain in weight can be fully corrected by insulin therapy [17], but only partially (~30% of insulin effect) by vanadate therapy, suggesting that future vanadium therapy may be used, in general, in insulin-dependent diabetic patients as a supplement (rather than a substitute) to insulin.

A more promising picture is, however, seen in diabetic experimental animals representing Type II diabetes. We refer here to rodents and other mammals which produce normal levels of insulin, but whose insulin-responsive tissues (i.e., muscle, liver, adipose) have fully or partially lost their capability to metabolize glucose in response to the hormone.

Vanadium therapy in insulin-independent diabetes mellitus

Four genetically well-studied rodent models of Type II diabetes are ob/ob mice, db/db mice, BB rats and fa/fa rats. The four models are characterized by hyperglycemia, hyperinsulinemia, sometimes obesity, and blunted response to insulin at the receptor and post-receptor levels. All four models demonstrate several metabolic defects consistent with peripheral insulin insensitivity. In simple words, exogenously administered insulin is incapable of reducing blood glucose levels or of ameliorating any of the defects associated with utilization of glucose. Various tissues derived from these rodents fail to metabolize glucose appreciably in response to insulin.

Naturally, several laboratories evaluated the potential beneficial effects of vanadium therapy in these insulin-resistant rodents. Oral administration of vanadate (0.25 mg/ml in drinking water) to ob/ob and db/db mice lowered blood glucose levels, and substantially improved glucose homeostasis in these mice models [18] and in BB rats [19]. Beneficial effects of vanadate were also observed in the fa/fa obese rat model. Long-term vanadate treatment corrected their characteristic hyperinsulinemia and induced a better tolerance to oral glucose loads [20]. Moreover, the insulin-mediated glucose utilization indices were 2- to 3-fold higher in all types of muscles derived from these insulin-resistant rats. Further studies indicated that chronic administration of vanadate markedly increased the insulin-mediated glucose uptake in the muscle, without altering the number of glucose transporters [21]. Altogether, these studies indicated that vanadate (but not insulin) is able to increase glucose metabolism in these rodents and, therefore, to induce normoglycemia in the whole animal model. Moreover, vanadium therapy markedly increased tissue responsiveness to insulin.

Sites of insulin and vanadate action

The insulin receptor (InsRTK) is a large transmembrane glycoprotein composed of two 135 kDa α subunits and two 95 kDa β subunits linked by disulfide bonds to form a β-α-α-β heterotetramer [22–24]. The α subunits are entirely extracellular and contain the insulin-binding domain, while the β subunits are transmembrane proteins that possess protein tyrosine kinase activity. Insulin binding to the extracellular domain of the α subunits facilitates phosphorylation of several tyrosyl residues in the β subunits, and this in turn increases the tyrosine kinase activity of the receptor many fold. It is currently believed that hormone mediated receptor autophosphorylation is an early and essential event for *all* the rapid responses of insulin, and that receptor-mediated phosphorylation of endogenous proteins is essential as well for activating glucose transport, and glucose metabolism but may not be essential for the effect of insulin in inhibiting lipolysis or in activating pyruvate dehydrogenase [25–27].

Tyrosyl phosphorylation is linked to serine/threonine phosphorylation state of key enzymatic systems controlling glucose and fat metabolism [22, 28]. Termination can occur at several levels, one of which is dephosphorylation of tyrosyl residues on the receptor β-subunit by endogenous protein phosphotyrosine phosphatases (PTPases) when insulin is removed.

Since vanadate is an inhibitor of protein phosphotyrosine phosphatases (PTPases; [29]) it was initially believed that the ion acts intracellularly by blocking this PTPase which dephosphorylates the insulin receptor and therefore activates it in an insulin independent manner. This may not be the case. Numerous groups have reported no significant increase in the phosphotyrosine content of the insulin receptor in vanadate pretreated* cells or tissues (reviewed in [5]). This issue was further substantiated when quercetin was found to inhibit InsRTK catalyzed phosphorylation of exogenous substrates ($ID_{50} = 2 \pm 0.2$ μM, ref. [26] and the stimulating effects of insulin in intact rat adipocytes (i.e. lipogenesis, Table 2). The same biological effects when triggered by vanadate were not inhibited at all by quercetin (Table 2). This left us with two alternative working hypotheses for vanadate signalling: a) vanadate bypasses the step of tyrosine phosphorylation, or b) vanadate effects can be signalled through an additional non-insulin receptor protein tyrosine kinase(s).

A cytosolic protein tyrosine kinase in rat adipocytes

The level of protein phosphorylation on tyrosyl residues is extremely low compared to phosphorylation on seryl and

* Unlike vanadate, peroxide of vanadate does activate the insulin receptor in rat adipocytes (i.e. ref. [42]). Peroxides of vanadate are to be discussed in other chapters.

Table 2. Effect of quercetin on activation of lipogenesis by insulin and vanadate[a]

Addition	[U-^{14}C]glucose conversion into lipids cpm (10^4 cells)$^{-1}$h^{-1}	% maximal stimulation
Basal	2700 ± 150	0
Insulin (17 nM)	8400 ± 560	100
+ quercetin	2900 ± 220	3.5
Vanadate (0.3 mM)	6800 ± 300	72
+ quercetin	7200 ± 280	79

Adipocytes (2 × 10^4 cells/ml) suspended in KRB buffer containing 0.7% BSA were preincubated in the absence and the presence of quercetin (110 μM) for 40 min at 37°C. Lipogenesis was measured over a period of 1 h in the presence of indicated concentrations of insulin or vanadate (from Shisheva and Shechter, Biochemistry 31, 8059–8063, 1992, ref. [26]).

threonyl residues in intact cellular systems. The amount of phosphotyrosine in cells comprises less than 0.1% of the combined amounts of phosphoserine and phosphothreonine [30]. The first protein tyrosine kinases identified were intrinsic plasma membrane transmembrane protein receptors such as the insulin and the EGF-receptors [22]. More recently cytosolic (soluble) protein-tyrosine kinases, have been identified in several mammalian tissues (reviewed in [31]). We found that the 40,000 × g supernatant fraction of rat adipocyte homogenate contains a soluble protein tyrosine kinase (CytPTK; [32]). The basic features of this enzyme are summarized in Table 3. Like the InsRTK, CytPTK readily phosphorylates PolyGlu$_4$Tyr, a random copolymer of about 30 kDa. However, the CytPTK is distinct from the insulin receptor kinase. The cytosolic enzyme has an estimated molecular weight of 53 ± 3 kDa and its activity is supported by Co^{2+} rather than by Mn^{2+}. Unlike the InsRTK, the cytosolic kinase is not inactivated by N-ethylmaleimide. Of the several protein tyrosine kinase inhibitors tested, staurosporine was the most potent inhibitor of the soluble kinase (IC_{50} = 3 ± 0.2 nM), but a poor inhibitor of the insulin receptor kinase (IC_{50} = 8 μM; [32]). A rough calculation seemed to indicate that in the cytosolic PolyGlu$_4$Tyr phosphorylating capacity exceeded that of the detergent-soluble plasma membrane fraction (i.e. InsRTK) by about one order of magnitude.

Activation of CYT-PTK by vanadate

When adipocytes were preincubated with vanadate (20–30 min at 37°C) CytPTK was activated about 3–5-fold (Fig. 2). Enzyme activation parallelled the dose-response curve for vanadate's insulin-like effects. The effect was found to be highly specific for vanadate. Neither insulin, isoproterenol, dibutyryl cyclic-AMP, okadaic acid, hydrogen peroxide nor the phorbol ester TPA, altered CytPTK activity in rat adipocytes (Fig. 2).

Table 3. Basic features of cytosolic protein tyrosine kinase of rat adipocytes

Divalent ions	Mg^{2+}, Co^{2+}
Exogenous substrate	Poly-[(GluNa:Tyr)4: 1], Angiotensin II
Phosphoryl donor	ATP
Specific activity	0.9 pmol PTyr/μg PolyGlu$_4$Tyr/min/,ug cytosolic protein
Estimated molecular weight	53 ± 3 kDa
Destroyed by extreme pH values, and boiling. Resistant to N-ethylmaleimide.	
Potently inhibited by staurosporine (IC_{50} = 3 ± 0.2 nM).	

From Shisheva and Shechter, FEBS Lett. 300, 93–96, 1991 and Shisheva and Shechter, J Biol Chem 268, 6463–6469, 1992, ref. [33]).

Staurosporine effects on lipogenesis

As mentioned earlier, staurosporine is a powerful inhibitor of adipocyte CytPTK but a poor inhibitor of InsRTK. In intact adipocytes, staurosporine inhibited selectively the effect of vanadate but only had a marginal effect on insulin-stimulated lipogenesis (ref. [32, 33] and Fig. 3).

Since staurosporine is an effective inhibitor of CytPTK, we could map the vanadate effects mediated by the cytosolic enzyme. This includes activation of lipogenesis and oxidation of glucose. The activating effects of vanadate on hexose transport are not quenched by staurosporine. The same is valid for the vanadate effect in inhibiting lipolysis (Fig. 4). Thus, vanadate can facilitate several additional insulin effects by mechanism(s) which are not dependent upon activating CytPTK, nor on activation of the insulin receptor.

Activation of CytPTK by vanadate in a cell-free system

The fact that most or all the activating effects of insulin are lost on cell disintegration was the predominant reason for the slow progress in studying the insulin-dependent mechanisms involved in signal transduction (reviewed in [11]). The steps linking vanadium pretreatment to activation of CytPTK can be significantly elucidated if the activating effect is preserved after cell disintegration. Following a large set of experiments such a cell-free system has been established. The assay is performed in Hepes buffer (pH 7.4) and includes the high speed (cytosolic) supernatant fraction from freshly prepared rat adipocytes (0.5–5 μg protein), 2 mM cobalt (II) acetate, 20 mM $MgCl_2$, 100 μM ATP, PolyGlu$_4$Tyr (0.7 mg/ml), and is performed in the presence or the absence of vanadate or other agents to be studied. The results of our cell-free studies are summarized in reference [37] and in Table 5. Vanadate activated CytPTK 3,5 fold, ED_{50} value was calculated to be 3 ± 0.7 μM. Other agents that activated the cytosolic enzyme were tungstate ($ID_{50} \cong 15$ μM), molybdate ($ID_{50} \cong 25$ μM) and phenylarsineoxide ($ID_{50} \cong 7$ μM). Sodium fluoride and va-

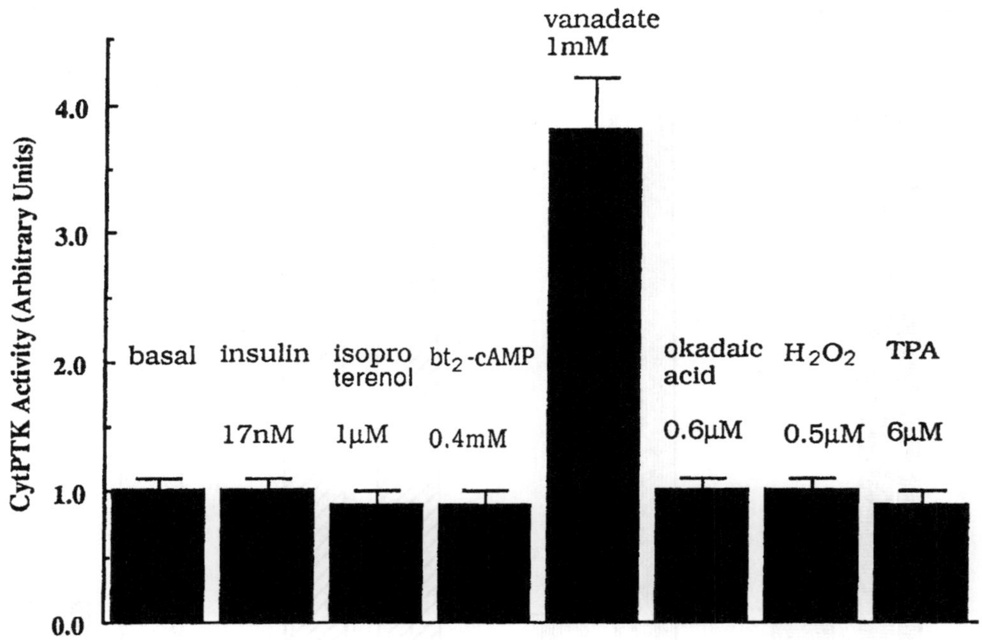

Fig. 2. Specific activation of cytosolic tyrosine kinase by preincubating adipocytes with vanadate (from Shisheva and Shechter, 1992, ref. [32]).

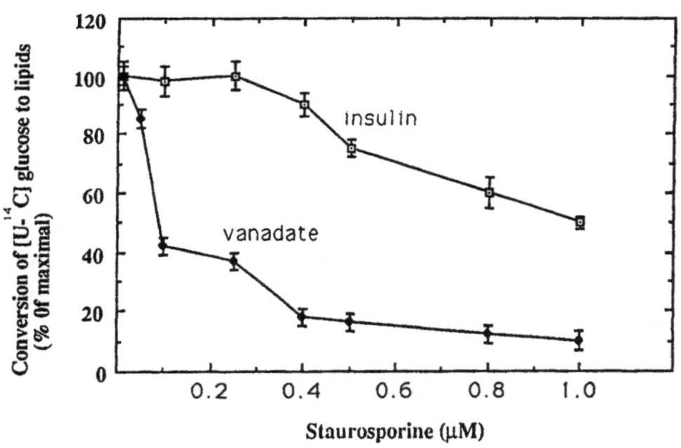

Fig. 3. Staurosporine inhibits selectively the stimulating effect of vanadate on lipogenesis (from Shisheva and Shechter, 1993, ref. [33]).

nadyl (IV) did not activate CytPTK (Table 5). Hydrogen peroxide at micromolar concentrations readily oxidizes stoichiometrically vanadyl to vanadate, and a link between vanadate, NADPH oxidase and endogenous activation of tyrosine phosphorylation has been noticed in several studies [35–37]. Based on this, a role for the intracellular vanadium pool implicating the latter in regulating the nonreceptor protein tyrosine kinases has bee suggested (illustrated schematically in Fig. 5).

Since tungstate, molybdate and phenylarsine oxide (PAO) are known to inhibit protein phosphotyrosine phosphatases (PTPases, ref. [38]) our cell free studies also suggest that CytPTK is activated by autophosphorylation and vanadate activation is secondary to inhibition of PTPase(s) by vanadate. That in vanadate pretreated adipocytes CytPTK (but not the InsRTK) is activated, may further suggest the existence of vanadate sensitive and vanadate insensitive species of PTPases (to be discussed in a subsequent paragraph).

Insulin-like effects of tungstate, molybdate and their permetalooxides

The observations made in cell-free experiments, demonstrating that tungstate and molybdate can also activate CytPTK, suggested to us that like vanadate, those metalooxides can facilitate insulin effects in intact adipocytes as well. Indeed, both molybdate and tungstate mimicked the actions of insulin. This, however, occurred only at high (millimolar) concentrations of molybdate and tungstate (Fig. 6).

Incubation of tungstate or molybdate with hydrogen peroxide (H_2O_2) followed by catalase (to destroy excess H_2O_2) resulted in the formation of pertungstate (peroxide of tungstate) and permolybdate (peroxide of molybdate). Pertungstate and permolybdate were found to be highly stable at aqueous solutions and to mimic most or all the biological effects of insulin. Moreover, these permetalooxides were 100–180 folds more potent in activating insulin responses than the corresponding metalooxides (Fig. 6). This shift in potency resembles that of pervanadate relative to vanadate in stimulating the same effects in rat adipocytes [39].

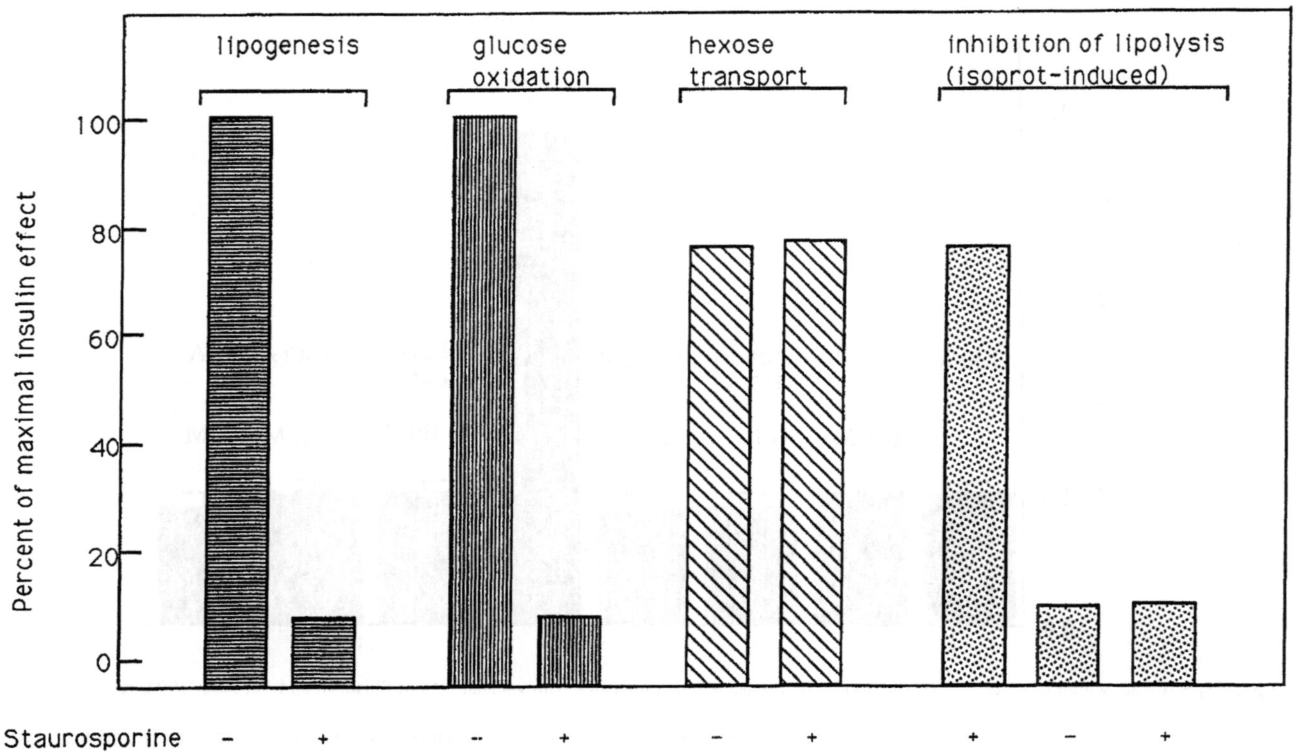

Fig. 4. Effect of staurosporine on several vanadate mediated effects in rat adipocytes (from Shisheva and Shechter, 1991, 1993, ref. [32, 33]).

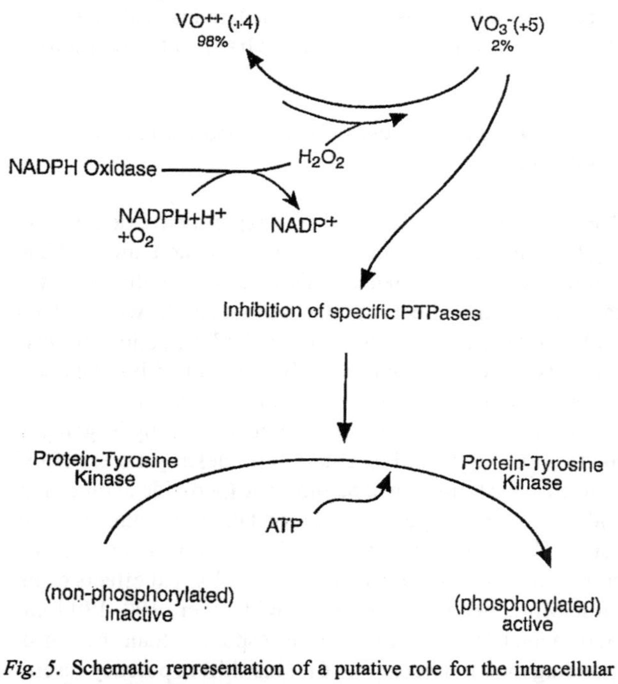

Fig. 5. Schematic representation of a putative role for the intracellular pool of vanadium in modulating the activity of non-receptor protein tyrosine kinases (from Elberg *et al.*, ref. [34]).

Vanadate-sensitive and vanadate-insensitive PTPases

Protein phosphotyrosine phosphatases (PTPases) are a structurally diverse family of transmembrane receptor-like and non transmembrane cytoplasmic enzymes (for review see [40]). As such, variations in properties are expected to exist among family members. A common denominator, however, that seems to unite most members is their features to be inhibited by micromolar concentrations of vanadate. Here we have studied whether indeed all PTPases are equally inhibited by vanadate and if negative (as suggested in previous paragraphs) to determine the common mechanistic features among vanadate-sensitive PTPases. Table 4 demonstrates several representative examples of PTPases exhibiting high sensitivity (IC_{50} = 1.5–3.5 μM), intermediate sensitivity (IC_{50} = 7.5–8.5 μM) and insensitivity (IC_{50} = >100 μM) to vanadate inhibition. The substrate used in these studies was p-nitrophenylphosphate. This low molecular weight high Km substrate was found to be most appropriate for evaluating PTPase sensitivity to vanadate inhibition (in preparation). Although PTPases are a structurally diversed large family, from an enzymological standpoint, they can be classified into three groups depending on whether seryl (i.e. alkaline phosphatases), histidyl (i.e. acid phosphatases) or cysteinyl

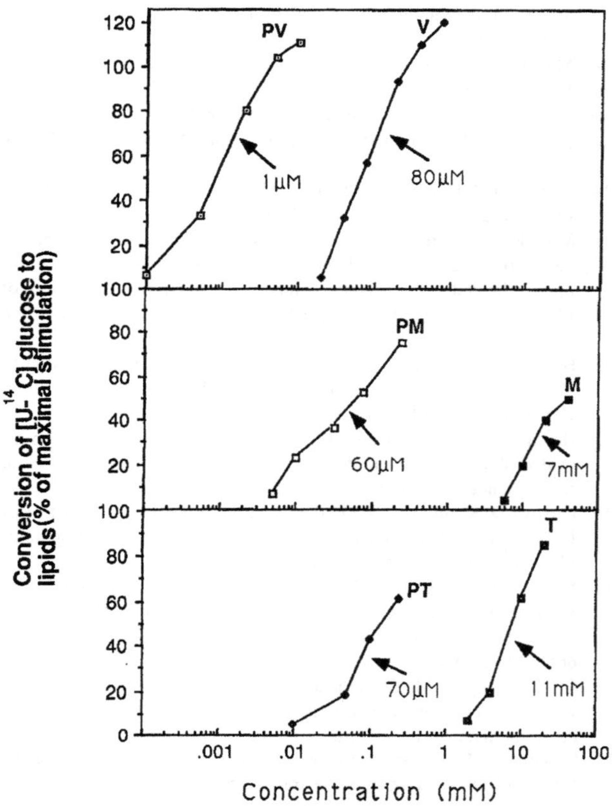

Fig. 6. Stimulation of lipogenesis by vanadate (V), tungstate (T), molybdate (M) and their permetalooxides. (PV, PM, and PT) (from Li, Elberg and Shechter, submitted manuscript).

Table 4. Activation of CytPTK in cell-free experiments by vanadate, tungstate, molybdate and phenylarsine oxide. Lack of vanadyl effect

Additions		Pmoles phosphotyrosine/ μg PolyGlu$_4$Tyr/40 min	Relative activity %
None	1.0 ± 0.1	100	
Vanadate	3 μM	2.4 ± 0.2	240
	10 μM	3.5 ± 0.2	350
	100 μM	2.7 ± 0.2	270
Tungstate	3 μM	1.4 ± 0.1	140
	10 μM	2.1 ± 0.12	210
	100 μM	3.7 ± 0.3	370
Molybdate	3 μM	1.1 ± 0.1	110
	10 μM	1.7 ± 0.15	170
	100 μM	3.05 ± 0.2	305
Phenylarsineoxide	3 μM	2.1 ± 0.1	210
	10 μM	3.7 ± 0.2	370
	100 μM	5.7 ± 0.3	570
Vanadyl	3 μM	1.1 ± 0.1	110
	10 μM	1.2 ± 0.1	120
	100 μM	1.05 ± 0.1	105

From Elberg, Li and Shechter, J Biol Chem 269, 9521–9527, 1994, ref. [34].

Table 5. Sensitivity of PTPases from various sources to inhibition by vanadate

Source of PTPase	Vanadate inhibition IC_{50} in μM
Adipose, liver, pancreas, brain (cytosolic)	1.5–3.5
Smooth muscle, cardiac muscle, Testis (cytosolic)	7.5–8.5
Liver (receptor like, glycoprotein)	2.5–3.5
Ovary (receptor like, glycoprotein)	>100

Elberg, Li and Shechter, submitted.

Table 6. Activity of the vanadate-sensitive, rat adipocytic PTPase under different experimental conditions

	μmoles pNPP hydrolyzed/mg protein/1 h at 37°C	% initial activity
No enzyme added	0.01	0
PTPase	0.13	100
PTPase + NH_2OH (50 mM)	0.29	223
PTPase pretreated with chloramine T (10 min, 5 mM)	0.03	23
PTPase pretreated with chloramine T followed by addition of NH_2OH (50 mM)	0.28	215
PTPase + vanadate (10 μM)	0.04	31
PTPase + vanadate (10 μM)+NH_2OH (50 mM)	0.28	215

Elberg, Li and Shechter, submitted

moieties participate in the formation of the enzyme phosphate intermediate during catalysis (reviewed in [41]).

Following chemical and enzymological studies the first unique feature of the vanadate-sensitive PTPases observed was their efficacy to be significantly activated by nucleophils such as hydroxyl amine. Of the numerous cysteinyl modifying reagents tested, chloramine T partially inactivated all PTPases examined, and hydroxylamine reactivated the vanadate sensitive PTPases. Finally, we have observed that PTPases are insensitive to vanadate inhibition in the presence of hydroxylamine. The results obtained with the vanadate-sensitive rat adipocytic PTPase are summarized in Table 6. Overall, the studies appear to indicate that vanadate interferes with the last step in catalysis, namely the dephosphorylation (by H_2O) of the phosphoenzyme intermediate. It also implicates an oxidation sensitive cysteinyl moiety of PTPase with vanadate inhibition.

Perspective

Fifteen years of research on the insulin mimetic actions of vanadium salts yielded interesting basic and clinical implications. From a basic point of view, the overall studies appear to indicate that vanadate signalling originates from a separate branch that is independent of insulin-receptor activation. More importantly, the studies illustrate that insulin-responsive tissues are equipped with the potency to metabolize glucose through a backup system, therefore enabling vanadate to act when the primary (insulin-dependent) system is faulty. This explains the potency of vanadium to induce normoglycemia in insulin resistant rodents.

It is still early to conclude whether vanadium salts themselves will be used therapeutically in the future care of diabetes. It is more likely that vanadium research will assist in a better understanding of the backup system(s) involved, and eventually lead to the development of more potent and less toxic vanadium substitutes. For example, it currently appears that the vanadium cascade is initiated by inhibiting specific vanadate sensitive PTPases. This may encourage further identification and characterization of the PTPases involved. Specific cell permeable inhibitors for these PTPases that will be effective in the nanomolar concentration range may substitute vanadium, and assist in the future care of diabetes in humans.

Acknowledgements

We thank Rona Levin for typing this manuscript, Rachel Benjamin for critical reading, Dr. Nava Nevo for technical assistance. This study was supported in part by grants from Minerva Foundation (Munich, Germany), The Israel Ministry of Health, Israel Academy of Sciences and Humanities, the Levine Fund, and by the Rowland and Sylvia Shaefer Program in Diabetes Research. Y.S. is the incumbent of C.H. Hollenberg Chair in Metabolic and Diabetes Research, established by the Friends and Associates of Dr. C.H. Hollenberg of Toronto, Canada. G.E. is the incumbent of the Eshkol Postdoctoral Fellowship.

References

1. Dubyak GR, Kleinzeller A: The insulin-mimetic effects of vanadate in isolated rat adipocytes. J Biol Chem 255: 5306–5312, 1980
2. Shechter Y, Karlish SJD: Insulin-like stimulation of glucose oxidation in rat adipocytes by vanadyl ions (IV) ions. Nature 284: 556–558, 1980
3. Shechter Y, Ron A: Effect of depletion of phosphate and bicarbonate ions on insulin action in rat adipocytes. J Biol Chem 261: 14945–14950, 1986
4. Degani H, Gochin M, Karlish SJD, Shechter Y: Electron paramagnetic studies and insulin-like effects of vanadium in rat adipocytes. Biochemistry 20: 5795–5799, 1981
5. Shechter Y: Insulin mimetic effects of vanadate; possible implications toward the future treatment of diabetes. Diabetes 'Perspective in Diabetes' 39: 1–5, 1990
6. Cantley LC, Aisen P: The fate of cytoplasmic vanadium. Implication of (Na^+, K^+) ATPase inhibition. J Biol Chem 254: 1781–1784, 1979
7. Simons TJB: Vanadate – a new tool for biologists. Nature 281: 337–338, 1979
8. Makara IG: Vanadium – an element in search of a role. Trends Biochem 5: 92–94, 1980
9. Cantley LC, Resh MD, Guidotti G: Vanadate inhibits the red cell (Na^+, K^+)ATPase from the cytoplasmic side. Nature 272: 552–554, 1978
10. Mondon EC, Dolkas CB, Olefsky JM, Reaven GM: Insulin sensitivity of isolated perfused rat liver. Diabetes 24: 225–229, 1974
11. Kahn RC, Shechter Y: In: A.G. Gilman, T.W. Rall, A.S. Nies, and P. Taylor (eds). Insulin, oral hypoglycemic agents, and the pharmacology of the endocrine pancreas. Goodman and Gilman Text Book of Pharmacology, Eighth edition, Chapter 61, Pergamon Press, 1990, pp 1463–1495
12. Bruck R, Prigozin H, Krepel Z, Rotenberg P, Shechter Y, Bar-Meir S: Vanadate inhibits glucose-output from isolated perfused rat liver. Hepatology 14: 540–544, 1991
13. Heyliger CE, Tahiliani AG, McNeill JH: Effect of vanadate on elevated blood glucose and depressed cardiac performance of diabetic rats. Science 227: 1474–1476, 1985
14. Hirsova D, Koldovsky O: On the question of the absorption of insulin from the gastrointestinal tract during postnatal development. Physiolog Bohemoslov 18: 281–284, 1969
15. Meyerovitch J, Farfel Z, Sack J, Shechter Y: Oral administration of vanadate normalizes blood glucose levels in streptozotocin-treated rats: Characterization and mode of action. J Biol Chem 262: 6658–6662, 1987
16. Gil J, Miralpeix M, Carreras J, Bartrons R: Insulin-like effects of vanadate on glucokinase activity and fructose 2.6 bisphosphate levels in the liver of diabetic rats. J Biol Chem 263: 1&68–1871, 1988
17. Kobayashi M, Olefsky JM: Effects of streptozotocin-induced diabetes on insulin binding, glucose transport, and intracellular glucose metabolism in isolated rat adipocytes. Diabetes 28: 87–95, 1979
18. Meyerovitch J, Rothenberg P, Shechter Y, Bonner-Weir SA, Kahn CR: Vanadate normalizes hyperglycemia in two mouse models of non-insulin-dependent diabetes mellitus. J Clin Invest 87: 1286–1294, 1991
19. Ramanadham S, Brownsey RW, Cross GH, Mongold JJ, McNeil, JH: Sustained prevention of myocardial metabolic abnormalities in diabetic rats following withdrawal from oral vanadyl treatment. Metabolism 38: 1022–1028, 1989
20. Brichard SM, Poltier AM, Henquin JC: Long term improvement of glucose homeostasis by vanadate in obese hyperinsulinemic fa/fa rats. Endocrinology 125: 2510–2516, 1989
21. Brichard SM, Assimacopoulos-Jeannet F, Jeanrenaud B: Vanadate treatment markedly increases glucose utilization in muscle of insulin-resistant fa/fa rats without modifying glucose transporter expression. Endocrinology 131: 311–317, 1992
22. Czech MP: The nature and regulation of the insulin receptor structure and function. Ann Rev Physiol 47: 357–381, 1985
23. Ebina Y, Ellis L, Jarmagin K, Standring D, Beaudoin J, Roth RA, Rutter WJ: The human insulin receptor cDNA: The structural basis for hormone-activated transmembrane signalling. Cell 40: 7447–7458, 1985
24. Kahn CR, White MF: The insulin receptor and the molecular mechanism of insulin action. J Clin Invest 82: 1151–1156, 1988
25. Shechter Y, Yaish P, Chorev M, Gilon C, Braun S, Levitzki A: Inhibition of insulin-dependent lipogenesis and anti-lipolysis by protein tyrosine kinase inhibitors. EMBO J 8: 1671–1676, 1989
26. Shisheva A, Shechter Y: Quercetin selectively inhibits insulin receptor function *in vitro* and the bioresponses of insulin and insulinomimetic agents in rat adipocytes. Biochemistry 31: 8059–8063, 1992
27. Gottschalk WK: The pathway mediating insulin's effects on pyru-

vate dehydrogenase bypasses the insulin receptor tyrosine kinase. J Biol Chem 266: 8814–8819, 1991

28. Shoelson SE, Kahn CR: Phosphorylation, the insulin receptor, and insulin action. In: M.Draznin, S. Melmed and D. LeRoith, (eds). Molecular and Cellular Biology of Diabetes Mellitus. Alan R. Liss, New York, Vol. II, 1989, pp. 23–33.
29. Swarup G, Speeg KV Jr, Cohen S, Garbers DL: Phosphotyrosyl protein phosphatase of TCRC-2 cells. J Biol Chem 257: 7298–7301, 1982
30. Hunter T, Sefton BM: Transforming gene product of rous sarcoma virus phosphorylates tyrosine. Proc Natl Acad Sci USA 77: 1311–1315, 1980
31. Srivastava A: Non-receptor protein tyrosine kinases of normal tissues. Intl J Biochem 11: 1229–1234, 1990
32. Shisheva A, Shechter Y: A cytosolic protein tyrosine kinase in rat adipocytes. FEBS Lett 300: 93–96, 1991
33. Shisheva A, Shechter Y: Role of cytosolic tyrosine kinase in mediating insulin-like actions of vanadate in rat adipocytes. J Biol Chem 268: 6463–6469, 1993
34. Elberg G, Li J, Shechter Y: Vanadium activates or inhibits receptor and non-receptor protein tyrosine kinases in cell-free experiments, depending on its oxidation state. Possible role of endogenous vanadium in controlling celllular protein tyrosine kinase activity. J Biol Chem 269: 9521–9527, 1994
35. Liochev S, Fridovich I: The oxidation of NADH by tetravalent vanadium. Arch Biochem Biophys 255: 274–278, 1987
36. Trudel S, Downey GP, Grienstein S, Paquet MC: Activation of permeabilized HL60 cells by vanadate. Evidence for diverged signal pathways. Biochem J 269: 127–131, 1990
37. Grinstein S, Furuya W, Lu DJ, Mills BG: Vanadate stimulates oxygen consumption and tyrosine phosphorylation in electropermeabilized human neutrophils. J Biol Chem 265: 318–327, 1990
38. Lau KHW, Farley JR, Baylink DJ: Review article: Phosphotyrosyl protein phosphatases. Biochem J 257: 23–36, 1989
39. Fantus IG, Kadota S, Deragon G, Foster D, Posner BI: Pervanadate [peroxide(s) of vanadate] mimics insulin action in rat adipocytes via activation of the insulin-receptor tyrosine kinase. Biochemistry 28: 8864–8871, 1989
40. Carbonneau H, Tonks NK: 1002 Protein phosphatases? Annu Rev Cell Biol 8: 463–493, 1992
41. Walton KM, Dixon JE: Protein tyrosine phosphatases. Ann Rev Biochem 62: 101–120, 1993

Molecular and Cellular Biochemistry **153:** 49–58, 1995.

Peroxovanadium compounds: Biological actions and mechanism of insulin-mimesis

A. Paul Bevan, Paul G. Drake, Jean-Francois Yale[1], Alan Shaver[2] and Barry I. Posner

Polypeptide Hormone Laboratory, [1]McGill Nutrition and Food Science Centre, and [2]Department of Chemistry, McGill University, Montreal, Quebec, H3A 2B2, Canada

Abstract

When used alone, both vanadate and hydrogen peroxide (H_2O_2) are weakly insulin-mimetic, while in combination they are strongly synergistic due to the formation of aqueous peroxovanadium species $pV_{(aq)}$. Administration of these $pV_{(aq)}$ species leads to activation of the insulin receptor tyrosine kinase (IRK), autophosphorylation at tyrosine residues and inhibition of phosphotyrosine phosphatases (PTPs). We therefore undertook to synthesize a series of peroxovanadium (pV) compounds containing one or two peroxo anions, an oxo anion and an ancillary ligand in the inner co-ordination sphere of vanadium, whose properties and insulin-mimetic potencies could be assessed. These pV compounds were shown to be the most potent inhibitors of PTPs yet described. Their PTP inhibitory potency correlated with their capacity to stimulate IRK activity. Some pV compounds showed much greater potency as inhibitors of insulin receptor (IR) dephosphorylation than epidermal growth factor receptor (EGFR) dephosphorylation, implying relative specificity as PTP inhibitors. Replacement of vanadium with either molybdenum or tungsten resulted in equally potent inhibition of IR dephosphorylation. However IRK activation was reduced by greater than 80% suggesting that these compounds did not access intracellular PTPs. The insulin-like activity of these pV compounds were demonstrable *in vivo. Intra venous* (i.v.) administration of bpV(pic) and bpV(phen) resulted in the lowering of plasma glucose concentrations in normal rats in a dose dependent manner. The greater potency of bpV(pic) compared to bpV(phen) was explicable, in part, by the capacity of the former but not the latter to act on skeletal muscle as well as liver. Finally administration of bpV(phen) and insulin led to a synergism, where tyrosine phosphorylation of the IR β-subunit increased by 20-fold and led to the appearance of four insulin-dependent *in vivo* substrates. The insulin-mimetic properties of the pV compounds raises the possibility for their use as insulin replacements in the management of diabetes mellitus. (Mol Cell Biochem **153:** 49–58, 1995)

Key words: insulin receptor, tyrosine kinase, peroxovanadium, insulin-mimetic, phosphotyrosine phosphatase

Introduction

It has previously been shown that vanadate has insulin-mimetic properties *in vitro* [1–8] although the underlying mechanism remains to be fully elucidated [4, 7, 8]. Strikingly, the oral administration of vanadate resulted in reduced hyperglycaemia in diabetic rats and raised the possibility of vanadate use as an oral insulin-mimetic [9–13].

More recently a combination of vanadate and hydrogen peroxide (H_2O_2; also a weak insulin-mimetic [14–18]) was shown to be strongly synergistic at both the level of insulin receptor tyrosine kinase (IRK) activation and insulin stimulated responses (e.g. recruitment of type-2 insulin-like growth factor receptors to the cell surface of adipocytes [19]. It was proposed that this synergy was due to the formation of aqueous peroxovanadium complexes ($pV_{(aq)}$) [19] since removal of free H_2O_2 with catalase did not diminish activity [19]. Subsequent studies demonstrated that $pV_{(aq)}$ resulted in a potent inhibition of IRK associated phosphotyrosine phosphatase (PTP) activity thus promoting IRK tyrosine phosphorylation

Address for offprints: B.I. Posner, Polypeptide Hormone Lab., Strathcona Medical Building, 3640 University Street; Rm W3.15, Montreal, Quebec, H3A 2B2, Canada

Table 1. Biological Action of peroxovanadiums

Effect	Efficacy compared to vanadate	Reference
Hypoglycaemic action		
Sprague Dawley rats[a]	+++[b]	24,38 [c]
Diabetic BB rats[a]	+++	[c]
Streptozotocin diabetic Wistar rats	++	39
Adipose		
↑IRK activity	++	7,19,21,22
↓PTP activity	++	21,22
↑IR tyrosine phosphorylation	+++	21,22
↑Tyrosine phosphorylation of substrates	ND[d]	22
↑IGF-II translocation	++	7
↑IGF-II binding	+++	7,19
↑Glucose transport	+++	22,40
↑Lipogenesis[e]	+++	21,24
↓Lipolysis	++	21,40
↑Protein synthesis	++	21
Muscle		
↑IRK activity[a]	++	38
↑Tyrosine phosphorylation of substrates	ND	20
↑Glucose transport[a]	ND	[f]
↑Glycogen synthase[a]	+++	41
↑Glycogen synthesis[e]	+++	38,41,42
↑Glucose oxidation	ND	42
↑Lactate formation	+++	41,42
Liver		
↑IRK activity[e]	+++	24
↓PTP activity[e]	+++	20,23,24,30
↑IR tyrosine phosphorylation[e]	+++	20,23,24
↑Insulin receptor substrate-1 tyrosine phosphoryiation[e]	ND	23,43 [f]
↑Phospholipase C tyrosine phosphorylation	ND	23
↑Phosphatidylinositol-3 kinase activity	ND	43
↑Tyrosine phosphorylation of substrates	+++	20,23,44 [f]
↑Na^+ transport by (Na^+ K^+)ATPase	ND	25
↑Na^+ transport by (Na^+ H^+)exchanger	ND	25
Chinese hamster ovary cells		
↑IR tyrosine phosphorylation	ND	45
↑Tyrosine phosphorylation of substrates	ND	45
Mononuclear cells		
↑$p56^{lck}$, $p59^{fyn}$ TK activities	ND	26
↑Phospholipase D activity	ND	46
↑Phospholipase C tyrosine phosphorylation	ND	26
↑Mobilization of Ca^{2+}	ND	26
↑Tyrosine phosphorylation of substrates	++	26,44,46,47
↑glucose transport	ND	48
Erythrocytes		
↑$p72^{syk}$ TK activity	ND	27
↑Band 3 tyrosine phosphorylation	ND	27
Platelets		
↑Tyrosine phosphorylation of P-selectin	ND	49
↑Tyrosine phosphorylation of substrates	ND	50
↑Platelet aggregation	ND	50

[a]Effect measured using crystallized, structurally defined pV compounds synthesized as described elsewhere [24]. [b]Degree of increased efficacy when compared to vanadate (+ = modest increase; ++ = large increase; +++ = substantial increase). [c]Yale, J.-.F., LaChance, D., Bevan, A.P., Vigeant, C., Shaver, A. and Posner, B.l. Hypoglycemic effects of peroxovanadium compounds in Sprague-Dawley and diabetic BB rats. Diabetes (accepted). [d]ND not determined. [e]Effect measured using both $pV_{(aq)}$ and structurally defined pV compounds. [f]Manuscript in preparation, Bevan, A.P., Burgess, J.W., Drake, P.G., Shaver, A., Bergeron, J.J.M. and Posner, B.l.

[20–23]. We then proceeded to synthesize stable, structurally defined pV compounds and assessed both their PTP inhibitory activities and insulin-mimetic properties [24].

Biological actions of pV compounds

The biological actions of pV compounds have been assessed in a number of studies (Table 1). A majority of these effects are insulin-mimetic in nature consistent with the view that pV compounds activate the insulin signalling pathway by inhibiting an IR associated PTP [24]. With the exception, perhaps, of the activation of Na^+ transport by the (Na^+ K^+)ATPase [25] all the responses described pertain to an alteration either directly or indirectly in the activity of tyrosine kinase(s) (TK) in the system under examination. Thus pV compounds appear to activate various TKs [24, 26, 27].

The rather broad specificity of $pV_{(aq)}$, the pV species formed upon mixing vanadate with H_2O_2, may reflect the fact that such a mixture contains a number of pV complexes [21]. However as will be discussed later the highly purified, stable, structurally defined pV compounds appear to exhibit relative PTP specificity.

Table 1 indicates that, where the comparisons have been made, the potencies of pV complexes are considerably greater than that of vanadate.

Biochemical effects of pV compounds

One of the most salient effects of incubating pV compounds with cultured cells is the activation of cellular IRK. Incubation of cultured hepatoma cells with 100 nM insulin led to a rapid activation of the IRK by 1 min which decreased to a plateau, approximately 50% of maximum, for the following 60 min (Fig. 1A). The compound bpV(phen)* and $pV_{(aq)}$ also induced IRK activation but with a considerably slower time course. Thus the level of IRK activation attained with the pVs equalled that of insulin by 20 min and exceeded it by 60 min before returning toward baseline by 120 min. The activation of IRK was dose dependent as shown in Fig. 1B. The ED_{50} for insulin was 10^{-8} M whereas that for bpV(phen) and $pV_{(aq)}$ was about 10^{-4} M. Even at concentrations of 10^{-3} M, vanadate or H_2O_2 alone did not activate the IRK which emphasizes the dramatic synergy achieved on combining them to yield pV complex formation.

The slower time course of IRK activation along with the prior demonstration, in adipocytes, that pV compounds promoted IR phosphorylation on tyrosine residues led to the hypothesis that their action resides at the level of PTP inhibition [21]. This would permit IRK phosphorylation not by activation as such but by the inhibition of an ongoing deactivation process. Previous work has established that the *in vivo* administration of insulin leads to activation of rat liver IRK and its internalization into endosomes (ENs) [28, 29]. Such ENs were shown to continue IRK phosphorylation *in vitro* [30]. Indeed studies by Faure *et al.* [30], in which ENs were incubated with [^{32}P]ATP *in vitro*, demonstrated that autophosphorylation of the IRK is a dynamic process involving a balance between TK activity on the one hand and

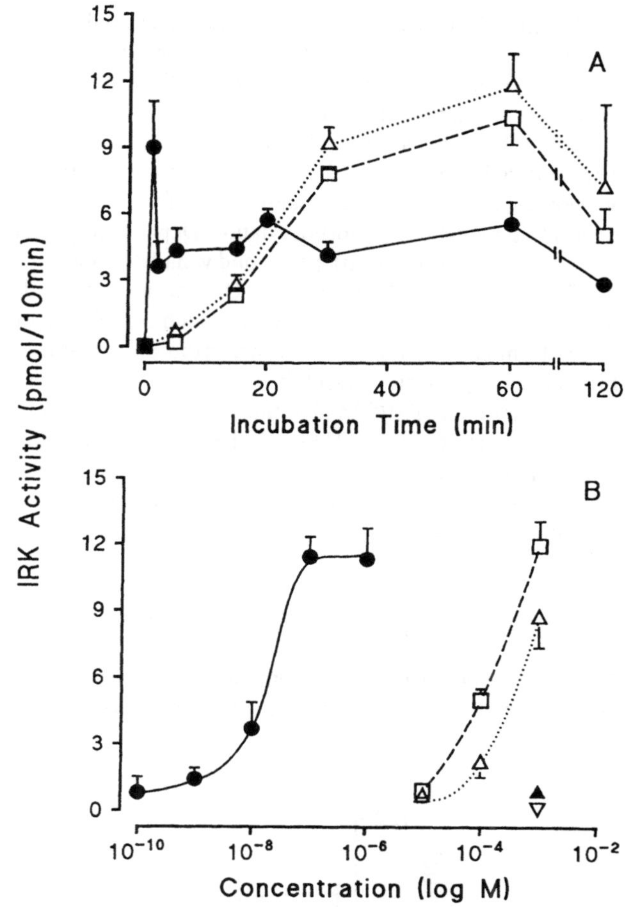

Fig. 1. Time course and dose response of IRK activation in hepatoma cells. H411EC3 hepatoma cells were incubated at 37°C in HEPES buffered Krebs-Ringer bicarbonate solution containing 1% BSA with the indicated additions and processed for determination of the exogenous tyrosine kinase activity measured as described elsewhere [24]. In panel A cells were incubated with 10^{-7} M insulin (●), 1 mM bpV(phen) (Δ) or a combination of 1 mM vanadate and 2 mM H_2O_2 (□) The results are the mean ± S.E. of 3–22 separate experiments. In Panel B cells were incubated with the above concentrations of insulin (●), bpV(phen) (Δ), vanadate plus H_2O_2 (□), vanadate (∇) or H_2O_2 (▲). The results are the mean ± S.E. of 3 to 19 separate experiments. Fig. 1A is taken from reference [24].

*The full structures for these compounds can be found in the paper by Shaver *et al.*, *ibid* and [24].

dephosphorylation by a receptor-associated PTP on the other. In these studies both the rate of autophosphorylation and dephosphorylation could be measured. At 37°C the endosomal IR underwent dephosphorylation with a $t_{½}$ of 1.6 min.

Figure 2A depicts the dephosphorylation of endosomal IRK and its inhibition by 100 μM $pV_{(aq)}$ but not by 100 μM vanadate. Figure 2B depicts a study of the effect of inhibitor concentration on IRK dephosphorylation. Dephosphorylation was inhibited by bpV(phen), $pV_{(aq)}$, and vanadate with ED_{50}s of 10^{-7}, 10^{-5} and 10^{-3} M, respectively. Okadaic acid was without inhibitory effect at concentrations which completely inhibited phosphoserine/phosphothreonine phosphatases. Furthermore pVs were non-inhibitory of such phosphatases at concentrations of 10^{-5} M [24] thus attesting to their relatively high specificity for PTPs. We therefore conclude that pV compounds are potent and quite specific PTP inhibitors and that their capacity to activate the IRK could well derive from this latter capacity.

Mechanism of insulin receptor kinase activation

To test the hypothesis that pV compounds activate the IRK by inhibiting an IRK associated PTP we evaluated the activities of synthetic pV compounds differing in their ancillary ligand and containing either one or two peroxo groups [24]. We assessed their ability to activate the IRK in hepatoma cells on the one hand and to inhibit dephosphorylation of the hepatic endosomal IRK on the other [24]. As seen in Fig. 3A there was a good correlation between the IRK activating and PTP inhibiting potencies of different synthetic pV compounds. Figure 3B quantifies this correlation for 12 different compounds ($r = 0.74$, $p < 0.003$). These data also indicate that the activity of a compound decreases with a decrease in the number of peroxo groups from 2 to 1 to 0 (Fig. 3B; cf compounds f→g→m). Furthermore, the ancillary ligand itself is an important determinant of activity. Increasing the bulkiness of the ancillary ligand (eg methylation of the phenanthroline; cf compounds a→b→c) decreased potency in both assays. The correlation between IRK activating and PTP inhibiting potencies is further supported by the data in Fig. 3C. In this study we observed that when a solution of bpV(phen) was allowed to stand at room temperature for 14 days there was a parallel loss of its NMR signal and its potencies to activate hepatoma IRK and inhibit endosomal PTP with time [24].

The catalytic domain of PTPs contains a well conserved sequence with a critical cysteine residue without which

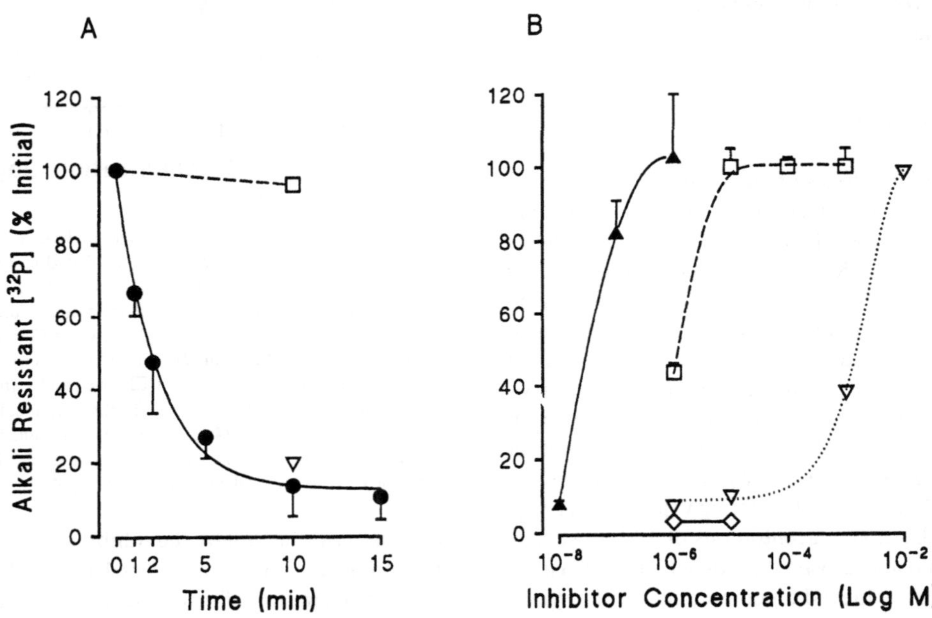

Fig. 2. Time course and dose response of IR dephosphorylation of [^{32}P]-labelled endosomal insulin receptors by different inhibitors. Rat liver ENs isolated 2 min following the *in vivo* injection of insulin (1.5 μg/100 g bwt) were autophosphorylated with 25 μM [γ-^{32}P]ATP before adding 5 mM unlabelled ATP (final concentration) to terminate phosphorylation and initiate dephosphorylation. At the noted times ENs were solubilized, IRs immunoprecipitated and then subjected to SDS-PAGE on 7.5% gels which were subsequently treated with alkali to remove [^{32}P]-serine and threonine to leave [^{32}P]-tyrosine. Following autoradiography, the 94 kDa band was quantitated by laser densitometry and the data plotted as a percent of the initial autophosphorylation observed as described elsewhere [30]. Panel A shows the time course of IR dephosphorylation in the absence of inhibitors (●) or the presence of 100 μM pV(aq) (□) or 100 μM vanadate (▽). The results are the mean ± S.D. of 3–4 separate experiments. Panel B shows the dose response to bpV(phen) (▲), combination of vanadate and H_2O_2 (□), vanadate (▽) or okadaic acid (◇). The results are the mean ± S.D. of 3–4 separate experiments. Fig. 2A and B are adapted from reference [30].

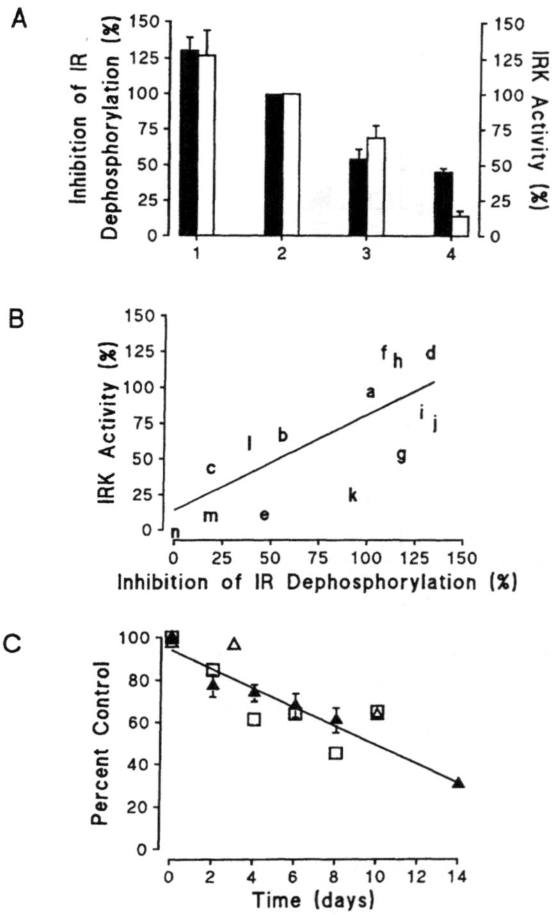

Fig. 3. The Relationship between PTP inhibition and IRK activation. IRK activation in H411EC3 hepatoma cells was measured in the presence of 1 mM pV compound and the inhibition of endosomal IR dephosphorylation was measured in the presence of 1 µM pV compound as described in Figs 1 and 2 respectively. Panel A depicts the relationship between PTP inhibition (solid bars) and IRK activation (open bars) for four of the pV compounds: 1, bpV(bipy); 2, bpV(phen); 3, bpV(Me$_2$phen); 4, bpV(bipy(CO$_2$)$_2$). Panel B depicts, in a more extensive study, a linear correlation between hepatoma cell IRK activation and inhibition of endosomal IR dephosphorylation by pV and related compounds ($r = 0.74$; $p < 0.003$). The compounds studied were: a, bpV(phen); b, bpV(Me$_2$phen); c, bpV(Me$_4$phen); d, bpV(bipy); e, bpV(bipy(CO$_2^-$)$_2$); f, bpV(ox); g, mpV(ox); h, bpV(NH$_3$); i, bpV(pic); j, bpV(HOpic); k, mpV(pic); I, mpV(2,6-pdc); m, V(O)$_2$(ox)$_2$; n, vanadate. Panel C delineates an experiment in which bpV(phen) was dissolved in 0.9% NaCl at a concentration of 3 mM and maintained at room temperature. At the noted times aliquots of solution were assayed as described previously for (▲) hepatoma cell IRK activation (dilution of 1 in 3; final concentration 1 mM); (□) inhibition of endosomal IR dephosphorylation (dilution of 1 in 3000; final concentration 1 µM) and (Δ) pV concentration using [^{51}V] NMR [24]. In Panels A and B all the data represent the mean ± S.E. of 3 separate experiments. In Panel C the IRK activation data represent the mean ± S.E. of 3 separate experiments whereas the other data were evaluated in parallel in one experiment. Fig. 3B and C are adapted from reference [24].

phosphatase activity is totally abolished [31, 32]. It appears highly likely that this critical cysteine residue forms a thiophosphate ester intermediate during catalysis [33, 34]. The pV compounds are known oxidants and have been shown to be much more effective than vanadate in oxidizing cysteine to cystine in solution [35]. Their mode of action may involve oxidation of the catalytic cysteine thus differing from the mode of action of vanadate which is thought to function as a phosphate analogue [36]. Information acquired from the recently deduced crystal structure of PTP 1B [37] may be of relevance to our observations. The recent structural data have shown that the depth of the catalytic cleft confers specificity for phosphotyrosine rather than phosphoserine or phosphothreonine since the former has a longer side chain allowing access to the bottom of the cleft wherein the catalytic cysteine is found. In a similar manner the bulkiness of a particular pV ancillary ligand may limit penetration by that compound into the cleft and hence reduce its inhibitory activity. This could be the physical basis for our observations above (Fig. 3).

The relationship between PTP inhibition and IRK activation deserves further comment. Various studies have observed that even in the complete absence of insulin, cultured cells still display a low level of basal IRK activity [24]. We envision a low level futile cycle in which phosphotyrosine is formed and degraded in the basal state at rates which prevent any significant net autophosphorylation. Upon administration of pVs this equilibrium is disturbed since dephosphorylation becomes inhibited allowing the low level of autophosphorylating activity to prevail and lead to IRK activation. This would explain the slower time course of activation observed with the pVs than with insulin.

The presence of vanadium is also of critical importance. In Fig. 4 the activities of bpV(phen) are compared to those of similar compounds in which the peroxo groups are coordinated to either molybdenum or tungsten. These latter compounds proved to be not only poor activators of the IRK in hepatoma cells but also unable to stimulate the incorporation of glucose into glycogen *in vivo* [38]. However with the exception of the one mono peroxo species (mpMo(pdc)) all were potent PTP inhibitors as measured by the endosomal IRK dephosphorylation assay. These observations suggest that the molybdenum and tungsten derivatives may be unable to access the interior of the cell in the way the pVs can thus reducing their capacity to inhibit PTPs *in situ* and hence activate the IRK. We suggest that another key property of the pV compounds may be their accessibility to the cellular interior.

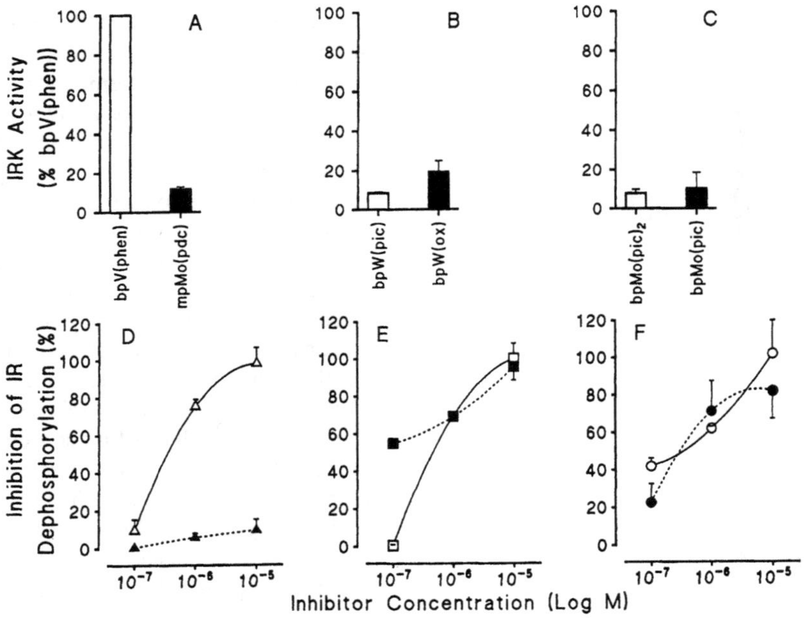

Fig. 4. Effect of peroxides of molybdenum and tungsten on IRK activation and PTP inhibition. IRK activation in H411EC3 hepatoma cells and inhibition of endosomal IR dephosphorylation were measured as described in Figs 1 and 2 respectively. Panels A, B and C show the activation of the IRK. The data represent the mean ± S.E. of 2–4 separate experiments. Panels D, E and F show the inhibition of endosomal IR dephosphorylation. The compounds studied were bpV(phen) (Δ), mpMo(pdc) (▲), bpW(pic) (□), bpW(ox) (■), bpMo(pic)$_2$ (O), bpMo(pic) (●), These data represent the mean + ½ the range of 2 separate experiments.

Specificity of the pV compounds

Previous work has suggested that several pV compounds are more potent inhibitors of IRK dephosphorylation than of EGFR tyrosine kinase (EGFRK) dephosphorylation [24]. Figure 5 depicts the inhibitory effects of two different pV compounds on the dephosphorylation of the IRK and the EGF receptor kinase (EGFRK). In Fig. 5A the compound bpV-(2,5pdc) is 4-fold more potent with regard to the IRK than the EGFRK at a concentration of 10^{-7} M. In contrast bpV-(phen) at 10^{-7} M, resulted in a similar inhibition of both EGFRK and IRK dephosphorylation (Fig. 5B). Therefore some compounds would appear to have a greater specificity for the IRK associated PTP. Clearly more work will have to be done with a range of PTPs to establish the range of this phenomenon. Furthermore it cannot be stated with certainty that these data indicate that the EGFRK and the IRK are dephosphorylated by different PTPs. Our observations may reflect the different behaviour of the same PTP acting on two quite different substrates.

In vivo activity

The pV compounds are active *in vivo* [24] (Fig. 6). As seen in Fig. 6A the i.v. administration of either bpV(pic) (0.15 μmol/100 g bwt) or bpV(phen) (0.6 μmol/100 g bwt) produced a 30% decrease in rat plasma glucose levels whereas a dose of vanadate of 0.6 μmol/100 g bwt was without effect. Indeed increasing the i.v. dose of vanadate produced toxic effects which were lethal before any significant hypoglycaemic action was observed†.

The two pV compounds were significantly different in potency (Fig. 6B). Thus bpV(pic) produced its maximal effect at a dose of 0.075 μmol/100 g bwt whereas bpV(phen) was comparably effective at a dose of 0.6 μmol/100 g bwt. At a dose of 0.2 μmol/100 g bwt bpV(phen) was without any significant hypoglycaemic action.

Studies of the effect of these agents on glycogen synthesis *in vivo* provided insight into the differences in hypoglycaemic potencies which have been observed. Thus bpV(pic) was a potent stimulator of the incorporation of [^{14}C]-glucose into glycogen in diaphragm muscle in contrast to the virtually total inactivity of bpV(phen) in this respect (Fig. 6C).

This different *in vivo* potency was puzzling in view of the comparable efficacy of these two compounds as IRK activa-

†Yale J-F, LaChance D, Bevan AP, Vigeant C, Shaver A and Posner BI, Hypoglycemic effects of peroxovanadium compounds in Sprague-Dawley and diabetic BB rats. Diabetes (accepted).

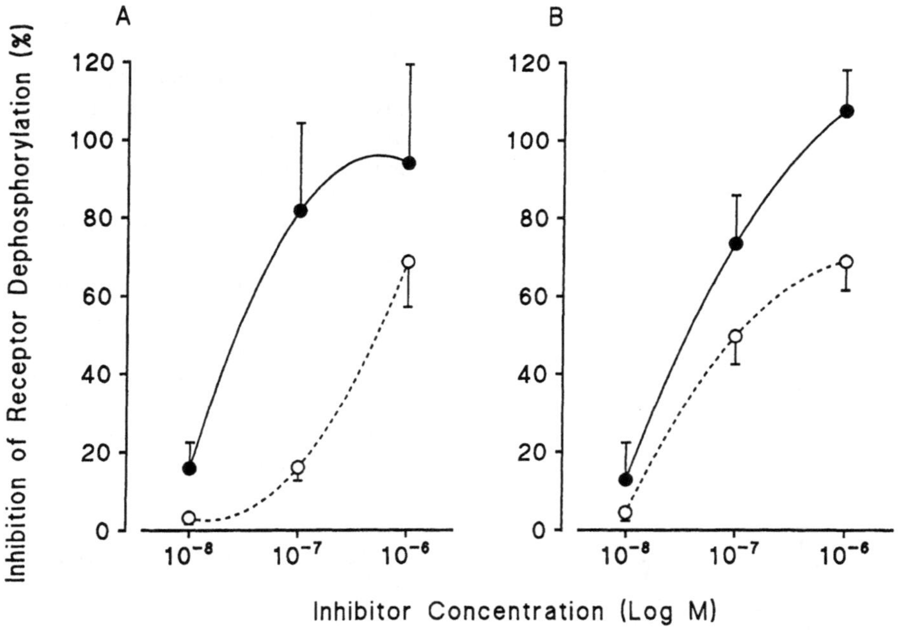

Fig. 5. Inhibition of endosomal IR or EFGR dephosphorylation by selected pV compounds: Relative specificity of action. Following the intrajugular injection of peptide, rat liver ENs were prepared 2 min after insulin (1.5 μg/100 g bwt) or 15 min following EGF (1.0 μg/100 g bwt) as described in Fig. 2. Panel A shows the inhibition of receptor dephosphorylation assayed at the noted concentrations of bpV(2,5pdc) for both the IR (●) and the EGFR (○). Panel B depicts the inhibition of receptor dephosphorylation at the noted concentrations of bpV(phen) for the IR (●) and EGFR (○). The data are the mean ± S.E. of 3 separate experiments.

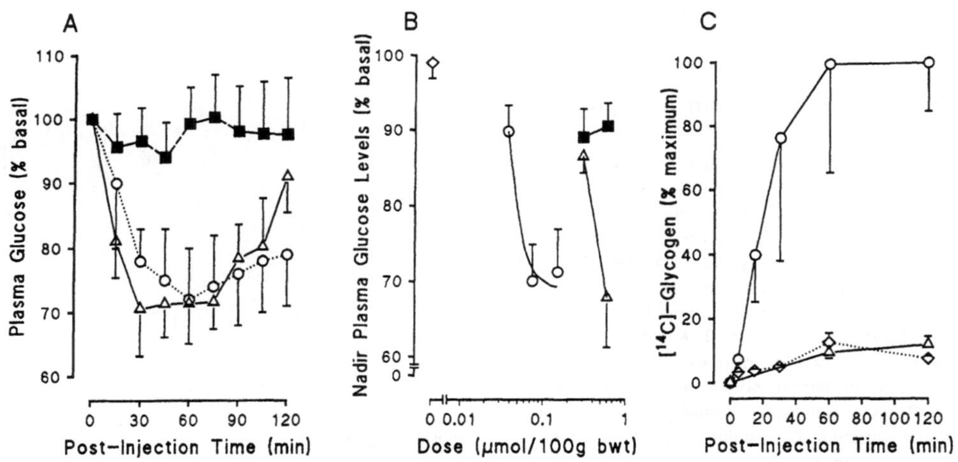

Fig. 6. In vivo metabolic effects of pVs. Fasted rats received an intrajugular injection of either vanadate (■), bpV(phen) (Δ), bpV(pic) (○) or phosphate buffered saline (PBS) (◊) at the indicated doses. Measurement of plasma glucose concentrations and the incorporation of [^{14}C]-glucose into diaphragm glycogen were performed as described elsewhere [24, 51]. Panel A shows the plasma glucose levels assayed every 15 min following vanadate (0.6 μmol/100 g bwt), bpV(phen) (0.6 μmol/100 g bwt) or bpV(pic) (0.15 μmol/100 g bwt) treatment. Each point is the mean ± S.E. of duplicate determinations performed on 4 different animals. Panel B shows the maximal reduction in plasma glucose level (nadir) achieved over a 2 h time course as a function of the administered dose of compound. The plasma glucose levels are expressed as a percent of pre-injection blood glucose levels (basal) and plotted as a function of the administered dose. Each point is the mean ± S.E. of duplicate determinations performed on 3–10 different animals. Panel C shows the incorporation of [^{14}C]-glucose into diaphragm glycogen following injection with [^{14}C]-glucose (1.25 μCi/100 g bwt) in PBS or in combination with 0.6 μmol/100 g bwt bpV(phen) or 0.6 μmol/100g bwt bpV(pic). Each point is the mean ± SE of determinations in 3–15 different animals.

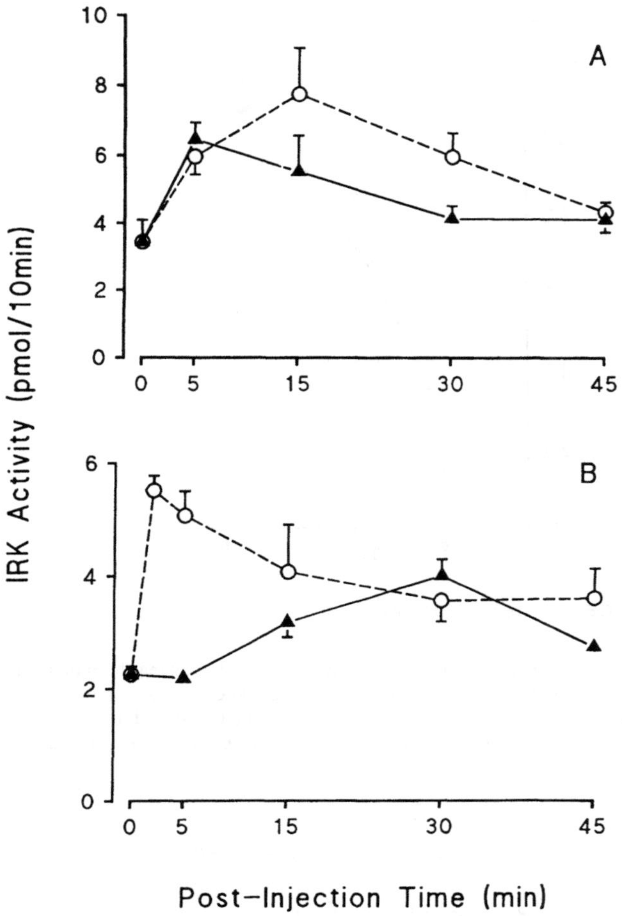

Fig. 7. Effect of bpV(phen) or bpV(pic) on IRK activity in rat diaphragm and hepatic microsomes. Fasted rats received an intrajugular injection of either bpV(phen) (0.6 µmol/100 g bwt) or bpV(pic) (0.6 µmol/100 g bwt) prior to sacrifice at the indicated times. IRs purified by lectin column chromatography from liver microsomes [29] and diaphragm [38] were assayed for exogenous kinase activity as described elsewhere [52]. Each assay was performed with an amount of IR which bound 10 fmol of [^{125}I]-insulin in the binding assay. Panel A shows the IRK activation in liver microsomes following bpV(phen) (▲) or bpV(pic) (○) administration. Each point is the mean ± S.E. of determinations on 3–10 different animals. Panel B shows the IRK activation in diaphragm following bpV(phen) (▲) or bpV(pic) (○) administration. Each point is the mean ± S.E. of determinations on 3–6 different animals.

tors in cultured hepatoma cells and as inhibitors of endosomal IRK dephosphorylation [24]. However the above observations made *in vivo* were clarified, at least in part, by studies performed to assess the ability of these compounds to activate the IRK in both liver and muscle - the two major target tissues for insulin's metabolic effects (Fig. 7). Upon i.v. administration of bpV(phen) maximal activation of liver microsome IRK was observed by 5 min (Fig. 7A). The administration of bpV(pic) also activated the liver microsomal IRK but to a greater extent with maximal activation being attained at 15 min post-injection. In contrast, whereas bpV(pic) rapidly activated the IRK of rat diaphragm, bpV(phen) produced modest activation and only by 30 min post-injection (Fig. 7B). The integrated response achieved with bpV(phen) was about 50% of that attained with bpV(pic). Thus the greater hypoglycaemic potency of bpV(pic) may reflect its capacity to act on both liver and muscle whereas the lesser potency of bpV(phen) may derive from its domain of action being relatively restricted to the liver. One possible explanation for these differences between the two compounds is that bpV(pic) may penetrate both liver and muscle tissue with comparable facility whereas bpV(phen) may readily penetrate liver but not muscle tissue. Pharmacodynamic studies will be required to establish whether ease of tissue accessibility can explain the above observations.

Synergy between insulin and peroxovanadium compounds in vivo

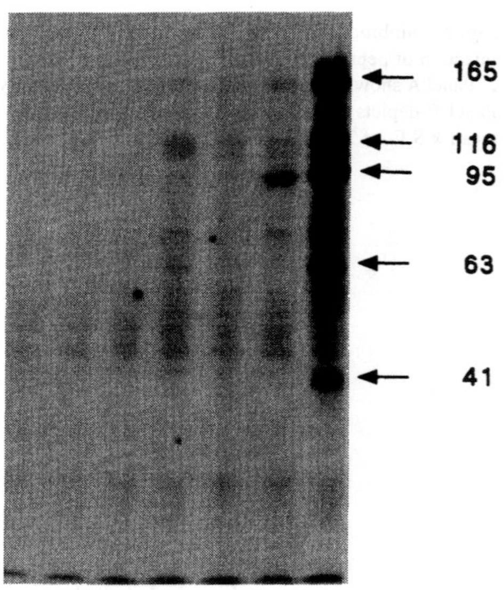

bpV(phen)	–	–	–	7	20	5	5
Insulin	–	2	15	–	–	2	15
Total time	–	2	15	7	20	7	20

Fig. 8. Rat liver phosphotyrosine protein profile: *in vivo* effect of bpV(phen) and Insulin. Fasted rats received an intrajugular injection of insulin (1.5 µg/100 g bwt), bpV(phen) (0.6 µmol/100 g bwt) or a combination of the two. At the indicated times (expressed in min) the rats were sacrificed, livers excised and liver microsomes were prepared as described elsewhere [29]. Aliquots of protein (100 µg) were subjected to SDS-PAGE on 7.5% gels and Western blotting with anti-phosphotyrosine antibodies as described elsewhere [29]. The figure depicts the phosphotyrosine profile following the various treatments.

If insulin perturbs the futile cycle limiting autophosphorylation in the basal state by directly activating the IRK, and the pVs promote activation by inhibiting receptor associated PTP(s) then we might anticipate that simultaneous administration of these two agents should promote a synergistic response. The results of one such study is depicted in Fig. 8. Here rats were injected with bpV(phen) 5 min prior to an injection of insulin and the animals killed either 2 or 15 min later. The level of various tyrosine phosphoproteins in subsequently prepared liver microsomes was assessed. Individually, insulin and bpV(phen) at these concentrations produced very modest effects on the level of tyrosine phosphorylated substrates (lanes 2–5). However in combination both the IR and other *in vivo* substrates became highly tyrosine phosphorylated (lanes 6 and 7). Furthermore IRK activity was also found to be highly augmented when bpV(phen) was administered prior to insulin [24]. Finally it has been observed that when insulin was administered 5 min after bpV(pic) the incorporation of [^{14}C]-glucose into diaphragm glycogen was synergistically stimulated [38]. Thus on the basis of several different measures of insulin action there is synergistic interaction between insulin and pV compounds.

Summary

We have described a number of properties of a new class of insulin-mimetic compounds, the peroxovanadate anions. Their ability to inhibit PTPs results in IRK activation leading to tyrosine phosphorylation of *in vivo* substrates. This in turn entrains normal insulin signalling steps culminating in insulin-dependent biological events. Both specificity and potency are affected by the ancillary ligand within the pV complex. This appears to influence both tissue and TK specificity of action. Their properties also raise the possibility of their use as insulin replacements or adjuvants in the management of diabetes mellitus.

Acknowledgements

These studies were supported by grants from the Medical Research Council of Canada and Nordic Merrell Dow Research, Laval, Quebec Canada. We also thank the Maurice Pollack Foundation, Montreal, Quebec, for their generous support. APB was previously supported by a fellowship from the Royal Victoria Hospital Research Institute and is currently the recipient of a fellowship from the Juvenile Diabetes Foundation. PGD was supported by a fellowship from the Royal Victoria Hospital Research Institute. We also thank Gerry Baquiran, Barbara Foster and Celyne Brule for their excellent technical assistance and finally we thank Robert Faure and James Burgess for their valuable contributions and help in these studies.

References

1. Tolman EL, Barris E, Burns M, Pansini A, Partridge R: Effects of vanadium on glucose metabolism *in vitro*. Life Sci 25: 1159–1164, 1979
2. Dubyak GR, Kleinzeller A: The insulin-mimetic effects of vanadate in isolated rat adipocytes. Dissociation from effects of vanadate as a (Na^+-K^+)ATPase inhibitor. J Biol Chem 255: 5306–5312, 1980
3. Shechter Y, Karlish SJ: Insulin-like stimulation of glucose oxidation in rat adipocytes by vanadyl (IV) ions. Nature 284: 556–558, 1980
4. Tamura S, Brown TA, Whipple JH, Fujita-Yamaguchi Y, Dubler RE, Cheng K, Larner J: A novel mechanism for the insulin-like effect of vanadate on glycogen synthase in rat adipocytes. J Biol Chem 259: 6650–6658, 1984
5. Clark AS, Fagan JM, Mitch WE: Selectivity of the insulin-like actions of vanadate on glucose and protein metabolism in skeletal muscle. Biochem J 232: 273–276, 1985
6. Kadota S, Fantus IG, Hersh B, Posner Bl: Vanadate stimulation of IGF binding to rat adipocytes. Biochem Biophys Res Commun 138: 174–178, 1986
7. Kadota S, Fantus IG, Deragon G, Guyda HJ, Posner Bl: Stimulation of insulin-like growth factor 11 receptor binding and insulin receptor kinase activity in rat adipocytes. Effects of vanadate and H_2O_2. J Biol Chem 262: 8252–8256, 1987
8. Strout HV, Vicario PP, Saperstein R, Slater EE: The insulin-mimetic effect of vanadate is not correlated with insulin receptor tyrosine kinase activity nor phosphorylation in mouse diaphragm *in vivo*. Endocrinology 124: 1918–1924, 1989
9. Heyliger CE, Tahiliani AG, McNeill JH: Effect of vanadate on elevated blood glucose and depressed cardiac performance of diabetic rats. Science 227: 1474–1477, 1985
10. Meyerovitch J, Farfel Z, Sack J, Shechter Y: Oral administration of vanadate normalizes blood glucose levels in streptozotocin-treated rats. Characterization and mode of action. J Biol Chem 262: 6658–6662, 1987
11. Brichard SM, Okitolonda W, Henquin JC: Long term improvement of glucose homeostasis by vanadate treatment in diabetic rats. Endocrinology 123: 2048–2053, 1988
12. Blondel O, Bailbe D, Portha B: *In vivo* insulin resistance in streptozotocin-diabetic rats-evidence for reversal following oral vanadate treatment. Diabetologia 32: 185–190, 1989
13. Gil J, Miralpeix M, Carreras J, Bartrons R: Insulin-like effects of vanadate on glucokinase activity and fructose 2,6-bisphosphate levels in the liver of diabetic rats. J Biol Chem 263: 1868–1871, 1988
14. Czech MP, Lawrence JCJ, Lynn WS: Evidence for the involvement of sulfhydryl oxidation in the regulation of fat cell hexose transport by insulin. Proc Natl Acad Sci USA 71 :4173–4177, 1974
15. May JM, de Haen C: The insulin-like effect of hydrogen peroxide on pathways of lipid synthesis in rat adipocytes. J Biol Chem 254: 9017–9021, 1979
16. Lawrence JCJ, Larner J: Activation of glycogen synthase in rat adipocytes by insulin and glucose involves increased glucose transport and phosphorylation. J Biol Chem 253: 2104–2113, 1978
17. Little SA, de Haen C: Effects of hydrogen peroxide on basal and hormone-stimulated lipolysis in perifused rat fat cells in relation to the mechanism of action of insulin. J Biol Chem 255: 10888–10895, 1980
18. Heffetz D, Zick Y: H_2O_2 potentiates phosphorylation of novel putative

substrates for the insulin receptor kinase in intact Fao cells. J Biol Chem 264: 10126–10132, 1989
19. Kadota S, Fantus IG, Deragon G, Guyda HJ, Hersh B, Posner Bl: Peroxide(s) of vanadium: a novel and potent insulin-mimetic agent which activates the insulin receptor kinase. Biochem Biophys Res Commun 147: 259–266, 1987
20. Heffetz D, Bushkin I, Dror R, Zick Y: The insulinomimetic agents H_2O_2 and vanadate stimulate protein tyrosine phosphorylation in intact cells. J Biol Chem 265: 2896–2902, 1990
21. Fantus IG, Kadota S, Deragon G, Foster B, Posner Bl: Pervanadate [peroxide(s) of vanadate] mimics insulin action in rat adipocytes via activation of the insulin receptor tyrosine kinase. Biochemistry 28: 8864–8871, 1989
22. Shisheva A, Shechter Y: Mechanism of pervanadate stimulation and potentiation of insulin-activated glucose transport in rat adipocytes: dissociation from vanadate effect. Endocrinology 133: 1562–1568, 1993
23. Hadari YR, Geiger B, Nadiv O, Sabanay I, Roberts CTJ, LeRoith D, Zick Y: Hepatic tyrosine-phosphorylated proteins identified and localized following in vivo inhibition of protein tyrosine phosphatases: effects of H_2O_2 and vanadate administration into rat livers. Mol Cell Endocrinol 97: 9–17, 1993
24. Posner Bl, Faure R, Burgess JW, Bevan AP, Lachance D, Zhang-Sun G, Fantus IG, Ng JB, Hall DA, Lum BS, Shaver A: Peroxovanadium compounds. A new class of potent phosphotyrosine phosphatase inhibitors which are insulin mimetics. J Biol Chem 269: 4596–4604, 1994
25. Jakubowski J, Jakob A: Vasopressin, insulin and peroxide(s) of vanadate (pervanadate) influence Na^+ transport mediated by (Na^+, K^+)ATPase or Na^+/H^+ exchanger of rat liver plasma membrane vesicles. Eur J Biochem 193: 541–549, 1990
26. Imbert V, Peyron JF, Farahi Far D, Mari B, Auberger P, Rossi B: Induction of tyrosine phosphorylation and T-cell activation by vanadate peroxide, an inhibitor of protein tyrosine phosphatases. Biochem J 297: 163–173, 1994
27. Harrison ML, Isaacson CC, Burg DL, Geahlen RL, Low PS: Phosphorylation of human erythrocyte band 3 by endogenous $p72^{syk}$. J Biol Chem 269: 955–959, 1994
28. Khan MN, Baquiran G, Brule C, Burgess J, Foster B, Bergeron JJ, Posner Bl: Internalization and activation of the rat liver insulin receptor kinase *in vivo*. J Biol Chem 264: 12931–12940, 1989
29. Burgess JW, Wada I, Ling N, Khan MN, Bergeron JJ, Posner Bl: Decrease in beta-subunit phosphotyrosine correlates with internalization and activation of the endosomal insulin receptor kinase. J Biol Chem 267: 10077–10086, 1992
30. Faure R, Baquiran G, Bergeron JJ, Posner Bl: The dephosphorylation of insulin and epidermal growth factor receptors. Role of endosome-associated phosphotyrosine phosphatase(s). J Biol Chem 267: 11215–11221, 1992
31. Guan KL, Dixon JE: Protein tyrosine phosphatase activity of an essential virulence determinant in Yersinia. Science 249: 553–556, 1990
32. Guan KL, Haun RS, Watson SJ, Geahlen RL, Dixon JE: Cloning and expression of a protein-tyrosine-phosphatase. Proc Natl Acad Sci USA 87: 1501–1505, 1990
33. Guan KL, Dixon JE: Evidence for protein-tyrosine-phosphatase catalysis proceeding via a cysteine-phosphate intermediate. J Biol Chem 266: 17026–17030, 1991
34. Pot DA, Dixon JE: Active site labeling of a receptor-like protein tyrosine phosphatase. J Biol Chem 267: 140–143, 1992
35. Shaver A, Ng JB, Hall DA, Lum BS, Posner Bl: Insulin mimetic peroxovanadium complexes: preparation and structure of potassium oxodiperoxo (pyridine-2-carboxylato) vanadate(V), $K_2[VO(O_2)_2(C_5H_4NCOO)]\cdot 2H_2O$, and potassium oxodiperoxo(3-hydroxypyridine-2-carboxylato)vanadate(V), $K_2[VO(O_2)_2(OHC_5H_3NCOO)].3H_2O$, and their reactions with cysteine. Inorg Chem 32: 3109–3113, 1993
36. Tracey AS, Gresser MJ: Interaction of vanadate with phenol and tyrosine: implications for the effects of vanadate on systems regulated by tyrosine phosphorylation. Proc Natl Acad Sci USA 83: 609–613, 1986
37. Barford D, Flint AJ, Tonks NK: Crystal structure of human protein tyrosine phosphatase 1 B. Science 263: 1397–1404, 1994
38. Bevan AP, Burgess JW, Yale J-F, Drake PG, Lachance D, Baquiran G, Shaver A, Posner Bl: *In vivo* insulin-mimetic effects of peroxovanadium compounds: A role for tissue targeting in determining potency. Am J Physiol 268: E60–E66, 1995
39. Shisheva A, Ikonomov O, Shechter Y: The protein tyrosine phosphatase inhibitor, pervanadate, is a powerful antidiabetic agent in streptozotocin-treated diabetic rats. Endocrinology 134: 507–510, 1994
40. Lonnroth P, Eriksson JW, Posner Bl, Smith U: Peroxovanadate but not vanadate exerts insulin-like effects in human adipocytes. Diabetologia 36: 113–116, 1993
41. Foot E, Bliss T, Fernandes LC, DaCosta C, Leighton B: The effects of orthovanadate, vanadyl and peroxides of vanadate on glucose metabolism in skeletal muscle preparations *in vitro*. Mol Cell Biochem 109: 157–162, 1992
42. Leighton B, Cooper GJS, DaCosta C, Foot EA: Peroxovanadates have full insulin-like effects on glycogen synthesis in normal and insulin-resistant skeletal muscle. Biochem J 276: 289–292, 1991
43. Hadari YR, Tzahar E, Nadiv O, Rothenberg P, Roberts CTJ, LeRoith D, Yarden Y, Zick Y: Insulin and insulinomimetic agents induce activation of phosphatidylinositol 3′-kinase upon its association with pp185 (IRS-1) in intact rat livers. J Biol Chem 267: 17483–17486, 1992
44. Zick Y, Sagi-Eisenberg R: A combination of H_2O_2 and vanadate concomitantly stimulates protein tyrosine phosphorylation and polyphosphoinositide breakdown in different cell lines. Biochemistry 29: 10240–10245, 1990
45. Heffetz D, Rutter WJ, Zick Y: The insulinomimetic agents H_2O_2 and vanadate stimulate tyrosine phosphorylation of potential target proteins for the insulin receptor kinase in intact cells. Biochem J 288: 631–635, 1992
46. Bourgoin S, Grinstein S: Peroxides of vanadate induce activation of phospholipase D in HL-60 cells. Role of tyrosine phosphorylation. J Biol Chem 267: 11908–11916, 1992
47. O'Shea JJ, McVicar DW, Bailey TL, Burns C, Smyth MJ: Activation of human peripheral blood T lymphocytes by pharmacological induction of protein-tyrosine phosphorylation. Proc Natl Acad Sci USA 89: 10306–10310, 1992
48. Estrada DE, Elliott E, Zinman B, Poon I, Liu Z, Klip A, Daneman D: Regulation of glucose transport and expression of GLUT3 transporters in human circulating mononuclear cells: studies in cells from insulin-dependent diabetic and nondiabetic individuals. Metabolism 43: 591–598, 1994
49. Modderman PW, von dem Borne AE, Sonnenberg A: Tyrosine phosphorylation of P-selectin in intact platelets and in a disulphide-linked complex with immunoprecipitated $pp60^{c\text{-}src}$. Biochem J 299: 613–621, 1994
50. Inazu T, Taniguchi T, Yanagi S, Yamamura H: Protein-tyrosine phosphorylation and aggregation of intact human platelets by vanadate with H_2O_2. Biochem Biophys Res Commun 170: 259–263, 1990
51. Rafaelsen OJ, Lauris V, Renold AE: Localized intraperitoneal action of insulin on rat diaphragm and epididymal adipose tissue *in vivo*. Diabetes 14:19–26, 1965
52. Burgess JW, Bevan AP, Bergeron JJM, Posner Bl: Pharmacological doses of insulin equalize insulin phosphotyrosine content but not tyrosine kinase activity in plasmalemmal and endosomal membranes. Biochem Cell Biol 70: 1151–1158, 1992

Molecular and Cellular Biochemistry **153:** 59–67, 1995.

Unique and selective mitogenic effects of vanadate on SV40-transformed cells

Hanlin Wang and Robert E. Scott
Department of Pathology, The University of Tennessee College of Medicine, 899 Madison Avenue, Memphis, Tennessee 38163, USA

Abstract

Vanadate and insulin both function as unique complete mitogens for SV40-transformed 3T3T cells, designated CSV3-1, but not for nontransformed 3T3T cells. The mitogenic effects induced by vanadate and insulin in CSV3-1 cells are mediated by different signaling mechanisms. For example, vanadate does not stimulate the tyrosine phosphorylation of the insulin receptor β-subunit nor the 170 kDa insulin receptor substrate-1. Instead, vanadate induces a marked increase in tyrosine phosphorylation of 55 and 64 kDa proteins that is not observed in insulin-stimulated CSV3-1 cells. Perhaps most interestingly, vanadate-induced mitogenesis is associated with the selective induction of *c-jun* and *junB* expression without significantly inducing *c-fos* or *c-myc*. Furthermore, treatment of CSV3-1 cells with genistein abolishes the effects of vanadate on protein tyrosine phosphorylation and *c-jun* induction. These and related data suggest that modulation of protein tyrosine phosphorylation and *c-jun* and *junB* expression may serve the critical roles in mediating vanadate-induced mitogenesis in SV40-transformed cells. (Mol Cell Biochem **153**: 59–67, 1995)

Key words: adipocyte differentiation, neoplastic transformation, insulin, *c-jun*, *junB*, tyrosine phosphorylation

Introduction

Vanadium compounds such as sodium orthovanadate (vanadate) are unique multifunctional agents that have been documented to mimic most of the metabolic effects of insulin in a variety of *in vitro* and *in vivo* experimental systems [1–6]. During the past decade, the possibility of vanadate as a potential therapeutic agent in the treatment of diabetes mellitus has been increasingly investigated [7, 8]. Numerous reports demonstrate that oral administration of vanadate to streptozotocin-induced diabetic rats can lead to normalization of hyperglycemia by modulating the activity and the expression of enzymes involved in glucose metabolism without causing detectable side effects [5–10]. These findings are significant not only because vanadate can be administered orally but more importantly because vanadate can bypass the insulin receptor and therefore can be effective in the treatment of insulin-resistant diabetes [7, 8].

Another important effect of vanadate is that it promotes mitogenesis and cell proliferation. Unlike the general growth-promoting effect of insulin, however, the mitogenic effect of vanadate seems more cell-type-selective. Under the appropriate culture conditions and at the optimal concentrations, vanadate by itself has been found to stimulate DNA synthesis in human fibroblasts [11], Swiss mouse 3T3 and 3T6 cells [12], a subpopulation of thymus cells [13], Nakano mouse lens epithelial cells [14], calvaria-derived osteoblast-like cells [15–17], bovine capillary endothelial cell [18], rat intestinal IEC-6 cells [19], and simian virus 40 (SV40)-transformed murine 3T3T cells [20]. In several different cell types, vanadate also synergistically enhances or potentiates the growth-promoting effects of other growth factors, such as insulin [12, 14, 16, 21, 22], insulin-like growth factor I [16], epidermal growth factor [11, 16], fibroblast growth factor [11], platelet-derived growth factor [17], estrogen [23], granulocyte-macrophage colony-stimulating factor [24], and interleukin-3 [24]. Furthermore, vanadate has been documented to promote neoplastic transformation of mouse and

Address for offprints: H. Wang, Department of Pathology, The University of Tennessee, Memphis, TN 38163, USA hamster fibroblasts [25, 26], normal rat kidney cells [25, 27], rabbit chondrocytes [28], a rat G5 cell line [29] and bovine capillary endothelial cells [18], and to potentiate the transforming activity of *c-fps/fes* and bovine papillomavirus [30, 31]. The effect of vanadate on cell transformation is reversible if vanadate is removed from the culture medium [25, 26] and is not always observed [32]. In addition, vanadate was found to have no influence on the incidence or histological type of mouse colon tumors induced by 1,2-dimethylhydrazine, although it increased thymidine incorporation [33]. It is therefore important to carefully evaluate the effects of vanadate on cell growth and transformation if vanadate is to be considered as a potential therapeutic agent for human diabetes.

The exact biochemical and molecular mechanisms by which vanadate mediates its metabolic, mitogenic and transforming effects remain to be established. Although vanadate can augment insulin binding and insulin-stimulated insulin receptor kinase activity and prolong insulin action in rat adipocytes [34, 35], it is generally accepted that vanadate by itself does not stimulate the tyrosine phosphorylation nor the kinase activity of the insulin receptor [7, 36–39]. Rather, it appears to selectively increase the tyrosine phosphorylation of specific cellular proteins including the mitogen-activated protein kinases [39, 40], which have been implicated to play a central role in the signaling pathways of insulin [41] and a variety of other growth factors and cytokines [42]. The mechanisms by which vanadate enhances cellular protein tyrosine phosphorylation may involve both an inhibition of tyrosine phosphatases, since vanadate is a tyrosine-specific phosphatase inhibitor [32, 43], and a direct stimulation of distinct tyrosine kinases [38–40, 44–47]. Accumulating studies have demonstrated that most of the vanadate effects are associated with an increase in tyrosine phosphorylation, suggesting that protein tyrosine phosphorylation may serve critical roles in mediating vanadate actions. This may be particularly true for its mitogenic and transforming effects because a number of growth factor receptors and oncogenes are associated with tyrosine kinase activity [48].

In the past four years, our studies, which will be reviewed next, have demonstrated that vanadate (and insulin) can independently induce mitogenesis in quiescent SV40-transformed murine 3T3T cells, i.e., CSV3-1 cells, but not in nontransformed 3T3T cells. The mitogenic effect of vanadate in these cells involves distinct protein tyrosine phosphorylation and selective induction of the expression of protooncogenes *c-jun* and *junB* [20, 38, 49, 50]. The focus of our studies is to establish the possible mechanisms by which vanadate has mitogenic activity in transformed cells, and more broadly, to understand how transformed cells escape the control mechanisms that regulate normal cell growth.

Growth and differentiation characteristics of 3T3T and CSV3-1 cells

Nontransformed 3T3T cells were derived from murine Balb/c 3T3 mesenchymal stem cells and have been used in the past years as an excellent *in vitro* model system to study adipocyte differentiation and to investigate the molecular mechanisms by which cell proliferation and differentiation are mutually regulated [51–55]. It has been well established that the process of adipocyte differentiation of 3T3T cells involves multiple steps [51]. In differentiation-inducing medium [52], 3T3T cells first arrest their growth at the predifferentiation growth arrest state, a distinct G_1 cell cycle restriction point at which cells possess the potentialities to undergo adipocyte differentiation or to reinitiate proliferation. Under defined culture conditions, cells at the predifferentiation growth arrest state next undergo nonterminal differentiation wherein they develop an adipocyte morphology while retaining their proliferative potential. Nonterminally differentiated cells can then either dedifferentiate and return to the proliferation state or they can be induced to irreversibly lose their proliferative ability and undergo terminal differentiation. An outstanding characteristic of this multiple differentiation process is that the differentiating cells progressively lose their mitogenic responsiveness to numerous growth factors [53, 55]. For example, nonterminally differentiated 3T3T adipocytes can be stimulated to proliferate by very high concentrations (e.g., 30%) of fetal bovine serum (FBS) but not by lower concentrations of FBS or purified growth factors. Terminally differentiated cells are not responsive to any known growth factors. The progressive repression of growth factor responsiveness in differentiating 3T3T cells is summarized in Fig. 1.

CSV3-1 cells used in our experiments were derived from 3T3T cells that were transformed by the SV40 large T antigen [56–58]. Typically, SV40-transformed cells show markedly reduced serum requirement for growth and the inability to differentiate. This makes it difficult to assay their growth factor responsiveness because of their inability to undergo growth arrest. From this perspective, CSV3-1 cells are valuable and unique in that they can differentiate into adipocytes efficiently and therefore can undergo growth arrest under differentiation-inducing conditions, although they cannot be growth-arrested by culture in serum-deficient medium or by high density culture conditions [58]. By using this differentiation approach, mitogenic assays on CSV3-1 cells at various biological states are feasible. The results presented in Fig. 1 further show that compared to 3T3T cells, the progressive loss of growth factor responsiveness during differentiation does not occur in CSV3-1 cells even though they express a fully differentiated phenotype. Figure 1 specifically shows that at the nonterminal differentiation state, 5% FBS is sufficient to induce the majority of CSV3-1 adi-

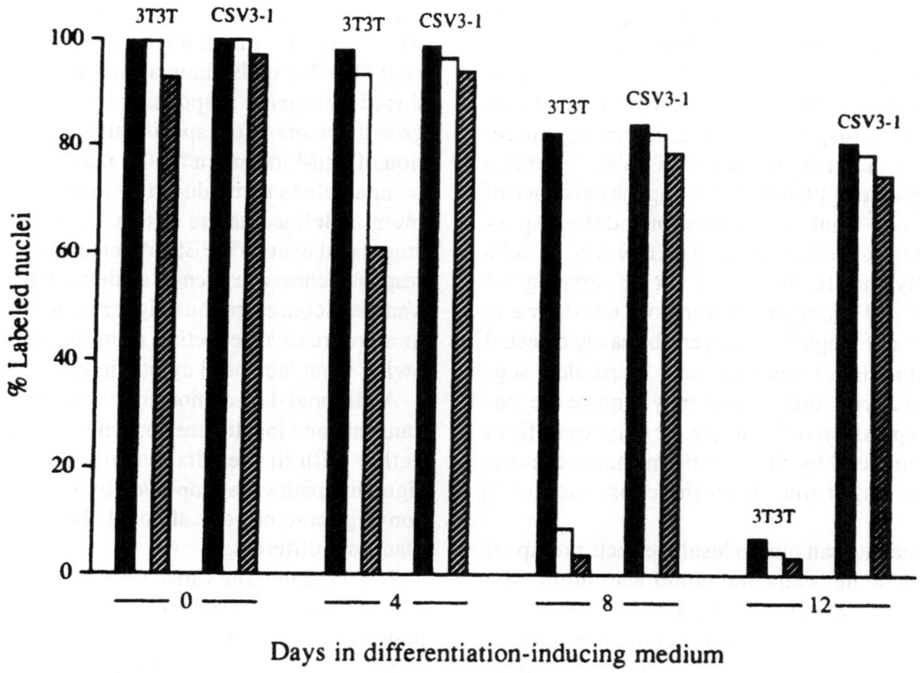

Fig. 1. Comparison of mitogenic responsiveness to various concentrations of FBS between CSV3-1 and 3T3T cells at the different biological states. Cells cultured in differentiation-inducing medium for various periods of time were treated with either 30% (closed bar), 10% (open bar), or 5% (hatched bar) FBS. Mitogenic response was determined as the percentage of labeled nuclei incorporating [^{3}H]thymidine after 24–40 h exposure to FBS. Days 0, 4, 8, and 12 represent rapid growing, predifferentiation growth arrest, nonterminal differentiation, and terminal differentiation states, respectively, in nontransformed 3T3T cells.

pocytes to undergo DNA synthesis and dedifferentiation, whereas 30% FBS is required for 3T3T cells. Under the conditions that induce terminal differentiation in 3T3T cells, CSV3-1 adipocytes can still be stimulated to proliferate by FBS concentrations of as low as 5%. Therefore, the terminal step of differentiation is blocked in transformed CSV3-1 cells [58, 59]. The steps involved in the process of adipocyte differentiation of CSV3-1 cells and their aberrant growth and differentiation characteristics, including the increased proclivity to reinitiate proliferation and the decreased ability to differentiate terminally, are further depicted in Fig. 2.

Mitogenic effects of insulin and vanadate on CSV3-1 cells

Since CSV3-1 cells show a dramatically increased sensitivity to serum growth factors even in a highly differentiated state, their mitogenic responsiveness to a variety of purified peptide growth factors was evaluated [49]. The results demonstrated that insulin is the only growth factor tested that selectively induces mitogenic response in CSV3-1 cells at both predifferentiation growth arrest state and nonterminal differentiation state. The mitogenic effect of insulin is also observed in SV40 large T antigen-transformed CSV3-35 cells that can be growth-arrested under the differentiation-inducing

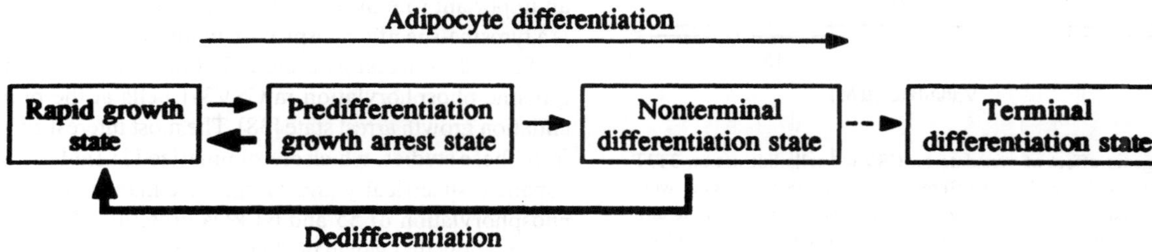

Fig. 2. A summary of the steps involved in the process of CSV3-1 adipocyte differentiation. The larger arrows represent an increased probability to reenter the proliferative cell cycle and the dashed arrow represents a decreased probability to undergo terminal differentiation.

conditions but is not observed in quiescent nontransformed 3T3T cells. Additional studies show that the insulin effect on CSV3-1 cells appears to be mediated by its own receptor rather than by the insulin-like growth factor I receptor [49, unpublished observations]. Furthermore, the mitogenic response to insulin is markedly decreased if SV40 T antigen expression is suppressed [49, 60], indicating the existence of a correlation between T antigen expression and the responsiveness to insulin-induced mitogenesis. However, in cells that are transiently transfected with a vector carrying the cDNA encoding SV40 T antigen, insulin does not show any mitogenic effect even though T antigen can be easily detected in these cells [unpublished observations]. These data suggest that insulin-induced mitogenesis may require the stable or extended expression of T antigen, or may result from cellular changes induced by the T antigen-induced transformation process rather than from the expression of T antigen *per se*.

The fact that vanadate can mimic insulin effects prompted us to investigate whether vanadate could also mimic the mitogenic effect of insulin in CSV3-1 cells [20]. To perform such studies, quiescent cells at the predifferentiation growth arrest state were used because they showed the highest insulin responsiveness. Figure 3 shows that whereas quiescent 3T3T cells show no mitogenic response to vanadate, quiescent CSV3-1 cells show a highly significant vanadate-induced mitogenic responsiveness in the absence of other growth factors. More specifically, vanadate at the concentration of 5 μM induces a > 50% maximum response, which is comparable to that induced by insulin [20, 49] (100% maximum is defined as the mitogenic response when cells are stimulated with 10% FBS). A detectable increase in mitogenic responsiveness can even be achieved with 50 nM of vanadate. Vanadate concentrations higher than 10 μM however result in a progressive reduction in mitogenic response probably owing to an increased cytotoxicity.

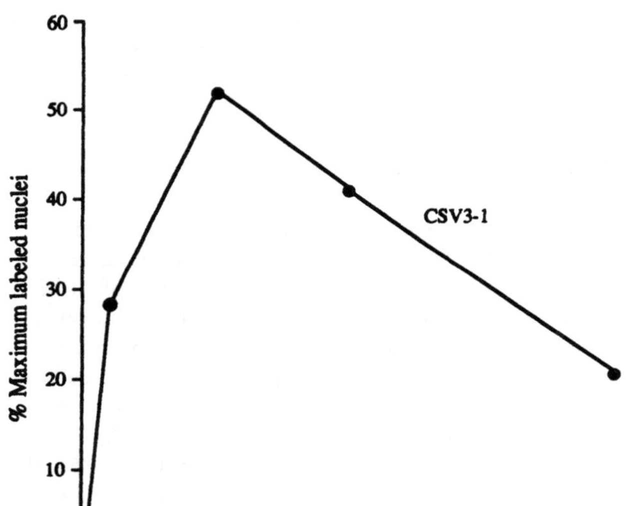

Fig. 3. Mitogenic effect of vanadate on CSV3-1 cells but not on 3T3T cells. Quiescent cells at the predifferentiation growth arrest state were exposed to vanadate for 24 h at the indicated concentrations in serum-free medium containing 2.5 μCi/ml of [*methyl*-^{3}H]thymidine (70–80 Ci/mmol; Amersham). The mitogenic response is reported as percentage of maximum labeled nuclei incorporating [^{3}H]thymidine where 100% maximum is defined as the mitogenic effect of 10% FBS.

Additional data demonstrate that the mitogenic effects of vanadate and insulin are not synergetic [20]. This result, together with the results described later, suggest that the signaling pathways employed by vanadate and insulin may converge at some point although their early signaling mechanisms are different.

To ascertain if the vanadate effect on CSV3-1 cells is specific and unique, a series of phosphoserine/phosphothreonine phosphatase inhibitors, including okadaic acid, molybdate, sodium pyrophosphate, and sodium fluoride, were examined over a wide range of concentrations. The results demonstrate that none of these inhibitors shows mitogenic effect on CSV3-1 cells at the predifferentiation growth arrest state. Another insulinomimetic agent, H_2O_2, which has been reported to enhance protein tyrosine phosphorylation and to potentiate the effect of vanadate on tyrosine phosphorylation in rat hepatoma Fao cells [61], also shows no mitogenic effect on CSV3-1 cells either alone or in combination with vanadate. Finally, a mitogenic response is not observed in CSV3-1 cells treated with calcium ionophore A23187, suggesting that the vanadate-induced mitogenesis is not primarily mediated through a mechanism involving an increase in intracellular calcium.

Distinct protein tyrosine phosphorylation by vanadate

Numerous studies have established that changes in protein phosphorylation, especially tyrosine phosphorylation, serve an important role in regulating cell growth. Since vanadate is a tyrosine-specific phosphatase inhibitor, experiments were performed to investigate the effect of vanadate on protein tyrosine phosphorylation in CSV3-1 cells at the predifferentiation growth arrest state [38]. The most interesting results from those studies, which are summarized in Table 1, are that vanadate specifically and markedly enhances the tyrosine phosphorylation of 55 and 64 kDa proteins. More detailed studies show that vanadate does not initially increase detectable tyrosine phosphorylation of any proteins, including neither the insulin receptor nor the insulin receptor substrate-

Table 1. Comparison of vanadate and insulin effects on protein tyrosine phosphorylation[a]

Protein	Vanadate	Insulin
95 kDa insulin receptor β-subunit[b]	–	+
170 kDa insulin receptor substrate-1[c]	–	+
55 kDa protein[d]	+	–
64 kDa protein[d]	+	–

[a]CSV3-1 cells were cultured in differentiation-inducing medium for 4 days to induce the quiescent state. They were then treated with 5 μM vanadate or 500 ng/ml insulin for 2 min–2 h. Soluble cellular proteins were subjected to immunoprecipitation using antibodies specified below and tyrosine phosphorylation was identified by probing Western blots with antiphosphotyrosine antibodies (Oncogene Science). [b]Immunoprecipitation using antibodies against the insulin receptor β–subunit (Santa Cruz Biotechnology, Inc.). [c]Immunoprecipitation using antibodies against the insulin receptor substrate-1 (Upstate Biotechnology, Inc.). [d]Immunoprecipitation using antiphosphotyrosine antibodies (Upstate Biotechnology, Inc.).

1, and does not stimulate the complex formation between the insulin receptor substrate-l and the 85 kDa subunit of phosphatidylinositol 3'-kinase. The marked increase in tyrosine phosphorylation of the 55 and 64 kDa proteins is only observed 60 min after vanadate stimulation and is not observed in insulin-stimulated CSV3-1 cells nor in vanadate-treated 3T3T cells. The effect of vanadate on tyrosine phosphorylation in quiescent CSV3-1 cells can be almost completely blocked by pretreatment of the cells with genistein [38], a general tyrosine-specific protein kinase inhibitor [62]. In contrast, insulin-induced tyrosine phosphorylation of the insulin receptor β-subunit is relatively insensitive to the inhibitory effect of genistein. Although the significance of increased tyrosine phosphorylation of the 55 and 64 kDa proteins in vanadate-induced mitogenesis remains unclear, these findings have further limited the possibility that the vanadate effect in CSV3-1 cells is mediated by activating the insulin receptor tyrosine kinase. Rather, vanadate may directly or indirectly activate a protein tyrosine kinase(s) that is distinct from those involved in the early insulin signaling pathways.

Enhanced protein tyrosine phosphorylation has also been reported in other vanadate-treated cells where vanadate may function as a growth- or transformation-promoting agent [18, 19, 25, 44]. In the reports by D'Onofrio *et al.* [39] and Nguyen *et al.* [40], the tyrosine phosphorylation of the 44 kDa and 42 kDa mitogen-activated protein kinases was markedly increased within 5 min after cells were exposed to 100–1000 μM of vanadate. At the dosages of less than 10 μM, however, the effect of vanadate was only minimal [39].

It remains to be determined why vanadate induces tyrosine phosphorylation selectively in transformed CSV3-1 cells but not in nontransformed 3T3T cells. It is our hypothesis that the activity of certain critical regulatory molecules, such as tyrosine phosphatases and/or kinases, may have been modified during SV40-induced transformation process so that they become more sensitive or 'primed' to the stimulatory effect of vanadate. One possibility is that CSV3-1 cells may contain higher levels of reactive H_2O_2 than nontransformed cells so that vanadate can be more effective as a tyrosine phosphatase inhibitor in the transformed cells by reacting more efficiently with intracellular H_2O_2.

Selective induction of c-jun *and* junB *by vanadate*

To further investigate the possible mechanisms by which mitogenesis is induced by vanadate in CSV3-1 cells, the effects of vanadate on the expression of growth-related genes were examined [20, 50]. The results demonstrate that while neither vanadate nor insulin has effect on the induction of the expression of *c-fos*, *c-myc*, *c-jun*, *junB*, or ornithine decarboxylase (ODC) activity in quiescent nontransformed 3T3T cells, both vanadate and insulin induce a marked increase in the expression of *c-jun* and *junB* mRNAs and ODC activity with an equivalent efficiency in quiescent CSV3-1 cells. Interestingly, insulin is also capable of inducing *c-fos* and *c-myc* expression in CSV3-1 cells but vanadate is not. Figure 4 presents this unique effect of vanadate on *c-jun* and *junB* induction and specifically shows that the *c-jun* and *junB* mRNA levels are significantly increased 30 min after vana-

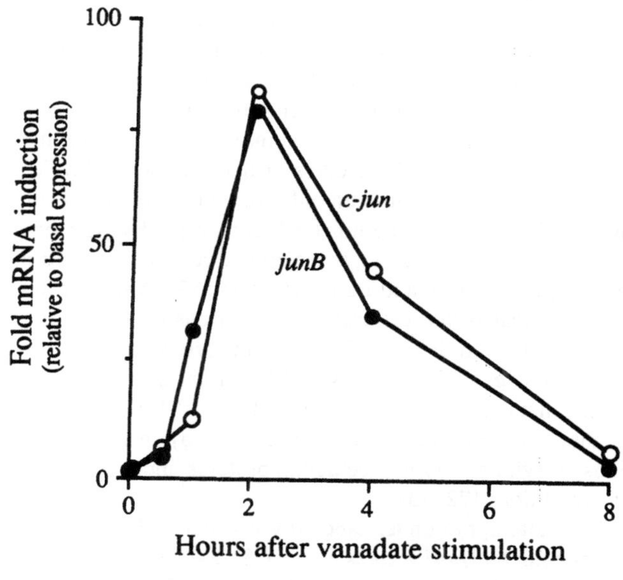

Fig. 4. Induction of *c-jun* and *junB* by vanadate in CSV3-1 cells. Quiescent CSV3-1 cells were treated with 5 μM vanadate for the indicated times. The relative *c-jun* and *junB* mRNA levels determined by Northern blot analysis were quantitated by densitometric analysis and normalized with glyceraldehyde-3-phosphate dehydrogenase mRNA levels. Data are reported as fold induction relative to the basal expression which are the *c-jun* or *junB* levels before vanadate treatment.

date stimulation, maximally induced by 2 h, and decline thereafter. More detailed studies designed to compare the early signal transduction pathways further demonstrated that vanadate and insulin use different mechanisms to induce *c-jun* expression [50]. For example, induction of *c-jun* by vanadate does not involve pertussis toxin-sensitive G protein(s) or the biosynthesis of polyamines (particularly spermidine), whereas induction of *c-jun* by insulin does. In addition, vanadate effect on *c-jun* induction is independent of protein kinase C but does require protein tyrosine kinase activity [50]. These data, together with the observations described previously, suggest that the early signaling mechanisms employed by vanadate and insulin to activate mitogenesis are distinct but that their effects may converge subsequently via their effects on *c-jun*, *junB*, and ODC.

Similar observations regarding vanadate effects on growth-related genes have also been reported in other cell types. Vanadate at the concentrations of 5–10 μM induces *c-jun* mRNA accumulation in cultured mouse C127 cells but does not induce *c-fos* or *c-myc* [63]. Vanadate at the concentration of 7 μM shows growth-promoting effect on mouse C2 myoblasts but shows no effect on *c-fos* induction [22]. Vanadate over the range of 0.05–5 μM also stimulates ODC activity in C3H/10T½ cells [64]. Furthermore, vanadate at the concentration of 100 μM enhances *c-myc* expression in human ovary carcinoma cells; but even at such a high concentration vanadate does not induce *c-fos* expression [65].

The *c-jun*, *junB* and *c-fos* genes are members of two related multigene families. The other members of the *fos* family include *fosB*, *fra-l*, and *fra-2* [66]. The products of *jun* and *fos* genes contain a leucine-zipper domain by which protein dimerization is mediated. Jun proteins are able to form either homodimers among themselves or heterodimers with Fos proteins but Fos proteins can only form heterodimers. Both Jun-Fos heterodimers and Jun-Jun homodimers, known as the AP-1 transcription factors, serve crucial roles in the control of cell proliferation and transformation [55, 66–70]. It has been suggested that different compositions of AP-1 complexes exist and different Fos proteins are required at the different stages of the cell cycle [71]. It has also been suggested that the expression of some of the AP-1 transcription factors, such as *c-jun*, can be autoregulated by changes in the phosphorylation of the preexisting proteins under appropriate conditions [72, 73].

In an attempt to understand why vanadate selectively induces *c-jun* and *junB* expression in CSV3-1 cells, a series of experiments have recently been performed. Preliminary results suggest that both Fos-Jun heterodimers and Jun-Jun homodimers exist in vanadate-stimulated CSV3-1 cells even though vanadate does not induce an increase in *fos* expression. The initial experiments employing gel shift assays demonstrated that vanadate markedly induces AP-1 DNA binding activity and that the AP-1 complex can be partially supershifted by a polyclonal antibody raised against all Fos proteins. Even prior to vanadate treatment, a substantial amount of AP-1 activity can be detected and can also be partially supershifted by the anti-Fos antibody, suggesting that low levels of Jun and Fos proteins are present in quiescent CSV3-1 cells before vanadate stimulation. This possibility is supported by data presented in Fig. 5 showing that the c-Jun and JunB protein levels are dramatically increased after vanadate stimulation. However, a significant amount of c-Jun and JunB proteins is readily detected in quiescent CSV3-1 cells before vanadate treatment (Fig. 5B and C). In addition, Fig. 5A demonstrates that a number of proteins that are recognized by the anti-Fos antibody are present in quiescent CSV3-1 cells. However, none of them shows an increase after vanadate stimulation but several of them can be induced by 10% FBS. The preexistence of low abundance of Jun and Fos or Fos-related proteins in CSV3-1 cells may facilitate a positive regulatory mechanism for *c-jun* and *junB* induction and therefore make cells more sensitive to the mitogenic effect of vanadate. Preliminary studies have also been performed to investigate the role *c-jun* and/or JunB may serve in vanadate-induced mitogenesis. In this regard, repression of *c-jun* or *junB* expression by antisense oligonucleotides results in a ~70% inhibition in vanadate-induced

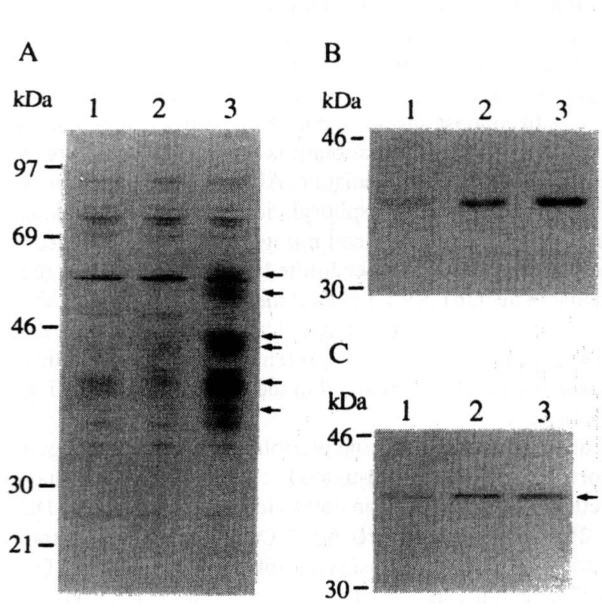

Fig. 5. Expression of c-Jun, JunB, and Fos proteins in quiescent CSV3-1 cells and the effects of vanadate. Quiescent CSV3-1 cells (lane 1) were treated with either 5 μM vanadate (lane 2) or 10% FBS (lane 3) for 2 h and 100 μg of soluble proteins from cell lysates was subjected to electrophoresis and Western blot analysis using antibodies specific for c-Jun (B), JunB (C), or an antibody raised against all Fos proteins (A). The increased expression of c-Jun (~39 kDa), JunB (~39 kDa), or Fos-related antigens in response to vanadate or FBS are indicated (arrows). All three antibodies were obtained from Santa Cruz Biotechnology, Inc.

mitogenesis, whereas sense oligonucleotides show little inhibitory effect [unpublished observations]. These results therefore provide strong evidence of *c-jun* and *junB* involvement in vanadate-induced mitogenesis in CSV3-1 cells.

Conclusion and future perspective

The mitogenic effect of vanadate has been documented in various cell types in culture and the molecular mechanisms that mediate this effect are being investigated in detail. These studies will broaden our understanding of vanadate actions as well as actions of other growth factors in general. It is important to emphasize here that like many other growth factors, vanadate-induced changes in protein tyrosine phosphorylation and *c-jun* and *junB* expression are only part of the requirement for mitogenesis. Changes in other critical signaling molecules and transcription factors are also absolutely required.

In the future, the effects of vanadate on cell proliferation and transformation need to be more extensively evaluated in *in vivo* systems. This is relevant to the fact that vanadate at the optimal dosages seems to preferentially act as a mitogen for transformed cells and may potentiate the effects of carcinogenic agents on certain cell types. The cytotoxic effects of vanadate at higher concentrations also need to be more carefully studied. The results of these studies are essential to further the ultimate goal of vanadate research to make the clinical use of vanadate as an alternative treatment for diabetes.

Acknowledgement

The authors thank Ms. Sherry Xie for excellent technical assistance and Dr. Michael M. Witte for critical reading of the manuscript. This research was supported by NIH grant CA51715 and the Muirhead Chair of Excellence to RES.

References

1. Shechter Y, Karlish SJD: Insulin-like stimulation of glucose oxidation in rat adipocytes by vanadyl (IV) ions. Nature 284: 556–558, 1980
2. Dubyak GR, Kleinzeller A: The insulin-mimetic effects of vanadate in isolated rat adipocytes: dissociation from effects of vanadate as a (Na^+-K^+)ATPase inhibitor. J Biol Chem 255: 5306–5312, 1980
3. Duckworth WC, Solomon SS, Liepnieks J, Hamel FG, Hand S, Peavy DE: Insulin-like effects of vanadate in isolated rat adipocytes. Endocrinology 122: 2285–2289, 1988
4. Strout HV, Vicario PP, Biswas C, Saperstein R, Brady EJ, Pilch PF, Berger J: Vanadate treatment of streptozotocin diabetic rats restores expression of the insulin-responsive glucose transporter in skeletal muscle. Endocrinology 126: 2728–2732, 1990
5. Bollen M, Miralpeix M, Ventura F, Toth B, Bartrons R, Stalmans W: Oral administration of vanadate to streptozotocin-diabetic rats restores the glucose-induced activation of liver glycogen synthase. Biochem J 267: 269–271, 1990
6. Pugazhenthi S, Khandelwal RL: Insulinlike effects of vanadate on hepatic glycogen metabolism in nondiabetic and streptozocin-induced diabetic rats. Diabetes 39: 821–827, 1990
7. Shechter Y: Insulin-mimetic effects of vanadate: possible implication for future treatment of diabetes. Diabetes 39: 1–5, 1990
8. Shechter Y, Shisheva A: Vanadium salts and the future treatment of diabetes. Endeavour 17: 27–31, 1993
9. Heyliger CE, Tahiliani AG, McNeill JH: Effect of vanadate on elevated blood glucose and depressed cardiac performance of diabetic rats. Science 227: 1474–1476, 1985
10. Valera A, Rodriguez-Gil JE, Bosch F: Vanadate treatment restores the expression of genes for key enzymes in the glucose and ketone bodies metabolism in the liver of diabetic rats. J Clin Invest 92: 4–11, 1993
11. Carpenter G: Vanadate, epidermal growth factor and the stimulation of DNA synthesis. Biochem Biophys Res Commun 102: 1115–1121, 1981
12. Smith JB: Vanadium ions stimulate DNA synthesis in Swiss mouse 3T3 and 3T6 cells. Proc Natl Acad Sci USA 80: 6162–6166, 1983
13. Ramanadham M, Kern M: Differential effect of vanadate on DNA synthesis induced by mitogens in T and B lymphocytes. Mol Cell Biochem 51: 67–71, 1983
14. Jones TR, Reid TW: Sodium orthovanadate stimulation of DNA synthesis in Nakano mouse lens epithelial cells in serum-free medium. J Cell Physiol 121: 199–205, 1984
15. Canalis E: Effects of sodium vanadate on deoxyribonucleic acid and protein syntheses in cultured rat calvariae. Endocrinology 116: 855–862, 1985
16. Lau KHW, Tanimoto H, Baylink DJ: Vanadate stimulates bone cell proliferation and bone collagen synthesis *in vitro*. Endocrinology 123: 2858–2867, 1988
17. Davidai G, Lee A, Schvartz I, Hazum E: PDGF induces tyrosine phosphorylation in osteoblast-like cells: relevance to mitogenesis. Am J Physiol 26: E205–E209, 1992
18. Maher PA: Stimulation of endothelial cell proliferation by vanadate is specific for microvascular endothelial cells. J Cell Physiol 151: 549–554, 1992
19. Raid A, Oliver B, Abdelrahman A, Sha'Afi RI, Hajjar J-J: Role of tyrosine kinase and phosphotyrosine phosphatase in growth of the intestinal crypt cell (IEC-6) line. Proc Soc Exp Biol Med 202: 435–439, 1993
20. Wang H, Wang J-Y, Johnson LR, Scott RE: Selective induction of *c-jun* and *jun-B* but not *c-fos* or *c-myc* during mitogenesis in SV40-transformed cells at the predifferentiation growth arrest state. Cell Growth Differ 2: 645–652, 1991
21. Hori C, Oka T: Vanadate enhances the stimulatory action of insulin on DNA synthesis in cultured mouse mammary gland. Biochim Biophys Acta 610: 235–240, 1980
22. Montarras D, Pinset C, Dubois C, Chenevert J, Gros F: High level of *c-fos* mRNA accumulation is not obligatory for renewed cell proliferation. Biochem Biophys Res Commun 153: 1090–1096, 1988
23. Sato B, Miyashita Y, Maeda Y, Noma K, Kishimoto S, Matsumoto K: Effects of estrogen and vanadate on the proliferation of newly established transformed mouse Leydig cell line *in vitro*. Endocrinology 120: 1112–1120, 1987
24. Kanakura Y, Druker B, Cannistra SA, Furukawa Y, Torimoto Y, Griffin JD: Signal transduction of the human granulocyte-macrophage colony-stimulating factor and interleukin-3 receptors involves tyrosine

phosphorylation of a common set of cytoplasmic proteins. Blood 76: 706–715, 1990

25. Klarlund JK: Transformation of cells by an inhibitor of phosphatases acting on phosphotyrosine in proteins. Cell 41:707–717,1985
26. Marchisio PC, D'Urso N, Comoglio PM, Giancotti FG, Tarone G: Vanadate-treated baby hamster kidney fibroblasts show cytoskeleton and adhesion patterns similar to their Rous sarcoma virus-transformed counterparts. J Cell Biochem 37: 151–159, 1988
27. Rijksen G, Voller MCW, Van Zoelen EJJ: Orthovanadate both mimics and antagonizes the transforming growth factor β action on normal rat kidney cells. J Cell Physiol 154: 393–401, 1993
28. Owada MK, Iwamoto M, Koike, T, Kato Y: Effects of vanadate on tyrosine phosphorylation and the pattern of glycosaminoglycan synthesis in rabbit chondrocytes in culture. J Cell Physiol 138: 484–492, 1989
29. Dessureault J, Weber JM: Retransformation of a revertant cell line with the adenovirus E1 oncogenes and vanadate. J Cell Biochem 43: 293–296, 1990
30. Feldman RA, Lowy DR, Vass WC: Selective potentiation of *c-fps/fes* transforming activity by a phosphatase inhibitor. Oncogene Res 5: 187–97, 1990
31. Kowalski LA, Tsang SS, Davison AJ: Vanadate enhances transformation of bovine papillomavirus DNA-transfected C3H/10T½ cells. Cancer Lett 64: 83–90, 1992
32. Gordon JA: Use of vanadate as protein-phosphotyrosine phosphatase inhibitor. Methods Enzymol 201: 477–482, 1991
33. Kingsnorth AN, LaMuraglia GM, Ross JS, Malt RA: Vanadate supplements and 1,2-dimethylhydrazine induced colon cancer in mice: increased thymidine incorporation without enhanced carcinogenesis. Br J Cancer 53: 683–686, 1986
34. Fantus IG, Ahmad F, Deragon G: Vanadate augments insulin binding and prolongs insulin action in rat adipocytes. Endocrinology 127: 2716–2725, 1990
35. Fantus IG, Ahmad F, Deragon G: Vanadate augments insulin-stimulated insulin receptor kinase activity and prolongs insulin action in rat adipocytes: evidence for transduction of amplitude of signaling into duration of response. Diabetes 43: 375–383, 1994
36. Green A: The insulin-like effect of sodium vanadate on adipocyte glucose transport is mediated at a post-insulin-receptor level. Biochem J 238: 663–669, 1986
37. Strout HV, Vicario PP, Saperstein R, Slater EE: The insulin-mimetic effect of vanadate is not correlated with insulin receptor tyrosine kinase activity nor phosphorylation in mouse diaphragm *in vivo*. Endocrinology 124: 1918–1924, 1989
38. Wang H, Scott RE: Distinct protein tyrosine phosphorylation during mitogenesis induced in quiescent SV40-transformed 3T3 T cells by insulin or vanadate. J Cell Physiol 158: 408–416, 1994
39. D'Onofrio F, Le MQU, Chiasson J-L, Srivastava AK: Activation of mitogen activated protein (MAP) kinases by vanadate is independent of insulin receptor autophosphorylation. FEBS Lett 340: 269–275, 1994
40. Nguyen TT, Scimeca J-C, Filloux C, Peraldi P, Carpentier J-L, Van Obberghen E: Co-regulation of the mitogen-activated protein kinase, extracellular signal-regulated kinase 1, and the 90-kDa ribosomal S6 kinase in the PC12 cells: distinct effects of the neurotrophic factors, nerve growth factor, and the mitogenic factor, epidermal growth factor. J Biol Chem 268: 9803–9810, 1993
41. White MF, Kahn CR: The insulin signaling system. J Biol Chem 269: 1–4, 1994
42. Pelech SL, Sanghera JS: Mitogen-activated protein kinases: versatile transducers for cell signaling. TIBS 17: 233–238, 1992
43. Swarup G, Cohen S, Garber DL: Inhibition of membrane phosphotyrosyl-protein phosphatase activity by vanadate. Biochem Biophys Res Commun 107: 1104–1109, 1982
44. Klarlund JK, Latini S, Forchhammer J: Numerous proteins phosphorylated on tyrosine and enhanced tyrosine kinase activities in vanadate-treated NIH 3T3 fibroblasts. Biochim Biophys Acta 971: 112–120, 1988
45. Shisheva A, Shechter Y: A cytosolic protein tyrosine kinase in rat adipocytes. FEBS Lett 300: 93–96, 1992
46. Shisheva A, Shechter Y: Role of cytosolic tyrosine kinase in mediating insulin-like actions of vanadate in rat adipocytes. J Biol Chem 268: 6463–6469, 1993
47. Elberg G, Li J, Shechter Y: Vanadium activates or inhibits receptor and non-receptor protein tyrosine kinases in cell-free experiments, depending on its oxidation state: possible role of endogenous vanadium in controlling cellular protein tyrosine kinase activity. J Biol Chem 269: 9521–9527, 1994
48. Cadena DL, Gill GN: Receptor tyrosine kinases. FASEB J 6: 2332–2337, 1992
49. Wang H, Scott RE: Insulin-induced mitogenesis associated with transformation by the SV40 large T antigen. J Cell Physiol 147: 102–110, 1991
50. Wang H, Scott RE: Induction of *c-jun* independent of PKC, pertussis toxin-sensitive G protein, and polyamines in quiescent SV40-transformed 3T3 T cells. Exp Cell Res 203: 47–55, 1992
51. Filipak M, Estervig DN, Tzen CY, Minoo P, Hoerl BJ, Maercklein PB, Zschunke MA, Edens M, Scott RE: Integrated control of proliferation and differentiation of mesenchymal stem cells. Environ Health Perspect 80: 117–125, 1989
52. Wang H, Sturtevant DB, Scott RE: Nonterminal and terminal adipocyte differentiation of murine 3T3 T mesenchymal stem cells. In: J.E. Celis (ed.) Cell Biology: A Laboratory Handbook. Academic Press, San Diego, 1994, pp 193–198
53. Hoerl BJ, Scott RE: Nonterminally differentiated cells express decreased growth factor responsiveness. J Cell Physiol 139: 68–75, 1989
54. Wang H, Scott RE: Inhibition of distinct steps in the adipocyte differentiation pathway in 3T3 T mesenchymal stem cells by dimethyl sulphoxide (DMSO). Cell Prolif 26: 55–66, 1993
55. Wang H, Scott RE: Adipocyte differentiation selectively represses the serum inducibility of *c-jun* and *junB* by reversible transcription-dependent mechanisms. Proc Natl Acad Sci USA 91: 4649–4653, 1994
56. Estervig DN, Minoo P, Tzen CY, Scott RE: Inhibition of simian virus 40 T-antigen expression by cellular differentiation. J Virol 63: 2718–2725, 1989
57. Scott RE, Estervig DN, Tzen CY, Minoo P, Maercklein PB, Hoerl BJ: Nonterminal differentiation represses the neoplastic phenotype in spontaneously and simian virus 40-transformed cells. Proc Natl Acad Sci USA 86: 1652–1656, 1989
58. Estervig DN, Minoo P, Tzen CY, Scott RE: Three distinct effects of SV40 T-antigen gene transfection on cellular differentiation. J Cell Physiol 142: 552–558, 1990
59. Wang H, Scott RE: Autocrine inhibitor of terminal differentiation secreted by SV40-transformed 3T3 T cells. Mol Cell Differ 1: 345–355, 1993
60. Witte MM, Parker RF, Wang H, Scott RE: Repression of SV40 T oncoprotein expression by DMSO. J Cell Physiol 151: 50–55, 1992
61. Heffetz D, Bushkin I, Dror R, Zick Y: The insulinomimetic agents H_2O_2 and vanadate stimulate protein tyrosine phosphorylation in intact cells. J Biol Chem 265: 2896–2902, 1990
62. Akiyama T, Ishida J, Nakagawa S, Ogawara H, Watanabe S-i, Itoh N, Shibuya M, Fukami Y: Genistein, a specific inhibitor of tyrosine-specific protein kinases. J Biol Chem 262: 5592–5595, 1987
63. Yin X, Davison AJ, Tsang SS: Vanadate-induced gene expression in

mouse C127 cells: roles of oxygen derived active species. Mol Cell Biochem 115: 85–96, 1992

64. Davison AJ, Stern A, Fatur DJ, Tsang SS: Vanadate stimulates ornithine decarboxylase activity in C3H/10T½ cells. Biochem Int 24: 461–466, 1991
65. Itkes AV, Imamova LR, Alexandrova NM, Favorova OO, Kisselev LL: Expression of *c-myc* gene in human ovary carcinoma cells treated with vanadate. Exp Cell Res 188: 169–71, 1990
66. Angel P, Karin M: The role of Jun, Fos and the AP-1 complex in cell-proliferation and transformation. Biochim Biophys Acta 1072: 129–157, 1991
67. Kovary K, Bravo R: The Jun and Fos protein families are both required for cell cycle progression in fibroblasts. Mol Cell Biol 11: 4466–4472, 1991
68. Hughes M, Sehgal A, Hadman M, Bos T: Heterodimerization with c-Fos is not required for cell transformation of chicken embryo fibroblasts by Jun. Cell Growth Differ 3: 889–897, 1992
69. Suzuki T, Murakami M, Onai N, Fukuda E, Hashimoto Y, Sonobe MH, Kameda T, Ichinose M, Miki K, Iba H: Analysis of AP-1 function in cellular transformation pathways. J Virol 68: 3527–3535, 1994
70. Dong Z, Birrer MJ, Watts RG, Matrisian LM, Colburn NH: Blocking of tumor promoter induced AP-1 activity inhibits induced transformation in JB6 mouse epidermal cells. Proc Natl Acad Sci USA 91: 609–613, 1994
71. Kovary K, Bravo R: Existence of different Fos/Jun complexes during the G_0-to-G_1 transition and during exponential growth in mouse fibroblasts: differential role of Fos proteins. Mol Cell Biol 12: 5015–5023, 1992
72. Angel P, Hattori K, Smeal T, Karin M: The *jun* protooncogene is positively autoregulated by its product, Jun/AP-1. Cell 55: 875–885, 1988
73. Pulverer BJ, Kyriakis JM, Avruch J, Nikolakaki E, Woodgett JR: Phosphorylation of *c-jun* mediated by MAP kinases. Nature 353: 670–674, 1991

Molecular and Cellular Biochemistry **153:** 69–78, 1995.

Vanadium salts stimulate mitogen-activated protein (MAP) kinases and ribosomal S6 kinases

Sanjay K. Pandey, Jean-Louis Chiasson and Ashok K. Srivastava
Research Group On Diabetes and Metabolic Regulation, Centre de Recherche/Hotel-Dieu de Montreal Hospital and Department of Medicine, University of Montreal, Montreal Quebec, H2W IT8, Canada

Abstract

Effect of several vanadium salts, sodium orthovanadate, vanadyl sulfate and sodium metavanadate on protein tyrosine phosphorylation and serine/threonine kinases in chinese hamster ovary (CHO) cells overexpressing a normal human insulin receptor was examined. All the compounds stimulated protein tyrosine phosphorylation of two major proteins with molecular masses of 42 kDa (p42) and 44 kDa (p44). The phosphorylation of p42 and p44 was associated with an activation of mitogen activated protein (MAP) kinase as well as increased protein tyrosine phosphorylation of $p42^{mapk}$ and $p44^{mapk}$. Vanadium salts also activated the 90 kDa ribosomal s6 kinase ($p90^{rsk}$) and 70 kDa ribosomal s6 kinase ($p70^{s6k}$). Among the three vanadium salts tested, vanadyl sulfate appeared to be slightly more potent than others in stimulating MAP kinases and $p70^{s6k}$ activity. It is suggested that vanadium-induced activation of MAP kinases and ribosomal s6 kinases may be one of the mechanisms by which insulin like effects of this trace element are mediated. (Mol Cell Biochem **153:** 69–78, 1995)

Key words: vanadium salts, MAP kinase, ribosomal s6 kinases ($p90^{rsk}$ and $p70^{s6k}$), insulinomimesis, protein tyrosine phosphatase

Abbreviations: eIF-4 – eukaryotic protein synthesis initiation factor-4; GRB-2 – growth factor receptor bound protein-2; GSK-3 – Glycogen Synthase Kinase-3; IRS-1 – insulin receptor substrate-1; ISPK – insulin stimulated protein kinase; MAPK – mitogen activated protein kinase, also known as: ERK – extracellular signal regulated kinase; MAPKK – mitogen activated protein kinase kinase, also known as – MEK, MAPK or ERK kinase; PHAS-1 – phosphorylated heat and acid stable protein regulated by insulin; PI3K – phosphatidyl inositol 3-kinase; PP1-G – protein phosphatase-glycogen bound form; PTK – protein tyrosine kinase; PTPase – protein tyrosine phosphatase; rsk – ribosomal s6 kinases; shc – src homology domain containing protein; SOS – son of sevenless

Introduction

Vanadium is a group V transition element and is an endogenous constituent of all or most mammalian tissues [1–3]. Vanadium salts such as sodium orthovanadate (Na_3VO_4), sodium metavanadate ($NaVO_3$) and vanadyl sulfate ($VOSO_4$) mimic several of the metabolic and growth promoting effects of insulin (4–24]; for a review see reference [25]). However, the most remarkable insulinomimetic effect of vanadium salts is their ability to normalise blood glucose in type I (insulin-dependent diabetes mellitus – IDDM) [26, 27] and type II (non insulin-dependent diabetes mellitus – NIDDM) [28–30] animal models of diabetes mellitus. Vanadium salts also improved the metabolic abnormalities associated with both type I as well as type II models of diabetes mellitus ([27, 31–43; for a review see reference [44]).

It is well established that the biological actions of insulin are initiated by binding of insulin to a specific receptor located on the membrane of target cells [45–46]. The insulin receptor is a heterotetrameric glycoprotein composed of two extracellular α-subunits and two transmembrane β-subunits. The α-subunit possesses the insulin binding activity and the β-subunit has an intrinsic protein tyrosine kinase (PTK) activity [44–48]. The binding of insulin to the α-

Address for offprints: Ashok K Srivastava, Centre de Recherche/Hotel-Dieu de Montreal, 3850 Saint-Urbain Street, Montreal, Quebec, H2W IT8, Canada subunit of its receptor activates the PTK activity of the β-subunit and results in autophosphorylation of the β-subunit in tyrosine residues as well as the tyrosyl phosphorylation of endogenous substrate [45–48]. Insulin receptor β-subunit autophosphorylation and activation of its PTK activity is believed to be a major pathway mediating the insulin action since cells with PTK deficient insulin receptors are unable to elicit several of the biological effects of insulin [49–50]. However, the mechanism by which vanadium compounds exert their insulin like effects remain to be clarified.

Vanadium salts are potent inhibitors of protein tyrosine phosphatases (PTPases) [51] and thus were initially thought to activate the insulin receptor PTK activity by preventing the dephosphorylation of insulin receptor β-subunit [11, 52–55]. However, recent work has suggested that the site of action of vanadium might not involve insulin receptor PTK [56–61].

A critical step in insulin signalling cascade appears to be activation of a group of protein serine/threonine kinases which include mitogen activated protein (MAP) kinases, 90 kDa ribosomal S6 kinase ($p90^{rsk}$) and 70 kDa ribosomal s6 kinase, ($p70^{s6k}$) [48, 62–66]. We have shown earlier that sodium orthovanadate stimulates the tyrosyl phosphorylation and activation of MAP kinases in an insulin receptor protein tyrosine kinase independent manner [60, 61]. Therefore, we were interested to examine if $p90^{rsk}$, which is immediate downstream of MAPK in insulin signalling cascade [48], could also be activated by vanadium salts. In addition, we also questioned whether $p70^{s6k}$ which is also activated by insulin by as yet undefined pathway [48], will be similarly activated by these salts. Furthermore, in the light of the suggestion that vanadyl sulfate may be more potent than sodium orthovanadate in eliciting hypoglycemic effects [67], we have compared the effects of various vanadium salts on MAPK, $p90^{rsk}$ and $p70^{s6k}$ activities in chinese hamster ovary cells overexpressing a normal human insulin receptor (CHO-HIRc).

Materials and methods

CHO cells overexpressing a normal human insulin receptor (CHO-HIRc) were a kind gift of Dr. Morris F. White (Joslin Diabetes Centre, Boston, MA, USA). Insulin was from Eli Lilly Co. Indianapolis, IN, USA). Myelin basic protein (MBP), sodium orthovanadate and sodium metavanadate were purchased from Sigma Chemical Company (St. Louis, MO, USA). Vanadyl sulfate was from Aldrich Chemical Co. (Milwaukee, WI, USA). Antiphosphotyrosine antibody, anti-MAP kinase antibody (raised against a peptide based upon residue 333–367 of the C-terminus of the rat 43 kDa ERK1) that detects several isoforms of MAP kinases including $p44^{mapk}$ and $p42^{mapk}$, antiphosphotyrosine agarose beads and S6 peptide RRRLSSLRA were from Upstate Biotechnology (Lake Placid, NY, USA). Goat anti-mouse or anti-rabbit IgG conjugated to alkaline phosphatase was from Bio-Rad (Mississauga, Ontario, Canada). The $p70^{s6k}$ antiserum (raised against a peptide corresponding to amino acids 2–30 of rat $p70^{s6k}$ [68] and $p90^{rsk}$ antiserum raised against bacterially produced chicken S6 kinase II [69] were generously provided by Drs. Frederic Hall of USC, School of Medicine, Los Angeles and John Blenis of Harvard Medical School, Boston respectively.

Cell culture

CHO cells were maintained on HAM's F-12 medium containing 10% fetal bovine serum. Cells were grown to confluence in 100 mm plates and incubated in serum-free F-12 medium for 20 h prior to the experiment [61].

Detection of phosphotyrosine-containing proteins

Tyrosine phosphorylation of cellular proteins stimulated in the absence or presence of insulin or vanadium compounds was assessed by immunoblotting using antiphosphotyrosine antibodies. Cells were stimulated with insulin or different vanadium compounds. The cells were lysed on ice in 400 μl of buffer A (25 mM Tris-Hcl pH 7.5, 25 mM NaCl, 1 mM Na orthovanadate, 10 mM Na fluoride, 10 mM Na pyrophosphate, 20 nM Okadaic acid, 0.5 mM EGTA, 1 mM phenylmethylsulphonyl fluoride (PMSF), 10 μg/ml aprotinin, 1% Triton X-100 and 0.1% SDS). The lysates were clarified by centrifugation for 12 min at 10,000 × g. Equal amounts of protein samples were electrophoresed on 10% SDS-polyacrylamide gels and transferred to polyvinilidine difluoride (PVDF) membranes and blotted with antiphosphotyrosine antibody (1:1000) and detected using goat-anti-mouse IgG conjugated to alkaline phophatase (1:3000).

Immunoaffinity purification of phosphotyrosyl proteins and MAP kinase immunoblotting

The cell lysates were affinity purified on antiphosphotyrosine-agarose column and immunoblotted using a MAP kinase antibody [61]. For affinity purification of phosphotyrosyl proteins, briefly, the clarified lysates from control or stimulated cells were incubated for 1 hour at 4°C with antiphosphotyrosine agarose beads equilibrated in buffer B (20 mM Tris-HCl pH 7.4, 1% Nonidet p-40, 10 mM EDTA, 0.2 mM Na vanadate, 0.01% Na azide and 100 mM NaCl). The beads were washed 3 times with buffer B and phosphotyrosine containing proteins were eluted with 1 mM phosphotyrosine in buffer B. The eluates were boiled in 3× Laemmli's sample buffer, electrophoresed on 10% SDS-polyacrylamide gels, transferred to PVDF membranes and blotted with anti-MAP kinase antibody (1:500) and detected using goat-anti-rabbit IgG conjugated to alkaline phosphatase.

MAP kinase assay

For MAP kinase assay cell lysates were prepared in buffer A devoid of Triton X-100 or SDS. To 5 µl of cell lysate (approximately 3–4 µg protein), 40 µl of kinase buffer (25 mM TrisHCl pH 7.4, 10 mM $MgCl_2$, 2 mM $MnCl_2$, 1 mM dithiothreitol, 40 µM ATP, 1 µM staurosporine, 15 µg MBP, 0.5 mM EGTA and 0.5 µCi (γ^{32}P)ATP) was added. After 12 min at 30°C, the reaction was stopped by spotting 20 µl of the reaction mixture onto P-81 filter papers (2 × 2 cm), washed 4 times in 0.5% phosphoric acid and counted for radioactivity [70].

Assay of $p70^{s6k}$ and $p90^{rsk}$

The clarified lysates prepared in buffer A were normalized to contain equal amounts of protein (100 µg) and incubated for 4 h at 4°C with 3 µl of either $p70^{s6k}$ or $p90^{rsk}$ antibody preadsorbed to protein A sepharose beads (Pharmacia Biotech Inc, Missisauga, Ontario, Canada). The immune complex was collected by centrifugation followed by washing 3 times with buffer A and once with buffer B (20 mM HEPES pH 7.4, 10 mM $MgCl_2$, 1 mM DTT, 10 mM β-glycerophosphate). The beads were resuspended in 20 µl of buffer B containing S6 peptide, RRLSSLRA and phosphotransferase reaction was initiated by adding 5 µl of 100 µM ATP containing 2 µCi (γ^{32}P)ATP, (Amersham, Canada). After 15 min at 30°C, the reaction was stopped by spotting on P-81 filter paper, washing in 0.5% phosphoric acid and counted for radioactivity.

Results

Effect of various vanadium salts on protein tyrosine phosphorylation

We have shown earlier that sodium orthovanadate (OV) causes the tyrosyl phosphorylation and activation of $p42^{mapk}$ and $p44^{mapk}$ in a concentration and time dependent manner [60, 61]. Therefore, to examine if a similar response is observed with other vanadium salts such as sodium metavanadate (MV) and vanadyl sulfate (VS) (Fig. 1), the CHO-HIRc cells were treated with 100 µM of OV, MV or VS and lysates were immunoblotted with antiphosphotyrosine antibodies. As shown in Fig. 2, all the three vanadium salts stimulated the tyrosyl phosphorylation of two major proteins having molecular masses of 44 kDa (p44) and 42 kDa (p42) respectively. However VS appeared to be about 40% more potent than either OV or MV. Furthermore, in contrast to insulin, none of the vanadium salts stimulated the tyrosyl phosphorylation of either the insulin receptor or any other proteins (Fig. 2).

Na_3VO_4

(Sodium Orthovanadate)

$VOSO_4.3H_2O$

(Vanadyl sulfate)

$NaVO_3$

(Sodium Metavanadate)

Fig. 1. Structural formula of vanadium salts used.

Effect of vanadium salts on MAP kinase activity

Since the molecular masses of p44 and p42 corresponded to that of MAP kinases $p44^{mapk}$ and $p42^{mapk}$ which are activated in response to insulin [71] it was of interest to examine if increased tyrosyl phosphorylation of these proteins by MV and VS correlated with an increase in MAP kinase activity. As shown in Fig. 3, stimulation of CHO-HIR cells with all the 3 vanadium compounds resulted in the activation of MAP kinase activity as judged by increased phosphorylation of myelin basic protein (MBP) as an exogenous substrate. The stimulatory effect of VS on MAP kinase activity was slightly more than that of insulin (5.8 fold for VS versus 4.5. fold for insulin). However, VS was about 2 times more potent than either OV or MV in stimulating the MAP kinase activity (Fig. 3). Vanadium salt-mediated activation of MAP kinase was accompanied by an increase in the phosphotyrosyl content of $p44^{mapk}$ (ERK 1) and $p42^{mapk}$ (ERK 2) as determined by immunoprecipitation using antiphosphotyrosine antibody followed by immunoblotting with a specific antibody to MAP kinase which recognises both ERK 1 and ERK 2 (data not shown, [61]).

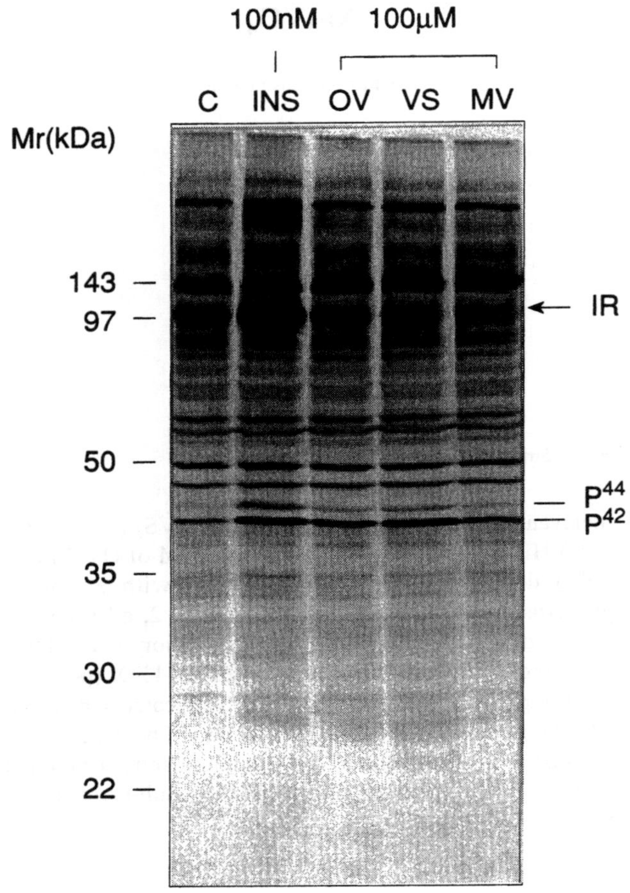

Fig. 2. Effect of various vanadium salts or insulin on protein tyrosine phosphorylation. CHO-HIRc cells were treated in the absence (C) or presence of insulin (INS, 100 nM) or 100 μM each of sodium orthovanadate (OV), vanadyl sulfate (VS) or sodium metavanadate (MV) for 5 min. Cell lysates were prepared and immunoblotted with antiphosphotyrosine antibodies as described in Materials and methods. The numbers in the left indicate the position of molecular weight standards. IR denotes the position of the insulin receptor β-subunit. This is a representative result from one of three experiments.

Effect of vanadium salts on 90 kDa ribosomal S6 kinase (p90^rsk^)

In insulin stimulated cells MAP kinase activation is associated with the activation of an immediate downstream serine/threonine kinase-a 90 kDa ribosomal S6 kinase ($p90^{rsk}$) [48, 63, 66]. Therefore, to evaluate if vanadium salts exerted a similar effect, the $p90^{rsk}$ activity was assessed in CHO-HIR cells stimulated with various vanadium compounds. The $p90^{rsk}$ activity was determined by using an immune complex kinase assay in which $p90^{rsk}$ protein from cell lysates was immunoprecipitated using a specific antiserum. This antiserum was prepared against a recombinant chicken rsk S6

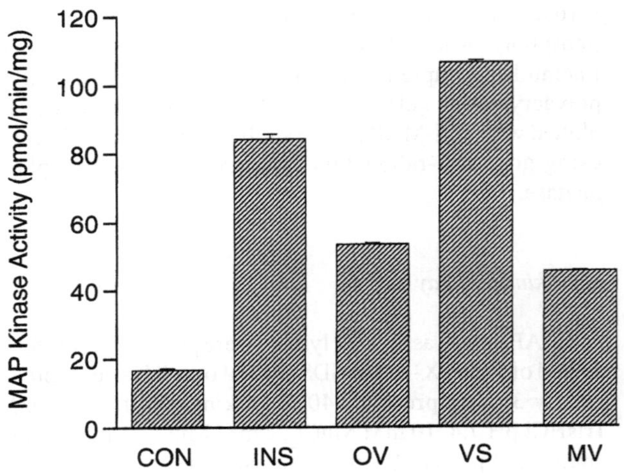

Fig. 3. Effect of insulin or vanadium salts on MAP kinase activity from CHO-HIRc cells. CHO-HIRc cells were treated in absence (CON) or presence of insulin or different vanadium compounds as mentioned in legend to Fig. 2, and MAP kinase activity was assayed in cell lysate using MBP (myelin basic protein) as exogenous substrate as described in Materials and methods. These results represent one of three independent experiments done in triplicate.

protein kinase which specifically immunoprecipitated $p90^{rsk}$ in various cell types [69]. The phosphotransferase reaction in the immune complex was performed by using a synthetic peptide substrate corresponding to amino acid 231–239 in human 40 S ribosomal protein S6. As shown in Fig. 4, incubation of CHO-HIR cells with all the vanadium salts (100 μM) activated the phosphotransferase activity up to 3-fold

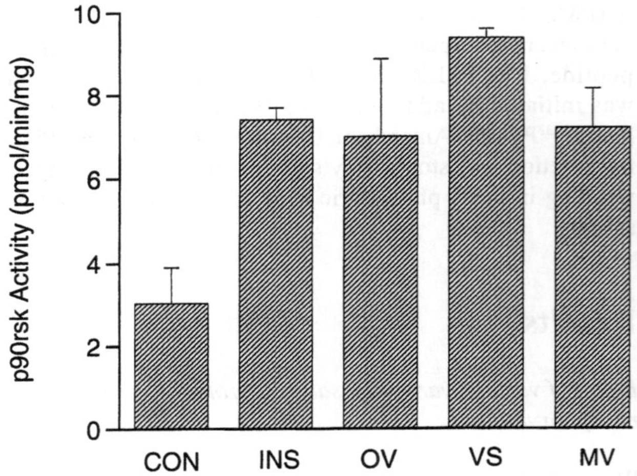

Fig. 4. Effect of insulin or various vanadium salts on 90 kDa ribosomal protein ($p90^{rsk}$) activity in CHO-HIRc cells. CHO-HIRc cells were treated as described in Fig. 2 and $p90^{rsk}$ activity was determined in immune complex using S6 peptide RRRLSSLRA as substrate as described in Materials and methods. Results are representative of one of three experiments performed in triplicate.

as compared to untreated control cells. All the salts tested were almost equipotent in activating the p90^{rsk} activity and the response was similar to that observed with insulin.

Effect of vanadium salts on 70 kDa ribosomal S6 kinase (p70^{s6k})

In addition to p90^{rsk}, insulin also stimulates a 70 kDa ribosomal S6 kinase termed as p70^{s6k} [48, 64, 65]. However, in contrast to p90^{rsk}, MAPK is not the upstream kinase mediating p70^{s6k} phosphorylation and activation, and the mechanism by which p70^{s6k} is activated remains to be clarified [48]. We therefore asked whether similar to insulin, vanadium salts could also activate p70^{s6k} in CHO-HIR cells. For these experiments also an immune complex kinase assay using an antiserum raised against a synthetic peptide corresponding to amino acid 2–30 of rat p70^{s6k} which specifically recognizes p70^{s6k} was performed. The results shown in Fig. 5 indicate that OV, VS and MV all caused a potent stimulation in the phosphotransferase activity of p70^{s6k}. The fold stimulation ranged between 2.5–4 fold as compared to untreated control. Among the 3 salts tested, the stimulatory response of VS was higher than that of either OV or MV. VS caused about 4 fold stimulation whereas only 2.5 fold stimulation was observed with OV or MV in comparison to control cells (Fig. 5). Furthermore the stimulatory response of VS was almost equal to that of insulin (Fig. 5).

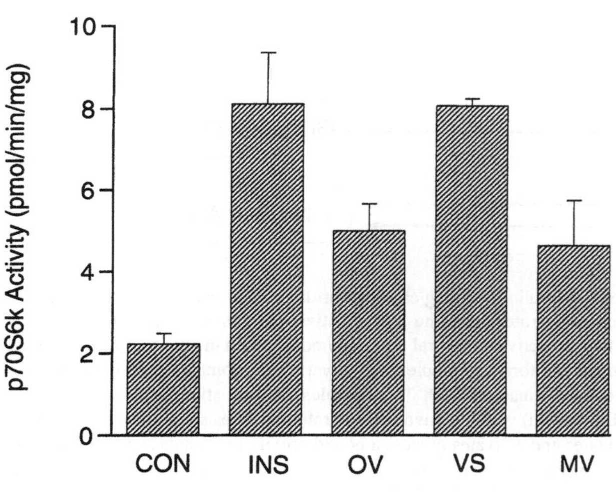

Fig. 5. Effect of insulin or vanadium compounds on 70 kDa ribosomal protein s6 kinase activity (p70^{s6k}) CHO-HIR cells were treated with vanadium compounds as described in Fig. 2 and p70^{s6k} activity was determined in immune complex using S6 peptide, RRRLSSLRA as substrate as described in Materials and methods. The results are one of three independent experiments performed in triplicate.

Discussion

In this study we have shown that various vanadium salts activate serine/threonine kinases – MAP kinase, p90^{rsk} and p70^{s6k} in CHO cells overexpressing a normal human insulin receptor. The activation observed by vanadium salts was comparable to that observed with insulin and occurred in the absence of detectable tyrosyl phosphorylation of any proteins other than p44^{mapk} and p42^{mapk}. Most notably, the tyrosyl phosphorylation of neither insulin receptor β-subunit nor the insulin receptor substrate I (IRS-I) could be detected even after 30 min incubation of cells with any of the vanadium salts. Vanadate-mediated activation of MAPK and p90^{rsk} has previously been shown in other cell types but in these studies no attempts were made to correlate it with tyrosyl phosphorylation of insulin receptor β-subunit [72, 73]. Our earlier observation that tyrosyl phosphorylation and activation of MAP kinases can occur in the absence of IRS-1 phosphorylation [60, 61] has recently been confirmed by Tamemoto *et al.* who demonstrated that in the livers of mice deficient in IRS-1, MAP kinase activation was not significantly altered as compared to normal mice [74].

Recent studies have implicated MAP kinase signalling pathways in the regulation of glycogen metabolism by insulin [75–78]. Dent *et al.* demonstrated that an insulin-stimulated protein kinase (ISPK) is able to phosphorylate and thereby activate the regulatory subunit of the glycogen bound form of protein phosphatases-1 (PP1-G) [75]. Activated PP1-G dephosphorylates glycogen synthase and phosphorylase kinase and thus stimulates glycogen synthesis [75]. ISPK has subsequently been identified as an isoform of p90^{rsk} or rsk II [79, 80]. Moreover, both p90^{rsk} and p70^{s6k} catalyze *in vitro* phosphorylation of glycogen synthase kinase-3 (GSK-3) [81–83]. GSK-3, which is able to phosphorylate and inhibit the activity of glycogen synthase is phosphorylated and inactivated in response to insulin, is believed to play an important role in the glycogen metabolism [83]. Recent studies with rapamycin, an immunosuppressant and specific inhibitor of p70^{s6k}, revealed that p70^{s6k} may not be involved in the phosphorylation and inactivation of GSK-3 [76, 77] However, based on experiments using wortmannin, an inhibitor of phosphatidyl-inositol kinase (PI3K), a possible role of MAP kinase/p90^{rsk} signalling pathway in insulin-mediated inactivation of GSK-3 has been suggested [76, 77]. Furthermore, insulin-stimulated glycogen synthesis and glycogen synthase activation was also blocked by wortmannin and rapamycin in 3T3-L1 adipocytes indicating the involvement of PI3K and p70^{s6k} signalling pathway in this process [78]. These results are however in contrast to the studies of Lin and Lawrence who demonstrated that rapamycin did not attenuate the stimulatory effect of insulin on glycogen synthase and suggested that activation of MAP kinase and ribosomal s6 kinases may not be sufficient for the activation of glycogen synthesis in adipocytes [84].

An involvement of PI3K in insulin-mediated glucose transport and antilipolytic effects has also been suggested [85, 86]. Inhibition of PI3K activity by wortmannin and LY294002, another specific inhibitor of PI3K, resulted in complete inhibition of insulin-stimulated p70^{s6k} as well as glucose uptake in 3T3-L1 adipocytes [85]. LY294002 treatment also inhibited the translocation of GLUT 4 glucose transporters to the plasma membrane [85]. Rapamycin, however, failed to exert any effect on insulin-stimulated glucose transport in 3T3-L1 adipocytes [87] suggesting the existence of alternate mechanisms.

Thus, the ability of vanadium salts to stimulate MAP kinase, p90^{rsk} and p70^{s6k} might be one of the mechanisms by which these compounds exert insulinomimetic effects in

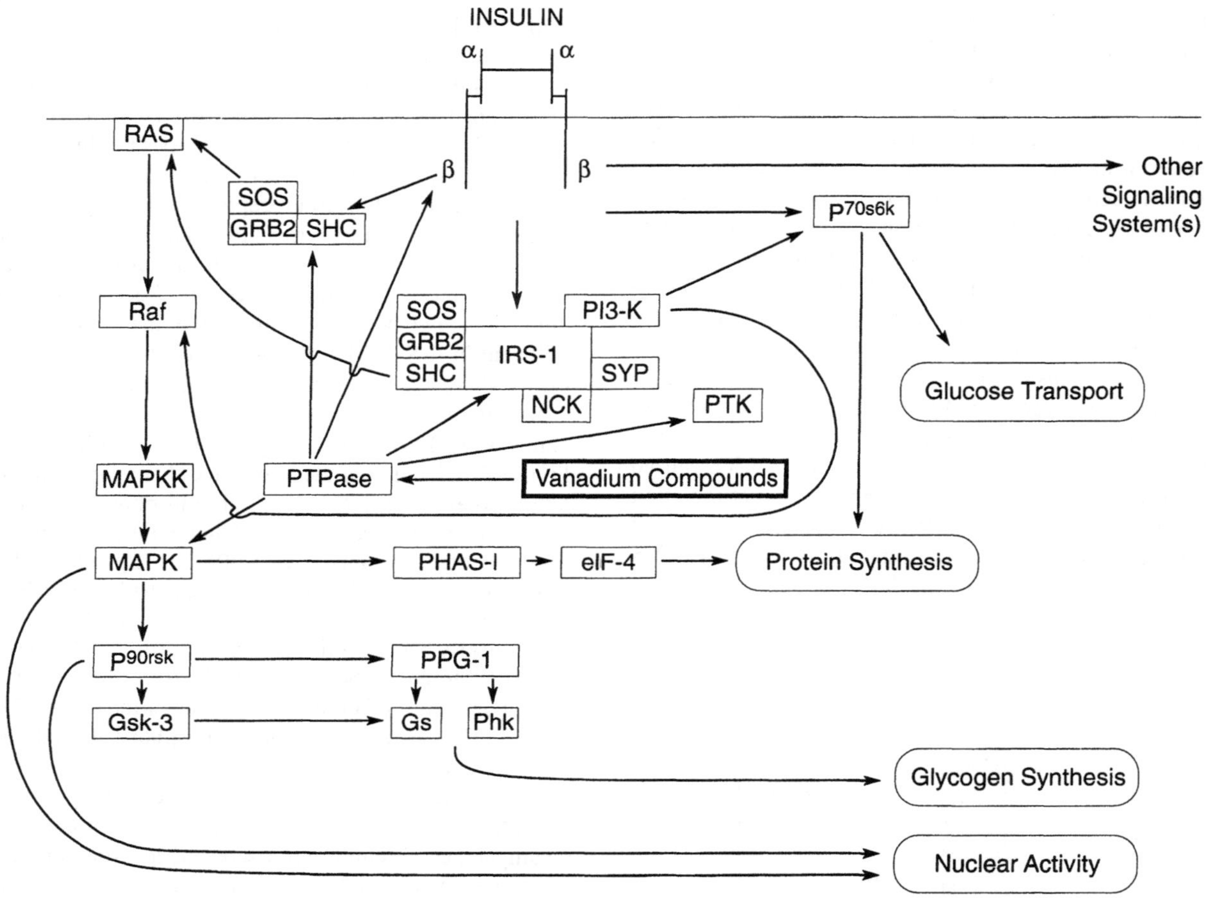

Fig. 6. Schematic model showing possible target sites of vanadium actions in relation to insulin signalling cascade: Insulin initiates its action by binding to α-subunit of its receptor, which is a heterotetrameric protein, and activates the intrinsic protein tyrosine kinase activity of β-subunit of the receptor by autophosphorylation. The activated β-subunit receptor in turn phosphorylates and thereby activates several proteins among which insulin receptor substrate-1 (IRS-l) is most widely characterized. The activated IRS-1 serves as docking protein and forms a complex with several SH2 domains containing proteins e.g. sos (son of sevenless), GRB2, Shc, Nck, Syp (PTPase) and phosphatidyl inositol 3-kinase (PI3K). This complex, in turn, stimulates Ras, MAPKK, MAPK and p90^{rsk} cascade. Activated MAPK phosphorylates PHAS-1 (a heat stable protein) which is involved in protein synthesis via activation of eIF-4 (elongation initiation factor-4). The stimulated ribosomal protein p90^{rsk} phosphorylates and activates glycogen bound protein phosphatase-1 (PP1-G). The activated PP1-G dephosphorylates glycogen synthase (GS) and phosphorylase kinase (Phk) and stimulates glycogen synthesis. P90rsk mediated phosphorylation of glycogen synthase kinase-3 (Gsk-3). MAPK and p90^{rsk} may regulate nuclear activity by phosphorylating transcription factors. The other form of ribosomal protein s6 kinase, the 70 kDa protein, p70^{s6k} lies on the other side of insulin signalling cascade and among other events is also implicated in modulating glucose transport and protein synthesis. Vanadium compounds are inhibitors of PTPases and by preventing dephosphorylation are able to increase tyrosine phosphorylation of key proteins involved in insulin signalling cascade. The potential target of this PTPase(s) could be IR-subunit, IRS-1, Shc, MAPK or a cytosolic PTK. Based on the results presented here and elsewhere the IR-β subunit/IRS-1 may not be the sites of vanadium action, however, MAPK-specific PTPase and/or Shc could be possible targets of vanadium compounds. With regard to p70^{s6k} activation by vanadium compounds a role of PI3-K may be suggested. The possibility that vanadium may act at other sites can not be excluded.

various systems. This notion is further supported by a recent study in which OV-stimulated glucose transport was partially inhibited by wortmannin treatment [86]. In this regard, it is noteworthy that, in circulating mononuclear leucocytes from diabetic subjects, the insulin-mediated activation of ribosomal S-6-kinases (s6k) and MAP kinase was completely diminished, as compared to non-diabetics. $NaVO_3$ therapy of diabetic subjects not only improved the glucose homeostasis but also stimulated the basal s6k and MAPK activity between 1.7 and 3.9-fold [88] further suggesting a role of these serine/threonine kinases in the insulinomimetic effects of vanadium. Involvement of additional mechanisms in insulinomimesis by vanadium compounds is also possible. For example, studies of Fantus *et al.* have demonstrated a possible role of vanadium (OV) in enhancing insulin sensitivity and prolonging insulin action at the level of insulin receptor PTK via PTPase [89, 90]. An involvement of OV-stimulated, staurosporine-inhibited cytosolic PTK in vanadium action has been proposed in rat adipocytes [91]. It has been suggested that cytosolic PTK activation is secondary to the inhibition of a PTPase [91].

The mechanism by which vanadium salts activate MAP kinase and $p90^{rsk}$ and $p70^{s6k}$ is not clearly understood. A tentative scheme with potential sites where vanadium compounds might act in insulin signalling cascade is shown in Fig. 6. Vanadium compounds are inhibitors of PTPases [51] and by preventing the dephosphorylation are able to increase the phosphotyrosyl content of key protein molecules in insulin signalling cascade [48] (Fig. 6) and thus could result in the activation of the MAP kinases as well as $p70^{s6k}$ activating pathway. The potential target could be insulin receptor β-subunit and/or IRS-I (Fig. 6); however results presented here and elsewhere [58–61] do not support this notion. Shc protein which is tyrosyl phosphorylated and activated in response to insulin [92] may serve as another target. Shc is able to interact with GRB2 in an IRS-1 dependent and independent manner and may activate the $p21^{ras}$ signalling pathway leading to the activation of MAPK and $p90^{rsk}$ [48] (Fig. 6). Alternatively, vanadium compounds might activate MAP kinase pathway by inhibiting a constitutively active MAP kinase-specific PTPase. A vanadium-inhibitable MAPK-specific protein tyrosine phosphatase (PTP) has recently been described in Xenopus-oocytes [93] and a similar PTPase might also be expressed in CHO cells. Since several lines of evidence link PI3K with the activation of $p70^{s6k}$ [85, 94] and recent studies have shown a modulation of PI3K activity by vanadate [86], it may be suggested that vanadium compounds utilize a similar pathway in activating $p70^{s6k}$ activity.

In conclusion, these results demonstrate that similar to insulin, vanadium compounds activate serine/threonine kinases, MAP kinases, $p90^{rsk}$ and $p70^{s6k}$. However, in contrast to insulin vanadium mediated activation is independent of the tyrosine phosphorylation of either the insulin receptor or IRS-I. It is suggested that vanadium-mediated activation of MAP kinases, $p90^{rsk}$ and $p70^{s6k}$ plays an important role in mimicking insulin-like effects.

Acknowledgement

We are grateful to Dr. John Blenis of Harvard Medical School and Dr. Frederic Hall of USC, School of Medicine, Los Angeles, for their generosity in providing the $p90^{rsk}$ and $p70^{s6k}$ antibodies, respectively. We would also like to thank Ms. Edith Giasson and Dr. Sylvain Meloch for help with S6 Kinasay assay and Mrs. Susanne Bordeleau-Chenier for secretarial help.

References

1. Ramasarma T, Crane FL: Does vanadium play a role in cellular regulation? Curr Topics Cell Regul 20: 247–301, 1991
2. Chasteen ND: The biochemistry of vanadium. In: Structure and bonding 53. Springer Verlag, Berling, Heidelberg, 1983, pp 105–138
3. Nechay BR: Mechanisms of action of vanadium. Ann Rev Pharmacol Toxicol 24: 501–524, 1984
4. Tolman EL, Barris E, Burns M, Pansini A, Partridge R: Effects of vanadium on glucose metabolism *in vitro*. Life Sci 25: 1159–1164, 1979
5. Shechter Y, Karlish SJD: Insulin-like stimulation of glucose oxidation in rat adipocytes by vanadyl (IV) ions. Nature 284: 556–558, 1980
6. Dubyak GR, Kleinzeller A: The insulin-mimetic effects of vanadate as (Na+, K+) ATPase inhibitor. J Biol Chem 255: 5306–5312, 1980
7. Singh J, Nordlis RC, Jorgenson RA: Vanadate: a potent inhibitor of multifunctional glucose-6-phosphatase. Biochem Biophys Acta 678: 477–482, 1981
8. Degani H, Gochin M, Karlish SJO, Shechter Y: Electron paramagnetic resonance studies and insulin-like effects of vanadium in rat adipocytes. Biochem 20: 5795–5799, 1981
9. Carpenter G: Vanadate, epidermal growth factor and the stimulation of DNAsynthesis. Biochem Biophys Res Commun 102: 1115–1121, 1982
10. Smith JB: Vanadium ions stimulated DNA synthesis in Swiss mouse 3T3 and 3T6 cells. Proc Natl Acad Sci USA 80: 6162–6166, 1983
11. Tamura S, Brown TA, Whipple JH, Yamaguchi YF, Dubler RE, Chen K, Larner J: A novel mechanism of the insulin-like effects of vanadate on glycogen synthase in rat adipocytes. J Biol Chem 259: 6650–6658, 1984
12. Clark AS, Fagan JM, Mitch WE: Selectivity of the insulin-like actions of vanadate on glucose and protein metabolism in skeletal muscle. Biochem J 232: 273–276, 1985
13. Klarlund JR: Transformation of cells by an inhibitor of phosphatases acting on phosphotyrosine in proteins. Cell 41: 707–717, 1985
14. Green A: The insulin like effect of sodium vanadate on adipocyte glucose transport is mediated at a post-insulin-receptor level. Biochem J 238: 663–669, 1986
15. Bosch F, Arino J, Gomez-Foix AM, Guinovart JJ: Glycogenolytic noninsulin-like effects of vanadate on rat hepatocyte glycogen synthase and phosphorylase. J Biol Chem 262: 218–222, 1987
16. Gomez-Foix AM, Rodriguez-Gil JE, Fillat C, Guinovart JJ, Boxch

F: Vanadate raises fructose 2,6-bisphosphate concentrations and activates glycolysis in rat hepatocytes. Biochem J 255: 507–512, 1988
17. Jackson TK, Salhanick AI, Sparks JD, Sparks CE, Bolognino M, Amatruda JM: Insulin mimetic effects of vanadate in primary cultures of rat hepatocytes. Diabetes 37: 1234–1240, 1988
18. Duckworth WC, Solomon SS, LiepnicksS J, Hamel FG, Hand S, Peaw DE: Insulin-like effects of vanadate in isolated rat adipocytes. Endocrinology (1988) 122: 2285–2289, 1988
19. Fantus IG, Kadota SI, Deragon G, Foster B, Posner BI: Pervanadate (peroxide(s) of vanadate) mimics insulin action in rat adipocytes via activation of the insulin receptor tyrosine kinase. Biochemistry 28: 8864–8871, 1989
20. Rodriguez-Gil JE, Gomez-Foix AM, Arino J, Guniovart JJ, Bosch F: Control of glycogen synthase and phosphorylase in hepatocytes from diabetic rats. Diabetes 38: 793–798, 1989
21. Miralpeix M, Gil J, Rosa JC, Carreras J, Bartrons R: Vanadate counteracts glucagon effects in isolated rat hepatocytes. Life Sci 44: 1491–1497, 1989
22. Rider MH, Bartrons R, Hue L: Vanadate inhibits liver fructose-2,6-bisphosphatase. Eur J Biochem 209: 53–56, 1990
23. Mountjoy KG, Flier JS: Vanadate regulates glucose transporter (Glut-1) expression in NIH 3T3 mouse fibroblasts. Endocrinology 26: 2778–2787, 1990
24. Miralpeix M, Decaux JF, Kahn A, Bartrons R: Vanadate induction of L-type pyruvate kinase mRNA in adult rat hepatocytes in primary culture. Diabetes 40: 462–464, 1991
25. Shechter Y: Insulin-mimetic effects of vanadate: Possible implications for future treatment of diabetes. Diabetes 39: 1–5, 1990
26. Heyliger CE, Tahiliani AG, McNeill JH: Effect of vanadate on elevated blood glucose and depressed cardiac performance of diabetic rats. Science 277: 1474–1477, 1985
27. Meyerovitch J, Farfel Z, Sack J, Shechter Y: Oral administration of vanadate normalizes blood glucose levels in streptozotocin-treated rats. J Biol Chem 262: 6658–6662, 1987
28. Brichard SM, Pottier AM, Henquin JC: Long term improvement of glucose homeostasis by vanadate in obese hyperinsulinemic fa/fa rats. Endocrinology 125: 2510–2516, 1989
29. Pugazhenthi S, Angel JF, Khandelwal RZ: Long-term effects of vanadate treatment on glycogen metabolizing and lipogenic enzymes of liver in genetically diabetic (db/db) mice. Metabolism 40: 941–946, 1991
30. Meyerovitch J, Rothenberg P, Shechter Y, Bonner-Weir S, Kahn CR: Vanadate normalizes hyperglycemia in two mouse models of noninsulin-dependent diabetes mellitus. J Clin Invest 87: 1286–1294, 1991
31. Gil J, Miralprix M, Carreras J, Bartrons R: Insulin-like effects of vanadate on glucokinase activity and fructose 2,6-biphosphate levels in the liver of diabetic rats. J Biol Chem 263: 1868–1871, 1988
32. Brichard SM, Okitolonda W, Henquin JC: Long term improvement of glucose homeostasis by vanadate treatment in diabetic rats. Endocrinology 123: 2048–2053, 1988
33. Pugazhenthi S, Khandelwal RL: Insulinlike effects of vanadate on hepatic glycogen metabolism in nondiabetic and streptozotocin-induced diabetic rats. Diabetes 39: 821–827, 1990
34. Rossetti L, Laughlin MR: Correction of chronic hyperglycemia with vanadate, but not with phlorizin, normalizes *in vivo* glycogen repletion and *in vitro* glycogen synthase activity in diabetic skeletal muscle. J Clin Invest 84: 892–899, 1989
35. Bollen M, Miralpeix M, Ventura F, Toth B, Bartrons R, Stalmans W: Oral administration of vanadate to streptozotocin-diabetic rats restores the glucose-induced activation of liver glycogen synthase. Biochem J 267: 269–271, 1990
36. Strout HV, Vicario PP, Biswas C, Superstein R, Brady EJ, Pilch PF, Berger J: Vanadate treatment of streptozotocin diabetic rats restores expression of the insulin responsive glucose transporter in skeletal muscle. Endocrin 126: 2728–2732, 1990
37. Sekar N, Kanthasamy A, William S, Subramanian S, Govindasamy S: Insulin actions of vanadate in diabetic rats. Pharmacol. Res. 22: 207–217, 1989
38. Saxena AK, Srivastava P, Bacquer NZ: Effects of vanadate on glycolytic enzymes and malic enzyme in insulin-dependent and independent tissues of diabetic rats. Eur J Pharmacol 216: 123–126, 1992
39. Valera A, Rodriguez-Gil JE, Bosch F: Vanadate treatment restores the expression of genes for key enzymes in the glucose and ketone bodies metabolism in the liver of diabetic rats. J Clin Invest 92: 4–11, 1993
40. Brichard SM, Desbuquois B, Girard J: Vanadate treatment of diabetic rats reverses the impaired expression of genes involved in hepatic glucose metabolism: effects on glycolytic and gluconeogenic enzymes and on glucose transporter GLUT2. Mol Cell Endocrinol 91: 91–97, 1993
41. Miralpeix M, Carballo E, Bartrons R, Crepin K, Hue L, Rousseau GG: Oral administration of vanadate to diabetic rats restores liver 6-phosphofructo-2-kinase content and mRNA. Diabetologia 35: 243–248, 1992
42. Pugazhenthi S, Khandelwal RL, Angel JF: Insulin like effects of vanadate on malic enzyme and glucose-6-phosphate dehydrogenase activites in streptozotocin-induced diabetic rat liver. Biochim Biophys Acta 1083: 310–312, 1991
43. Brichard SM, Ongemba LN, Girard J, Henquin JC: Tissue specific correction of lipogenic enzyme gene expression in diabetic rats given vanadate. Diabetologia 37: 1065–1072, 1994
44. Srivastava AK: Potential use of vanadium compounds in the treatment of diabetes mellitus. Exp Opin Invest Drugs 4: 525–536, 1995
45. Rosen O: After insulin binds. Science 237: 1452–1458, 1987
46. Olefsky JM: The insulin receptor: a multifunctional protein. Diabetes 3A: 1009–1116, 1990
47. Myers MG, White MF: The new elements of insulin signaling insulin receptor substrate-1 and proteins with SH2 domains. Diabetes 42: 643–650, 1993.
48. White MF, Kahn CR: The insulin signaling system. J Biol Chem 269: 14, 1994
49. Chou CK, Dull TJ, Russel DS, Ghezri R, Lebwohl D, Ullrich A, Rosen OM: Human insulin receptors mutated at the ATP-binding site lack protein tyrosine kinase activity and fail to mediate post-receptor effects of insulin. J Biol Chem 262: 1842–1847, 1987
50. McClain DA, Maegawa H, Lee J, Dull TJ, Ullrich A, Olefsky JM: A mutant insulin receptor with defective tyrosine kinase displays no biological activity and does not go endocytosis. J Biol Chem 262: 14663–14671, 1987
51. Swarup G, Cohen S, Garbers DL: Inhibition of membrane phosphotyrosyl protein phosphatase activity by vanadate. Biochem Biophys Res Commun 107: 1104–1109, 1982
52. Tamura S, Brown TA, Dubler RE, Larner J: Insulin like effects of vanadate on glycogen synthase and on phosphorylation of 95,000 dalton subunit of the insulin receptor. Biochem Biophys Res Commun 113: 80–86, 1984
53. Bernier M, Laird DM, Lane MD: Effect of vanadate on the cellular accumulation of pp15, an apparent product of insulin receptor tyrosine kinase action. J Biol Chem 263: 13626–13634, 1988
54. Gherzi R, Caratti C, Andraghetti G, Bertolini S, Montemurrd A, Sesti G, Cordera R: Direct modulation of insulin receptor protein tyrosine kinase by vanadate and anti-insulin receptor monoclonal antibodies. Biochem Biophys Res Commun 152: 1474–1480, 1988
55. Pugazhenthi S, Khandelwal RL: Does the insulin-mimetic action of

vanadate involve insulin receptor kinase? Mol Cell Biochem 217–218: 211–218, 1993
56. Strout HV, Vicario PP, Saperstein R, Slater EE: The insulinmimetic effect of vanadate is not correlated with insulin receptor tyrosine kinase activity nor phosphorylation in mouse diaphragm *in vivo*. Endocrinology 124: 1918–1924, 1989
57. Shisheva A, Shechter Y: Quercetin selectively inhibits insulin receptor function *in vitro* and the bioresponse of insulin and insulinomimetic agents in rat adipocytes. Biochemistry 31: 8059–8063, 1992
58. Mooney RA, Bordwell KL, Luhowsky S, Casnelli JE: The insulin-like effect of vanadate on lipolysis in rat adipocytes is not accompanied by an insulin-like effect on tyrosine phosphorylation. Endocrinology 124: 422–429, 1989
59. Blondel O, Simon J, Chevalier B, Portha B: Impaired insulin action but normal insulin receptor activity in diabetic rat liver: effect of vanadate. Am J Physiol 258: E459–E467, 1990
60. D'Onofrio F, Le MQ, Chiasson J-L, Srivastava AK: Vanadate dependent activation of mitogen activated protein (MAP) kinase in Chinese hamster ovary cells overexpressing a wild type human insulin receptor (CHO-HIRc). The Pharmacologist 35: 109, 1993
61. D'Onofrio F, Le MQ, Chiasson J-L, Srivastava AK: Activation of mitogen activated protein (MAP) kinases by vanadate is independent of insulin receptor autophosphorylation. FEBS Lett 340: 269–275, 1994
62. Ray LB, Sturgill TW: Rapid stimulation by insulin of a serine/threonine kinase in 3T3-L1 adipocytes that phosphorylates microtubule associated protein-2 *in vitro*. Proc Natl Acad Sci USA 1502–1506, 1987
63. Sturgill, TW, Ray, LB, Erkson, E, Maller, JL.: Insulin-stimulated MAP-2 kinase phosphorylates and activates ribosomal s6 kinase II. Nature 334: 715–718, 1988
64. Banerjee P, Ahmad MF, Grove JR, Kozlosky C, Price DJ, Avruch J: Molecular structure of a major insulin/mitogen-activated 70kDa s6 protein kinase. Proc Natl Acad Sci USA 87: 8550–8554, 1990
65. Cobb MH: An insulin stimulated ribosomal protein kinase in 3T3-L1 cells. J Biol Chem 261: 12994–12999, 1986
66. Blenis J: Transduction via the MAP kinases: Proceed at your own RSK. Proc Natl Acad Sci (USA) 90: 5889–5892, 1993
67. Pederson RA, Ramanadham S, Buchan A, McNeill JH: Long term effects of vanadyl treatment on streptozotocin-induced diabetes in rats. Diabetes 38: 1390–1395, 1989
68. Kozma SC, Ferrari S, Bassand P, Siegmann M, Totty N, Thomas G: Cloning of the mitogen-activated s6 kinase from rat liver reveals an enzyme of the second messenger subfamily. Proc Natl Acad Sci USA 87: 7365–7369, 1990
69. Chen R-H, Blenis J: Identification of Xenopus S6 protein kinase homologus ($pp90^{rsk}$) in somatic cells: phosphorylation and activation during the initiation of cell proliferation. Mol Cell Biol 10: 3204–3215, 1. Ramasarma T, Crane FL: Does vanadium play a role in cellular regulation? Curr Topics Cell Regul 20: 247–301, 1991
70. Srivastava AK, Chiasson J-C, Chiasson J-L, Lacroix A, Windisch L: Biochemical characterstics of cytosolic and particulate forms of protein tyrosine kinases from N-methyl-N-nitrosourea (MNU)-induced rat mammmary carcinoma. Mol Cell Biochem 106: 87–97, 1991
71. Cobb MH, Boulton TG, Robbins DJ: Extracellular signal-regulated kinases: ERKS in progress. Cell-Regul 2: 965–978, 1991
72. Chen R.-H, Chung J, Blenis J: Regulation of $pp90^{rsk}$ phosphorylation and s6 phosphotransferase activity in swiss 3T3 cells by growth factor – phorbol ester – and cyclic AMP-mediated signal transduction. Mol Cell Biol 11: 1861–1867, 1991
73. Nguyen TT, Scimeca J-C, Filloux C, Peraldi P, Carpentier J-L, Van Obberghen E: Co-regulation of the mitogen activated protein kinase, extracellular signal-regulated kinase 1 and the 90 kDa ribosomal s6 kinase in PC12 cells. Distinct effects of the neurotrophic factor, nerve growth factor and the mitogenic factor, Epidermal growth factor. J Biol Chem 268: 9803–9810, 1993
74. Tamemoto H, Kadowaki T, Tobe K, Yagi T, Sakura H, Hyakawa T, Terauchi Y, Ueki K, Kaurage Y, Satoh S, Sekihara H, Yoshioka S, Horikoshi H, Furuta Y, Ikawa Y, Kasuga M, Yazaki Y, Aizawa S: Insulin resistance and growth retardation in mice lacking insulin receptor substrate-1. Science 372: 182–186, 1994
75. Dent P, Lavoinne A, Nakielny S, Caudwell FB, Watt P, Cohen P: The molecular mechanism by which insulin stimulates glycogen synthesis in mammalian skeletal muscle. Nature 348: 302–308, 1990
76. Cross DAE, Alessi DR, Vadenheede JR, McDowell HE, Hundal HS, Cohen P: The inhibition of glycogen synthase kinase-3 by insulin or insulin-like growth factor 1 in the rat skeletal muscle cell line L6 is blocked by worthmannin, but not by rapamycin: evidence that wortmannin blocks activation of the mitogen activated protein kinase pathway in L6 cells between ras and raf. Biochem J 303: 21–26, 1994
77. Welsh GI, Foulstone EJ, Young SW, Tavare JM, Proud CG: Wortmannin inhibits the effect of insulin and serum on the activities of glycogen synthase kinase-3 and mitogen activated protein kinase. Biochem J 303: 15–20, 1994
78. Shepherd PR, Nave BT, Siddle K: Insulin stimulation of glycogen synthesis and glycogen synthase activity is blocked by wortmannin and repamycin in 3T3-L1 adipcoytes: evidence for the involvement of phosphoinositde 3-kinase and p70 ribosomal protein S6 kinase. Biochem J 305: 25–28, 1995
79. Donella-Dean A, Lavoinne A, Marin O, Pinna LA, Cohen, P: An analysis of the substrate specificity of insulin-stimulated protein kinase-1. Biochem Biophys Acta 1178: 189–193, 1993
80. Sutherland C, Cambell DG, Cohen P: Identification of insulin stimulated protein kinase-1 as the rabbit equivalent of rsk-2 identification of two threonine phophorylation during activation by mitogen activated protein kinase. Eur J Biochem 212: 581–588, 1993
81. Sutherland C, Leighton IA, Cohen P: Identification of glycogen synthase kinase-3 by phosphorylation: new kinase connections in insulin and growth factor signalling. Biochem J 296: 15–19, 1993
82. Sutherland C, Cohen P: The alpha-isoform of glycogen synthase kinase-3 from rabbit skeletal muscle is inactivated by p70s6 kinase or MAP kinase-activated protein kinase-1 *in vitro*. FEBS Lett 338: 37–42, 1994
83. Woodgett JR: A common denominator linking glycogen metabolism, nuclear oncogene and development. Trends Biochem Sci 16: 177–181, 1991
84. Lin TA, Lawrence JC Jr: Activation of ribosomal protein s6 kinase does not increase glycogen synthesis or glucose transport in rat adipocytes. J Biol Chem 269: 21255–21261, 1994
85. Cheatham B, Vlahos CJ, Cheatham L, Wang L, Blenis J, Kahn CR: Phosphatidyl inositol 3-kinase activation is required for insulin stimulation of $pp70^{s6k}$, DNA synthesis, and glucose transporter translocation. Mol Cell Biol 14: 4902–4911, 1994
86. Berger J, Hayes N, Szalkowski DM, Zhang B: PI3-kinase activation is required for insulin stimulation of glucose transport into L6 myotube. Biochem Biophys Res Commun 205: 570–576, 1994
87. Finger DC, Hausdorff SF, Blenis J, Birnbaum MJ: Dissociation of pp70 ribosomal protein s6 kinase from insulin-stimulated glucose transport in 3T3-L1 adipocytes. J Biol Chem 268: 3005–3008, 1993
88. Goldfine AB, Folli F, Patti ME, Simonson DC, Kahn CR: Clinical trials of vanadium compounds in human diabetes mellitus. Can J Physiol Pharmacol 72 (suppl 3): 11, 1994
89. Fantus IG, Ahmad F, Deragon G: Vanadate augments insulin binding and prolongs insulin action in rat adipocytes. Endocrinology 127: 2716–2725, 1990
90. Fantus IG, Ahmad F, Deragon G: Vanadate augments insulin stimu-

lated tyrosine kinase activity and prolongs insulin action in rat adipocytes: evidence for transduction of amplitude of signaling into duration of response. Diabetes 43: 375–383, 1994
91. Shisheva A, Shechter Y: Role of cytosolic tyrosine kinase in mediating insulin-like actions of vanadate in rat adipocytes. J Biol Chem 268: 6463–6469, 1993
92. Skolnik EY, Lee CH, Batzer A, Vincentine LM, Zhou M, Daly R, Myers MJ Jr, Backer JM, Ullrich A, White MF, Schlessinger J: The SH2/SH3 domain containing protein GRB2 interacts with tyrosine-phosphorylated IRS-1 and shc: implications for insulin control of ras signalling. The EMBO J 12: 1429–1436, 1993
93. Sarcevic B, Erikson E, Maller JL: Purification and characterization of a mitogen-activated protein kinase tyrosine phosphatase from Xenopus eggs. J Biol Chem 268: 25075–25083, 1993
94. Chung J, Grammer TC, Lemon KP, Kazlauskar A, Blenis J: PDGF- and insulin-independent pp70^{s6k} activation mediated by phosphatidylinositol-3-OH kinase. Nature (London) 370: 71–75, 1994

Molecular and Cellular Biochemistry **153**: 79–85, 1995.

Protective effect of vanadate on oxyradical-induced changes in isolated perfused heart

Taku Matsubara, Sorin Musat-Marcu, Hara P. Misra[1] and Naranjan S. Dhalla

Division of Cardiovascular Sciences, St. Boniface General Hospital Research Centre, Faculty of Medicine, University of Manitoba, Winnipeg, Manitoba, R2H 2A6, Canada and [1]Department of Biomedical Sciences, Virginia Polytechnic Institute and State University, Blacksburg, Virginia 24060-0442, USA

Abstract

In order to examine the mechanisms of the beneficial effects of vanadate on cardiac dysfunction in chronic diabetes, rat hearts were perfused with xanthine plus xanthine oxidase, an oxyradical generating system in the absence or presence of vanadate. The heart failed to generate contractile force and increased the resting tension markedly within 5 min of perfusion with xanthine plus xanthine oxidase. These changes were prevented by the addition of 4 μM vanadate in the perfusion medium. The protective effects of vanadate on the loss of developed tension and increased resting tension due to xanthine plus xanthine oxidase were dose-dependent (0.1–5 μM). Perfusion of the hearts with glucose-free medium did not abolish the protective actions of vanadate. The sarcolemmal Ca^{2+}-pump (ATP-dependent Ca^{2+} uptake and Ca^{2+}-stimulated ATPase) and Na^+-dependent Ca^{2+} uptake activities were decreased upon perfusing the hearts with a medium containing xanthine plus xanthine oxidase for 5 min; these effects were prevented by the addition of 2–4 μM vanadate in the perfusion medium. The signals for superoxide radicals produced by xanthine plus xanthine oxidase, as detected by electron paramagnetic resonance spectroscopic technique, were inhibited by 5–100 μM vanadate. These results suggest that vanadate is an oxyradical scavenger and thus may prevent heart dysfunction under some pathological conditions by its antioxidant action. (Mol Cell Biochem **153:** 79–85, 1995)

Key words: oxygen free radicals, vanadate, rat heart, sarcolemmal calcium pump activity, sarcolemmal sodium-calcium exchange

Introduction

By virtue of their ability to enhance the uptake and oxidation of glucose in the cell, different vanadium compounds including vanadate are considered to exert insulin-like action in the body [1–6]. The insulin mimetic effects of vanadium compounds have been suggested to be due to either stimulation of tyrosine kinase and/or inhibition of phosphotyrosine phosphatase activities [7, 8]. Although the beneficial effects of vanadium compounds in preventing heart dysfunction are associated with their hypoglycemic action in chronic diabetes [9–12], the exact mechanisms of their action on the heart are not clear at present. This view is supported by the fact that vanadate has been shown to inhibit Na^+-K^+ ATPase activity and is thus considered to exert actions on the heart like digitalis glycosides [13–15]. Furthermore, vanadate is known to exert positive inotropic action by acting on Ca^{2+} pools involved in contractile activation in the myocardium [16]. Since diabetes is an internal stress associated with increased oxidative processes and free radical formation [17], it is likely that the beneficial effects of vanadate on heart dysfunction in diabetes are due to its antioxidant property. This was found to be the case since the treatment of diabetic animals with vanadate was found to prevent depressions in the activities of glutathione peroxidase, catalase and superoxide dismutase as well as glutathione content, which are known to serve as cellular defense mechanisms [18]. In this study we have tested the antioxidant actions of vanadate by examining its ability to prevent oxyradical-induced changes in cardiac function and membrane Ca^{2+}-transport in the isolated perfused rat

Address for offprints: N.S. Dhalla, Division of Cardiovascular Sciences, St. Boniface General Hospital Research Centre, 351 Tache Avenue, Winnipeg, Manitoba, R2H 2A6, Canada hearts. In addition, the effect of vanadate on oxyradical signals generated by xanthine plus xanthine oxidase in the solution was tested by electron paramagnetic resonance (EPR) spectroscopic technique.

Materials and methods

Isolated heart preparations

Adult male Sprague-Dawley rats (250–300 g) were killed by decapitation and their hearts quickly excised and immersed in ice-cold buffer. The adherent connective tissue was removed and the aorta of the heart was attached to a cannula. The heart was perfused by non-recirculating Langendorff technique with Krebs-Henseleit solution containing (in mM) NaCl 120, $NaHCO_3$ 25, KCl 4.7, KH_2PO_4 1.2, $MgSO_4$ 1.2, $CaCl_2$ 1.25 and glucose 11.0. This perfusion solution was continuously oxygenated with a mixture of 95% O_2 and 5% CO_2 (pH 7.4) and its temperature maintained at 37°C. The heart was electrically stimulated (Phipps and Bird stimulator) at 300 beats/min via a square wave of 1.5 ms duration at twice the threshold voltage. The coronary flow was maintained at 10 ml/min by a Harvard peristaltic pump. Cardiac contractile force and resting tension were recorded on a Beckman recorder via a force-displacement transducer (Grass FT.03). A resting tension of 2 g was applied to the heart at the beginning of each experiment.

To generate oxygen radicals, xanthine (2 mM) and xanthine oxidase (10 U/l) were dissolved in the perfusion buffer and mixed for 60 min. The hearts were allowed to equilibrate for 20–30 min with normal oxygenated Krebs-Henseleit buffer before use. In the first set of experiments, hearts were perfused with the buffer solution containing xanthine (X) and xanthine oxidase (XO) for 20 min to examine changes in contractile parameters. A second set of experiments was performed to study the effects of sodium vanadate on changes induced by X + XO. Immediately after the experimental period, the hearts were rapidly removed from the perfusion apparatus. The ventricles were trimmed and frozen using Wollenberger clamps precooled in liquid nitrogen. All hearts were stored at –90°C before isolating sarcolemmal membranes. No change was observed for Ca^{2+} pump and Na^+-Ca^{2+} exchange activities of the sarcolemmal membranes isolated from the frozen hearts for 2 months.

Isolation of sarcolemmal membranes

Membrane fraction enriched with sarcolemma was isolated from a pool of three to four hearts according to the method of Pitts [19]. The frozen hearts were crushed to a fine powder in a stainless steel percussion mortar precooled with liquid nitrogen. This frozen tissue was allowed to thaw on ice and then homogenized in ice-cold 0.6 M sucrose, 10 mM imidazole-HCl, pH 7.0 (3.5 ml/g tissue) using a Polytron PT-20 (5 × 20 sec, setting 5). The resulting homogenate was centrifuged at 12,000 × g for 30 min, and the pellet was discarded. After diluting (5 ml/g tissue) with 140 mM KCl, 20 mM (*N*-morpholino)propanesulfonic acid (MOPS), pH 7.4 (KCl-MOPS buffer), the supernatant was centrifuged at 95,000 × g for 60 min. The resulting pellet was suspended in the KCl-MOPS buffer and layered over a 30% sucrose solution containing 0.3 M KCl, 50 mM $Na_4P_2O_7$ and 0.1 M tri(hydroxymethyl)aminomethane (Tris)-HCl, pH 8.3. After centrifugation at 95,000 × g for 90 min (utilizing a Beckman swinging bucket rotor), the band at the sucrose-buffer interface was taken and diluted with 3 volumes of KCl-MOPS buffer. A final centrifugation at 95,000 × g for 30 min resulted in a pellet rich in sarcolemma. All these steps for the isolation of sarcolemma were carried out at 0–4°C.

For the determination of Na^+-Ca^{2+} exchange activity the pellet was resuspended in 140 mM NaCl, 20 mM MOPS, pH 7.4 and quickly frozen in liquid nitrogen. Samples were then stored at –90°C for about 3 weeks before carrying out biochemical studies. When the purified vesicles were used for any other enzyme assay, the pellet was resuspended in 0.25 M sucrose, 10 mM histidine, pH 7.0, and was stored as described above. The freeze-thaw process had a minimal effect (< 5% reduction in Ca^{2+} pump and Na^+-Ca^{2+} exchange activities compared with membrane assayed immediately after isolation. The protein concentration was measured using the method of Lowry *et al* [20]. The purity of the membrane preparations was examined by measuring the activities of marker enzymes such as Na^+-K^+-ATPase (sarcolemmal marker), rotenone-insensitive NADPH cytochrome *c* reductase (sarcoplasmic reticular marker) and cytochrome *c* oxidase (mitochrondrial marker) in both homogenates and sarcolemmal membrane fractions according to the procedures used in our laboratory [21, 22]. Sarcolemmal Na^+-K^+-ATPase activity was enriched about 18-fold with respect to the homogenate values and the marker enzyme activities showed negligible contamination of sarcoplasmic reticular and mitochondrial membranes in the sarcolemmal fraction.

Measurement of Na^+-Ca^{2+} exchange

The method for Na^+-dependent Ca^{2+} uptake measurement has been described in detail elsewhere [21]. Briefly, 5 µl of sarcolemmal vesicles (1.5 mg/ml; 7.5 µg protein/tube) preloaded with NaCl-MOPS buffer at 37°C for 30 min were rapidly diluted 50 times with Ca^{2+} uptake medium containing 140 mM KCl, 20 mM MOPS, 0.4 µM valinomycin, 0.3 µCi $^{45}Ca^{2+}$ and 20 µM Ca^{2+}, pH 7.4. After the appropriate time of incubation, the reaction was stopped by the addition of ice-cold 0.03 ml stopping solution containing 140 mM KCl, 1 mM $LaCl_3$, 20 mM MOPS, pH 7.4. Samples (0.25 from 0.28 ml of the total reaction mixture) were filtered through Millipore filters (pore size 0.4 μm) and washed twice with 2.5 ml of ice-cold washing solution containing 140 mM KCl, 0.1 mM $LaCl_3$, 20 mM MOPS, pH 7.4. Radioactivities of filters were measured with a Beckman LS 1701 counter. In parallel to these samples, nonspecific Ca^{2+} uptake was measured by placing Na^+-loaded vesicles in an equimolar medium containing 140 mM NaCl and 20 mM MOPS. Na^+-dependent Ca^{2+} uptake activity was corrected by subtracting nonspecific Ca^{2+} uptake from the total Ca^{2+} uptake values.

Measurement of Ca^{2+} pump activities

The experimental conditions were the same as reported elsewhere [22]. For the estimation of Mg^{2+}-ATPase, sarcolemmal vesicle (20–40 μg protein) were preincubated at 37°C for 5 min in 0.5 ml of medium containing 140 mM KCl, 10 mM MOPS-Tris (pH 7.4), 2 mM $MgCl_2$, 5 mM sodium azide and 0.1 mM ethylene glycol-bis (β-aminoethyl ether)-N,N,N′,N′ tetraacetic acid (EGTA). The reaction was started by the addition of 4 mM Tris-ATP, pH 7.4, and terminated 5 min later with 0.5 ml of ice-cold 12% trichloroacetic acid; the liberated phosphate was measured by the method of Taussky and Shorr [23]. Estimation of total (Ca^{2+}-Mg^{2+})-ATPase was made in medium containing 140 mM KCl, 10 mM MOPS-Tris, pH 7.4, 2 mM $MgCl_2$, 5 mM sodium azide, and 10 μM free Ca^{2+}; the concentration of free Ca^{2+} in the medium was adjusted by using EGTA. The Ca^{2+}-stimulated ATPase activity was the difference between the total ATPase and Mg^{2+}-ATPase activities. ATP-dependent Ca^{2+} accumulation assay was carried out by a method described in detail elsewhere [21] with some modifications. Sarcolemmal vesicles (50 μg protein) were preincubated at 37°C for 5 min in 0.25 ml of medium containing 140 mM KCl, 10 mM MOPS-Tris, pH 7.4, 2 mM $MgCl_2$, and $^{45}CaCl_2$-EGTA, which contained 10 μM free Ca^{2+}. Ca^{2+} accumulation was initiated by adding 4 mM Tris-ATP, pH 7.4. After 5 min of incubation at 37°C, the reaction was stopped by the addition of ice-cold 0.03 ml stopping solution containing KCl-MOPS and 1 mM $LaCl_3$, pH 7.4. Samples (0.25 ml from 0.28 ml of the total reaction mixture) were filtered through Millipore filters (pore size 0.45 μm), washed twice with 2.5 ml ice-cold KCl-MOPS and 1 mM $LaCl_3$, and the radioactivity was determined for calculating the total Ca^{2+} binding. Nonspecific Ca^{2+} binding was measured in the absence of ATP for each set of experiments. The ATP-dependent Ca^{2+} binding was calculated by subtracting nonspecific Ca^{2+} binding from the total Ca^{2+} binding.

Xanthine and xanthine oxidase were obtained from Sigma Chemical Company (St. Louis, MO) and 5,5-dimethyl-1-pyrroline-N-oxide (DMPO) was obtained from Aldrich Chemical Inc. (Milwaukee, WI). The DMPO was further purified by eluting through an activated charcoal column anaerobically. The purified DMPO solution (0.1 M) gave no EPR signal when tested at pH 7.0 using 0.05 M phosphate buffer containing 10% DMSO. The DMPO was quantitated as previously described [24]. EPR measurements were made on a Bruker D-200 X-band spectrometer using a magnetic field modulation frequency of 100 KHz. The microwave power was maintained at 20 milliwatts and the scans were traced with the modulation amplitude of 0.8 G, time constant of 0.32 sec, scan rate of 500 sec, and receiver gain of 5×10^5. In the kinetic studies, the EPR spectrometer was set at a field setting to monitor the height of the first line of the EPR signal doublet as a function of time using the above parameters, except that the scan was off.

Statistical analysis

Results are presented as mean ± S.E. The data were subjected to analysis of variance followed by the Bonferroni corrected *t* test for multiple comparison. Probability (P) value of less than 0.05 was considered to be statistically significant.

Results

In one series of experiments, the effects of different concentrations of vanadate on changes in cardiac contractile force development (developed tension) and resting tension due to X + XO were studied by employing isolated rat hearts perfused with a medium containing glucose. Perfusion of hearts with medium containing X + XO was found to produce a biphasic effect (a slight increase followed by a marked depression) on developed tension whereas the resting tension was increased in a time-dependent manner (Fig. 1). These X + XO induced changes in cardiac developed tension and resting tension were prevented in hearts perfused with a medium containing 4 μM vanadate (Fig. 1). The data in Fig. 2 show that the depression in cardiac developed tension and increase in resting tension by X + XO were prevented by vanadate in a dose-dependent manner. The results in Fig. 2 also reveal that complete prevention of X + XO induced increase in resting tension was seen at 1 μM concentration of vanadate whereas no significant X + XO induced depression in developed tension was observed in the presence of 4 μM vanadate.

In order to test if the ability of vanadate to prevent X + XO induced changes in heart function was due to its ability to enhance glucose uptake and subsequent oxidation [1–6], the hearts were perfused in glucose-free medium. The results in Fig. 3 indicate that the biphasic effects of X + XO were seen

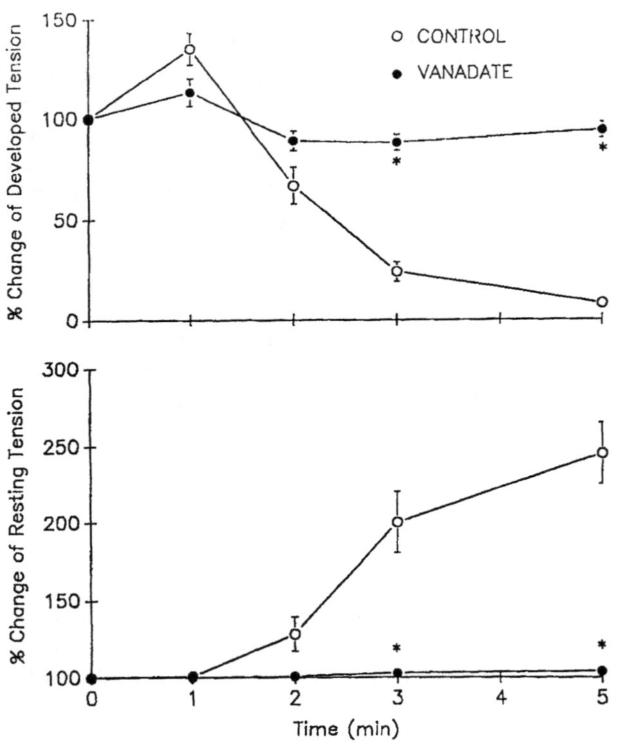

Fig. 1. Effect of 4 μM vanadate on xanthine plus xanthine oxidase (X + XO)-induced changes in developed tension and resting tension in isolated perfused rat hearts. Values are mean ± S.E. of 5 experiments and are expressed as percent changes from values obtained just before starting perfusion with a medium containing X + XO. *$P < 0.05$ compared to hearts perfused without vanadate (control group).

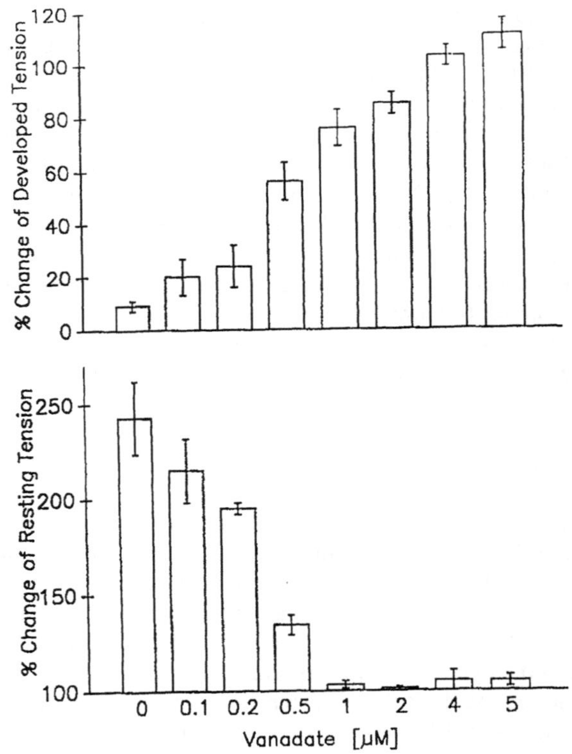

Fig. 2. Effect of different concentrations of vanadate on changes in developed tension and resting tension induced by 5 min perfusion with xanthine plus xanthine oxidase (X + XO). Values are mean ± S.E. of 4–6 experiments and are expressed as percent changes from values obtained just before starting perfusion with a medium containing X + XO.

in hearts perfused with glucose-free medium. Furthermore, X + XO induced depression in cardiac developed tension and increase in resting tension were prevented to a great extent in the presence of 4 μM vanadate (Fig. 3).

Another series of experiments was designed to test if the X + XO induced changes in sarcolemmal Ca^{2+}-pump and Na^+-Ca^{2+} exchange activity were prevented by vanadate. For this purpose the hearts were perfused with a medium containing X + XO for 5 min in the absence or presence of vanadate. The results in Tables 1 and 2 indicate that hearts perfused with X + XO in the absence of vanadate exhibited depressed sarcolemmal ATP-dependent Ca^{2+}-uptake, Ca^{2+}-stimulated ATP-ase and Na^+-dependent Ca^{2+}-uptake activities. These depressant effects of X + XO on sarcolemmal Ca^{2+}-pump and Na^+-Ca^{2+} exchange activities were prevented when hearts were perfused with 2 or 4 μM vanadate (Tables 1 and 2). It should be noted that perfusion of control hearts with 2 or 4 μM vanadate depressed the sarcolemmal Ca^{2+}-pump activities without affecting the Na^+-Ca^{2+} exchange activities (Tables 1 and 2).

The effect of vanadate on X + XO induced changes in sarcolemmal ATP-dependent Ca^{2+} uptake and Na^+-dependent Ca^{2+} uptake activities were also examined by incubating sarcolemmal preparations with X + XO in the absence or presence of 1 μM vanadate for 10 min. The results in Table 3 indicate that incubation of sarcolemma with X + XO under in vitro conditions depressed both ATP-dependent Ca^{2+} uptake and Na^+-dependent Ca^{2+} uptake activity; these effects were prevented by the presence of vanadate in the incubation medium.

It should be pointed out that xanthine oxidase is known to catalyze the conversion of xanthine to uric acid and during this process molecular oxygen is reduced to generate superoxide radicals ($O_2^{\cdot-}$). The $O_2^{\cdot-}$ so generated can react with DMPO, a spin trap, to form DMPO.$O_2^{\cdot-}$ spin adduct. Superoxide dismutase inhibits this reaction by effectively competing with DMPO for the available flux of $O_2^{\cdot-}$. As shown in Fig. 4 (upper panel), xanthine and xanthine oxidase produced DMPO.$O_2^{\cdot-}$ spin adduct and superoxide dismutase at 0.1 μg/ml inhibited this reaction, indicating that the formation of DMPO.$O_2^{\cdot-}$ spin adduct was dependent on $O_2^{\cdot-}$. We tested the effects of Na_3VO_4 in this system as a possible $O_2^{\cdot-}$ scavenger. The Na_3VO_4 at 5–100 μM inhibited

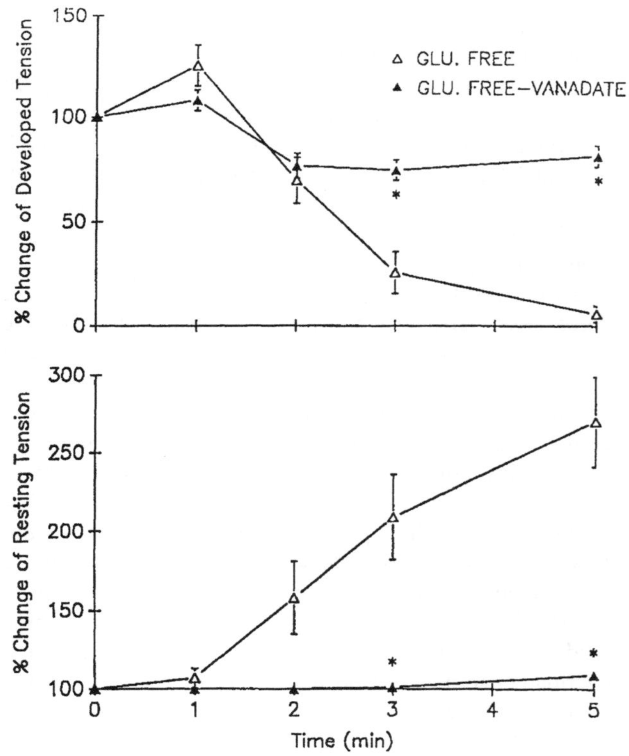

Fig. 3. Effect of vanadate (4 μM) on the changes in developed tension and resting tension induced by xanthine plus xanthine oxidase (X + XO) in rat hearts perfused with a glucose-free medium. Values are mean ± S.E. of 5 experiments and are expressed as percent changes from values obtained just before starting perfusion with a glucose-free medium containing X + XO. *P < 0.05 compared to hearts perfused without vanadate.

Table 1. Effect of vanadate on xanthine plus xanthine oxidase induced changes in sarcolemmal Ca^{2+}-pump activities in the isolated perfused rat heart

	ATP-dependent Ca^{2+} uptake (nmol Ca^{2+}/mg/5 min)	ATPase activities (μmol Pi/mg/h)	
		Mg^{2+}-ATPase	Ca^{2+}-stimulated ATPase
A. *Hearts perfused in the absence of vanadate*			
Control	24.7 ± 1.0	113 ± 5.8	12.2 ± 0.8
X + XO	6.3 ± 1.0*	88 ± 5.6	5.2 ± 1.1*
B. *Hearts perfused with 2 μM vanadate*			
Control	14.2 ± 0.7	117 ± 6.9	9.4 ± 0.5
X + XO	12.8 ± 0.9	110 ± 5.2	8.6 ± 0.5
C. *Hearts perfused with 4 μM vanadate*			
Control	8.0 ± 0.8	111 ± 7.5	6.4 ± 0.6
X + XO	7.4 ± 0.9	102 ± 6.3	5.6 ± 0.5

Each value is a mean ± S.E. of 4–5 experiments. X – xanthine (2 mM); XO – xanthine oxidase (10 U/l). *Significantly (p < 0.05) different from their respective control value.

Table 2. Effect of vanadate on xanthine plus xanthine oxidase induced changes in sarcolemmal Na^+-Ca^{2+} exchange activity in the isolated perfused rat heart

	Na^+-dependent Ca^{2+}-uptake (nmol Ca^{2+}/mg/2 sec)	Nonspecific Ca^{2+} uptake (nmol Ca^{2+}/mg/2 sec)
A. *Hearts perfused in the absence of vanadate*		
Control	3.31 ± 0.20	1.01 ± 0.09
X + XO	1.66 ± 0.13*	0.99 ± 0.10
B. *Hearts perfused with 2 μM vanadate*		
Control	3.25 ± 0.23	1.17 ± 0.12
X + XO	3.08 ± 0.16	1.04 ± 0.08
C. *Hearts perfused with 4 μM vanadate*		
Control	3.14 ± 0.28	1.03 ± 0.14
X + XO	2.89 ± 0.35	1.03 ± 0.11

Each value is a mean ± S.E. of 4–5 experiments. X – xanthine (2 mM); XO – xanthine oxidase (10 U/1). *Significantly (P < 0.05) different from their respective control value.

Table 3. Effect of vanadate on xanthine (X) plus xanthine oxidase (XO) induced changes in ATP-dependent Ca^{2+}-uptake and Na^+-dependent Ca^{2+}-uptake activities in sarcolemma isolated from control unperfused rat hearts

	ATP-dependent Ca^{2+}-uptake (nmol Ca^{2+}/mg/min)	Na^+-dependent Ca^{2+}-uptake (nmol Ca^{2+}/mg/2 sec)
A. *In the absence of vanadate*		
–	5.2 ± 0.25	4.8 ± 0.22
X + XO	3.0 ± 0.12*	1.6 ± 0.13*
B. *In the presence of 1 μM vanadate*		
–	4.1 ± 0.26	4.7 ± 0.27
X + XO	3.9 ± 0.25	4.4 ± 0.19

Each value is a mean ± S.E. of 4 experiments. Sarcolemmal preparations obtained from unperfused hearts were incubated with 2 mM X and 0.03 U/ml XO for 10 min in the absence or presence of 1 μM vanadate. *P < 0.05 compared to respective values in sarcolemmal preparations without X + XO.

the EPR signal intensity of the spin adduct in a dose-dependent manner. Thus, as shown in Fig. 4 (lower panel), Na_3VO_4 did powerfully inhibit the EPR signal intensity and as little as 35 μg Na_3VO_4 caused 50% inhibition.

Since there exists a competition between DMPO and Na_3VO_4 for the available $O_2^{\cdot-}$, the rate constant of Na_3VO_4 in the above system can be obtained as derived earlier [24]:

$$K_{VO_4^{3-}} = \frac{(I_0/I - 1) \cdot (K_{DMPO}[DMPO] + K_d)}{[VO_4^{3-}]}$$

Under a given condition, item (K_{DMPO}[DMPO] + K_d) is constant and a plot of ($I_0/I - 1$) against [VO_4^{3-}] should give a straight line. That this was the case is shown in Fig. 4 (lower panel), where a simple Stern-Volmer behavior for scavenging $O_2^{\cdot-}$ by VO_4^{3-} was observed with a correlation coefficient

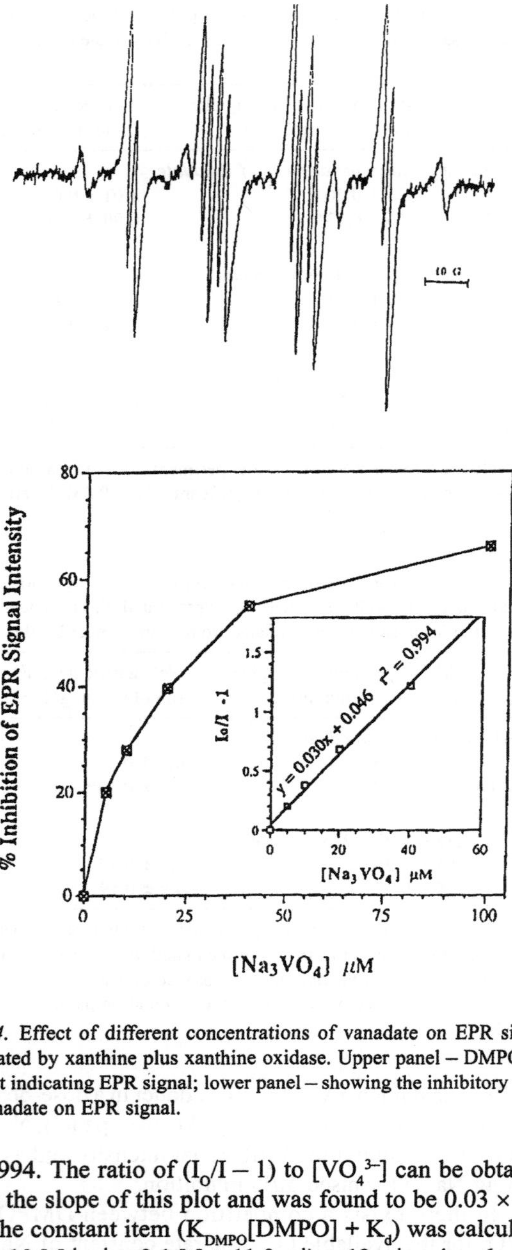

Fig. 4. Effect of different concentrations of vanadate on EPR signals generated by xanthine plus xanthine oxidase. Upper panel – DMPO.$O_2^{\cdot -}$ adduct indicating EPR signal; lower panel – showing the inhibitory effect of vanadate on EPR signal.

of 0.994. The ratio of $(I_0/I - 1)$ to $[VO_4^{3-}]$ can be obtained from the slope of this plot and was found to be 0.03×10^{-6} M. The constant item ($K_{DMPO}[DMPO] + K_d$) was calculated to be $10\ M^{-1}.s^{-1} \times 0.1\ M + 11.0\ s^{-1}) = 12\ s^{-1}$, using the rate constant, K_{DMPO}, the rate of DMPO trapping of $O_2^{\cdot -}$ of $10\ M^{-1}s^{-1}$ [25] and K_d, the rate $O_2^{\cdot -}$ decay at pH 7.13 of $11\ s^{-1}$ [26]. The rate constant, $K_{VO_4^{3-}}$, for VO_4^{3-} scavenging $O_2^{\cdot -}$ was found to be $3.6 \times 10^5\ M^{-1}s^{-1}$.

Discussion

By incubating sarcolemmal preparations with xanthine plus xanthine oxidase, a well known free radical generating system, free radicals have been reported to exert depressant effects on Ca^{2+}-pump and Na^+-Ca^{2+} exchange activities [22, 27, 28]. In this study we have observed that perfusing the heart with xanthine plus xanthine oxidase decreased the sarcolemmal ATP-dependent Ca^{2+} uptake, Ca^{2+}-stimulated ATPase and Na^+-dependent ATPase activities. Thus the results described in this study extend our previous observations [22, 27, 28] and support the conclusion that sarcolemmal membrane is one of the sites for the action of free radicals. Likewise, our results concerning the depression of contractile force development and increase in resting tension of isolated heart upon perfusion with xanthine plus xanthine oxidase confirm the findings of other investigators [29, 30]. In view of the role of sarcolemmal Ca^{2+}-pump and Na^+-Ca^{2+} exchange system in the efflux of Ca^{2+} from cardiomyocytes [31, 32], it is possible that the free radical induced increase in the resting tension of the heart may be due to the inhibitory effect of xanthine plus xanthine oxidase on the sarcolemmal Ca^{2+}-transport systems and subsequent occurrence of intracellular Ca^{2+}-overload. Whether the depression in contractile force development due to xanthine plus xanthine oxidase is also due to the occurrence of intracellular Ca^{2+}-overload or some other mechanisms cannot be described at the present time.

The major finding of this study concerns the ability of vanadate to prevent the xanthine plus xanthine oxidase induced changes in heart function and sarcolemmal Ca^{2+}-transport activities. Although vanadate is known to promote glucose uptake and oxidation [1–6], its protective effect on xanthine plus xanthine oxidase on heart function cannot be explained on the basis of increased glycolysis. This view is consistent with our observations that vanadate was capable of preventing xanthine plus xanthine oxidase induced depression in contractile force development and increase in resting tension in hearts perfused with glucose-free medium. On the other hand, it is likely that the beneficial effects of vanadate with respect to xanthine plus xanthine oxidase induced changes in heart function and sarcolemmal Ca^{2+} transport systems may be due to its ability to serve as a free radical scavenger. In this regard it should be noted that vanadate was found to exert an inhibitory effect on the EPR signals generated during the xanthine plus xanthine oxidase reaction. Furthermore, vanadate was observed to prevent sarcolemmal Ca^{2+}-pump and Na^+-Ca^{2+} exchange activity upon incubating the sarcolemmal membranes with xanthine plus xanthine oxidase system in a manner similar to that reported for some commonly used free radical scavengers [22, 27, 28]. Thus it appears that vanadate may possess an antioxidant property and its beneficial effects on diabetic heart dysfunction [9–12] may be related to its ability to serve as a scavenger of free radicals. It should be pointed out that under some special conditions in vitro, vanadate has been shown to generate oxygen free radicals [33]; however, other investigators [34] failed to observe any action of reductants, metal chelators or antioxidants on vanadate-stimulated expression of actin and c-Ha-ras mRNA. Perhaps the antioxidant or oxidant property of vanadium compound depends upon the form of vanadate species as well as cell type.

Acknowledgements

The work reported here was supported by the Medical Research Council of Canada (MRC Group in Experimental Cardiology). Dr. T. Matsubara was a postdoctoral fellow of the Medical Research Council during the tenure of this study.

References

1. Tolman EL, Barris E, Burns M, Pansini A, Partridge R: Effects of vanadium on glucose metabolism in vitro. Life Sci 25: 1159–1164, 1979
2. Shecter Y, Karlish SJD: Insulin-like stimulation of glucose oxidation in rat adipocytes by vanadyl (IV) ions. Nature 284: 556–558, 1980
3. Dubyak GR, Kleinzeller A: The insulin-mimetic effects of vanadate in isolated rat adipocytes. Dissociation from effects of vanadate as a (Na^+-K^+)-ATPase inhibitor. J Biol Chem 255: 5306–5312, 1980
4. Duckworth WC, Solomon SS, Liepnieks J, Hamel FG, Hand S, Peavy DE: Insulin-like effects of vanadate in isolated rat adipocytes. Endocrinology 122: 2285–2289, 1988
5. Fantus IG, Ahmad F, Deragon G: Vanadate augments insulin binding and prolongs insulin action in rat adipocytes. Endocrinology 127: 2716–2725, 1990
6. Clark AS, Fagan JM, Mitch WE: Selectivety of insulin-like actions of vanadate on glucose and protein metabolism in skeletal muscle. Biochem J 232: 273–276, 1985
7. Tracey AS, Gresser MJ: Interaction of vanadate with phenol and tyrosine. Implications for the effects of vanadate on systems regulated by tyrosine phosphorylation. Proc Natl Acad Sci USA 83: 609–613, 1986
8. Chernoff J, Li H-C: A major phosphotyrosyl-protein phosphatase from bovine heart is associated with a low molecular weight acid phosphatase. Arch Biochem Biophys 240: 135–145, 1985
9. Heyliger CE, Tahiliani A, McNeill JH: Effects of vanadate on blood glucose and depressed cardiac performance of diabetic rats. Science 227: 1474–1477, 1985
10. Ramanadham S, Mongold JJ, Brownsey RW, Cros GH, McNeill JH: Oral vanadyl sulfate in treatment of diabetes mellitus in rats. Am J Physiol 257: H904–H911, 1989
11. Pederson RA, Ramanadham S, Buchan AMJ, McNeill JH: Long term effects of vanadyl treatment on streptozotocin-induced diabetes in rats. Diabetes 38: 1390–1395, 1989
12. Paulson DJ, Kopp SJ, Tow JP, Peace DG: Effects of vanadate on *in vivo* myocardial reactivity to norepinephrine in diabetic rats. J Pharmacol Exp Therap 240: 529–534, 1987
13. Cantley LC Jr, Aisen P: The fate of cytoplasmic vanadium. Implications on (Na,K)-ATPase inhibition. J Biol Chem 254: 1781–1784, 1979
14. Erdmann E, Werdan K, Krawietz W, Schimitz W, Scholz H: Vanadate and its significance in biochemistry and pharmacology. Biochem. Pharmacol. 33: 945–950, 1984
15. Nechay BR: Mechanism of action of vanadium. Ann Rev Pharmacol Toxicol 24: 501–524, 1984
16. Takeda K, Temma K, Akera T: Inotropic effects of vanadate in isolated rat and guinea-pig heart under conditions which may modify calcium pools involved in contractile activation. J Pharmacol Exp Therap 222: 132–139, 1982
17. Dhalla NS, Pierce GN, Innes IR, Beamish RE: Pathogenesis of cardiac dysfunction in diabetes mellitus. Can J Cardiol 1: 263–284, 1985
18. Sekar N, Kanthasamy A, William S, Balasubramaniyan N, Govindasamy S: Antioxidant effects of vanadate on experimental diabetic rats. Acta Diabetol Lat 27: 285–293, 1990
19. Pitts BJR: Stoichiometry of sodium-calcium exchange in cardiac sarcolemmal vesicles. J Biol Chem 254: 6232–6235, 1979
20. Lowry OH, Rosenbrough NJ, Farr AL, Randall RJ: Protein measurement with the Folin phenol reagent. J Biol Chem 193: 265–275, 1951
21. Dixon IMC, Eyolfson DA, Dhalla NS: Sarcolemmal Na^+-Ca^{2+} exchange activity in hearts subjected to hypoxia reoxygenation. Am J Physiol 253: H1026–H1034, 1987
22. Kaneko M, Elimban V, Dhalla NS: Mechanism for depression of heart sarcolemmal Ca^{2+}-pump by oxygen free radicals. Am J Physiol 257: H804–H811, 1989
23. Tausky H, Shorr E: A microcolorimetric method for the estimation of inorganic phosphorus. J Biol Chem 202: 678–685, 1953
24. Zang LY, Misra HP: EPR kinetic studies of superoxide radicals generated during the autooxidation of 1-methyl-4-phenyl-2,3-dihydro-pyridinium, a bioactivated intermediate of Parkinsonian-inducing neurotoxin 1-methyl-4-phenyl-1,2,3,6-tetrahydropyridine. J Biol Chem 267: 23601–23608, 1992
25. Finkelstein E, Rogen GM, Rauckman EJ: Spin trapping of superoxide and hydroxyl radical: Practical aspect. Arch Biochem Biophys 200: 1–16, 1980
26. Rabani J, Nielsen SO: Absorption spectrum and decay kinetics of $O_2^{\cdot -}$ and $HO_2\cdot$ in aqueous solution by pulse radiolysis. J Phys Chem 73: 3736–3744, 1969
27. Kaneko M, Beamish RE, Dhalla NS: Depression of heart sarcolemmal Ca^{2+}-pump activity by oxygen free radicals. Am J Physiol 256: H368–H374, 1989
28. Hata T, Kaneko M, Beamish RE, Dhalla NS: Influence of oxygen free radicals on heart sarcolemmal Na^+-Ca^{2+} exchange. Coronary Artery Disease 2: 397–407, 1991
29. Burton KP, McCord JM, Ghai G: Myocardial alterations due to free radical generation. Am J Physiol 246: H776–783, 1984
30. Gupta M, Gameiro A, Singal PK: Reduced vulnerability of the hypertrophied rat heart to oxygen-radical injury. Can J Physiol Pharmacol 65: 1157–1164, 1987
31. Reuter H: Exchange of calcium ions in the mammalian myocardium: Mechanisms and physiological significance. Circ Res 34: 599–605, 1974
32. Dhalla NS, Pierce GN, Panagia V, Singal PK, Beamish RE: Calcium movements in relation to heart function. Basic Res Cardiol 77: 117–139, 1982
33. Kalyani P, Vijaya S, Ramasarma T: Characterization of oxygen free radicals generated during vanadate-stimulated NADH oxidation. Mol Cell Biochem 111: 33–40, 1992
34. Yin X, Davison AJ, Tsang SS: Vanadate induced gene expression in mouse C127 cells: Roles of oxygen derived active species. Mol Cell Biochem 115: 85–96, 1992

Molecular and Cellular Biochemistry **153:** 87–94, 1995.

In vivo effects of vanadate on hepatic glycogen metabolizing and lipogenic enzymes in insulin-dependent and insulin-resistant diabetic animals

Ramji L. Khandelwal and Subbiah Pugazhenthi

Department of Biochemistry, College of Medicine, University of Saskatchewan, 107 Wiggins Road, Saskatoon, Saskatchewan, S7N 5E5, Canada

Abstract

The insulin-mimetic action of vanadate is well established but the exact mechanism by which it exerts this effect is still not clearly understood. The role of insulin in the regulation of hepatic glycogen metabolizing and lipogenic enzymes is well known. In our study, we have, therefore, examined the effects of vanadate on these hepatic enzymes using four different models of diabetic and insulin-resistant animals. Vanadate normalized the blood glucose levels in all animal models. In streptozotocin-induced diabetic rats, the amount of liver glycogen and the activities of the active-form of glycogen synthase, both active and inactive-forms of phosphorylase, and lipogenic enzymes like glucose 6-phosphate dehydrogenase and malic enzyme were decreased and vanadate treatment normalized all of these to near normal levels. The other three animal models (db/db mouse, sucrose-fed rats and fa/fa obese Zucker rats) were characterized by hyperinsulinemia, hypertriglyceridemia, increases in activities of lipogenic enzymes, and marginal changes in glycogen metabolizing enzymes. Vanadate treatment brought all of these values towards normal levels. It should be noted that vanadate shows differential effects in the modulation of lipogenic enzymes activities in type I and type II diabetic animals. It increases the activities of lipogenic enzymes in streptozotocin-induced diabetic animals and prevents the elevation of activities of these enzymes in hyperinsulinemic animals. The insulin-stimulated phosphorylation of insulin receptor β subunit and its tyrosine kinase activity was increased in streptozotocin-induced diabetic rats after treatment with vanadate. Our results support the view that insulin receptor is one of the sites involved in the insulin-mimetic actions of vanadate. (Mol Cell Biochem **153**: 87–94, 1995)

Key words: vanadate, diabetes, glycogen synthase, phosphorylase, lipogenic enzymes, liver

Introduction

In the past 15 years, the insulin-mimetic effects of vanadium have been well documented by both *in vivo* and *in vitro* studies [1–33]. Salts of vanadium normalize metabolic aberrations in insulin-dependent [1–18] and insulin-resistant [19–22] diabetic animals . Orthovanadate has been shown to exert insulin-like effects in adipose tissue, muscle and liver on metabolic pathways such as glucose transport, glycolysis, glycogen synthesis and lipogenesis [23–33]. This review will focus our studies on the *in vivo* effects of vanadate on the glycogen metabolizing and lipogenic enzymes of liver in animal models for type I and type II diabetes.

The mammalian liver plays a pivotal role in glucose homeostasis. Hepatic glycogen metabolism is one of the processes that plays an important role in maintenance of normal blood glucose [34, 35]. Glycogenesis and glycogenolysis pathways are regulated by two key enzymes, glycogen synthase and phosphorylase. The activities of these enzymes are regulated by interconversion between active and inactive forms involving phosphorylation-dephosphorylation of enzymes and by allosteric effectors [34, 35]. Insulin activates glycogenesis by stimulating the formation of glycogen synthase a (active form) through an increase in activity of synthase phosphatase, and this process is counteracted by anti-insulin hormones (for e.g. glucagon, catecholamines,

Address for offprints: R.L. Khandelwal, Department of Biochemistry, University of Saskatchewan, 107 Wiggins Road, Saskatoon, Saskatchewan, S7N 5E5, Canada vasopressin etc.) [34–40]. On the other hand, insulin inhibits glycogenolysis but the exact mechanism for this is still unknown [39, 41–43]. It is well known that insulin stimulates lipogenesis by stimulating the activities of lipogenic enzymes such as glucose 6-phosphate dehydrogenase, malic enzyme, ATP-citrate lyase and acetyl CoA carboxylase [44–46].

Although, it is now well established that metabolic abnormalities associated with insulin-dependent and insulin-resistant diabetic animals can be normalized by oral administration of vanadate, the mechanism by which vanadate exerts its insulin-like effect is not clearly understood. There are conflicting reports in the literature whether insulin receptor is involved in vanadate action. Some reports have implicated the involvement of insulin receptor for vanadate action through the activation of tyrosine kinase activity [25, 26, 28, 47, 48] whereas others have reported no effect of this metal ion on insulin receptor [18, 29, 49]. The results on vanadate's action on insulin receptor properties is also included in this paper.

Animal models for insulin-dependent and insulin-resistant diabetes mellitus

In this study, four different animal models were employed. The general characteristics of these animals are given in Table 1. The administration of streptozotocin specifically destroys pancreatic islet β cells and within 1–3 days produces insulin deficiency, hyperglucagonemia and hyperglycemia. Because of the ease with which this insulin-dependent (type I, IDDM) diabetic state can be induced, this animal model has been routinely used in most studies. The genetically diabetic C57BL/KsJ-db/db (commonly referred as db/db mouse) is a good model for non-insulin dependent diabetes mellitus (type II, NIDDM) and is characterized by hyperglycemia, hyperinsulinemia and obesity. These animals are almost normal till the age of 3–4 weeks and then start to develop diabetic characteristics [50–53]. A model for insulin resistance can easily be produced by feeding a high sucrose (or fructose) diet to rats [22]. These animals are characterized by obesity, hypertriglyceridemia, hyperinsulinemia, mild hyperglycemia with an impaired glucose tolerance test. Genetic fa/fa obese Zucker rat is a model for obesity with mild diabetes [54–56]. These animals are hyperinsulinemic, hypertriglyceridemic, mildly hyperglycemic with impaired glucose tolerance. The two latter animal models provide characteristics similar to those found at early stages of type II diabetes. Since no animal model reflects all the features of human diabetes, there is a need to use several animal models especially for type II diabetes. These four animal models, therefore, provide the characteristics generally found in human beings in both types of diabetes.

Table 1. General characteristics of animal models

Animals	General characteristics
Streptozotocin-induced diabetic rats	Model for type-1 diabetes (IDDM) Hyperglycemic Hypoinsulinemic Ketotic
db/db mouse	Model for type-2 diabetes (NIDDM) Hyperglycemic Hyperinsulinemic Obese Insulin resistant
High-sucrose fed rats	Model for insulin resistance Hyperinsulinemic Hypertriglyceridemic Obese Impaired glucose tolerance
fa/fa obese Zucker rats	Model for obesity Insulin resistant Hyperinsulinemic Hypertriglyceridemic Impaired glucose tolerance

Plasma levels of glucose, insulin and triacylglycerols

The results on plasma levels of glucose, insulin and triacylglycerols in different animal models with and without vanadate treatment are presented in Table 2. As expected, all models of diabetic animals were hyperglycemic and vanadate treatment of these animals almost normalized plasma glucose levels similar to respective control animals. In type I diabetic animals, plasma insulin levels are lower whereas in type II diabetic animals plasma insulin levels are several fold higher than their respective control animals. Vanadate treatment of type I diabetic animals caused a slight but insignificant increase in plasma insulin level. The treatment of type II diabetic hyperinsulinemic animals by vanadate lowered plasma insulin levels to varying degrees in different animal models. The vanadate treatment of normal (or lean) control animals of all groups lowered plasma insulin with little or no effect on plasma glucose (Table 2), [8, 9]. This is significant since under these conditions no changes in any insulin-sensitive metabolic processes have been observed. There is so far no evidence that vanadate decreases insulin release and, therefore, it would indicate that this metal ion potentiates the actions of endogenous insulin. Similar *in vivo* observation on levels of plasma glucose and insulin after vanadate treatment of type I diabetic animals has been made by many investigators [3–6, 8]. Meyerovitch *et al.* [4] indicated that such effect was reversible but another study by Ramanadham *et al.* [57] pointed out that oral administration of vanadyl sulphate to diabetic rats for 3 weeks sustained improvement in glu-

Table 2. Effect of vanadate on plasma levels of glucose, insulin and triacylglycerols in insulin-deficient and insulin-resistant animals

Animals	Glucose (mM)	Insulin (pM)	Triacylglycerols (mM)	Reference Number
Normal control rats	6.7 ± 0.2	283.6 ± 25.6	N.D.	8, 9
Vanadate-treated normal rats	6.3 ± 0.5	89.5 ± 7.8**	N.D.	
STZ-diabetic rats	31.7 ± 0.8**	48.4 ± 4.3**	N.D.	
Vanadate-treated diabetic rats	10.9 ± 1.1[c]	75.7 ± 14.7	N.D.	
Normal control (starch-fed) rats	7.8 ± 0.2	349.0 ± 23.7	2.8 ± 0.1	22
Vanadate-treated normal rats	7.3 ± 0.4	139.0 ± 21.6**	1.8 ± 0.1*	
Sucrose-fed rats	10.3 ± 0.5**	755.0 ± 48.8**	6.6 ± 0.8**	
Vanadate-treated sucrose-fed rats	7.5 ± 0.6[b]	156.0 ± 30.8[c]	2.3 ± 0.3[c]	
Lean control rats	8.4 ± 0.6	567.0 ± 60.0	2.0 ± 0.2	75, 76
Vanadate-treated lean rats	6.2 ± 0.6*	127.0 ± 8.0**	0.8 ± 0.1**	
Obese Zucker rats	9.4 ± 0.8	2804.0 ± 449.0**	77.7 ± 16.8**	
Vanadate-treated obese Zucker rats	7.1 ± 0.4[b]	682.0 ± 165.0[c]	25.6 ± 6.4[c]	
Lean control mice	7.5 ± 0.6	162.8 ± 16.9	N.D.	20
Vanadate-treated lean mice	6.1 ± 0.5	88.3 ± 11.8*	N.D.	
Diabetic db/db mice	24.2 ± 1.1**	1763.3 ± 234.7**	N.D.	
Vanadate-treated diabetic db/db mice	7.2 ± 1.1[c]	918.7 ± 137.7[c]	N.D.	

N.D. = not determined. Values are means ± S.E. Comparisons were made by student's *t* test. *$p < 0.01$; **$p < 0.001$ vs. normal (or lean) control animals. [b]$p < 0.01$; [c]$p < 0.001$ vs. diabetic (or sucrose-fed or obese Zucker) animals.

cose homeostasis at least for 13 weeks after withdrawal of vanadyl sulphate. It is still debatable which of the salt of vanadium orthovanadate, metavanadate or vanadyl sulphate is better for observing these effects with minimal toxic effects. Ramanadha *et al.* [15, 57] and Cros *et al.* [16] have strongly argued in favor of treatment with vanadyl sulphate. However, these investigators have always used mild diabetic animals by starting the treatment 3 days after streptozotocin injection. Other investigators who have tested vanadate started treatment after 7–10 days of streptozotocin injection. It is, therefore, still to be determined whether this sustained treatment is due to better protection of β cells by initiating an early treatment.

A comparison of effect of vanadate treatment on two types of diabetic animals is interesting. In type I diabetics, vanadate has little or no effect on insulin levels whereas in type II diabetics, insulin levels are significantly reduced. The mechanisms for such diverse effects by vanadate in two different types of animals might be different and has to be determined by future studies. It is probable that in type I diabetics, vanadate normalizes metabolic processes by potentiating the actions of residual insulin whereas in type II diabetics, it could be due to increased sensitivity of insulin leading to decreased insulin level. A decrease in food consumption leading to decreased body weight has been also observed [3–9]. Meyerovitch *et al.* [58] have suggested that anorexigenic effect of vanadate could be due to an increase in glucose uptake by the central nervous system.

The type II diabetic animals are associated with hyperinsulinemia and hypertriglyceridemia (Table 2). Vanadate treatment significantly lowered levels of plasma triacylglycerols in two animal models examined. This again could be related to decrease in plasma insulin levels and a decrease in food consumption (Table 2). As will be described later in this paper, these animals also had decreased levels of lipogenic enzymes as compared to untreated animals.

Glycogen storage levels and activities of glycogen synthase and phosphorylase

Insulin is one of the important hormones involved in the synthesis and maintenance of liver glycogen. In type I diabetes, hepatic glycogen levels are decreased because of low circulating insulin (Table 3). In type II diabetic animals, however, glycogen levels are maintained closer to normal levels in spite of the increased rates of glycogenolysis (Table 3). This is because of huge reservoir of substrate supplied by voracious eating habits of animals and also due to an active gluconeogenic pathway in these animals. The treatment of type I diabetic animals by insulin or vanadate normalizes hepatic glycogen stores [6, 8], (Table 3). Insulin or vanadate treatment of type II diabetic animals has no effect on glycogen storage levels but does normalize rate of glycogen turn over [20–22, 59]. In ob/ob mouse, Brichard *et al.* [59] observed an increase in liver glycogen levels after vanadate treatment but these glycogen levels were still lower than those found in other animal models.

Table 3. Effect of vanadate on hepatic levels of glycogen, glycogen synthase and phosphorylase in insulin-deficient and insulin-resistant animals[#]

Animals	Glycogen (mg/g)	Glycogen synthase		Phosphorylase		Reference Number
		Total	% active	Total	% active	
Normal control rats	47.7 ± 5.7	1.23 ± 0.13	42 ± 2	138.8 ± 9.3	54 ± 2	8, 9
STZ-diabetic rats	6.7 ± 1.3**	1.16 ± 0.11	14 ± 1**	35.2 ± 2.0**	61 ± 3	
Vanadate-treated diabetic rats	28.4 ± 5.0[c]	1.01 ± 0.08	34 ± 3[c]	85.9 ± 13.5[c]	59 ± 3	
Normal control (starch-fed) rats	42.1 ± 1.5	1.19 ± 0.07	36 ± 4	117.6 ± 7.1	54 ± 3	22
Sucrose-fed rats	37.1 ± 2.8	1.20 ± 0.08	36 ± 4	127.9 ± 8.7	54 ± 4	
Vanadate-treated sucrose-fed rats	38.5 ± 3.0	1.22 ± 0.08	34 ± 3	116.1 ± 7.9	57 ± 3	
Lean control rats	36.6 ± 2.5	1.75 ± 0.10	29 ± 1	184.0 ± 9.0	23 ± 1	
Obese Zucker rats	34.8 ± 4.4	2.83 ± 0.14**	24 ± 1	254.0 ± 13.0**	23 ± 1	
Vanadate-treated obese Zucker rats	42.0 ± 3.0	2.26 ± 0.19[a]	24 ± 2	178.0 ± 10.0[c]	23 ± 1	
Lean control mice	45.8 ± 2.5	1.26 ± 0.06	24 ± 1	133.0 ± 10.8	46 ± 3	20
Diabetic db/db mice	48.3 ± 1.7	1.85 ± 0.10**	20 ± 2	271.9 ± 14.9**	42 ± 3	
Vanadate-treated diabetic db/db mice	45.0 ± 3.8	1.98 ± 0.11	21 ± 2	193.6 ± 10.5[c]	50 ± 3	

[#]Vanadate treatment of normal (or lean) control animals did not significantly change values for these parameters and therefore are not included in this table. These values are reported in original articles. All values are means ± S.E. Comparisons were made by student's *t* test. **$p < 0.001$ vs. normal (or lean) control animals. [c]$p < 0.001$ vs. diabetic (or sucrose-fed or obese Zucker) animals.

The activity of glycogen synthase, the rate limiting enzyme for glycogen synthesis, is primarily regulated by phosphorylation-dephosphorylation mechanism by glucose and insulin. There is some evidence that this regulation is through the control of synthase phosphatase activity [34–40]. Glycogen synthase, a highly phosphorylated enzyme, is phosphorylated on many sites by several known protein kinases. However, it is still debatable whether one or more of these protein kinases are regulated by insulin and/or glucose. In type I diabetic animals, the activity of glycogen synthase active form is low and this can be normalized by insulin or vanadate [8, 39], (Table 3). There is no effect of diabetes on the total glycogen synthase activity. There is, however, parallel decrease in synthase phosphatase activity in livers of type I diabetic animals which is also normalized by insulin and vanadate [8, 39]. In type II diabetic animals, there is either no change or an increase in total glycogen synthase activity without any change in percentage active form of the enzyme (Table 3). There were ~45 and ~60% increases in total synthase activity in db/db mouse and fa/fa rats without any change in percentage active form of the enzyme. There have been some contradictory reports on glycogen synthase activity in db/db mouse. Chan *et al.* [60] have reported a similar change in glycogen synthase activity in these animals, whereas Sterns and Benzo [61] did not find any change. We have previously reported that this could be due to a shift in the time of peak of glycogen synthase diurnal rhythm [62]. Vanadate treatment did not alter total synthase activity in db/db mouse but slightly decreased the enzyme activity in fa/fa rats. The percentage active forms were similar in both types of animals after vanadate treatment.

The activity of phosphorylase, the key regulatory enzyme for glycogenolysis, is also regulated by phosphorylation–dephosphorylation. The enzyme is activated by hormones such as glucagon, which activate the cyclic AMP cascade system, and by those such as epinephrine, vasopressin and angiotensin II, which cause an increase in intracellular calcium concentration [36]. Insulin antagonizes the activation of phosphorylase by these hormones. In type I diabetes, activity (active form and total) of phosphorylase decreases without any change in percentage activity form of the enzyme [8, 39], (Table 3). In all animal models of type II diabetes, activity of total and the active-form are increased, also without any change in percentage active form of the enzyme. We have previously demonstrated that the total phosphorylase activity is mainly governed by changes in enzyme protein concentration and that glucagon to insulin molar ratio to values higher than 1.0 causes a significant decrease in the liver phosphorylase protein level [42]. The exact mechanism by which glucagon, insulin and/or both regulate the total phosphorylase activity (and total protein concentration) has to be still worked out. Vanadate treatment of all types of diabetic animals, however, moves the total phosphorylase activity towards the levels found in their respective control animals again without any change in percentage active form of the enzyme (Table 3).

Other carbohydrate metabolizing enzymes

In insulin-dependent diabetes, liver normally switches from a net glucose utilizing organ to one which produces glucose. This is accomplished, in addition to a decrease in net synthesis of glycogen, with a decrease in glycolytic flux and an increase in gluconeogenic flux. Four additional enzymes are affected in this process. The activities of glucokinase and phosphofructokinase are decreased and activities of phosphoenolpyruvate carboxykinase and glucose-6-phosphatase are increased. Vanadate normalizes activities of all four enzymes in STZ-diabetic rats [2, 5, 7, 31, 32]. Activation of glycolysis by vanadate has been observed in hepatocytes from normal and STZ-diabetic rats [31, 32]. Expression of the gene for phosphoenolpyruvate carboxykinase is inhibited by this agent in rat hepatoma cells [63].

Lipogenic enzymes

The role of insulin in the regulation of lipogenic enzymes is well known [44–46]. Glucose-6-phosphate dehydrogenase and malic enzyme activities are decreased in type I diabetic animals and vanadate treatment restores these levels back to normal levels (Table 4). In db/db mouse, the alteration in activities of these enzymes showed different patterns, as indicated by a 50% decrease in glucose-6-phosphate dehydrogenase and 160% increase in malic enzyme activities. Such differential changes in ATP-citrate lyase activity (similar to malic enzyme in our study) and glucose-6-phosphate dehydrogenase activity in livers of db/db mouse were previously reported by Coleman and Hummel [51]. The decrease of glucose-6-phosphate dehydrogenase indicates that less glucose is being metabolized via hexose monophosphate pathway in these animals. It is probable that glucose-6-phosphate dehydrogenase serves a function different from those of malic enzyme and ATP-citrate lyase in db/db mouse. Surprisingly, vanadate treatment led to normalization of both enzyme activities. Furthermore, vanadate has opposite effects on hepatic lipogenic enzymes in insulin-deficient and insulin-resistant animals. In insulin-deficient animals, it stimulates lipogenesis by activating (or increase in amounts of) lipogenic enzymes whereas in insulin-resistant animals it inhibits the enhanced lipogenesis by decreasing the amounts of (or inhibiting) lipogenic enzymes. However, these effects are to bring enzyme activities towards levels found in normal animals. In insulin-dependent diabetic animals, stimulation of lipogenesis by vanadate is accompanied with a decrease in lipolysis [27, 29].

Table 4. Effect of vanadate on hepatic levels of lipogenic enzymes in insulin-deficient and insulin-resistant animals[#]

Animals	Glucose-6-P dehydogenase	Malic enzyme	Ref. No.
Normal control rats	7.8 ± 0.7	3.9 ± 0.4	9
STZ-diabetic rats	3.8 ± 0.4**	1.5 ± 0.2**	
Vanadate-treated diabetic rats	7.0 ± 1.3[a]	4.5 ± 0.5[c]	
Lean control mice	4.9 ± 0.5	19.0 ± 2.6	20
Diabetic db/db mice	2.5 ± 0.3**	48.4 ± 6.2*	
Vanadate-treated diabetic db/db mice	5.1 ± 0.9[b]	19.7 ± 1.4[c]	

[#]Vanadate treatment of normal (or lean) control animals did not significantly change values for these parameters and therefore are not included in this table. These values are reported in original articles. All values are means ± S.E. Comparisons were made by student's *t* test. *$p < 0.01$; **$p < 0.001$ vs. normal (or lean) control animals. [a]$p < 0.05$; [b]$p < 0.01$; [c]$p < 0.001$ vs. diabetic animals.

Mechanism for vanadate action

It is debatable whether vanadate exerts its insulin-mimetic effects through altering the properties of insulin receptor or at some post-receptor level or both. In our laboratory, we have examined the *in vivo* effects of vanadate on properties of insulin receptor in streptozotocin-induced diabetic rat livers. In diabetic animals, insulin binding increased to 195% and the autophosphorylation of the β subunit of insulin receptor and its tyrosine kinase activity decreased by approximately 60%. The vanadate treatment of diabetic animals caused a significant decrease in insulin binding and partially restored the autophosphorylation of β subunit of insulin receptor and its tyrosine kinase activity. These results support the view that insulin receptor is not the sole but is one of the sites involved in the insulin-mimetic actions of vanadate. There are reports which have correlated the insulin-like effects of vanadate to the stimulation of insulin receptor kinase [25, 26, 28, 47, 48]. Vanadate has been shown to increase the autophosphorylation of β subunit and activation of tyrosine kinase activity in isolated adipocytes and in 3T3-L1 adipocytes [25, 26]. Fantus *et al.* [28] showed that pervanadate, a more potent derivative of vanadate, mimics insulin action by stimulating the insulin receptor kinase activity. Tyrosine kinase activity of the partially purified insulin receptor from human placenta is

Table 5. Effect of vanadate on insulin receptor properties in streptozotocin-induced diabetic rat livers[#]

Animals	Insulin binding	Autophosphorylation	Tyrosine kinase activity
Normal control rats	100	100	100
STZ-diabetic rats	195 ± 15*	43 ± 2**	36 ± 3**
Vanadate-treated diabetic rats	140 ± 11[a]	68 ± 2[c]	54 ± 2[b]

Measurements of autophosphorylation and tyrosine kinase activity in different groups were carried out in the presence of 100 nM insulin using receptor preparation with equal insulin binding capacities. For all parameters, values for normal control rats were taken as 100. The data are taken from reference 48. [#]Vanadate treatment of normal control animals did not significantly change values for these parameters and therefore are not included in this table. All values are means ± S.E. Comparisons were made by student's *t* test. *$p < 0.01$; **$p < 0.001$ vs. normal control rats. [a]$p < 0.05$; [b]$p < 0.01$; [c]$p < 0.001$ vs. diabetic rats.

enhanced *in vitro* by vanadate [47]. The mechanism of activation of tyrosine kinase is probably through the inhibition of phosphotyrosine phosphatase [64]. Phosphotyrosine phosphatase dephosphorylates the β subunit of insulin receptor and terminates its action [65]. In diabetic rat, there is elevation in hepatic phosphotyrosine phosphatase activity which is decreased after treatment with vanadate [66, 67]. The control of phosphotyrosine phosphatase could, therefore, play an important role in the regulation of insulin receptor activity.

There are, however, some reports which indicate that vanadate can exert insulin-like effects without stimulating the receptor kinase suggesting that vanadate might be acting at some post-receptor level [18, 29, 49, 68, 69]. Blondel *et al.* [49] did not find any change in the autophosphorylation in diabetic rats induced by neonatal STZ after vanadate treatment. In partially pancreatectomized rats, there was an increase in basal but not in insulin-induced tyrosine kinase activity after chronic administration of vanadate [70]. However, in these two studies the animal models had mild diabetes and there was no impairment of receptor kinase when compared to the nondiabetic animals. When such models are used, it is difficult to conclude that vanadate does not act at the insulin receptor level *in vivo* since this agent is known to normalize the abnormalities in diabetic animals [5, 8, 9, 19–21]. Venkatesan *et al.* [18] using similar diabetic rats did not observe significant improvement in receptor kinase activity which could be due to the fact that a different salt of vanadium, i.e. vanadyl sulfate was used.

Vanadate mimics insulin action in diverse metabolic pathways. It is, therefore, probable that sites of action of vanadate could both be at the insulin receptor as well as at some points in the cascade of insulin-induced signal transduction. Evidence for both of these sites are presently reported in the literature. Further work is needed to be carried out to precisely understand the mechanism of vanadate action in manifesting *in vivo* insulin-like actions.

Body weight gain

In vivo administration of vanadate leads to a decrease in body weight gain in animals [3, 4, 8, 9]. The beneficial effects of vanadate could be partially due to reduction in body weight. However, Brichard *et al.* [21] and our recent data (Pugazhenthi *et al.*, unpublished observation) have demonstrated by using pair-fed control experiments that vanadate-induced improvement in glucose homeostasis is not secondary to decreased food intake. Also, in mild diabetic animals, low doses of vanadate exert the insulin-mimetic effects without altering the body weight gain [49]. This insulin-mimetic agent vanadate has been generally administered through drinking water. Water and food consumption are less in these animals. In one of our studies, we overcame this problem partially by mixing vanadate with the diet [14]. In these treated rats fluid intake was normal and food intake improved considerably in the later part of the treatment period.

Conclusions

There is an overwhelming evidence that oral administration of vanadate normalizes abnormalities associated with insulin-dependent and insulin-resistant diabetes mellitus. In this paper, we have reviewed information on the enzymes involved in the storage of glycogen and fat in liver. There is still on-going debate on possible toxic effects of this metal ion [2, 71, 72] and its precise mechanism of action. For its mechanism of action, it is probable that vanadate manifests its actions through insulin receptor as well as at some post-receptor level in cascade of insulin signal transduction pathway. Further studies are needed to investigate whether lower doses of vanadate in combination with insulin or other hypoglycemic drugs can be administered for normalization of glucose homeostasis and other abnormalities associated with diabetes mellitus. Rosetti *et al.* [73] have reported that vanadate administration in combination with other trace elements like lithium, zinc and magnesium provides an effective therapy in diabetic rats. The use of a chelator could also reduce the tissue accumulation and the toxicity of vanadate without altering its insulin-mimetic action [74].

Acknowledgment

The work in the authors' laboratory was supported by the Medical Research Council of Canada.

References

1. Shechter Y: Insulin-mimetic effects of vanadate. Possible implications for future treatment of diabetes. Diabetes 39: 1–5, 1990
2. Brichard SM, Lederer J, Henquin JC: The insulin-like properties of vanadium: A curiosity or a perspective for the treatment of diabetes? Diabete & Metabolisme 17: 435–440, 1991
3. Heyliger CE, Tahiliani AG, McNeill JH: Effect of vanadate on elevated blood glucose and depressed cardiac performance of diabetic rats. Science 227: 1474–1477, 1985
4. Meyerovitch J, Farfel Z, Sack J, Shechter Y: Oral administration of vanadate normalizes blood glucose levels in streptozotocin-treated rats. J Biol Chem 262: 6658–6662, 1987
5. Gil J, Miralpeix M, Carreras J, Bartrons R: Insulin-like effects of vanadate on glucokinase activity and fructose 2,6-bisphosphate levels in the liver of diabetic rats. J Biol Chem 263: 1868–1871, 1988
6. Brichard SM, Okitolanda W, Henquin JC: Long term improvement of glucose homeostasis by vanadate treatment in diabetic rats.

Endocrinology 123: 2048–2053, 1988

7. Schulz LO: Suppression of the hepatic glucose 6-phosphate system in diabetic rats by vanadate. Ann Nutr Metab 32: 289–296, 1988
8. Pugazhenthi S, Khandelwal RL: Insulinlike effects of vanadate on hepatic glycogen metabolism in nondiabetic and streptozotocin-induced diabetic rats. Diabetes 39: 821–827, 1990
9. Pugazhenthi S, Khandelwal RL, Angel JF: Insulin-like effect of vanadate on malic enzyme and glucose 6-phosphate dehydrogenase activities in streptozotocin-induced diabetic rat liver. Biochim Biophys Acta 1083: 310–312, 1991
10. Bollen M, Miralpeix M, Ventura F, Toth B, Bartrons R, Stalmans W: Oral administration of vanadate to streptozotocin-diabetic rats restores the glucose-induced activation of liver glycogen synthase. Biochem J 267: 269–271, 1990
11. Miralpeix M, Carballo E, Bartrons R, Crepin K, Hue L, Rosseau GG: Oral administration of vanadate to diabetic rats restores liver 6-phosphofructo-2-kinase content and mRNA. Diabetologia 35: 243–248, 1992
12. Rosseti L, Laughlin MR: Correction of chronic hyperglycemia with vanadate, but not with phlorizin, normalizes *in vitro* glycogen synthase activity in diabetic skeletal muscle. J Clin Invest 84: 892–899, 1989
13. Strout HV, Vicario PP, Biswas C, Saperstein R, Brady EJ, Pilch PF, Berger J: Vanadate treatment of streptozotocin diabetic rats restores expression of the insulin-responsive glucose transporter in skeletal muscle. Endocrinology 126: 2728–2732, 1990
14. Pugazhenthi, S, Angel JF, Khandelwal RL: Effects of high sucrose diet on the insulin-like effects of vanadate in diabetic rats. Mol Cell Biochem 122: 77–84, 1993
15. Ramanadham S, Mongold JJ, Brownsey RW, Cros GH, McNeill JH: Oral Vanadyl sulfate in treatment of diabetes mellitus in rats. Am J Physiol 257: H904–H911, 1989
16. Cros G, Mongold JJ, Serrano JJ, Ramanadham S, McNeill JH: Effects of vanadyl derivatives on animal models of diabetes. Mol Cell Biochem 109: 163–166,1992
17. Sakurai H, Tsuchiya K, Nukatsuda M, Sofue M, Kawada J: Insulin-like effects of vanadyl ion on streptozotocin-induced diabetic rats. J Endocrinol 126: 451–459, 1990
18. Venkatesan N, Avidan A, Davidson MB: Antidiabetic action of vanadyl in rats independent of *in vivo* insulin-receptor kinase activity. Diabetes 40: 492–498, 1991
19. Meyerovitch J, Rothenberg P, Shechter Y, Bonner-Weir S, Kahn CR: Vanadate normalizes hyperglycemia in two mouse models of non-insulin-dependent diabetes mellitus. J Clin Invest 87: 1286–1294, 1991
20. Pugazhenthi S, Angel JF, Khandelwal RL: Long-term effects of vanadate treatment on glycogen metabolizing and lipogenic enzymes of liver in genetically diabetic (db/db) mice. Metabolism 40: 941–946, 1991
21. Brichard SM, Pottier AM, Henquin JC: Long-term improvement of glucose homeostasis by vanadate in obese hyperinsulinemic fa/fa rats. Endocrinology 125: 2510–2516, 1989
22. Pugazhenthi S, Angel JF, Khandelwal RL: Effects of vanadate administration on the high sucrose diet-induced aberrations in normal rats. Mol Cell Biochem 122: 69–75, 1993
23. Tolman EL, Barris E, Burns M, Pansini A, Partridge R: Effects of vanadium on glucose metabolism *in vitro*. Life Sci 25: 1159–1164, 1979
24. Dubyak GR, Kleinzeller A: The insulin-mimetic effects of vanadate in isolated rat adipocytes. J Biol Chem 255: 5306–5312, 1980
25. Tamura S, Brown TA, Whipple JH, Fujita-Yamaguchi Y, Dubler RE, Cheng K, Larner J: A novel mechanism for the insulin-like effect of vanadate on glycogen synthesis in rat adipocytes. J Biol Chem 259: 6650–6658, 1984
26. Bernier M, Laird DM, Lane MD: Effect of vanadate on the cellular accumulation of pp 15, an apparent product of insulin receptor tyrosine kinase action. J Biol Chem 263: 13626–13634, 1988
27. Duckworth WC, Solomon SS, Liepnieks J, Hamel FG, Hand S and Peavy DE: Insulin-like effects of vanadate, in isolated rat adipocytes. Endocrinology 122: 2285–2289, 1988
28. Fantus IG, Kadota SI, Deragon G, Foster B, Posner BI: Pervanadate (peroxides of vanadate) mimics insulin action in rat adipocytes via activation of the insulin receptor tyrosine kinase. Biochemistry 28: 8864–8871, 1989
29. Mooney RA, Bordwell KL, Luhowskyi S, Casnellie JE: The insulin-like effect of vanadate on lipolysis in rat adipocytes is not accompanied by an insulin-like effect on tyrosine phosphorylation. Endocrinology 124, 422–429, 1989
30. Clausen T, Andersen TL, Sturup-Johansen M, Petkova O: The relationship between the transport of glucose and cations across cell membranes in isolated tissues. The effect of vanadate on ^{45}Ca-efflux and sugar transport in adipose tissue and in skeletal muscle. Biochim Biophys Acta 646: 261–267, 1981
31. Gomez-Foix AM, Rodriguez-Gil JE, Fillat C, Guinovart JJ, Bosch F: Vanadate raises fructose 2,6-bisphosphate concentrations and activates glycolysis in rat hepatocytes. Biochem J 255: 507–512, 1988
32. Rodriguez-Gil JE, Gomez-Foix AM, Fillat C, Bosch F, Guinovart JJ: Activation by vanadate of glycolysis in hepatocytes from diabetic rats. Diabetes 40: 1355–1359, 1991
33. Miralpeix M, Decaux JF, Kahn A, Bartrons R: Vanadate induction of L-type pyruvate kinase mRNA in adult rat hepatocytes in primary culture. Diabetes 40: 462–464, 1991
34. Hers HG: The control of glycogen metabolism in the liver. Ann Rev Biochem 45: 167–189, 1976
35. Van de Werve G, Jeanrenaud B: Liver glycogen metabolism: An overview. Diabetes/Metabolism Reviews 3: 47–78, 1987
36. Cohen P: The role of protein phosphorylation in the hormonal control of enzyme activity. Eur J Biochem 151: 439–448, 1985
37. Roach P: Control of glycogen synthase by hierarchal protein phosphorylation. FASEB J 4: 2961–2968, 1990
38. Roesler WJ, Pugazhenthi S, Khandelwal RL: Hepatic glycogen metabolism in the db/db mouse. Mol Cell Biochem 92: 99–106, 1990
39. Khandelwal RL, Zinman SM, Zebrowski EJ: The effect of streptozotocin-induced diabetes and of insulin supplementation on glycogen metabolism in rat liver. Biochem J 168: 541–548, 1977
40. Lavoie L, Bollen M, Stalmans W, Van de Werve G: Increased synthase phosphatase activity is responsible for the super-activation of glycogen synthase in hepatocytes from fasted obese Zucker rats. Endocrinology 129: 2674–2678, 1991
41. Hems DA, Whitton PD: Control of hepatic glycogenolysis. Physiol Rev 60: 1–50, 1980
42. Roesler WJ, Khandelwal RL: Regulation of rat liver glycogen phosphorylase concentration by *in vivo* relative levels of glucagon and insulin. Endocrinology 121, 227–232, 1987
43. Roesler WJ, Nijjar MS, Khandelwal RL: The rate of degradation of liver glycogen phosphorylase is specifically decreased in the C57BL/KsJ-db/db mouse. Mol Cell biochem 87: 147–152, 1989
44. Spence JT, Pitot HC: Induction of lipogenic enzymes in primary cultures of rat hepatocytes. Relationship between lipogenesis and carbohydrate metabolism. Eur J Biochem 128: 15–20, 1982
45. Katsurada A, Iritani N, Fukuda H, Matsumura Y, Noguchi T, Tanaka T: Effects of insulin and fructose on transcriptional and post-transcriptional regulation of malic enzyme synthesis in diabetic rat liver. Biochim Biophys Acta 1004: 103–107, 1989
46. Katsurada A, Iritani N, Fukuda H, Matsumura Y, Noguchi T, Tanaka T: Effects of nutrients and insulin on transcriptional and post-

transcriptional regulation of glucose-6-phosphate dehydrogenase synthesis in rat liver. Biochim Biophys Acta 1006: 104–111, 1989
47. Gherzi R, Caratti Ca, Andraghetti G, Bertolini S, Montemurro A, Sesti G, Cordera R: Direct modulation of insulin-receptor protein tyrosine kinase by vanadate and anti-insulin receptor monoclonal antibodies. Biochem Biophys Res Commun 152: 1474–1480, 1988
48. Pugazhenthi S, Khandelwal RL: Does the insulin-mimetic action of vanadate involve insulin receptor kinase? Mol Cell Biochem 217/218: 211–218,1993
49. Blondel O, Simon J, Chevalier B, Portha B: Impaired insulin action but normal insulin receptor activity in diabetic rat liver: effect of vanadate. Am J Physiol 258: E459–E467, 1990
50. Hummel KP, Dickie MM, Coleman DL: Diabetes, a new mutation in the mouse. Science 153: 1127–1128, 1966
51. Coleman, DL, Hummel KP: Studies with the mutation, diabetes, in the mouse. Diabetologia 3: 238–248, 1967
52. Wyse BM, Dulin WE: The influence of age and dietary conditions on diabetes in the db/db mouse. Diabetologia 6: 268–273, 1970
53. Roesler WJ, Khandelwal RL: Age-related changes in hepatic glycogen metabolism in the genetically diabetic (db/db) mouse. Diabetes 34: 395–402, 1985
54. Bray GA: The Zucker-fatty rat: a review. Federation Proc. 36:148–153, 1977
55. Triscari J, Stern JS, Johnson PR, Sullivan AC: Carbohydrate metabolism in lean and obese Zucker rats. Metabolism 28: 183–189, 1979
56. Martin RJ: *In vivo* lipogenesis and enzyme levels in adipose and liver tissues from pair-fed genetically obese and lean rats. Life Sci 14:1447–1453, 1974
57. Ramanadham S, Brownsey RW, Cros GH, Mongold JJ, McNeill JH: Sustained prevention of myocardial abnormalities in diabetic rats following withdrawl from oral vanadyl treatment. Metabolism 38: 1022–1028, 1989
58. Meyerovitch J, Shechter Y, Amir S: Vanadate stimulates *in vivo* glucose uptake in brain and arrests food intake and body weight gain in rats. Physiol Behav 45: 1113–1116, 1989
59. Brichard SM, Bailey CJ, Henquin JC: Marked improvement of glucose homeostasis in diabetic ob/ob mice given oral vanadate. Diabetes 39: 1326–1332, 1990
60. Chan TM, Young KM, Hutson NJ, Brumley FT, Exton JH: Hepatic metabolism of genetically diabetic (db/db) mice. I. Carbohydrate metabolism. Am J Physiol 229: 1702–1712, 1975
61. Sterns SB, Benzo CA: Structural and chemical alterations associated with hepatic glycogen metabolism in genetically diabetic (db/db) and in streptozotocin-induced diabetic mice. Lab Invest 37: 180–187, 1977
62. Roesler WJ, Helgason C, Gulka M, Khandelwal RL: Aberrations in the diurnal rhythms of plasma glucose, plasma insulin, liver glycogen and hepatic glycogen synthase and phosphorylase activities in genetically diabetic (db/db) mice. Horm Metab Res 17: 572–575, 1985
63. Bosch F, Hatzoglou M, Park EA, Hanson RW: Vanadate inhibits expression of the gene for phosphoenolpyruvate carboxykinase (GTP) in rat hepatoma cells. J Biol Chem 265: 13677–13682, 1990
64. Swarup G, Cohen S, Garbers DL: Inhibition of membrane phosphotyrosyl protein phosphatase activity by vanadate. Biochem Biophys Res Commun 107: 1104–1109, 1982
65. King MJ, Sale GJ: Dephosphorylation of insulin-receptor autophosphorylation sites by particulate and soluble phosphotyrosyl protein phosphatases. Biochem J 266: 251–259, 1990
66. Meyerovitch J, Backer JM, Kahn CR: Hepatic phosphotyrosine phosphatase activity and its alterations in diabetic rats. J Clin Invest 84: 976–983, 1989
67. Begum N, Sussman KE, Draznin B: Differential effects of diabetes on adipocyte and liver phosphotyrosine and phosphoserine phosphatase activities. Diabetes 40: 1620–1629, 1991
68. Donofrio F, Le MQU, Chiasson JL, Srivastava AK: Activation of mitogen activated protein (Map) kinases by vanadate is independent of insulin receptor autophosphorylation. FEBS Lett 340: 269–275, 1994
69. Green A: The insulin-like effect of sodium vanadate on adipocyte glucose transport is mediated at a post-insulin-receptor level. Biochem J 238: 663–669, 1986
70. Cordera R, Andraghetti G, DeFronzo RA, Rosseti L: Effect of *in vivo* vanadate treatment on insulin receptor tyrosine kinase activity in partially pancreatectamized diabetic rats. Endocrinology 126: 2177–2183, 1990
71. Paternain JL, Domingo JL, Gomez M, Ortega A, Corbella J: Developmental toxicity of vanadium in mice after oral administration. J Appl Toxicol 10: 181–186, 1990
72. Domingo JL, Gomez M, Llobet JM, Corbella J, Keen CL: Improvement of glucose homeostasis by oral vanadyl or vanadate treatment in diabetic rats is accompanied by negative side effects. Pharmacol and Toxicol 68: 249–253, 1991
73. Rosseti L, Giaccari A, Klein-Robbenhaar E, Vogel LR: Insulinomimetic properties of trace elements and characterization of their *in vivo* mode of action. Diabetes 39: 1243–1250, 1990
74. Domingo JL, Gomez M, Sanchez DJ, Llobet JM, Keen CL: Tiron administration minimizes the toxicity of vanadate but not its insulinmimetic properties in diabetic rats. Life Sci 50: 1311–1317, 1992
75. King MJ, Pugazhenthi S, Khandelwal RL, Sharma RK: Membrane-associated N-myristoyltransferase activity is reduced in obese (fa/fa) Zucker rat liver. Biochem Biophys Res Commun 196: 665–670, 1993
76. Pugazhenthi S, Hussain A, Yu B, Brownsey RW, Angel JF, Khandelwal RL: Vanadate induces normolipidemia and a reduction in levels of hepatic lipogenic enzymes in obese Zucker rat. Mol Cell Biochem 153: 125–129, 1995

Molecular and Cellular Biochemistry **153**: 95–102, 1995.

The relationship between insulin and vanadium metabolism in insulin target tissues

Frederick G. Hamel and William C. Duckworth
Veterans Administration Medical Center, Omaha, NE 68105, USA

Abstract

Vanadium (V) is an orally effective treatment for diabetes, but relatively little is known about the mechanisms controlling its normal metabolism nor the long term pharmacokinetics of oral administration. We have examined the accumulation of V in various organs from rats fed liquid diet for up to 18 days, containing no additional V, 1.6, 80, or 160 μmole/kg/day as either sodium orthovanadate (SOV) or vanadyl sulfate (VS). V content was assayed using a sensitive neutron activation analysis method. The organs of the nonsupplemented animals contained widely varying concentrations (ng of V/g dry tissue weight) with brain < fat < blood < heart < muscle < lung < liver < testes < spleen < kidney. All organs accumulated V in a dose dependent manner. Not all organs showed steady state amount of V at 18 days, so additional rats were fed SOV or VS, switched to control diet, and assayed at 0, 4 and 8 days. From this data we calculated organ half lives of V. Insulin sensitive tissue tissues, such as liver and fat, had shorter half-lives than tissues that are relatively less insulin sensitive, such as spleen, brain and testes. SOV and VS fed rats showed similar patterns, but VS had somewhat shorter t½'s. Additional studies of old and young rats fed control diet for 45 days show accumulation of V in spleen and testes. These results indicate that vanadium metabolism varies widely among different organs, and that insulin, either directly or indirectly has effects on the retention of vanadium. This may have impact on the therapeutic use of vanadium in Type I diabetics with no insulin, or Type II patients who may be relatively hyperinsulinemic. (Mol Cell Biochem **153**: 95–102, 1995)

Key words: vanadium, insulin, radiochemical neutron activation analysis

Introduction

A consensus is forming that vanadium is likely an essential element [1–4], but definitive proof and the assignment of an exact function are still lacking. The majority of studies have found that its function is related to proper growth and development of the organism. Of related interest is the fact that vanadium has been shown to have growth factor-like effects. McKeehan *et al* [5] first showed that vanadium was required for optimal growth of fibroblasts in culture. Vanadium has been shown to have mitogenic responses in human fibroblasts and 3T3 and 3T6 cell lines [6–8]. Vanadium can have effects similar to fibroblast growth factor, epidermal growth factor and insulin [9–12]. Thus it appears that whatever vanadium's mechanism of action, it is related to similar actions of growth factors.

Vanadium's similarity to growth factors is probably best documented by its ability, at pharmacological doses, to mimic the actions of insulin. These effects and its important potential clinical use [13–17] have been demonstrated in studies with diabetic animals. Orally administered vanadate can normalize blood glucose levels in streptozotocin diabetic rats as well as restore cardiac function and increase lipogenesis [18, 19]. Glucose metabolism in general is normalized in diabetes [20, 21]. The vanadyl form of vanadium has also been shown to be effective in 'treating' diabetes [22–28].

However, oral administration of vanadium is not without its drawbacks. Toxic side effects have been noted, including some deaths, decreased weight gain, increased serum urea and creatine concentrations [29, 30]. Development and reproductive problems have also been noted with vanadium administration [31–33]. Vanadium has toxic effects on the immune system as well [34, 35] possibly mediated through alteration in microfilaments [36], cellular energy metabolism [37] or cytokine production [38]. Thus, as with any potential drug, the metabolism of this normally trace metal will be important if any significant clinical use is to be made of vanadium in humans.

Address for offprints: F.G. Hamel, Veterans Administration Medical Center, Research 151, 4101 Woolworth Avenue, Omaha, NE 68105, USA

We have demonstrated (unpublished data) that a liquid diet is an effective and convenient means of administering vanadium compounds to experimental animals. We have used this method to examine the accumulation of vanadium in various tissues when administered at different doses, including one that would be effective at lowering blood glucose levels in diabetic rats. We have also demonstrated that the various organs retain vanadium at different rates. This work demonstrates the complex nature of vanadium metabolism, and identifies some of the organs which are most likely to be affected by long term administration of vanadium.

Materials and methods

Liquid Rat Diet, Original Lieber-DeCarli Control (product #F0005) was purchased from Bio-Serv (Frenchtown, NJ). Vanadyl sulfate was purchased from Fisher (Plano, TX). Cupferron and sodium orthovanadate were obtained from Sigma (St. Louis, MO). Male Sprague-Dawley rats were purchased from Sasco (Omaha, NE) and maintained and used in accordance with the NIH Guide for the Care and Use of Laboratory Animals. The rats used in these studies were 140–160 g at the time of purchase, except for those used in comparing vanadium content of young versus old animals. The young animals were 39–41 days old (120–140 g) and the old animals were retired breeders 5–6 months old (approximately 500 g).

Rats were fed Purina Rat Chow and water *ad lib* for 4 days after their arrival at our facility. The rats were then switched to 90 ml per day Liquid Diet for 3 days prior to starting the vanadium treatment protocol. Sodium orthovanadate (SOV) or vanadyl sulfate (VS) was then added to 90 ml Liquid Diet at 1.6, 80, or 160 μmoles of vanadium/kg body weight/day. Control animals were given liquid diet alone without added vanadium. Animals were fed for 4, 11, or 18 days, with sacrifice at that time to determine organ vanadium content. The amount of liquid diet was recorded daily, and weights taken at the start of feeding and at 4, 11, and 18 days. Determination of vanadium content of the tissues was done by radiochemical neutron activation analysis as described [39] and as detailed below.

The rats used in the determination of the tissue half-lives were fed vanadium (either SOV or VS at 160 μmol V/kg*day) for one week, and then fed liquid diet alone without added vanadium until organ harvest. The animals were killed and organs taken at 0, 4, and 8 days and tissue levels determined as described below. The half lives were determined by curve fitting the data using the InPlot computer program (GraphPad Software Inc., San Diego, CA). The data were fitted to the equation

$$Y = A \times e^{-BX} + E$$

where Y is the vanadium concentration at time X (in days), A+E is the vanadium concentration at day 0, and E is the ending vanadium concentration. For our purposes, E was assumed to be the organ concentration of untreated control animals.

At the time of sacrifice the rats were anesthetized using 60 mg/kg sodium pentobarbital, and killed by pneumothorax and the subsequent removal of the organs. The following organs were routinely analyzed, though data is not presented for all the organs at all time points: brain; blood; epidymal fat pads; heart; kidney; liver; lungs; skeletal muscle (hind limb); spleen; and testes. The organs were removed using a Teflon-coated scalpel to assure no contamination of the sample from the stainless steel, which contains vanadium. The organs were placed in 50 ml Nalgene beakers and lyophilized to dryness. The organs were weighed, and digested with 10 ml of concentrated nitric acid at 56°C until dry. The samples had 10 ml of nitric acid added to them, were sonicated for 15 min and redigested as before. The sample was then dissolved in 5 ml 1 N nitric acid, sonicated for 15 min, and transferred to an irradiation vial. Samples were irradiated for 5 min at thermal neutron flux of 3.2×10^{11} n·cm^{-2} and a gamma-ray flux of 3×10^{17} eV/g·min in the Omaha Department of Veterans Affairs Medical Center TRIGA Mark I nuclear reactor by means of a pneumatic transfer tube. The sample was transferred to a separatory funnel 3 ml of fresh 6% cupferron solution were added and the funnel shaken for 30 sec. Five ml of chloroform were added, shaken for an additional 30 sec and the layers allowed to separate. The organic layer was withdrawn and 2 ml counted 5 min after removal of the sample from the reactor. The samples were counted for a live time of 200 sec in a 53 cm^3 closed end coaxial well Ge detector (Canberra, IN) with a relative efficiency of 10.4% for the 1332 keV gamma photon emitted by ^{60}Co. The gamma photons from the $^{51}V(n,\gamma)^{52}V$ reaction with an energy of 1.434 MeV were counted with Nuclear Data (Schaumberg, IL) ND 680 4096-channel analyzer employing ND 'Peak' software. The counts were compared to a known standard run on the same day to determine the vanadium content of the sample, with appropriate dilution factors taken into account for preparation and organic extraction of the sample. The vanadium value was divided by the dry weight of the sample to determine ng V/g dry tissue.

It should be noted here that fat tissue presented a difficult challenge in that it would not acid digest the way other tissues did. The reason for this was apparently the high lipid content which was insoluble in the aqueous nitric acid. To solve this problem we homogenized the wet tissue with chloroform:methanol (2:1) and separated the aqueous layer. The aqueous layer was dried down, and digested with nitric acid similar to the other tissues. Samples spiked with vanadium standard prior to homogenization reveal a loss of about 50% using this procedure. Thus the fat data underestimate the vanadium content by half, and the results are not as consistent. However, using this method meant we could at least obtain some measure of the vanadium content of fat, which we were unable to do using the standard procedure.

For the study of vanadium effects on insulin binding and degradation, rats were fed Liquid Diet with or without 160 µmole V/kg body weight/day added in SOV form for 14 days. Hepatocytes were then prepared as described previously [40] and used to examine cell association and degradation of ^{125}I iodoinsulin labeled at the A- 14 position.

Results

Table 1 shows the vanadium content in control animals for the organs tested. The levels vary widely with kidney being the highest and brain the lowest. While the control groups varied somewhat over time and among the treatments, no significant pattern of change in vanadium content was seen for any organ. Thus these values represent the normal tissue levels against which the SOV and VS treated groups should be compared.

Figures 1 and 2 show the time course (Panel A) and the dose response (Panel B) curves for liver and testes, respectively. Note that the accumulation of vanadium is essentially linearly dependent on dose over the range tested. All organs tested showed similar dose response curves. Note that for liver their was no significant further accumulation from day 11 to day 18 (Fig. 1, Panel A). Kidney showed a similar pattern (data not shown). However testes (Fig. 2, Panel A) showed no tendency to plateau at later time points. Similar data was obtained for brain and spleen (not shown).

Figure 3 shows the binding and degradation of ^{125}I iodoinsulin to hepatocytes isolated from control rats and from rats fed SOV for 14 days. Essentially no difference is found in the affinity or number of insulin receptors, nor in degradation, as measured by trichloroacetic acid solubility. The time course of insulin association to hepatocytes was the same in

Table 1. Tissue vanadium levels

Organ	ng V/g Dry Tissue Weight	
Blood	11 ± 2	n = 26
Brain	2 ± 1	n = 29
Fat	5 ± 1	n = 23
Heart	16 ± 3	n = 25
Kidney	133 ± 14	n = 27
Leg Muscle	18 ± 5	n = 27
Liver	30 ± 2	n = 26
Lung	27 ± 3	n = 27
Spleen	64 ± 8	n = 26
Testes	38 ± 5	n = 24

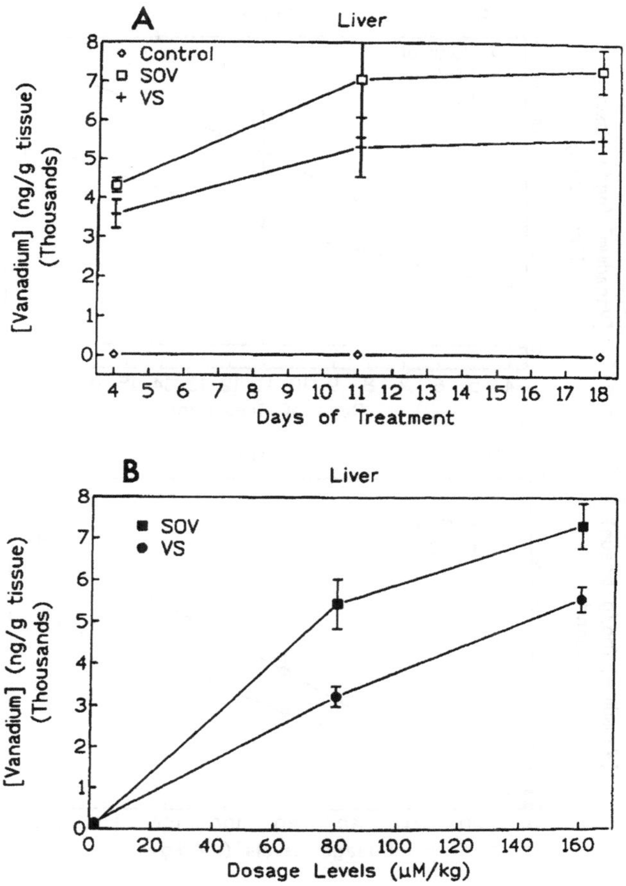

Fig. 1. Time course (Panel A) and dose-accumulation (Panel B) curves of vanadium for liver. Panel A shows the amount of vanadium (ng V/ gram dry tissue weight) in the organs of animals fed liquid diet alone (Cont) or with 160 µmole Vlkg/day as SOV or VS at days 4, 11 and 18. Panel B shows the amount of vanadium (ng V/gram dry tissue weight) in the organs after 18 days of feeding vanadium as SOV or VS at the dose indicated on the x-axis.

control and treated animals (data not shown).

Table 2 shows the half-lives in various organs from animals treated with SOV or VS. Note that the values vary greatly among the organs, with liver and kidney having the shortest and testes the longest. Almost all of the tissues show a slightly longer half-life of vanadium for SOV fed rats compared to the VS fed rats.

Table 3 shows the vanadium content of the organs listed of young and old rats shortly after arrival at our facility and after 45 days of feeding control Liquid Diet. Brain and leg muscle showed no difference between animal groups before or after feeding. All other organs showed an increase over time in their vanadium content, suggesting the Liquid Diet may contribute more vanadium to the animal's diet than the chow fed by the supplier. Preliminary tests of the supplier's chow found them to be relatively low in vanadium content

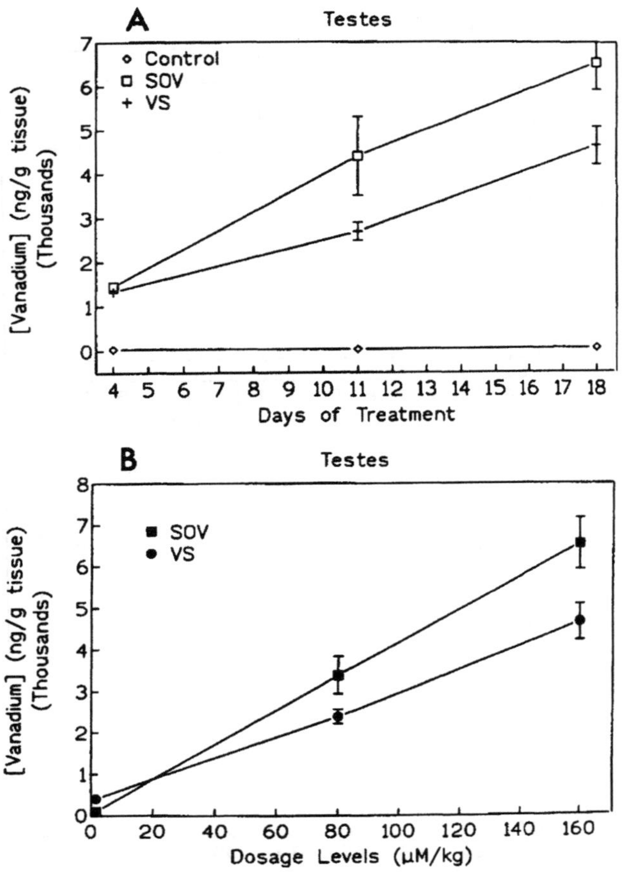

Fig. 2. Time course (Panel A) and dose accumulation (Panel B) curves of vanadium for testes. Panels are the same as described in the legend for Fig. 1.

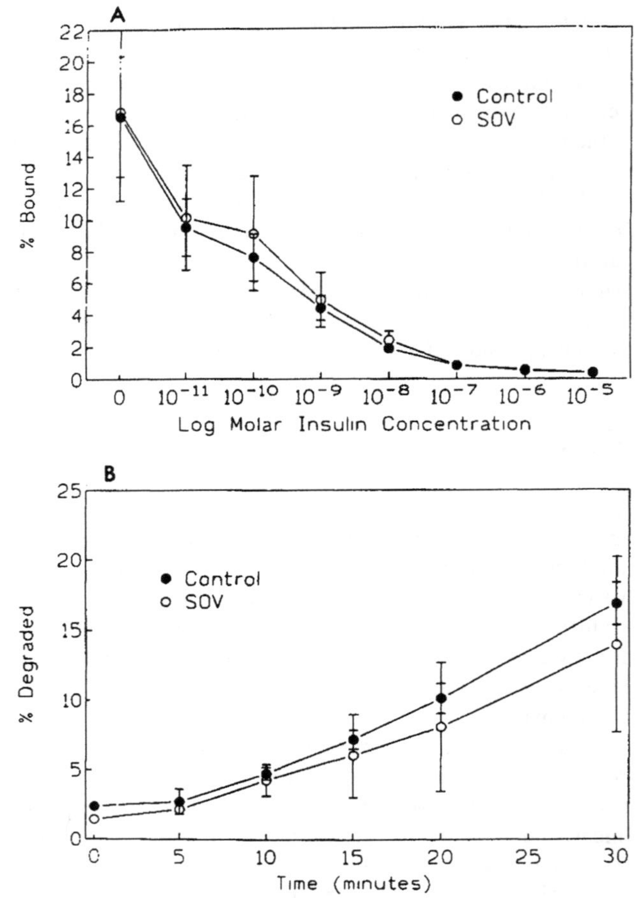

Fig. 3. Insulin binding and degradation in hepatocytes of control and SOV fed rats. Panel A shows displacement of tracer ^{125}I iodoinsulin by unlabeled insulin normalized to 10^6 cells/ml. Panel B shows tracer ^{125}I-iodoinsulin degradation in the media of cells incubated for the times shown.

(data not shown). This probably also explains why the values in Table 3 are lower than those in Table 1. Note that at the end of the age study the vanadium content of most of the organs is similar to that reported in Table 1. The slightly lower values probably reflect the lot to lot variability in the vanadium content of the diet that we have noted (unpublished observation). However spleen and testes showed a difference between age groups at the start and end of the feeding study. These results suggest these organs accumulate vanadium over time to a greater extent than the other organs tested.

Discussion

Vanadium is becoming increasingly important for its potential use in diabetes [13–17], its possible role as an essential element [1–4], and as an environmental toxicant [41]. To help understand its metabolic function and toxicology, it is important to know where the element resides in the body and where, under increased absorption, it may be concentrated. Because of its ability to decrease blood glucose concentrations we have examined vanadium levels in a variety of organs with varying sensitivities to insulin. Our hypothesis was that tissues in which glucose transport is sensitive to insulin may show greater levels of, or ability to concentrate, vanadium relative to those that are not insulin sensitive.

The data shown in Table 1 indicate that with respect to vanadium levels this hypothesis is false. Tissues such as fat and muscle, which have insulin sensitive glucose transport systems have relatively low levels of vanadium, while tissues that relatively insensitive to insulin, such as brain and kidney have widely disparate levels. The tissues with the lowest levels: brain; fat; heart; and muscle; have a large portion of their mass made up of components with which vanadium may not be expected to associate, i.e. lipid for fat and brain, and contractile proteins for heart and muscle. Thus vanadium

Table 2. Organ half lives (days)

Organ	SOV	r^2	VS	r^2
Liver	3.57	.997	3.18	.966
Kidney	3.93	.996	3.27	.999
Fat	4.06	.998	5.04	.983
Lung	5.52	.999	4.45	.998
Muscle	6.11	.999	4.49	.997
Heart	7.03	.994	5.05	.995
Spleen	9.13	.999	5.15	.993
Brain	11.17	.985	9.17	1.00
Testes	15.95	.987	13.50	.997

r^2 indicates the correlation coefficient of the data to the equation.

Table 3. Vanadium concentration in old and young rats

Organ /Rat# (Vng/gTis)	Young Day 0 (Vng/gTis)	Old Day 0 (Vng/gTis)	Young Day 45 (Vng/gTis)	Old Day 45 (Vng/gTis)
Brain	0	0	2 ± 2	0
Heart	3 ± 2	8 ± 3	12 ± 3	16 ± 1
Kidney	23 ± 2	18 ± 8	89 ± 13**	100 ± 8***
Leg muscle	6 ± 1	2 ± 2	6 ± 1	7 ± 1
Liver	6 ± 3	11 ± 1	18 ± 1*	24 ± 3*
Lung	5 ± 3	10 ± 1	17 ± 2**	21 ± 1**
Spleen	5 ± 4	40 ± 3^{++}	46 ± 6**	88 ± 6**
Testes	5 ± 5	23 ± 2^{+}	24 ± 3*	50 ± 3^{***++}

Values are mean ± S.E.M. Day 45 Different from Day 0: * = $p < 0.05$; ** = $p < 0.01$; *** = $p < 0.001$. Old different from Young: $^{+}$ = $p < 0.05$; $^{++}$ = $p < 0.01$. As determined by a Tukey-Kramer Multiple Comparisons test post ANOVA.

levels appear to correlate with the available 'space' it occupies in the tissue. A possible exception is the kidney, which had the highest levels of vanadium. However, this is not unexpected given the role the kidney plays in electrolyte metabolism. Thus the high levels probably represent vanadium which is being processed for excretion.

We have examined the accumulation of vanadium in animals orally treated with sodium orthovanadate (SOV) and vanadyl sulfate (VS) at three different doses. Liver and testes were typical (Panel B of Figs 1 and 2) of all organs tested, showing a linear dose-dependent increase in vanadium content. The organs did not accumulate vanadium uniformly, with kidney and liver increasing 150–200 fold at the highest dose by day 18 while fat only increased 10 fold at the same dose.

Panel A of Figs 1 and 2 show the time course of vanadium accumulation differs among the organs. Liver showed a plateau between day 11 and day 18, whereas testes appear to continue to accumulate vanadium at a relatively steady rate. Kidney behaved similarly to liver, while brain and spleen were more like testes. The other organs showed intermediate behavior, with a decrease in accumulation rate between day 11 and day 18 compared to between day 4 and day 11.

These differences in accumulation rate could result from varying assimilation rates, or from different elimination rates. We therefore determined the half-life of vanadium for the various organs.

While vanadium levels do not correlate with insulin sensitive glucose transport, our studies of the organ half-lives reveals another general relationship. The organs with the shorter half-lives tend to be more insulin sensitive in general, while those with the longest are relatively insensitive. For example, testes, brain and spleen are not considered to be exquisitely sensitive target tissues of insulin, and they have the longest half-lives, indicating a relatively slow metabolic turnover of vanadium. Fat and liver however, are two model tissues for insulin sensitivity, and they have short half-lives. This suggests that vanadium metabolism may be controlled by or otherwise related to insulin or the metabolic effects under the influence of insulin.

We have recently shown [42] in a Type I diabetes model, (BB/Wor-UTM Wistar rats from the colony at University of Tennessee), that the vanadium content of livers from diabetic rats is decreased from their non-diabetic littermates. If vanadium is involved in insulin sensitive metabolic events in the cell, it is logical that the tissue levels would be influenced by insulin. However, we have also recently discovered (unpublished data) that streptozotocin diabetic rats have increased vanadium levels in liver, muscle and spleen. Heart and brain also showed increases, although they were not statistically significant. Kidney showed a decrease in vanadium content. Insulin treatment of the diabetic animals returned the vanadium values to normal. We believe the discrepancy in these findings is the result of insulin treatment of the BB/Wor-UTM increasing 150–200 fold at the highest dose by day 18 while fat only increased 10 fold at the same dose.

Panel A of Figs 1 and 2 show the time course of vanadium accumulation differs among the organs. Liver showed a plateau between day 11 and day 18, whereas testes appear to continue to accumulate vanadium at a relatively steady rate. Kidney behaved similarly to liver, while brain and spleen were more like testes. The other organs showed intermediate behavior, with a decrease in accumulation rate between day 11 and day 18 compared to between day 4 and day 11. These differences in accumulation rate could result from varying assimilation rates, or from different elimination rates. We therefore determined the half-life of vanadium for the various organs.

While vanadium levels do not correlate with insulin sensitive glucose transport, our studies of the organ half-lives reveals another general relationship. The organs with the shorter half-lives tend to be more insulin sensitive in general, while those with the longest are relatively insensitive. For example, testes, brain and spleen are not considered to be exquisitely sensitive target tissues of insulin, and they have the longest half-lives, indicating a relatively slow metabolic turnover of vanadium. Fat and liver however, are two model tissues for insulin sensitivity, and they have short half-lives. This suggests that vanadium metabolism may be controlled by or otherwise related to insulin or the metabolic effects under the influence of insulin.

We have recently shown [42] in a Type I diabetes model, (BB/Wor-UTM Wistar rats from the colony at University of Tennessee), that the vanadium content of livers from diabetic rats is decreased from their non-diabetic littermates. If vanadium is involved in insulin sensitive metabolic events in the cell, it is logical that the tissue levels would be influenced by insulin. However, we have also recently discovered (unpublished data) that streptozotocin diabetic rats have increased vanadium levels in liver, muscle and spleen. Heart and brain also showed increases, although they were not statistically significant. Kidney showed a decrease in vanadium content. Insulin treatment of the diabetic animals returned the vanadium values to normal. We believe the discrepancy in these findings is the result of insulin treatment of the BB/Wor-UTM spontaneously diabetic rat. These rats must be treated with insulin for survival, although they were taken off insulin treatment 2 days prior to analysis. As we now know from our half-life study, two days would be insufficient time for the insulin-treated animal organs to return to their 'normal' levels. Thus, we believe vanadium levels in some organs are under the influence of insulin, with tissues tending to accumulate vanadium in the absence of insulin. This effect may be a direct control of the element's levels, or may be the result of more general metabolic changes. Our current data are insufficient to distinguish between those possibilities. However, our results do suggest that vanadium turnover in insulin sensitive tissues would be expected to be greater than that in tissues whose metabolism is less dependent on insulin.

It has been proposed that once vanadium, in either the SOV or VS form, enters the blood, it is converted to the vanadyl form within red blood cells, and that this is the form that resides in the tissues where it accumulates. If this were so, the organ half-life should be independent of the administered form. However our results suggest an alternative explanation. Whether administered as the orthovanadate or vanadyl species, equal vanadium concentrations are achieved in the blood. Even if there is some interconversion, some proportion of each remains in the blood, and the anionic vanadate as well as the cationic vanadyl are transported into the tissue. Once inside the cell there may be additional interconversion of the vanadium forms, but the vanadate species is either less metabolically active, preferentially converted to a more stable form, sequestered into a compartment which limits its turnover, or some combination of the above. Regardless of the exact mechanism, the vanadium involved is no longer able to be lost or transported out of the cell in the same manner as vanadium originally accumulated in the vanadyl form. Thus the administered form of vanadium can have an effect on its subsequent metabolism, which could have important consequences in increasing it effectiveness and decreasing toxicity if used therapeutically for diabetes mellitus. Furthermore, alternative forms of vanadium that have been proposed for therapetltic use, such as pervanadate [43, 44] or coordinated oxovanadium compounds [45–47], may exhibit different pharmacokinetic behaviors and should be examined with regard to organ ll accumulation.

These results may have important consequences with regard to the mechanism of vanadium's actions. Vanadate's activation of MPAP kinases [48] and glucose transport [49] are insulin receptor tyrosine kinase independent events. Pervanadate, however, appears to work by increasing insulin receptor tyrosine kinase activity by relatively specifically inhibiting the receptor's dephosphorylation by phosphotyrosine phosphatase(s) [44]. If the various vanadium compounds are handled differently by cells, and if the metabolism of the various forms of vanadium is affected differently by insulin, therapeutic use of specific compounds may be able to target and/or be limited by specific conditions or tissues. For example, differential sequestration of vanadate may mean it is more useful in some tissues than others, and its metabolic dependence on insulin may alter its usefulness between Type I and Type II diabetics. Furthermore, if vanadium compounds that work via the insulin receptor correct vanadium metabolism in diabetes, it may give insight into the mechanisms controlling vanadium levels.

Our results indicate a relationship between insulin and vanadium metabolism in some tissues. We also examined whether the converse were true, that vanadium could alter insulin metabolism. Figure 3 demonstrates that insulin binding and degradation were not altered in hepatocytes by oral treatment of rats with sodium orthovanadate. This indicates that insulin receptors are not affected by vanadium, and that subsequent processing of bound insulin by hepatocytes is also unchanged. Thus, the insulin-like effects of vanadium are unrelated to changes in insulin metabolism, at least in hepatocytes.

Vanadium content of the old and young animals organs before and after feeding liquid diet indicates two things. First, diet is very important in determining the levels of vanadium in some of the tissues. Forty-five days of consuming the liquid diet resulted in an increase in most, but not all organs. Thus, studies of vanadium levels depend on a consistent source of diet. Second, before and after liquid diet feeding, the older animals showed significantly higher levels of vanadium in spleen and testes. These data, combined with the long half-life of vanadium in these organs suggest they may be particular targets of vanadium toxicity. The continued accumulation of vanadium in tissues such as spleen, testes, or brain, may be a limiting factor in the use of vanadium for the treatment of diabetes mellitus. This deserves further study.

In conclusion, we have shown that organ concentrations

of vanadium vary widely, with tissues whose mass is made of largely a functional entity (e.g. contractile protein in muscle) tend to have less vanadium on a dry weight basis. The metabolism of vanadium, that is its turnover in organs, is roughly approximated by the tissue's sensitivity to insulin, with sensitive tissues having shorter half-lives. Two organs with longer half-lives, spleen and testes, tend to accumulate vanadium over time. These findings may relate to the immunotoxicity [34–38] and reproduction effects [31–33] found in animals. Finally, the nature of the vanadium compound exposed or administered to the animal can affect the metabolism of the element. Since the chemistry of vanadium is so complex, it will require more study to determine exactly which species are in the various organs. However, this study suggests there is a relationship between vanadium and insulin-stimulated metabolism, and the specific vanadium oxidation state and/or form is important in its metabolism.

Acknowledgements

The Authors acknowledge the technical assistance of Sally Payne and the secretarial assistance of Carol Plamondon.

References

1. Nielsen F, Uthus E: The essentiality and metabolism of vanadium. In: N.D. Chasteen (ed.). Vanadium in Biological Systems. Kluwer Academic Publishers, Dordrecht. p. 51–62, 1990
2. Nielsen F: Nutritional requirements for boron, silicon, vanadium, nickel and arsenic: Current knowledge and speculation. Biological and Trace Element Research. 26(27): 599–611, 1990
3. Nielsen F: Nutritional requirements for boron, silicon, vanadium, nickel, and arsenic: Current knowledge and speculation. FASEB. 5(12): 2661–2667, 1991
4. Neve J: Clinical implications of trace elements in endocrinology. Biological and Trace Element Research 32: 173–185, 1992
5. McKeehan W, McKeehan K, Hammond S, Ham R: Improved medium for clonal growth of human diploid fibroblasts at low concentrations of serum protein. In Vitro. 3: 399–416, 1977
6. Carpenter G, Vanadate EGF: The stimulation of DNA synthesis. Biochem Biophys Res Communications. 102: 1115–1121, 1981
7. Mountjoy K, Flier J: Vanadate regulates glucose transporter (Glut-1) expression in NIH3T3 mouse fibroblasts. Endocrinology 127(4): 2025–2034, 1990
8. Smith J: Vanadium ions stimulate DNA synthesis in Swiss mouse 3T3 and 3T6 cells. PNAS 80: 6162–6166, 1983
9. Canalis E: The effect of sodium vanadate on deoxyribonucleic acid and protein synthesis in cultured rat calvaraie. Endocrinology 116: 855–862, 1985
10. Kato Y, Iwamoto M, Koike T, Suzuki F: Effect of vanadateoncartilage-matrix proteoglycan synthesis in rabbit costal chondrocyte cultures. Journal of Cell Biology. 104: 311–319, 1987
11. Lau K, Tanimoto H, Baylink D: Vanadate stimulates bone cell proliferation and bone collagen synthesis in vitro. Endocrinology 123: 2858–2867, 1988
12. Nechay B, Norcross-Nechay K, Nechay P, Jena D: Presented at the Spurenelement Symposium, 1989
13. Shechter Y: Insulin-mimetic effects of vanadate. Possible implications for future treatmentof diabetes. Diabetes.39(1): 1–5, 1990
14. Robertson R, Klein O: Treatment of diabetes mellitus. Diabetologia 35(Sup. 2): S8–S17, 1992
15. Brichard S, Lederer J, Henquin J: The insulin-like properties of vanadium: A Curiosity or a perspective for the treatment of diabetes? Diabetes Metabolism. 17(5): 435–440, 1991
16. Vialettes B, Silvestre P: Pharmacological approach in the treatment of insulin resistance. Horm Res 38(1–2): 51–56, 1992
17. Shechter Y, Shisheva A: Vanadium salts and the future treatment of diabetes. Endeavour 17(1): 27–31, 1993
18. Heyliger C, Tahiliani A, McNeill J: Effect of vanadate on elevated blood glucose and depressed cardiac performance of diabetic rats. Science 227: 1474–1477, 1985
19. Meyerovitch J, Farfel Z, Sack J, Schechter Y: Oral administration of vanadate normalizes blood glucose levels in streptozotocin-treated rats: Characteristics and mode of action. Journal of Biological Chemistry 262: 6658–6662, 1987
20. Blondel O, Bailbe D, Portha B: *In vivo* insulin resistance in streptozotocin-diabetic rats: Evidence for reversal following oral vanadate treatment. Diabetologia. 32(3): 185–190, 1989
21. Brichard, S., Okitolonda, W., and Henquin, J. Long term improvement of glucose homeostasis by vanadate treatment in diabetic rats. Endocrinology. 123(4): 2048–2053, 1988.
22. Ramanadham S, Mongole J, Brownsey R, Cros G, MeNeill J: Oral vanadyl sulfate in the treatment of diabetes mellitus in rats. Am J Physiol 257(3):H904–911, 1989
23. Pederson, R., Ramanadham, S., Buchan, A., and McNeill, J. Long-term effects of vanadyl treatment on streptozotocin-induced diabetes in rats. Diabetes. 38(11):1390–1395, 1989
24. Ramanadham S, Brownsey R, Cros G, Mongold J, McNeill J: Sustained prevention of myocardial and metabolic abnormalities in diabetic rats following withdrawal from oral vanadyl treatment. Metabolism. 38(10): 1022–1028, 1989
25. Sakurai H, Tsuchiya K, Nukatsuke M, Sofue M, Kawada J: Insulin-like effect of vanadyl ion on streptozotocin-induced diabetic rats. J. Endocrinol. 126(3): 451–459, 1990
26. Ramanadham S, Cros G, Mongold J, Serrano J, McNeill J: Enhanced in vivo sensitivity of vanadyl-treated diabetic rats to insulin. Can J Physiol-Pharmacol 68(4): 486–491, 1990
27. Venkatesan N, Avidan A, Davidson M: Antidiabetic action of vanadyl in rats independent of in vivo insulin-receptor kinase activity. Diabetes 40(4): 492–498, 1991
28. Ozansoy G, Karasu C, Ozcelikay A: The effect of oral vanadyl treatment on the reactivity of tracheal smooth muscle obtained from insulin dependent diabetic rats. Gen Pharmacol 24(1): 115—119, 1993
29. Domingo J, Gomez M, Llobet J, Corbella J, Keen C: Oral vanadium administration to streptozotocin-diabetic rats has marked negative side-effects which are independent of the form of vanadium used. Toxicology 66(3): 279–287, 1991
30. Domingo J, Gomez M, Llobet J, Corbella J, Keen C: Improvement of glucose homeostasis by oral vanadyl or vanadate treatment in diabetic rats is accompanied by negative side effects. Pharmacology-Toxicology 68(4): 249–253, 1991
31. Sanchez D, Ortega A, Domingo J, Corbella J: Developmental toxicity evaluation of orthovanadate in the mouse. Biological and Trace Element Research.30: 219–226, 1991
32. Llobet J, Colomina M, Sirvent J, Domingo J, Corbella J: Reproductivetoxicity evaluation of vanadium in mab mice. Toxicology 80: 199–206, 1993
33. Leonard A, Gerber G: Mutagenicity, carcinogenicity and terato-

genicity of vanadium compounds. Mutation Research 317(1): 81–88, 1994
34. Cohen, M., Wei, C., Tom, H., and Kao, K. Effect of ammonium meta vanadate on the murine immune response. J Tox Envir Health 19: 279–288, 1986
35. Cohen M, Chen C, Wei C: Decreased resistance to *Listeria monocytogenes* in mice following vanadate exposure and effects upon the function of macrophages. Int J Immunopharmac 11: 185–292, 1989
36. Wang E, Choppin P: Effect of vanadate on intracellular distribution and function of 10 nm filaments. Proc Natl Acad Sciences USA 78: 2363–2367, 1981
37. Cohen M, Wei C: Effects of ammonium metavanadate treatment upon macrophage glutations redox cycle activity, superoxide production, and intracellular glutation status. J Leukocyte Biol 44: 122–129, 1988
38. Cohen M, Parsons E, Schlesinger R, Zelikoff J: Immunotoxicity of *in vitro* vanadium exposures: Effects on interleukin-1, tumor necrosis factor-α and prostaglandin E2 production by WEHI-3 macrophages. Int J Immunopharmac 15(3): 437–446, 1993
39. Blotcky AJ, Duckworth WC, Ebrahim A, Hamel FG, Rack EP, Sharma RB: Determination of vanadium in serum by pre-irradiation and post-irradiation chemistryand neutron activation analysis. Journal of Radioanalytical and Nuclear Chemistry. 134(1): 151–160, 1989
40. Peavy DE, Edmondson JW, Duckworth WC: Selective effects of inhibitors of hormone processing on insulin action in isolated hepatocytes. Endocrinology 114: 753–760, 1984
41. World Health Organization. International Programme on Chemical Safety, Environmental Health Criteria 81 Vanadium 1988
42. Hamel F, Solomon S, Jespersen A, Blotcky A, Rack E, Duckworth W: Alteration of tissue vanadium content in diabetes. Metabolism 42(12): 1503–1505, 1993
43. Shisheva A, Ikonomow O, Schechter Y: The Protein tyrosine phosphatase inhibitor, pervanadate, is a powerful antidiabetic agent in streptozotocin-treated diabetic rats. Endocrinology 134(1): 507–10, 1994
44. Posner BI, Faure R, Burgess JW, Bevan AP, Lachance D, Zhang-Sun G, Fantus IG, Ng, JB, Hall DA, Lum BS: Peroxovanadium compounds. A new class of potent phosphotyrosine phosphatase inhibitors which are insulin mimetics. J Biol Chem 269(6): 4596–4604, 1994
45. Yuen VG, Orvig C, Thompson KH, McNeill JH: Improvement in cardiac dysfunction in streptozotocin-induced diabetic rats following chronic oral administraiton of bis (maltolato) oxovanadium (IV). Can J Physiol Pharmacol 71(3–4): 270–276, 1993
46. Yuen V, Orvig C, McNeill J: Glucose-lowering effects of a new organic vanadium complex, bis(maltolato) oxovanadium (IV). Can J Physiol Pharmacol 71(3-4): 263–269, 1993
47. Cam M, Cros G, Serrano J, Lazaro R, McNeill J: *In vivo* antidiabetic actions of naglivan, an organic vanadyl compound in streptozotocin-induced diabetes. Diabetes Res Clin Pract 20(2): 111–121, 1993
48. D'Onofrio F, Le MQ, Chiasson JL, Srivastava AK: Activation of mitogen activated protein (MAP) kinases by vanadate is independent of insulin receptor autophosphorylation. FEBS Letters 340(3): 269–75, 1994
49. Shisheva A, Shechter Y: Mechanism of Pervanadate stimulation and potentiation of insulin-activated glucose transport in rat adipocytes: Dissociation from vanadate effect. Endocrinology 133(4): 1334–1562, 1993

Molecular and Cellular Biochemistry **153**: 103–112, 1995.

Modulation of insulin action by vanadate: evidence of a role for phosphotyrosine phosphatase activity to alter cellular signaling

I.G. Fantus, G. Deragon, R. Lai and S. Tang

Department of Medicine, Mount Sinai Hospital, Banting and Best Diabetes Centre and Department of Physiology, University of Toronto, Canada

Abstract

A number of vanadium compounds (vanadate, vanadyl sulfate, metavanadate) have insulin-mimicking actions both *in vitro* and *in vivo*. They have multiple biological effects in cultured cells and interact directly with various enzymes. The inhibitory action on phosphoprotein tyrosine phosphatases (PTPs) and enhancement of cellular tyrosine phosphorylation appear to be the most relevant to explain the ability to mimic insulin. We demonstrated that in rat adipocytes both acute insulin effects, e.g. stimulation of IGF-II and transferrin binding and a chronic effect, insulin receptor downregulation, were stimulated by vanadate. Vanadate also enhanced insulin binding, particularly at very low insulin concentrations, associated with increased receptor affinity. This resulted in increased adipocyte insulin sensitivity. Finally vanadate augmented the extent of activation of the insulin receptor kinase by submaximal insulin concentrations. This was associated with a prolongation of the insulin biological response, lipogenesis, after removal of hormone.

In conclusion: in rat adipocytes vanadate promotes insulin action by three mechanisms, 1) a direct insulin-mimetic action, 2) an enhancement of insulin sensitivity and 3) a prolongation of insulin biological response. These data suggest that PTP inhibitors have potential as useful therapeutic agents in insulin-resistant and relatively insulin-deficient forms of diabetes mellitus. (Mol Cell Biochem **153**: 103–112, 1995)

Key words: insulin resistance, adipocytes, insulin receptor, vanadate, tyrosine kinase

Introduction

Since the first descriptions of several insulin-like biological effects of vanadium compounds [1–3] a large number of studies have documented that sodium orthovanadate (vanadate, Na_3VO_4) and vanadyl sulfate ($VOSO_4$) mimic many of the actions of insulin in its three major target tissues, liver, muscle and fat [4–10]. The demonstration that oral administration of these compounds to insulin-deficient diabetic rats can lead to normal blood glucose concentrations and improved metabolism [11–17] without apparent serious toxicity gave rise to the concept of vanadium compounds as potential therapeutic agents [18]. More recently the successful use of oral vanadate or vanadyl sulfate in rodent models of insulin resistance with or without overt diabetes [19–25] has given this idea great impetus because, in contrast to merely substituting for subcutaneous insulin injections as may occur in insulin-dependent (IDDM/Type 1) forms of diabetes, vanadate may actually be more effective than insulin in non-insulin-dependent (NIDDM)/Type 2) forms. In addition the reduction of endogenous and exogenous hyperinsulinemia may be advantageous in view of the postulated contributing role of hyperinsulinemia to the pathophysiology of diseases associated with NIDDM such as hypertension and atherosclerosis [26].

In this paper we review briefly the biological effects and possible biochemical mechanisms of vanadate action and then focus on our work in isolated rat adipocytes, a classical insulin target tissue. We demonstrate that vanadate mimics insulin directly, enhances sensitivity to insulin and prolongs

Address for offprints: I. G. Fantus, Mount Sinai Hospital, 600 University Avenue, Suite 780, Toronto, Ontario, M5G 1X5 Canada insulin biological effects, thus promoting insulin action by at least three physiological mechanisms. These observations support the notion that vanadate or related vanadium compounds may be particularly effective therapeutic agents for insulin-resistant forms of diabetes mellitus.

Actions of vanadate

Some of the biological actions of vanadate have been reviewed [18, 27, 28]. The most prominent of these include the enhancement of protein tyrosine phosphorylation [29, 30], growth factor action [31] and cellular transformation [29, 32]. Vanadate has also been demonstrated to alter the regulation of several cell surface receptors [33, 34] as well as gene expression [31, 35, 36]. A wide variety of insulin-mimetic actions have been well documented *in vitro* [30, 37–39] and *in vivo* [35, 36, 40] and see Introduction [1–25]. However the effects of vanadate or vanadyl sulfate administration *in vivo* on cellular growth, transformation, oncogenesis, tumor growth and receptor expression have not been examined in detail.

Many *in vitro* studies have been published documenting the interactions of vanadate with multiple enzyme systems [18, 27, 28]. The general biological effects, in particular the insulin-mimetic and growth factor enhancing actions suggest specific biochemical mechanisms. Thus the inhibition of phosphoprotein tyrosine phosphatases appears to be of primary importance. However, with respect to the insulin-like effects the direct interaction of vanadate with a variety of enzymes involved in glucose metabolism, in particular the inhibition of glucose-6-phosphatase [41, 42] and fructose-2,6-bisphosphatase [35, 36, 43], may also be relevant. To gain a better understanding of the possible sites at which vanadium compounds may exert their metabolic effects it is helpful to review the mechanism of insulin action.

The mechanism of insulin action is incompletely understood. The initial steps involve the stimulation by insulin of an autophosphorylation reaction on tyrosine residues of the intracytoplasmic portion of its specific plasma membrane spanning receptor (reviewed in [44–46]). This activates the receptor as a tyrosine protein kinase resulting in the phosphorylation of multiple cellular protein substrates [45, 47]. At least one of these, insulin receptor substrate-1 (IRS-1) has been found to function as a 'docking' protein, such that a number of proteins containing SH2 (src homology 2) domains can bind to several phosphorylated tyrosine sites within the protein [47]. These SH2 containing proteins may in turn mediate many, if not all of the actions of insulin [45, 46]. Clearly the regulation of tyrosine phosphorylation of the insulin receptor itself and its substrate(s) and the balance between tyrosine phosphorylation and dephosphorylation are potential key mechanisms for altering the actions of insulin [48–50]. The phosphoprotein tyrosine phosphatase (PTP) inhibitory action of vanadate makes this an attractive hypothesis to explore. Although the action to inhibit a number of PTPs at various cellular sites may contribute to the general and insulin-specific bioeffects of vanadium compounds (Fig. 1), we have focused on the insulin receptor and the control of its tyrosine phosphorylation and intrinsic tyrosine kinase activity to investigate the insulin-mimetic actions of vanadate.

Vanadate mimics the rapid actions of insulin

Several years ago we examined the increase in insulin-like growth factor II (IGF-II) binding as a convenient marker for a rapid bioeffect of insulin. The stimulation by insulin of an increase in IGF-II binding had been documented to involve a translocation of receptors from an intracellular compartment to the plasma membrane [51]. This was analogous to the translocation process demonstrated for the glucose transporter involved in the stimulation of glucose transport by insulin [52] as well as by vanadate [53]. Vanadate was able to stimulate IGF-II binding in freshly isolated rat adipocytes (Table 1) and this effect was inhibited by amiloride [9]. Amiloride is an effective tyrosine kinase inhibitor and also inhibited the action of insulin suggesting that tyrosine kinase activation was involved in the vanadate response. Further support for this hypothesis was evident when we correlated the ability of insulin, vanadate, hydrogen peroxide (H_2O_2) and pervanadate (a novel insulin-mimetic agent generated upon mixing of vanadate with H_2O_2 which is discussed in detail in other sections of this symposium) to stimulate IGF-II binding with their respective extent of activation of the insulin receptor kinase. The correlation was highly significant [39]. Apart from glucose uptake and IGF-II binding, a similar translocation process stimulated by insulin is apparent for both LDL (low density lipoprotein) [54] and transferrin receptors [55]. We have demonstrated that vanadate increases both ^{125}I-LDL and ^{125}I-transferrin binding in cultured human fibroblasts and isolated rat adipocytes respectively (Table 1). The latter effect in adipocytes is time- and dose-dependent and associated with an increase in receptor number without a change in receptor affinity. This was the same as the effect of insulin. Furthermore, both insulin and vanadate stimulated ^{59}Fe uptake from ^{59}Fe-diferric transferrin in rat adipocytes associated with the increase in transferrin receptor binding capacity (ms in preparation).

Vanadate mimics a long term action of insulin

The actions of insulin have been classified chronologically by Rosen [44] as rapid (seconds), intermediate (minutes) and chronic (hours). The relationship of the rapid tyrosine autophosphorylation of the insulin receptor and activation of its intrinsic tyrosine kinase to the chronic effects of insulin, particularly the downregulation of its receptor, was not clear.

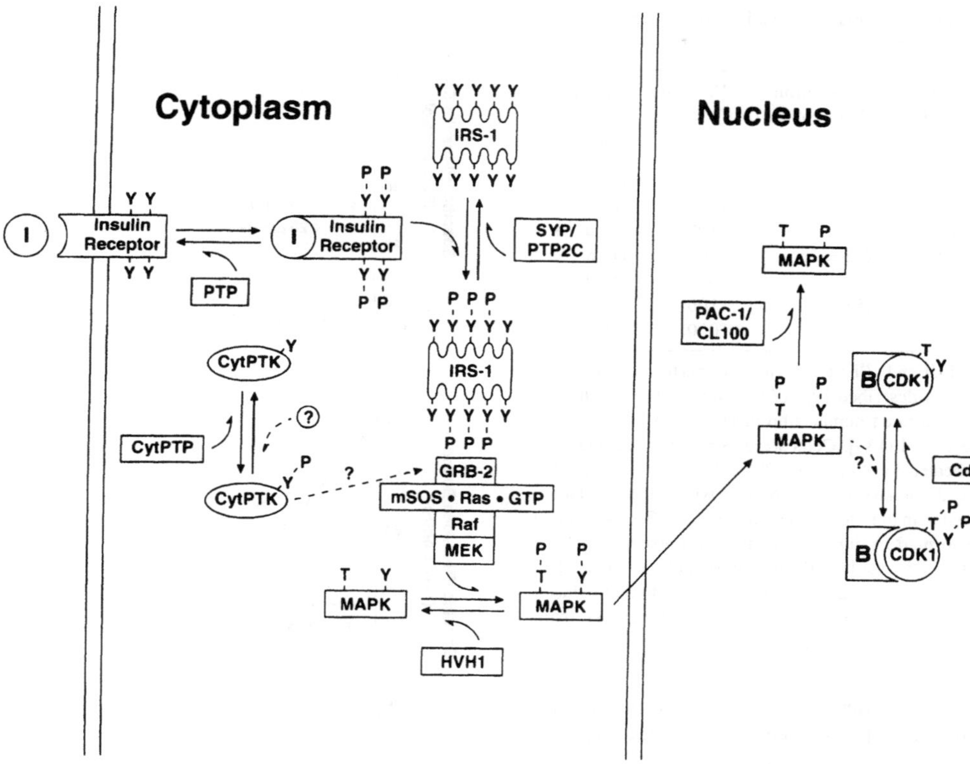

Fig. 1. Possible sites of protein tyrosine phosphatase (PTP) inhibition by vanadium compounds. PTPs that may play a role in modulating insulin signal transduction include (1) the PTP which dephosphorylates the insulin receptor itself (which remains unknown) designated here as PTP. (2) SYP or PTP2C, an SH2 domain containing cytosolic PTP [68] which binds to tyrosine phosphorylated IRS-1 and may be involved in either (or both) forward signal transduction or IRS-1 dephosphorylation and signal termination [69]. (3) HVH1, a dual specificity PTP which dephosphorylates both tyrosine and threonine residues of MAPK (mitogen-activated protein kinase) is inhibited by vanadate [70]. The activation of MAPK by vanadate has been demonstrated (see Pandey *et al.* in this issue) and requires tyr and thr phosphorylation [71]. (4) PAC-1 or CL100 are dual specificity phosphatases localized to the nucleus [72]. MAPK has been demonstrated to translocate to the nucleus [73]. (5) Cdc 25 is another nuclear dual specificity phosphatase shown to play a role in cell cycle progression, activating cyclin dependent kinase-1 (cdc 2) by dephosphorylation [74]. In this case, vanadate may be predicted to block cell cycle progression. In fact there is evidence that vanadate inhibits cell proliferation [75]. (6) The demonstration that a cytosolic tyrosine kinase is activated by vanadate (cyt PTK) [76] suggests that inhibition of a cyt PTP (as yet undefined) may be important in mediating some of insulin's bioeffects.

To explore this relationship we used vanadate as a probe to inhibit PTPs and examine whether the insulin receptor could be downregulated in the absence of ligand. In cultured IM-9 human lymphocytes, the classical and first model used to demonstrate insulin-induced receptor downregulation [56], vanadate stimulated only the first step of the downregulation process, i.e. receptor internalization [33]. We found that receptors accumulated intracellularly and their degradation was inhibited. The cause of this inhibitory action is not known but does not appear to be part of the insulin-mimetic repertoire of vanadate's actions. We turned to the rat adipocyte model in which other insulin-mimetic actions of vanadate were well documented ([39, 57] and ref therein).

In this insulin target tissue vanadate did stimulate downregulation in a manner similar to that induced by insulin. This was suggested by Marshall [58] who demonstrated a decrease in cell surface receptors. We have now found a time and dose-dependent downregulation of insulin receptors by both PTP inhibitors, vanadate and pervanadate (Fig. 2). In rat adipocytes both agents mimicked insulin completely such that ^{125}I-insulin binding to intact cells (Fig. 3) as well as to solubilized cells, i.e. the total cellular complement of insulin receptors (Fig. 4), was decreased. Thus both internalization and receptor degradation were stimulated by vanadate. These studies taken together with the demonstration of defects in insulin-stimulated downregulation of mutated, kinase-defective receptors [59, 60] and of receptors in which tyrosine autophosphorylation was inhibited [61] indicate the requirement of insulin receptor tyrosine autophosphorylation for both the long-term as well as short-term insulin biological effects. A recent model proposed by Carpentier *et al.* suggests that the early phase of

Table 1. Insulin-mimetic effects of vanadate on receptor protein translocation: Dose and time dependence

Receptor	Insulin (nM)	Stimulation (% of control)	Vanadate (μM)	Stimulation (% of control)
IGF-II	2	289 ± 16	100	142 ± 6
			1000	206 ± 22
LDL	10	124 ± 14	100	131 ± 16
Transferrin	10	273 ± 26	100	118 ± 7
			1000	159 ± 19
Transferrin (4 h preincubation)	10	181 ± 7	100	153 ± 12
			1000	290 ± 50

The binding of ^{125}I-IGF-II and ^{125}I-transferrin were performed in freshly isolated rat adipocytes as described [9, 106]. Binding of ^{125}I-LDL was performed in confluent cultured human fibroblasts grown in lipoprotein deficient serum as described [139]. Cells were exposed to the agents at the concentrations indicated for 30 min at 37°C followed by cooling and the binding assays. An additional set of experiments with 4 h preincubation was performed for transferrin binding. Values are mean ± S.E. of 4 separate experiments each performed in duplicate or triplicate except for LDL binding for which n = 5. The results indicate a dose and time-dependent stimulation of binding of all 3 ligands by vanadate similar to that induced by insulin and consistent with the translocation of receptors from intracellular sites to the plasma membrane.

downregulation, the movement of insulin receptors from the microvillus surface of cells to coated pits is the process stimulated by tyrosine autophosphorylation [62].

Modulation of insulin action by vanadate

Enhancement of insulin sensitivity

During our experiments examining downregulation, we noted that short term incubation [30 min to 2 h) of freshly isolated rat adipocytes with vanadate (≤ 200 uM) resulted in an increase in ^{125}I-insulin binding associated with an increase in receptor affinity. This effect of vanadate was dose-dependent such that the increase in tracer ^{125}I-insulin binding was 12 ± 7.7% at 10 μM vanadate and reached maximum stimulation of 44 ± 5.8% at 200 μM. This increase in binding was most marked at lower insulin concentrations (≤ 1 nM) with minimal effect (6.6%) at 10 ng/ml (1.7 nM) consistent with the increase in receptor affinity (see Fig. 2 ref [63]). Relative internalization of tracer ^{125}I-insulin was not altered but the greater absolute amount of cell-associated radioactivity was associated with a slower rate of release and a greater percentage of insulin released as intact hormone [63]. Although the mechanism by which vanadate causes these alterations in receptor binding affinity and insulin processing is not clear, the consequence of these changes was an increase in insulin sensitivity (Fig. 5). Thus lipogenesis in rat adipocytes was more sensitive to insulin in the presence of vanadate; the concentration of insulin required to stimulate half of the

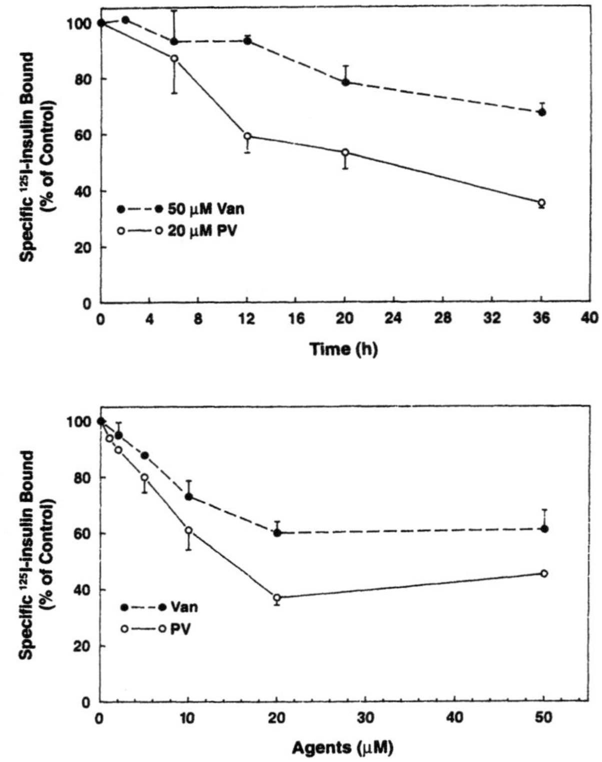

Fig. 2. Vanadate and pervanadate downregulate insulin receptors in rat adipocytes. Freshly isolated rat adipocytes were prepared as described [63] and incubated in DMEM supplemented with 25 mM HEPES, 1% BSA (DMEM-HA), 1% FBS, pH 7.6, in the absence and presence of vanadate (●) or pervanadate (○) (prepared as previously described [39] for the times and at the concentrations indicated at 37°C. The adipocytes were then washed and resuspended in DMEM-HA buffer, pH 7.6, at 15°C for 90 min with 0.2 ng/ml ^{125}I-insulin without and with 20 ug/ml unlabeled insulin to determine nonspecific binding as described [63]. Upper panel – Time course of insulin receptor downregulation by 50 μM vanadate and 20 μM pervanadate. Lower Panel – Dose-dependent downregulation of cell surface insulin receptors by vanadate and pervanadate after 20 h exposure.

maximum effect decreased from 0.2–0.08 ng/ml.

Prolongation of insulin action

The knowledge that tyrosine phosphorylation was critical for insulin signaling and the subsequent prediction that tyrosine dephosphorylation would be important in terminating the cellular response led us to test the hypothesis that vanadate, by inhibiting PTP activity would prolong an insulin biological response. We thus examined the duration of insulin-stimulated lipogenesis in the presence and absence of vanadate after the removal of insulin. We found that after thorough washing of cells to remove hormone and allowing control cells to return to basal activity, there was a significant residual insu-

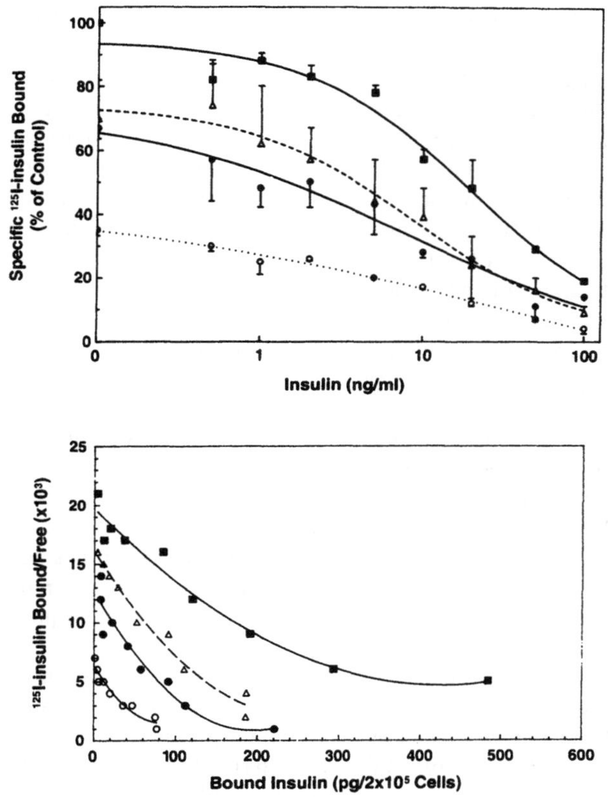

Fig. 3. Vanadate and pervanadate mimic insulin to downregulate insulin receptors in rat adipocytes. Freshly isolated rat adipocytes were prepared and cultured in the absence (■) and presence of 50 μM vanadate (●), 20 μM pervanadate (O) or 10^{-7} M insulin (Δ) for 36 h at 37°C as described in the legend to Fig. 2. The cells were then washed and resuspended in DMEM-HA and insulin binding assayed as in Fig. 2 with additional determinations performed in the presence of 0–20 μg/ml of unlabeled insulin. Upper panel – Competition binding curves. Lower panel – Scatchard analyses. The decrease in ^{125}I-insulin binding by all agents was associated with a decrease in receptor binding capacity.

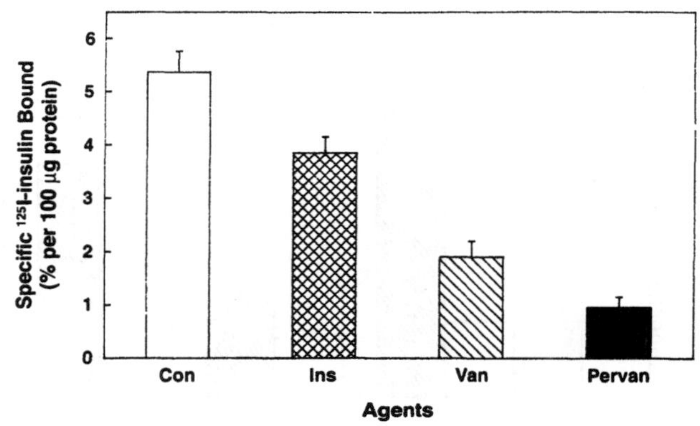

Fig. 4. Vanadate and pervanadate mimic insulin to downregulate total cellular adipocyte insulin receptors. Freshly isolated rat adipocytes were prepared and cultured in the presence of 50 μM vanadate (Van), 20 μM pervanadate (Pervan), 10^{-7} M insulin (Ins) or medium alone (Con) for 36 h at 37°C as described in the legend to Fig. 2. Subsequently the adipocytes were washed and solubilized in 1% Triton X-100, 50 mM HEPES, pH 7.6 buffer with protease and phosphatase inhibitors at 4°C as described [33]. ^{125}I-insulin binding was performed by incubating 50–100 ul (equal amounts of protein) of total cell solubilized supernatant with 0.2 ng/ml ^{125}I-insulin in the presence and absence of excess 10 μg/ml unlabeled insulin at 4°C for 18 h. Bound insulin-receptor complexes were separated from free ^{125}I-insulin by addition of 1 ml of 0.125% rabbit γ-globulin and 1.5 ml of 25% polyethylene glycol and centrifugation for 30 min as described [33]. The pellets were washed once with 25% PEG and counted. Total solubilized cell specific ^{125}I-insulin binding (mean ± S.E., n = 4) was decreased by all 3 agents.

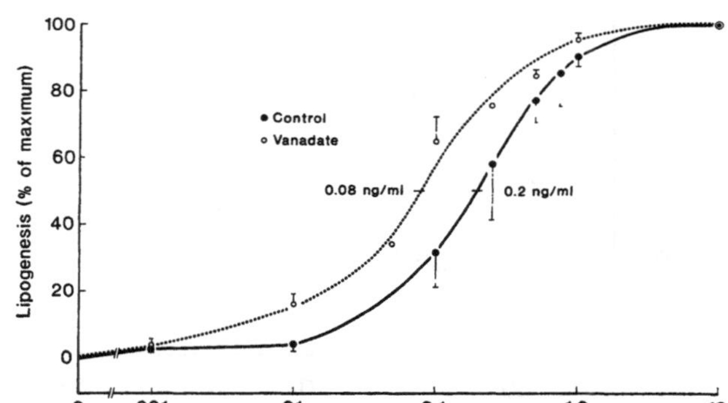

Fig. 5. Effect of vanadate on insulin action in rat adipocytes: insulin dose-response. Isolated rat adipocytes were preincubated in the absence (●) and presence (O) of 50 μM vanadate for 2 h at 37°C and subsequently exposed to various concentrations of insulin (0–10 ng/ml). The stimulatory effect of vanadate alone was taken as basal activity for the vanadate-preincubated cells. Values are the mean ± S.E. of three separate experiments performed in duplicate. The dose-response curve of the vanadate-treated cells was shifted to the left, with a lower concentration of insulin required for half-maximum response (0.08 ng/ml) compared to the control (0.2 ng/ml), indicating an increased sensitivity to insulin. Reproduced with permission [63].

lin-stimulated biological response in the presence of vanadate, which could not be accounted for by the weak insulin-mimetic action of the low vanadate concentration (< 200 uM) employed (Fig. 6). It should be noted that the concentration of insulin used in these initial experiments (10 ng/ml, 1.7 nM) achieved a maximum stimulation of lipogenesis but did not fully saturate all insulin receptors. A time course of the decay of insulin-stimulated lipogenesis in the presence and absence of vanadate revealed that vanadate prolonged the decay rate about 5 fold, such that the t½ of decay of the bioeffect was increased from 11–55 min in the presence of 200 uM vanadate [64]. To determine the potential relevance of this effect to *in vivo* actions of vanadate in rodent models of diabetes the dose-dependence of the prolongation was examined. As depicted in Fig. 7 vanadate sig-

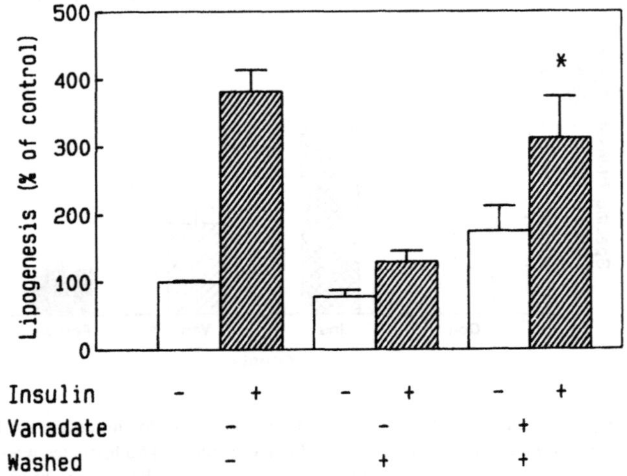

Fig. 6. Effect of vanadate on insulin action in rat adipocytes: preservation of insulin action. After 2 h of preincubation with or without 200 μM vanadate, adipocytes were incubated in the presence (▨) or absence (☐) of insulin (10 ng/ml, 1.7 nM) for 15 min at 37°C. Lipogenesis was assayed at 37°C either immediately or after repeated washing in pH 7.0 insulin-free buffer at 22°C, followed by reequilibration in pH 7.6 buffer. Each bar represents the mean ± S.E. (n = 11). *p < 0.05, washed vanadate plus insulin vs. washed insulin. Reproduced with permission [63].

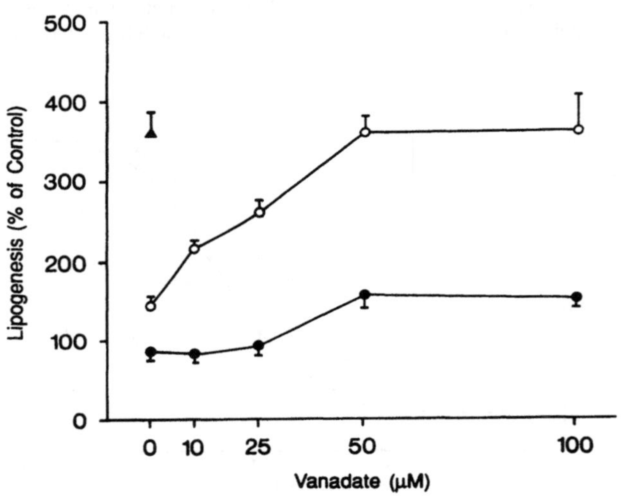

Fig. 7. Dose-dependent effect of vanadate on the maintenance of insulin-stimulated lipogenesis in rat adipocytes. Freshly isolated rat adipocytes were incubated in the absence and presence of the indicated concentrations of vanadate for 1 h at 37°C. The cells were further treated (O) or not treated (●) with 1.7 nM insulin for 15 min. The cells were then washed and lipogenesis was subsequently assayed as described in Fig. 6. Lipogenesis is expressed as a percentage of that measured in the absence of insulin and vanadate before the wash (100%). Maximum lipogenesis in the presence of 1.7 nM insulin, i.e. without washing, are represented (▲). Data are mean ± S.E. of 4–10 separate experiments each performed in triplicate. The residual insulin-stimulated lipogenesis after 30 min of wash was significantly augmented by vanadate (10 μM, p < 0.01; 25, 50 and 100 μM, p < 0.001). Reproduced with permission [64].

nificantly maintained insulin action after removal of hormone at concentrations in the range of 10 uM, the levels measured in the plasma of diabetic rats treated with vanadium compounds [12]. To elucidate the mechanism of this action two possibilities were tested. First our previous data suggested that insulin binding or dissociation of insulin from its receptor might be altered. However, at these relatively high (2–10 ng/ml) insulin concentrations used compared to tracer ^{125}I-insulin (0.1 ng/ml), there was no effect of vanadate to alter binding or dissociation [64]. Second the effect of vanadate on the duration of insulin-stimulated insulin receptor kinase activity was determined. The t½ of the decay of the insulin receptor tyrosine kinase upon removal of the hormone was the same in the absence and presence of vanadate; that is, 6 and 6.8 min respectively (see Fig. 3, ref [64]). Under the conditions of these experiments the decay rate of the activity of the insulin receptor tyrosine kinase is limited by the rate of dissociation of insulin from its receptor. Thus while receptor tyrosine dephosphorylation has now been documented to occur in intact rat adipocytes at a rapid rate, < 2 min [65], hormone-receptor dissociation requires several minutes [64].

Although vanadate did not alter the rate of decay of the insulin-stimulated receptor kinase, the extent of its activation by insulin was increased 3-fold in the presence of vanadate to 319 ± 18.7% of insulin alone [64]. This suggested that a partial inhibition of a PTP which would alter the balance between receptor tyrosine phosphorylation and dephosphorylation could be responsible. To determine the extent of insulin receptor tyrosine phosphorylation, adipocytes were preincubated as in the previous experiments in the absence and presence of vanadate 100 μM for 60 min at 37°C followed by the addition of insulin, 2 ng/ml for 15 min. The cells were separated from the medium by centrifugation through oil and the reaction terminated by immediate freezing in liquid nitrogen or addition of ice-cold solubilization buffer containing protease inhibitors and immediate cooling to –70°C. The adipocytes were thawed and solubilized in 1% Triton X-100, all in the presence of 1 mM vanadate, following which the insulin receptors were purified by WGA (wheat germ agglutinin) chromatography. The receptors were immunoprecipitated with human anti-insulin receptor antibodies and separated by SDS-PAGE (sodium dodecyl sulfate polyacrylamide gel electrophoresis). Proteins were transferred to nitrocellulose membranes and blotted with anti-phosphotyrosine antibodies. In the presence of vanadate insulin stimulation of phosphate incorporation onto tyrosine in the B-subunit was increased to 246% of insulin alone, very similar to the extent of stimulation of kinase activity [64]. This supported the concept that vanadate's action was mediated via inhibition of a PTP that was active to dephosphorylate the insulin receptor.

The fact that the insulin receptor kinase activity was not prolonged but rather augmented in the presence of vanadate,

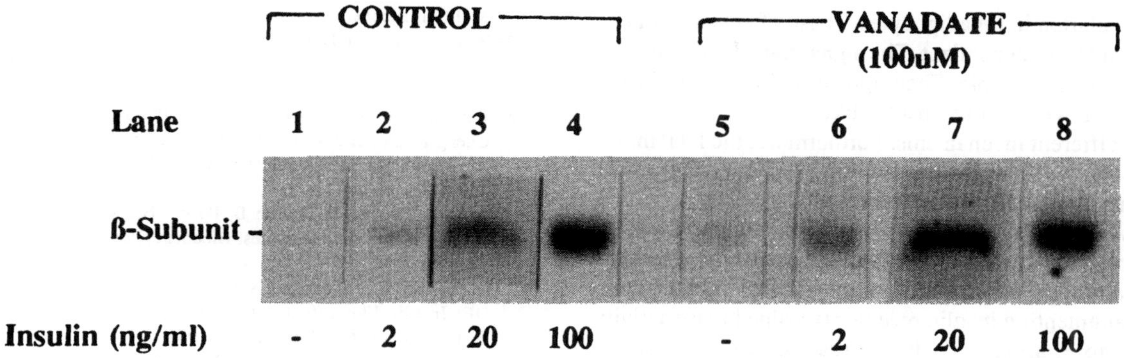

Fig. 8. Effect of vanadate on phosphate incorporation onto tyrosine in the insulin receptor β-subunit. Adipocytes were incubated in the absence or presence of vanadate for 60 min and with and without insulin 0.34 nM (2 ng/ml), 3.4 nM (20 ng/ml) and 17.2 nM (100 ng/ml) for 15 min. At the end of the incubation, the reaction was terminated by immediate freezing to –70°C. Cells were thawed, homogenized, and receptors solubilized as described [64]. Insulin receptors were immunoprecipitated and separated by SDS-PAGE. Proteins were transferred to nitrocellulose membranes and immunoblotted with anti-phosphotyrosine antibodies. The intensities of the 95 kD bands were determined by densitometry. The effect of vanadate on insulin receptor P-tyr content was similar to that observed on tyrosine kinase activity. The relative intensities in arbitrary units were: 15.1 (lane 1), 49.5 (lane 2), 207.4 (lane 3), 261.3 (lane 4), 24.7 (lane 5), 95.7 (lane 6) 244.9 (lane 7) and 317.3 (lane 8). Similar results were obtained in two separate experiments.

while under identical experimental conditions lipogenesis was prolonged suggested a possible relationship between the amplitude of the signal and the duration of the biological response. This concept had been proposed a number of years ago [66]. The demonstration of 'spare' receptors for insulin biological effects was consistent with at least two physiological consequences. First the number of cell surface insulin receptors could be a determinant of target cell hormone sensitivity [67]. Second, the binding and activation by insulin of receptors above that required to achieve a maximum biological response could result in a longer duration of action. To test whether the latter mechanism could explain our observations, it was required to determine whether vanadate would prolong insulin action in the presence of maximum activation of the insulin receptor kinase. Our previous dose-response studies suggested that maximum activation could be achieved by stimulation with 100 ng/ml (17.2 nM) of insulin [57]. We would expect that in this situation addition of vanadate would not significantly augment tyrosine kinase activity. We confirmed this by assessing the extent of tyrosine phosphorylation of the insulin receptor β-subunit in intact cells that had been exposed to increasing concentrations of insulin in the presence and absence of vanadate as described above. Figure 8 demonstrates that while vanadate markedly enhanced the insulin-stimulated incorporation of phosphate onto β-subunit tyrosine at 2 ng/ml (93% above insulin alone) similar to our previous results [64], this effect was minimal at 20 and 100 ng/ml insulin (18 and 21% above insulin alone respectively). Furthermore, at the latter insulin concentration (100 ng/ml, 17.2 nM) vanadate did not prolong the decay in insulin-stimulated lipogenesis (Table 2). Thus the data strongly support the hypothesis that the amplitude of the initial signal, i.e. the insulin receptor kinase, is a major determinant of the duration of the biological response. The mechanism of this signal transduction process, which likely involves intermediary molecules remains to be defined.

Table 2. Effect of insulin concentration on the decay of insulin action

		Insulin-stimulated lipogenesis (% of initial insulin)		
Vanadate(100 μM)	Insulin (nM)	20 min	30 min	40 min
Without	0.34	39.8 ± 18	26.5 ± 14	ND
With		95.5 ± 26*	70.8 ± 23*	ND
Without	17.2	98.5 ± 29	76.9 ± 24	57.7 ± 20
With		55.7 ± 14†	57.8 ± 24†	43.8 ± 13†

Freshly isolated rat adipocytes were incubated in the presence and absence of 100 μM vanadate for 60 min in KRBH buffer. Insulin (0.34 or 17.2 nM) was then added for 15 min. Lipogenesis was assayed immediately, or the cells were washed and incubated as described in Fig. 7. Lipogenesis was then assayed and expressed as the percentage of insulin-stimulated activity measured at the end of incubation before removal of hormone. Data are means ± S.E. (n = 4). Vanadate significantly increased lipogenesis with 0.34 nM insulin but not with 17.2 nM insulin. ND, not determined. *P < 0.05 in the presence compared with absence of vanadate. † = NS in the presence compared with absence of vanadate.

Summary

We have demonstrated that vanadate and similar PTP inhibitors mimic both acute insulin biological effects, e.g. receptor autophosphorylation, tyrosine kinase activation, glucose transport activation, stimulation of IGF-II and transferrin receptor binding, as well as a chronic effect, insulin receptor downregulation in rat adipocytes. These effects are concentration and time-dependent. The signal transduction mechanism for these distal bioeffects appears to involve the insulin receptor. However other insulin-like actions may be mediated by different mechanisms. Furthermore, the PTP inhibitors are less specific in their actions than insulin and may well have unwanted or adverse effects.

Our work has also shown that vanadate has important interactions with insulin to enhance insulin sensitivity, at least in part by altering insulin binding at low insulin concentrations and by augmenting insulin receptor tyrosine kinase activity. Furthermore, apparently by increasing insulin receptor autophosphorylation and tyrosine kinase activity stimulated by concentrations of insulin which do not maximally activate its receptor, vanadate can prolong the duration of the biological response. In this way the inhibition of PTP activity was used to demonstrate the physiological principle that the amplitude of the receptor kinase signal (above that required to achieve a maximum biological response) is a key determinant of the duration of the biological response. This is an important consideration *in vivo* because insulin secretion is pulsatile, nutrient-stimulated and often diminished in diabetes. These data support the potential utility of vanadate in insulin-resistant forms of diabetes mellitus.

Acknowledgements

We would like to thank Dr. B. Posner for helpful discussions, Ms. C. Walker for technical assistance and Mrs. B. Baubinas for excellent secretarial assistance. This work was supported by the Medical Research Council of Canada (Grant No. MT-7658).

References

1. Tolman EL, Barris E, Burns M, Pansini A, Partridge R: Effects of vanadium on glucose metabolism *in vitro*. Life Sciences 25: 1159–1164, 1979
2. Shechter Y, Karlish SJD: Insulin-like stimulation of glucose oxidation in rat adipocytes by vandyl (IV) ions. Nature (London) 284: 556–558, 1980
3. Dubyak GR, Kleinzeller A: The insulin-mimetic effects of vanadate in isolated rat adipocytes. Dissociation from effects of vanadate as a (Na^+/K^+) ATPase inhibitor. J Biol Chem 255: 5306–5312, 1980
4. Jackson TK, Salhanick AI, Sparks JD, Sparks CE, Bolognino M, Amatruda JM: Insulin-mimetic effects of vanadate in primary cultures of rat hepatocytes. Diabetes 37: 1234–1240, 1988.
5. Clark AS, Fagan JM, Mitch WE: Selectivity of the insulin-like action of vanadate on glucose and protein metabolism in skeletal muscle. Biochem J 232: 273–276, 1985
6. Haugaard N, Torbati A, Smithgall T, Wildey G: Stimulation of the phosphorylation of uridine in skeletal muscle by insulin and vanadate. Mol Cell Biochem 93: 13–19, 1990
7. Bruck R, Prigozin H, Krepel Z, Rotenberg P, Shechter Y, Bar-Meir S: Vanadate inhibits glucose output from isolated perfused rat liver. Hepatology 14: 540–544, 1991
8. Tamura S, Brown TA, Whipple JH, Fujita-Yamaguchi Y, Dubler RE, Cheng K, Larner J: A novel mechanism for the insulin-like effect of vanadate on glycogen synthase in rat adipocytes. J Biol Chem 259: 6650–6658, 1984
9. Kadota S, Fantus IG, Hersh B, Posner BI: Vanadate stimulation of IGF binding to rat adipocytes. Biochem Biophys Res Commun 138: 174–178, 1986
10. Duckworth WC, Solomon SS, Liepnieks J, Hamel FG, Hand S, Peavy DE: Insulin-like effects of vanadate in isolated rat adipocytes. Endocrinology 122: 2285–2289, 1988
11. Heyliger CE, Tahiliani AG, McNeill JH: Effect of vanadate on elevated blood glucose and depressed cardiac performance of diabetic rats. Science 227: 1474–1476, 1985
12. Meyerovitch J, Farfel Z, Sack J, Shechter Y: Oral administration of vanadate normalizes blood glucose levels in streptozotocin-treated rats. Characterization and mode of action. J Biol Chem 262: 6658–6662, 1987
13. Brichard SM, Okitolonda W, Henquin JC: Long term improvement of glucose homeostasis by vanadate treatment in diabetic rats. Endocrinology 123: 2048–2053, 1988
14. Rossetti L, Laughlin MR: Correction of chronic hyperglycemia with vanadate, but not with phlorizin, normalizes *in vivo* glycogen repletion and *in vitro* glycogen synthase activity in diabetic skeletal muscle. J Clin Invest 84: 892–899, 1989
15. Blondel O, Bailbe D, Portha B. *In vivo* insulin resistance in streptozotocin-diabetic rats – evidence for reversal following oral vanadate treatment. Diabetologia 32: 185–190, 1989
16. Ramanadham S, Mongold JJ, Brownsey RW, Cros GH, McNeill JH: Vanadyl sulfate in the treatment of diabetes mellitus in rats. Am J Physiol 257: H904–H911, 1989
17. Bendayan M, Gingras D: Effect of vanadate administration on blood glucose and insulin levels as well as on the exocrine pancreatic function in streptozotocin-diabetic rats. Diabetologia 32: 561–567, 1989
18. Shechter Y: Insulin-mimetic effects of vanadate: possible implications for future treatment of diabetes. Diabetes 39: 1–5, 1990
19. Brichard SM, Pottier AM, Henquin JC: Long term improvement of glucose homeostasis by vanadate in obese hyperinsulinemic fa/fa rats. Endocrinology 125: 2510–2516, 1989
20. Brichard SM, Bailey CJ, Henquin JC: Marked improvement of glucose homeostasis in diabetic ob/ob mice given oral vanadate. Diabetes 39: 1326–1332, 1990
21. Meyerovitch J, Rothenberg P, Shechter Y, Bonner-Weir S, Kahn CR: Vanadate normalizes hyperglycemia in two mouse models of non-insulin-dependent diabetes mellitus. J Clin Invest 87: 1286–1294, 1991
22. Pugazhenti S, Angel JF, Khandelwal RL: Long term effects of vanadate treatment on glycogen metabolizing and lipogenic enzymes of liver in genetically diabetic (db/db) mice. Metabolism 40: 941–946, 1991
23. Brichard SM, Ongemba LN, Henquin JC: Oral vanadate decreases muscle insulin resistance in obese fa/fa rats. Diabetologia 35: 522–527, 1992
24. Eriksson JW, Lönnroth P, Smith U: Vanadate increases cell surface insulin binding and improves insulin sensitivity in both normal and insulin-resistant rat adipocytes. Diabetologia 35: 510–516, 1992
25. Bhanot S, Bryer-Ash M, Cheung A, McNeill JH: Bis (maltolato) oxovanadium (IV) attenuates hyperinsulinemia and hypertension in spontaneously hypertensive rats. Diabetes 43: 857–861, 1994
26. DeFronzo RA, Ferrannini E: Insulin resistance: a multifaceted syndrome responsible for NIDDM, obesity, hypertension, dyslipidemia

and atherosclerotic cardiovascular disease. Diabetes Care 14: 173–194, 1991
27. Macara IG: Vanadium – an element in search of a role. Trends Biochem Sci 5: 92–94, 1980
28. Nechay BR, Nanninga LB, Nechay PSE, Post RL, Granthan JJ, Macara IG, Kubena LF, Phillips TD, Nielsen FFH: Role of vanadium in biology. Fed Proc 45: 123–132, 1986
29. Klarlund J: Transformation of cells by an inhibitor of phosphatases acting on phosphotyrosine proteins. Cell 41: 707–717, 1985
30. Bernier M, Laird DM, Lane MD: Effect of vanadate on the cellular accumulation of pp15, an apparent product of insulin receptor tyrosine kinase action. J Biol Chem 263: 13626–13634, 1988
31. Chen Y, Chan TM: Orthovanadate and 2,3-dimethoxy-1,4-naphthoquinone augment growth factor-induced cell proliferation and C-fos gene expression in 3T3-L1 cells. Arch Biochem Biophys 305: 9–16, 1993
32. Kowalski LA, Tsang SS, Davison AJ: Vanadate enhances transformation of bovine papillomavirus DNA-transfected C3H/10T½ cells. Cancer Lett 64: 83–90, 1992
33. Torossian K, Freedman D, Fantus IG: Vanadate downregulates cell surface insulin and growth hormone receptors and inhibits insulin receptor degradation in cultured human lymphocytes. J Biol Chem 263: 9353–9359, 1988
34. Marshall S, Monzon R: Down-regulation of cell surface insulin receptors in primary cultured rat adipocytes by sodium vanadate. Endocrinology 121: 1116–1122, 1987
35. Valera A, Rodriguez-Gil JE, Bosch F: Vanadate treatment restores the expression of genes for key enzymes in the glucose and ketone bodies metabolism in the liver of diabetic rats. J Clin Invest 92: 4–11, 1993
36. Miralpeix M, Carballo E, Bartrons R, Crepin K, Hue L, Rousseau GG: Oral administration of vanadate to diabetic rats restores liver 6-phosphofructo-2-kinase content and mRNA. Diabetologia 35: 243–248, 1992
37. Green A: The insulin-like effect of sodium vanadate on adipocyte glucose transport is mediated at a post-insulin-receptor level. Biochem J 238: 663–669, 1986
38. Foot E, Bliss T, Fernandes LC, Da Costa C, Leighton B: The effects of orthovanadate, vanadyl and peroxides of vanadate on glucose metabolism in skeletal muscle preparations *in vitro*. Mol Cell Biochem 109: 157–162, 1992
39. Kadota S, Fantus IG, Deragon G, Guyda HJ, Posner BI: Stimulation of insulin-like growth factor II receptor binding and insulin receptor kinase activity in rat adipocytes: Effects of vanadate and H_2O_2. J Biol Chem 262: 8252–8256, 1987
40. Cordera R, Andraghetti G, DeFronzo RA, Rosetti L: Effect of *in vivo* vanadate treatment on insulin receptor tyrosine kinase activity in partially pancreatectomized diabetic rats. Endocrinology 126: 2177–2183, 1990
41. Singh J, Nordlie RC, Jorgensen RA: Vanadate: a potent inhibitor of multifunctional glucose-6-phosphatase. Biochim Biophys Acta 678: 477–482, 1981
42. Schulz LO: Suppression of the hepatic glucose-6-phosphatase system in diabetic rats by vanadate. Ann Nutr Metab 32: 289–296, 1988
43. Rider MH, Bartrons R, Hue L: Vanadate inhibits liver fructose-2,6-bisphosphatase. Eur J Biochem 190: 53–56, 1990
44. Rosen O: After insulin binds. Science 237: 1452–1458, 1987
45. White MF, Kahn CR: The insulin signaling system. J Biol Chem 269: 1–5, 1994
46. Kahn CR: Insulin action, diabetogenes, and the cause of Type II diabetes. Diabetes 43: 1066–1084, 1994
47. Sun XJ, Crimmins DL, Myers MG, Miralpeix M, White MF: Pleiotropic insulin signals are engaged by multisite phosphorylation of IRS-1. Mol Cell Biol 13: 7418–7428, 1993
48. Tonks NK, Cicirelli MF, Diltz CD, Krebs EG, Fischer EH: Effect of microinjection of a low-Mr human placenta protein tyrosine phosphatase on induction of meiotic cell division in Xenopus oocytes. Mol Cell Biol 10: 458–463, 1990
49. Mooney RA, Bordwell KL: Differential dephosphorylation of the insulin receptor and its 160 kDa substrate (pp 160) in rat adipocytes. J Biol Chem 267: 14054–14060, 1992
50. Goldstein BJ: Protein-tyrosine phosphatases and the regulation of insulin action. J Cell Biochem 48: 33–42, 1992
51. Oka Y, Mottola C, Oppenheimer CL, Czech MP: Insulin activates the appearance of insulin like growth factor II receptors on the adipocyte cell surface. Proc Natl Acad Sci USA 81: 4028–4032, 1984
52. Simpson IA, Cushman SW: Hormonal regulation of mammalian glucose transport. Ann Rev Biochem 55: 1059–1089, 1986
53. Paquet MR, Romanek RJ, Sargeant RJ: Vanadate induces the recruitment of glut-4 glucose transporter to the plasma membrane of rat adipocytes. Mol Cell Biochem 109, 149–155, 1992
54. Salter AM, Fisher SC, Brindley DN: Binding of low-density lipoprotein to monolayer cultures of rat hepatocytes is increased by insulin and decreased by dexamethasone. FEBS Lett 220: 159–162, 1987
55. Davis RJ, Corvera S, Czech MP: Insulin stimulates cellular iron uptake and causes the redistribution of intracellular transferrin receptors to the plasma membrane. J Biol Chem 261: 8708–8711, 1986
56. Gavin JR, Roth J, Neville DM Jr, DeMeyts P, Buell DN: Insulin-dependent regulation of insulin receptor concentrations: a direct demonstration in cell culture. Proc Natl Acad Sci USA 71: 84–188, 1974
57. Fantus IG, Kadota S, Deragon G, Foster B, Posner BI: Pervanadate (Peroxide(s) of vanadate) mimics insulin action in rat adipocytes via activation of the insulin receptor tyrosine kinase. Biochemistry 28: 8864–8871, 1989
58. Marshall S, Monzon R: Down-regulation of cell surface insulin receptors in primary cultured rats adipocytes by sodium vanadate. Endocrinology 121: 1116–1122, 1987
59. Russell DS, Gherzi R, Johnson EL, Chou CK, Rosen OM: The protein-tyrosine kinase activity of the insulin receptor is necessary for insulin-mediated receptor down-regulation. J Biol Chem 262: 11833–11840, 1987
60. McClain DA, Maegawa H, Lee J, Dull TJ, Ullrich A, Olefsky JM: A mutant insulin receptor with defective tyrosine kinase displays no biologic activity and does not undergo endocytosis. J Biol Chem 262: 14663–14671, 1987
61. Torossian K, Nower P, Schwartz T, Fantus IG: Phorbol esters inhibit insulin-induced receptor downregulation in cultured human lymphocytes: association with diminished insulin receptor autophosphorylation. Biochem J 290: 151–158, 1993
62. Carpentier J-L, Paccaud J-P, Backer J, Gilbert A, Orci LM, Kahn CR: Two steps of insulin receptor internalization depend on different domains of the B-subunit. J Cell Biol 122: 1243–1252, 1993
63. Fantus IG, Ahmad F, Deragon G: Vanadate augments insulin binding and prolongs insulin action in rat adipocytes. Endocrinology 127: 2716–2725, 1990
64. Fantus IG, Ahmad F, Deragon G: Vanadate augments insulin-stimulated insulin receptor kinase activity and prolongs insulin action in rat adipocytes. Evidence for transduction of amplitude of signaling into duration of response. Diabetes 43: 375–383, 1994
65. Mooney RA, Anderson DL: Phosphorylation of the insulin receptor in permeabilized adipocytes in coupled to a rapid dephosphorylation reaction. J Biol Chem 264: 6850–6857, 1989
66. Roth J, Grunfeld C: Mechanism of action of peptide hormones and catecholamines. In: J.D. Wilson, D.W. Foster (eds). Williams Textbook of Endocrinology. Philadelphia, Saunders, pp 76–122, 1985

67. Kahn CR: Insulin resistance, insulin insensitivity, and insulin unresponsiveness: a necessary distinction. Metabolism 27: 1893–1902, 1978
68. Feng G-S, Hui C-C, Pawson T: SH2-containing phosphotyrosine phosphatase as a target of protein-tyrosine kinases. Science 259: 1607–1611, 1993
69. Kuhne MR, Zhao Z, Rowles J, Lavan BE, Shen S-H, Fischer EH, Lienhard GE: Dephosphorylation of insulin receptor substrate 1 by the tyrosine phosphatase PTP2C. J Biol Chem 269: 15833–15837, 1994
70. Kwak SP, Hakes DJ, Martell KJ, Dixon JE: Isolation and characterization of a human dual specificity protein-tyrosine phosphatasegene. J Biol Chem 269: 3596–3604, 1994
71. Nishida E, Gotoh Y: The MAP kinase cascade is essential for diverse signal transduction pathways. Trends Biochem Sci 18: 128–131, 1993
72. Nebreda AR: Inactivation of MAP kinases. Trends Biochem Sci 19: 1–2, 1994
73. Chen R-H, Sarnecki C, Blenis J: Nuclear localization and regulation of erk- and rsk-encoded protein kinases. Mol Cell Biol 12: 915–927, 1992
74. Norbury C, Nurse P: Animal cell cycles and their control. Ann Rev Biochem 61: 441–470, 1992
75. Cruz TF, Morgan A, Min W: *In vitro* and *in vivo* anti-neoplastic effect of orthovanadate. Mol Cell Biochem (In press)
76. Shisheva A, Shechter Y: Role of cytosolic tyrosine kinase in mediating insulin-like actions of vanadate in rat adipocytes. J Biol Chem 268: 6463–6469, 1993

Molecular and Cellular Biochemistry **153**: 113–119, 1995.

Reversal of defective G-proteins and adenylyl cyclase/cAMP signal transduction in diabetic rats by vanadyl sulphate therapy*

Madhu B. Anand-Srivastava[1]**, John H. McNeill[2] and Xiao-Ping Yang[1]
[1]Department of Physiology, University of Montreal, Montreal, Quebec; [2]Faculty of Pharmaceutical Sciences, University of British Columbia, Vancouver, BC, Canada

Abstract

Vanadium salts exhibit a wide variety of insulinomimetic effects. In the present studies, we have examined the modulation of G-protein levels and adenylyl cyclase activity in the liver of streptozotocin-induced chronic diabetic rats (STZD) by vanadyl sulfate treatment and compared it with that of insulin. The basal enzyme activity, as well as the stimulatory effects of guanine nucleotides, glucagon, N-Ethylcarboxamideadenosine (NECA), isoproterenol, forskolin and sodium fluoride (NaF) on adenylyl cyclase were significantly increased in STZ-D rat liver as compared to control. In addition, the levels of stimulatory (Gsα) as well as inhibitory ($Gi_{\alpha-2}$ and $Gi_{\alpha-3}$) as determined by immunoblotting techniques were also significantly higher in the STZ-D rat liver, however, the inhibitory effects of oxotremorine and low concentrations of GTPγS on adenylyl cyclase were not different in the two groups. Vanadyl sulfate and insulin treatments restored the augmented basal enzyme activity, the stimulations exerted by stimulatory inputs on adenylyl cyclase and the G-protein levels to various degrees, however, vanadyl sulfate was more effective than insulin. In addition, unlike vanadyl sulfate, insulin was unable to improve the stimulation exerted by glucagon and isoproterenol on adenylyl cyclase activity in STZD rats. These results suggest that vanadyl sulfate mimics the effects of insulin to restore the defective levels of G-proteins and adenylyl cyclase activity. From these results it may be suggested that one of the mechanisms by which vanadyl sulfate improves the glucose homeostasis in STZ-D rats may be through its ability to modulate the levels of G-proteins and adenylyl cyclase signal transduction system. (Mol Cell Biochem **153**: 113–119, 1995)

Key words: Adenylyl cyclase, G-proteins, diabetes, insulin, vanadate

Abbreviations: NECA – N-ethylcarboxamideadenosine; Iso – Isoproterenol; Glu – Glucagon; FSK – forskolin; GTPγS – guanosine 5'-[γ-thio]triphosphate; Gs – stimulatory guanine nucleotide regulatory protein; Gi – inhibitory guanine nucleotide regulatory protein; STZ – streptozotocin.

Introduction

Guanine nucleotide regulatory proteins (G-proteins) play an important role in the regulation of variety of signal transduction mechanisms which mediate the physiological responses of various hormones. One of these signal transduction systems is adenylyl cyclase/cAMP system, that has been shown to be altered in diabetes [1, 2]. The adenylyl cyclase/cAMP system is composed of three components: receptor, catalytic subunit and guanine nucleotide regulatory proteins (G-proteins). The stimulation and inhibition of adenylyl cyclase by hormones are mediated by two distinct G-proteins, Gs (stimulatory) and Gi (inhibitory), that couple the extracellular receptor to the catalytic subunit [3]. G proteins are heterotrimeric and composed of three distinct subunits: α, β, and γ [4]. The α-subunit binds and hydrolyzes GTP and

*This work was supported by grants from Medical Research Council and Canadian Diabetes Association.
**M.B.A-S is a recipient of the Medical Research Council Scientist Award from the Medical Research Council of Canada.
Address for offprints: M.B. Anand-Srivastava, Department of Physiology, University of Montreal, C.P. 6128, Succursale A, Montreal, H3C 3J7, Quebec, Canada

confers specificity in receptor and effector interactions [5]. Two different forms of the $G_{s\alpha}$ protein with a molecular size of 45 and 52 kDa have been characterized [6]. However, a third form of the $G_{s\alpha}$ protein having a molecular size of 47 kDa has also been detected in heart [7]. These different forms of $G_{s\alpha}$ arise from several species of mRNA that appear to be the products of alternative splicing of a common precursor [8, 9]. On the other hand, three distinct forms of $G_{i\alpha}$ ($G_{i\alpha-1}$, $G_{i\alpha-2}$, and $G_{i\alpha-3}$) have been identified and characterized and have been shown to be the products of different genes [10–12]. The defective G-proteins and adenylyl cyclase signal transduction system in STZ-induced acute diabetic rat liver has been demonstrated by several investigators [1, 2]. However, the results of these studies have been conflicting. An increased, decreased, or unaltered basal and stimulated adenylyl cyclase activity as well as the G-protein levels have been reported in the liver from STZ-D rats [1, 2, 13, 14].

Vanadium, a trace element, exerts a wide variety of insulinomimetic effects, such as stimulation of glucose oxidation, glucose transport in adipocytes and glycogen synthesis in liver and diaphragm [15–17]. The interest for the use of vanadium in the diabetic state was brought by the first demonstration from Heyliger and co-workers [18] about its antidiabetic activity in the STZ-D rats. Vanadate has also been shown to markedly decrease blood glucose levels in STZ-D rats [18–21]. The mechanism by which vanadium elicits improvement of glucose homeostasis in STZ-D is still obscure. Vanadate has been reported to increase glucose transport, cell surface insulin binding and to improve insulin sensitivity in both normal and insulin resistant rat adipocytes [22]. Recent reports suggested that the insulin-like effects of vanadium are known to occur independently of tyrosine kinase activation [23, 24]. The involvement of G-proteins in the regulation of glucose transport has also been reported [25]. In the present studies we have examined if G-proteins and adenylyl cyclase activity is also altered in chronic STZ-D liver and whether vanadium, like insulin could also interact with G-proteins and improve their levels as well as of adenylyl cyclase/cAMP system in STZ-induced diabetes.

Methods

Preparation of animals

Wistar rats weighing about 200 g were injected intravenously with a bolus of streptozotocin (STZ, 60 mg/kg) to induce diabetes. Age-matched control rats were injected with an equal volume of saline. The diabetic state was verified 48 h after the injection by measuring the blood glucose concentration using a glucometer. The rats were kept for two months before carrying out the biochemical analysis. In the study, vanadyl sulphate treatment was initiated at a concentration of 0.5 mg/ml in the drinking water and the concentration was gradually increased to 1 mg/ml. The actual dose of vanadyl sulphate fluctuated during the period of treatment due to variations in the amount of water the rats drank. On average, the dose was about 0.45 mmol/kg/day, based on calculations made according to the concentration of vanadyl sulphate in the drinking water and the amount of fluid consumed by the rats. The blood glucose levels were monitored throughout the study by sampling blood from the tail of the rat and measuring glucose with a glucometer. At the end of the treatment, the rats were fasted overnight and about half of the rats in each group received a bolus injection of insulin (10 U/kg, iv). The rats were anesthetized with pentobarbital at a dose of 100 mg/kg, livers were dissected out and were frozen at –70° until used. Serum samples from all the rats were prepared at the termination of the experiments for the determination of glucose concentration using a glucose assay kit from Boehringer-Mannheim.

Preparation of liver membranes

The frozen livers were pulverized to a fine powder in a percussion mortar cooled in liquid nitrogen. The powder was homogenized using a teflon-glass homogenizer in a buffer containing 10 mM Tris-HCl and 1 mM-EDTA (pH 7.5) for 10–15 strokes. The homogenate was centrifuged at 1500 g for 10 min. The supernatant fraction was discarded and pellet was washed once with Tris-EDTA buffer. The pellet was finally suspended in 10 mM-Tris/HCl, 1 mM EDTA and used for determination of adenylyl cyclase activity and for immunoblotting studies.

Adenylyl cyclase activity determination

Adenylyl cyclase activity was determined by measuring [^{32}P]-cAMP formation from [α-^{32}P] ATP as described previously [26–28]. The assay medium containing 50 mM-glycylglycine, pH 7.5, 0.5 mM Mg-ATP, [α-^{32}P]ATP[$(1–1.5) \times 10^6$ cpm], 5 mM $MgCl_2$ 0.5 mM cAMP, 1 mM-3-isobutyl-1-methylxanthine, 0.1 mM EGTA and an ATP-regenerating system consisting of 2 mM phosphocreatine, 0.1 mg of creatine kinase/ml, and 0.1 mg of myokinase/ml in total volume of 200 µl. For measuring inhibitory hormone-regulated adenylyl cyclase activity, NaCl (100 mM) and GTPγS (10 µM) were included in reaction mixture. The reactions were initiated by addition of reaction mixture to the membrane preparations which had been equilibrated for 2 min at 37°C. The reactions were conducted in triplicate for 10 min at 37°C and were terminated by addition 0.6 ml of 120 mM-Zinc acetate. cAMP was purified by co-precipitation of other nucleotides with $ZnCO_3$ by addition of 0.5 ml of 144 mM Na_2CO_3 and subsequent chromatography by double-column system as described by Salomon *et al.* [29]. Under the assay conditions used, adenylyl cyclase activity was linear with respect to protein concentration and time of incubation. Total proteins were determined as described by Lowry *et al.* [30].

Immunoblotting and Northern blotting analysis

Immunoblotting and Northern blotting analysis was performed as described previously [27, 31, 32].

Analysis of data

The data are expressed as the means + S.E.M. Comparisons between means were performed using student's *t*-test or analysis of variance as appropriate. Results were considered significantly different if $p < 0.05$.

Results

Blood glucose concentration

Blood glucose concentration was significantly higher in chronic STZ-induced diabetic rats as compared to the controls (control 6.78 ± 0.51 mM; n = 5 and diabetic rats (D = 16.42 ± 4.06 mM; n = 4, $p < 0.001$). Insulin treatment did not alter the blood glucose levels in diabetic rats but decreased it in control rats (control = 3.18 ± 0.24 mM; D = 16.40 ± 2.83 mM, n = 5), however, vanadate treatment reduced the blood glucose levels in diabetic rats but not in control rats (control 6.40 ± 0.38 nM, diabetic; 5.16 ± 0.66 nM), whereas insulin + vanadate treatment reduced the blood glucose levels in both control and diabetic rats (control 2.79 ± 0.38 mM, diabetic 2.20 ± 0.22 mM). However the reduction is significantly higher in diabetic rats as compared to control rats (87% reduction in diabetic rats as compared to 59% in control rats).

G-proteins and adenylyl cyclase in chronic STZ-induced diabetic liver

G-proteins play an important role in the regulation of variety of signal transduction systems including adenylyl cyclase/cAMP. Since the levels of G-proteins (Giα and Gsα) have been shown to be altered in short-term diabetic liver, it was of interest to investigate if the levels of G-protein in chronic liver remain altered or normalized by some compensatory mechanism. We therefore determined the levels of G-proteins; Giα and Gsα at protein and mRNA level by immunoblotting and Northern blotting techniques using specific antisera against different G-proteins and cDNA probes of G-proteins respectively. The RM/1 antibody specific for Gsα recognized the 42 kDa and 45 kDa forms of Gsα in both STZ-D and control rats, however the levels of both the forms were increased by 186 ± 45 and 156 ± 50% respectively as determined by densitometric scanning in STZ-D rats as compared to control rats. The AS/7 antibody specific for $Gi_{\alpha-1}$ and $Gi_{\alpha-2}$ recognized a single protein of 40 kDa referred to as $Gi_{\alpha-2}$ and the EC/2 antibody recognized a single protein of 41 KDa referred to as $Gi_{\alpha-3}$. The levels of $Gi_{\alpha-2}$ and $Gi_{\alpha-3}$ were also increased by 100 ± 24% and 50 ± 25% respectively in STZ-D rats as compared to control. However the mRNA levels of Gsα, $Gi_{\alpha-2}$ and $GiO_{\alpha-3}$ were not significantly different in control and STZ-D rats.

To investigate the relationship between the enhanced levels of Gs and functions, the effect of GTPγS on adenylyl cyclase was studied in livers from control and STZ-D rats. The results shown in Table 1 demonstrate that GTPγS stimulated adenylyl cyclase activity in a concentration dependent manner in STZ-D and control rats, however the net stimulation of adenylyl cyclase (stimulation with GTPγS-basal activity) was significantly increased in STZ-D rats. Similarly GTP, N-Ethylcarboxamide adenosine (NECA) and glucagon, which mediate their stimulatory responses through Gs, also stimulated adenylyl cyclase activity to a greater extent by about 150,300 and 90% respectively in STZ-D rats as compared to control rats. In addition NaF and FSK-stimulated adenylyl cyclase activity was also higher in STZ-D rats as compared to control rats.

We also examined the Gi-functions by determining the receptor-mediated and non-receptor-mediated inhibition of adenylyl cyclase, Gi receptor-independent functions of Gi were

Table 1. Effect of GTPγS on adenylyl cyclase activity in liver membranes from control and chronic STZ-D rats

	Adenylyl cyclase activity pmol.cAMP $(\text{mg protein.10 mn})^{-1}$		Net stimulation (stimulated activity – basal activity)	
$GTP_{\gamma}S$ (M)	Control	Diabetes	Control	Diabetes
None	35 ± 6	46 ± 7	—	—
10^{-9}	31 ± 4	63 ± 14	0	17
10^{-8}	131 ± 7	146 ± 22	96	100
10^{-7}	372 ± 22	535 ± 24**	337	489
10^{-6}	495 ± 24	735 ± 89**	460	689
10^{-5}	532 ± 49	796 ± 114**	497	750

Adenylyl cyclase activity was determined as under 'Methods'. Values are means ± S.E.M. of four separate experiments. **P < 0.01 between STZ-D and control.

examined by studying the effect of various concentrations of GTPγS on FSK-stimulated activity, whereas receptor dependent inhibition was examined by studying the effect of oxotremorine as an inhibitory hormone on adenylyl cyclase activity. GTPγS at lower concentrations (up to 10^{-9} M) inhibited the FSK-stimulated enzyme activity in a concentration dependent manner in both groups, however the percent inhibition was not significantly different in STZ-D rats as compared to control rats. On the other hand, at higher concentrations (10^{-8} M), the percent inhibition was decreased more in STZ-D rats as compared to control rats and at 10^{-7} M, GTPγS augmented the FSK-stimulated activity in control and STZ-D rats by about 35 and 50% respectively (data not shown), which may be due to the activation of Gs proteins (2). In addition, oxotremorine which inhibits adenylyl cyclase through Gi (33) (receptor-dependent inhibition) inhibited the enzyme activity in STZ-D and control rats to a similar extent (data not shown).

Effect of vanadium and insulin treatment on adenylyl cyclase and G-protein levels in STZ-D rat liver

Insulin and vanadium, an insulinomimetic agent, has been used in the present studies to investigate if altered levels of G-protein and altered adenylyl cyclase activity in STZ-D liver could be normalized to control levels in diabetic rats by such treatments. Insulin as well as vanadyl sulphate treatments restored the enhanced stimulation of adenylyl cyclase by GTPγS, GTP, NECA, glucagon, NaF and forskolin to various degrees, however, the extent of recovery was significantly higher in vanadyl sulphate-treated group as compared to insulin-treated group (control, (C); GTP, 48 ± 3; diabetes (D), 87 ± 9; D + insulin (I), 66 ± 12; D + vanadate (V), 64 ± 7; D + I + V, 46 ± 7; C; NECA + GTP, 67 ± 2; D, 166 ± 3; D + I, 135 ± 8, D + V, 96 ± 9, D + V + I, 73 ± 3 pmol cAMP (mg protein.10 min)$^{-1}$. The combined treatment of insulin and vanadyl sulphate improved further the GTP-, NECA- and glucagon-(data not shown) stimulated adenylyl cyclase activity but not of FSK- and NaF-(data not shown) stimulated enzyme activity (Control FSK, 1143 ± 182, D 1688 ± 240; D + I, 1051 ± 150; D + V, 942 ± 189; D + I + V, 1102 ± 256 pmol cAMP (mg protein.10 min)$^{-1}$.

In addition, the enhanced levels of $Gi_{\alpha-2}$ and $Gs_{\alpha42}$ and $Gs_{\alpha45}$ in diabetic liver were restored to control levels to different extents by vanadyl sulphate, insulin and combined treatments of both the agents. For example vanadyl sulphate treatment reversed the enhanced levels of $Gs_{\alpha45}$ and $Gs_{\alpha42}$ by about 80 and 60% respectively, whereas the percent recovery by insulin treatment was about 60%, however the combined treatments further improved the percentage recovery of $Gs_{\alpha42}$ but not of $Gs_{\alpha45}$ (data not shown). On the other hand, insulin treatment was less effective (~ 25%) in restoring the enhanced levels of $Gi_{\alpha-2}$ and $Gi_{\alpha-3}$ whereas vanadyl sulphate treatment restored these levels to control levels, and combination of the treatments did not further improve the recovery (data not shown).

Discussion

The present studies demonstrate that the chronic STZ-induced diabetes results in an enhanced levels of hepatic $Gi_{\alpha-2}$ and $Gi_{\alpha-3}$ proteins without any changes in their functions, whereas $Gs\alpha_{42}$ and $Gs\alpha_{45}$ protein levels and associated functions are augmented and vanadyl sulphate as well as insulin treatment were able to restore the altered levels of G-proteins and adenylyl cyclase responsiveness to various hormones to various degrees, however vanadyl sulphate was more effective than insulin.

An increased stimulation of adenylyl cyclase by GTPγS in livers from chronic STZ-D rats as compared to control rats may be due to the increased levels of $Gs_{\alpha45}$ and $Gs_{\alpha42}$ as demonstrated by immunoblotting experiments. An increased levels of Gsα proteins and increased stimulation of adenylyl cyclase by GTP has also been reported by Lynch *et al.* [1] in liver plasma membranes from acute 5–7 days STZ-D rats. In addition, Bushfield *et al.* [34] have also reported that short term diabetes resulted in an enhanced expression of $Gs_{\alpha45}$ protein as well as mRNA in plasma membranes of whole liver. Thus, it may be suggested that the enhanced expression of Gsα in liver membranes from acute (short term) as well as from chronic (long term) is independent of the duration of diabetes. However, in contrast to liver membranes, hepatocytes from acute STZ-D rats have been reported to have decreased levels of $Gs_{\alpha42}$ protein [34] and Gsα transcripts [35], whereas adipocytes from acute STZ-D rats did not show any alterations in the expression of Gsα levels [35]. These apparent discrepancies in the levels of Gsα in STZ-D rats appears to be attributed to the tissue differences.

The increased responsiveness of NECA and glucagon to adenylyl cyclase stimulation in chronic STZ-D rats may be explained by the upregulation of hormone receptors, however, several reports have shown a down regulation or no change [36–38] in the number of glucagon receptors in experimental short term diabetes. Thus, it appears that the increased levels of Gsα in chronic STZ-D rats may be responsible for the observed augmented responsiveness of adenylyl cyclase to glucagon stimulation. However, the failure of isoproterenol to stimulate adenylyl cyclase to a greater extent in STZ-D rats in spite of having higher levels of Gsα as compared to control rats, may be due to the possibility that β-adrenergic receptors are down regulated in chronic STZ-D rats. However the increased absolute enzyme activity in the presence of isoproterenol in STZ-D rats may be the reflection of the higher basal enzyme activity in the presence of GTP. In this regard, Bushfield *et al.* [34] by immunoblot analysis have shown enhanced levels of catalytic subunit of adenylyl cyclase in both hepatocytes and liver plasma membranes from acute STZ-D rats. The types VI and VII isoforms of adenylyl cyclase have been reported to be abundant in liver [39]. However, whether these isoforms are overexpressed in STZD rats and contribute to the enhanced basal activity in STZ-D rat is not known and remains to be explored.

Enhanced stimulation of adenylyl cyclase by FSK in STZ-D rat may be due to the enhanced levels of catalytic subunit of adenylyl cyclase in STZ-D rat [34]. The requirement of Gs and guanine nucleotides for the FSK activation of adenylyl cyclase has also been reported [40]. Since the present studies demonstrate an enhanced levels of Gsα, it may be possible that the increased sensitivity of adenylyl cyclase to FSK stimulation in STZ-D rats may be attributed to the augmented levels of Gsα.

A correlation between the levels of Gi protein and adenylyl cyclase inhibition has also been reported by several investigators in various pathophysiological conditions including diabetes [1, 27, 32]. In the present studies, we have also shown that the levels of $Gi_{\alpha-2}$ and $Gi_{\alpha-3}$ as well as the functions were unaltered in liver membranes from acute STZ-D rats. These results are in agreement with the studies reported by Lynch *et al.* [1] who by using pertussis-toxin-catalyzed ADP-ribosylation and immunological techniques also showed that the levels of Giα proteins as well as their functions were not altered in liver membranes from acute STZ-D rats. Similarly, Allard *et al.* [14] were unable to detect any changes in the levels of Giα-2 protein in liver plasma membranes from acute STZ-D rats, but showed a complete attenuation of Gi-functions. In addition, Bushfield *et al.* [34] also did not observe any alterations in the levels of Giα protein in liver membranes but showed a decreased expression of Giα-2 mRNA levels in livers from acute STZD rats. In contrast, a decreased levels of Giα-2 protein and a complete attenuation of Gi-mediated functions were observed in hepatocytes from acute STZ-D rats [34]. These apparent discrepancies on the levels of Giα-2 in liver and hepatocytes from acute STZ-D rats were suggested to be attributed to the heterogeneity of the cell population used for these studies [34]. For example whole liver membranes have been reported to have heterogenous population of cells containing parenchymal and non parenchymal cells, whereas hepatocytes represent only parenchymal cells [34]. However, Kirkham *et al.* [13] have shown that hepatocytes as well as whole liver membranes from acute STZ-D rats behave in a similar way as far as Gi-protein expression and functions are concerned. Similar to the results observed by other investigators [1, 14] these investigators have reported that the levels of Giα protein as well as Gi-function were unaltered in both hepatocytes (parenchymal cells) and whole liver membranes (parenchymal + non parenchymal cells) [13]. On the other hand, in adipocytes from STZ-D rats, the levels of Giα-3 protein and mRNA were significantly enhanced without any changes in the levels of Giα-2 and Gsα protein or mRNA, and the Gi functions were attenuated [2]. In the present studies, we also demonstrate that the Gi functions; receptor-independent-(GTPγS-mediated inhibition of forskolin-stimulated adenylyl cyclase) or receptor-dependent (oxotremorine-mediated inhibition of adenylyl cyclase) inhibition of adenylyl cyclase was not different in whole liver membranes from acute as well as chronic STZ-D rats as compared to control rats which suggests that the Gi functions may not be altered in both acute and chronic STZ-induced diabetes. However, we report enhanced levels of Giα-2 and Giα-3 proteins in chronic but not in acute STZ-D rats as determined by immunoblotting techniques, which were not reflected in the increased functions. The reason for this increase in $Gi_{\alpha-2}$ and Giα–3 levels in chronic STZ-D rats is not clear and may be attributed to the duration of diabetes. A lack of relationship between Gi functions and Gi levels has also been reported [2, 14]. However, the reason for this in chronic STZ-D rats is not clear and remains to be investigated. One possible explanation may be that the levels of Giα-2 and Giα-3 proteins which are increased in chronic STZ-D rats may be in the phosphorylated form and therefore are incapable of eliciting the enhanced Gi-mediated functions. In this context, an enhanced phosphorylation of Giα-2 protein in hepatocytes from diabetic rats has been demonstrated [34]. However, it would be interesting to explore if dephosphorylation of the liver membranes from chronic STZ-D rats by alkaline phosphatase treatment would also result in an enhanced Gi functions.

On the other hand, we were unable to detect any alterations in the levels of G-protein mRNAs, suggesting that the chronic STZ-induced diabetes may not alter the gene expression of G-proteins. However, it may be possible that the alterations in the expression of G-protein levels may be at translational or post-translational levels, and may involve an increased efficiency of translation of G-protein mRNA or decreased degradation rate of G-proteins. Our results are in contrast with the studies reported earlier [34] where acute STZ-induced diabetes resulted in the reduction of the levels of $Gi_{\alpha-2}$ mRNA and increase in Gs transcripts in liver and hepatocytes [34, 35]. However in agreement to our results, these investigators have also not been able to observe any significant changes in the levels of $Gi_{\alpha-3}$ mRNA in livers from STZ-D rats [34], whilst the $Gi_{\alpha-3}$ protein levels have been shown to be decreased. Similarly, no significant changes in the protein levels of $Gi_{\alpha-1}$, $Gi_{\alpha-2}$ and Gs_{α} have been shown in adipocytes from STZ-D rats, whereas mRNA coding for $Gi_{\alpha-1}$ and $Gi_{\alpha-3}$ were dramatically increased [35].

Insulin, a hypoglycemic hormone exerts some of the effects by interacting with G-proteins [25]. Insulin has also been shown to inhibit adenylyl cyclase in a GTP-dependent manner [41]. Insulin therapy has been shown to reverse the enhanced responsiveness of adenylyl cyclase to cholera toxin in liver membranes from short term STZ-D rats [1]. In addition, the augmented labelling of $Gs\alpha_{43}$ and $Gs\alpha_{47}$ in STZ-D rats was also shown to be partially reversed by insulin treatment [1]. In the present studies, we have also shown that insulin as well as vanadyl sulphate therapy of the STZ-D rats restored significantly the increased sensitivity of guanine nucleotides (GTP, GTPγS) to stimulate adenylyl cyclase activity, however vanadyl sulphate was more effective than insulin. The restoration of the enhanced GTP or GTPγS-sensitive adenylyl cyclase activity by these treatments appear to be associated with their ability to restore the levels of $Gs_{\alpha 42}$ and $Gs_{\alpha 45}$ protein as has been demonstrated in the present studies by using immunoblotting techniques. The restoration of the levels of Gsα as well as of $Gi_{\alpha-2}$ and $Gi_{\alpha-3}$ by vanadyl sulphate was also greater than that observed by insulin treatment.

In addition, vanadyl sulphate as well as insulin treatment was also able to restore the augmented stimulations of adenylyl cyclase exerted by NECA and glucagon, which may partly be due to the restoration of Gsα protein, that mediates the stimulatory effects of hormones on adenylyl cyclase. Since glucagon receptors have been shown to be decreased or unaltered but not increased in STZ-D rats [36–38], the restoration of glucagon-sensitive adenylyl cyclase may thus not be at the receptor level and may involve the Gsα proteins. The adenosine receptor have also been shown to be unaltered in adipocytes from STZ-D rats [42] however no studies have been conducted on adenosine receptors in liver membranes. Taken together, it may be possible that the reversal of NECA-sensitive adenylyl cyclase by insulin and vanadyl sulphate may also be associated with the restoration of Gsα protein. A further improvement in the restoration of the levels of Gsα as of well as of glucagon- and NECA-sensitive adenylyl cyclase activity by combined treatments of insulin and vanadyl sulphate may be due to the ability of vanadyl sulphate to increase the sensitivity of insulin to exert its effects. Vanadyl sulphate induced augmentation of insulin sensitivity in various actions of insulin has been reported previously [22].

In conclusion, we have shown that the chronic (long term, 10 week) STZ-diabetes like acute STZ-induced diabetes resulted in the overexpression of Gsα protein and its functions in whole liver membranes. In contrast, the levels of $Gi_{\alpha-2}$ and $Gi_{\alpha-3}$ proteins are increased without any alterations in Gi function. Vanadyl sulphate mimics the effect of insulin and is more potent than insulin in restoring the enhanced levels of G-proteins as well as the altered stimulations exerted by stimulatory hormones. It may thus be suggested that one of the mechanisms by which vanadyl sulphate improves the glucose homeostasis in STZ-D rats may be through its ability to modulate the levels of Gs proteins and adenylyl cyclase signal transduction system.

Acknowledgment

We would like to thank Christiane Laurier for her valuable secretarial help.

References

1. Lynch CJ, Blackmore PF, Johnson EH, Wange RL, Krone PK, Exton JH: Guanine nucleotide binding regulatory proteins and adenylate cyclase in livers of streptozotocin- and BB/Wor-diabetic rats. Immunodetection of Gs and Gi with antisera prepared against synthetic peptides. J Clin Invest 83: 2050–2062, 1989
2. Bushfield M, Griffiths SL, Strassheim D, Tang E, Shakur Y, Lavan B, Houslay MD: Guanine-nucleotide-binding proteins in diabetes and insulin-resistant states. Biochem Soc Symp 56: 137–154, 1990
3. Gilman AG: G-protein and dual control of adenylate cyclase. Cell 36: 577–579, 1984
4. Gilman AG: G-proteins: transducers of receptor-generated signals. Ann Rev Biochem 56: 615–649, 1987
5. Stryer L, Bourne HR: G-proteins: a family of signal transducers. Ann Rev Cell Biol 2: 391–419, 1986
6. Sternweis PC, Northup JK, Smigel MD, Gilman AG: The regulatory component of adenylate cyclase. J Biol Chem 256: 11517–11526, 1981
7. Murakami T, Yasuda H: Rat heart cell membranes contain three substrates for cholera toxin-catalyzed ADP-ribosylation and a single substrate for pertussis toxin-catalyzed ADP-ribosylation. Biochem Biophys Res Commun 138: 1355–1361, 1986
8. Robishaw JD, Smigel MD, Gilman AG: Molecular basis for two forms of the G-protein that stimulates adenylyl cyclase. J Biol Chem 261: 9587–9590, 1986
9. Bray P, Caster A, Simons C, Guo V, Puckett C, Hamholz J, Spiegel A, Nirenberg H: Human cDNA clones for four species of Gsα signal transduction protein. Proc Natl Acad Sci USA 83: 8893–8897, 1986
10. Itoh H, Kozaka T, Magata S, Nakamura S, Katada T, Ui M, Iwai S, Ohtsuka E, Kawasaki H, Suzuki K, Kaziro Y: Molecular cloning and sequence determination of cDNAs for a subunits of the guanine nucleotide-binding proteins Gs, Gi and Go from brain. Proc Natl Acad Sci USA 83: 3776–3780, 1986
11. Jones DT, Reed RR: Molecular cloning of five GTP-binding protein cDNA species from rat olfactory neuroepithelium. J Biol Chem 262: 14241–14249, 1987
12. Itoh H, Toyama R, Kozasa T, Tsukamoto T, Matsuoka M, Kaziro Y: Presence of three distinct molecular species of Gi protein: a subunit structure of rat cDNA and human genomic DNAs. J Biol Chem 263: 6656–6664, 1988
13. Kirkham DM, Murphy CJ, Yong Y: Demonstration of inhibitory guanine nucleotide regulatory protein (Gi) function in liver and hepatocyte membranes from streptozotocin-treated rats. Biochem J 284: 301–304, 1992
14. Allard WJ, Vicasio PP, Saperstein R, Slater EE, Strout HV: The functions but not the expression of rat liver inhibitory guanine nucleotides binding protein is altered in streptozotocin-induced diabetes. Endocrinology 129: 169–175, 1991
15. Tolman EL, Barris E, Burns M, Pansini A, Partridge R: Effects of vanadium on glucose metabolism *in vitro*. Life Sci 25: 1159–1164, 1979
16. Shechter Y, Karlish SJD: Insulin-like stimulation of glucose oxidation in rat adipocytes by vanadyl ions (IV) ions. Nature 284: 556–558, 1980

17. Shechter Y: Insulin mimetic effects of vanadate possible implications for future treatment of diabetes. Diabetes 39: 1–5, 1990
18. Heyliger CE, Tahiliani AG, McNeill JH: Effect of vanadate on elevated blood glucose and depressed cardiac performance of diabetic rats. Science 227: 1474–1477, 1985
19. Bendayan M, Gingras D: Effect of vanadate administration on blood glucose and insulin levels as well as on exocrine pancreatic function in streptozotocin-diabetic rats. Diabetologia 32: 561–567, 1989
20. Ramanadham S, Mongold JJ, Browsey W, Cros GH, McNeill JH: Oral vanadyl sulphate in treatment of diabetes mellitus in rats. Am J Physiol 257: 11904–11911, 1989
21. Pederson RA, Ramadaham S, Buchan AMJ, McNeill JH: Long term effects on vanadyl treatment on streptozotocin-induced diabetes in rats. Diabetes 38: 1390–1395, 1989
22. Erickson JW, Lonnroth P, Smith U: Vanadate increases cell surface insulin binding and improves insulin sensitivity in both normal and insulin-resistant adipocytes. Diabetologia 35: 510–516, 1992
23. Mooney RA, Boodwell KL, Luhowskyj S, Casneillie JE: The insulin-like effect of vanadate on lipolysis in rat adipocytes is not accompanied by an insulin-like effect on tyrosine phosphorylation. Endocrinology 124: 422–429, 1989
24. Strout HV, Vicario P, Saperstein R, Slater EE: The insulin-mimetic effect of vanadate is not correlated with insulin receptor tyrosine kinase activity nor phosphorylation in mouse diaphragms *in vivo*. Endocrinology 124: 1918–1924, 1989
25. Eckel J, Gerlack-Eskchen E, Reinauer H: G-protein mediated regulation of the insulin-responsive glucose transporter in isolated cardiac myocytes. Biochem J 272: 691–696, 1990
26. Srivastava AK, Srivastava MB: Streptozotocin-induced diabetes and hormone sensitivity of adenylate cyclase in rat myocardial sarcolemma, aorta and liver. Biochem Pharmacol 34: 2013–2017, 1985
27. Anand-Srivastava MB: Enhanced expression of inhibitory guanine nucleotide regulatory protein in spontaneously hypertensive rats: relationship to adenylate cyclase inhibition. Biochem J 288: 79–85, 1992
28. Anand-Srivastava MB: Platelets from spontaneously hypertensive rats exhibit decreased expression of inhibitory guanine nudeotide regulatory protein; relation with adenylyl cyclase activity. Circ Res 73: 1032–1039, 1993
29. Salomon Y, Londos C, Rodbell M: A highly sensitive adenylate cyclase assay. Anal Biochem 58: 541–548, 1974
30. Lowry OH, Rosebrough NJ, Farr AL, Randall RJ: Protein measurement with the folin phenol reagent. J Biol Chem 193: 265–275, 1951
31. Thibault C, Anand-Srivastava MB: Altered expression of G-protein mRNA in spontaneously hypertensive rats. FEBS Lett 313: 160–164, 1992
32. Anand-Srivastava MB, de Champlain J, Thibault C: DOCA-salt hypertensive rat hearts exhibit altered expression of G-proteins. Am J Hypertens 6: 72–75, 1993
33. Wei JW, Wang MC: Muscarinic M_2 receptors coupled to inhibition of adenylate cyclase in rat heart. Chinese J Physiol 33: 315–326, 1990
34. Bushfield M, Griffiths SL, Murphy GJ, Pyne NJ, Knowler JT, Milligan G, Parker PJ, Mollner S, Houslay MD: Diabetes-induced alterations in the expression, functioning and phosphorylation state of the inhibitory guanine nucleotide regulatory protein Gi-2 in hepatocytes. Biochem J 271: 365–372, 1990
35. Griffiths SL, Knowler JT, Houslay MD: Diabetes-induced changes in guanine nucleotide-regulatory protein mRNA detected using synthetic oligonucleotide probes. Eur J Biochem 193: 367–374, 1990
36. Bachmann W, Bottger I, Allgayer H, Haslbeck M, Hepp KD: [Insulin, glucagon binding and adenylate cyclase activity of liver plasma membranes in streptozotocin-diabetic rats] (trans.). Verhandlungen Deutschen Gesellschaft Inn Med 84: 1217–1220, 1978
37. Srikant CB, Freeman D, McCorkle K, Unger RH: Binding and biological activity of glucagon in liver cell membranes of chronically hyperglucagonemic rats. J Biol Chem 252: 7434–7436, 1977
38. Bhathena SJ, Voyles NR, Smith S, Recant L: Decreased glucagon receptors in diabetic rat hepatocytes. Evidence for regulation of glucagon receptors by hyperglucagonemia. J Clin Invest 51: 1488–1497, 1978
39. Krupinski J, Lehman TC, Prankenfield CD, Zwaagstra JC, Watson PA: Molecular diversity in the diversity in the adenylyl cyclase family: evidence for eight forms of the enzyme and cloning of type VI. J Biol Chem 267: 24858–24862, 1992
40. Hilderbrandt JD, Hanoune J, Birnbaumer J: Guanine nucleotide inhibition of Cyc 549 mouse lymphoma cell membrane adenylyl cyclase. 257: 14723–14725, 1982
41. Heyworth CM, Houslay MD: Insulin exerts actions through a distinct species of guanine nucleotide regulatory protein: inhibition of adenylyl cyclase. Biochem J 214: 547–552, 1983
42. Green A, Johnson JL: Evidence for impaired coupling of receptor to Gi protein in adipocytes from streptozotocin-induced diabetic rats. Diabetes 40: 89–94, 1991

Molecular and Cellular Biochemistry **153:** 121–124, 1995.

Effects of vanadate on the expression of genes involved in fuel homeostasis in animal models of Type I and Type II diabetes

S.M. Brichard

Unité d'Endocrinologie et Métabolisme, University of Louvain, Faculty of Medicine, UCL 55.30, Avenue Hippocrate 55, B-1200 Brussels, Belgium

Abstract

Vanadium is a trace element that has raised increasing interest in diabetology since the discovery of its insulin-like properties *in vitro* and *in vivo*. This brief article reviews the most recent data concerning the beneficial effects of vanadium compounds on fuel homeostasis in animal models of insulinopenic (Type I) or insulin-resistant (Type II) diabetes. These studies open obvious therapeutic possibilities in diabetes, and more particularly, in states of insulin resistance. (Mol Cell Biochem **153:** 121–124, 1995)

Key words: vanadium compounds, insulin-dependent diabetes, insulin-independent diabetes, gene expression in liver

Vanadium (V) salts are found to influence fuel homeostasis in animal models of Type I and Type II diabetes.

Insulin-deficient rats

The *in vivo* action of vanadium salts appears to be slower than that of insulin, even though they are considered as insulin-mimickers. In rats made insulin-deficient and diabetic by streptozotocin (STZ) injection (models of Type I diabetes), an acute intravenous infusion of vanadate for two hours (2–12 mg/kg body weight) exerted little or no effect on hyperglycemia and insulinemia . However, the levels of vanadium reached in the plasma of these animals were similar (1 μg/ml) (rev. in [1]) or higher than those measured after chronic administration of the compound, from the first hour onward (Fig. 1). These *in vivo* concentrations remained, however, ≈ 5–20-fold lower than those usually required to elicit acute insulin-like effects *in vitro* (rev. in [1]).

On the other hand, chronic administration of oral V compounds to STZ-diabetic rats induces a spectacular fall in blood glucose levels [2–6]. This fall is evidenced within the first week of the treatment and can be maintained for several months. It is often accompanied by a slowing of body weight gain [5–9], that can be regarded as an adverse reaction in insulin-deficient and catabolic rats, but as a positive effect in obese Type II animals (cfr infra). This reduction in body weight gain is explained by a decrease in food consumption induced by V compounds [5]. It has recently been claimed that the hypoglycemic action of vanadate was entirely due to its anorexigenic effect [10]. This is absolutely untrue for several reasons. First, vanadate hypoglycemic action is detectable in fasted as well as in fed rats [5]. Second, it is also evidenced during standardized glucose loads (oral or iv glucose tolerance tests) [5]. Third, oral vanadate given at low doses to mildly diabetic rats improves their glucose homeostasis without any repercussions on food intake and body weight gain [11]. Fourth, vanadate given to STZ-rats reverses the impaired expression of genes involved in hepatic fuel metabolism (cfr infra), a finding hardly compatible with mere anorexia or food restriction which produce opposite effects [12].

Vanadate has first been shown to correct the abnormal expression of genes coding for key enzymes of *glucose metabolism* in the liver of diabetic rats. This action contributes to the lowering of hyperglycemia. Liver mRNA levels of two major glycolytic enzymes, glucokinase (GK) and L-type pyruvate kinase (L-PK), which are blunted in diabetic rats, were partially (GK) or totally (L-PK) restored by vanadate treatment (Fig. 2) [13, 14]. This was also the case for the bifunctional enzyme, 6-phosphofructo-2-kinase [15]. In contrast to the glycolytic enzymes, mRNA levels of the gluconeogenic enzyme, phosphoenolpyruvate carboxykinase

Address for offprints: S.M. Brichard, Unité d'Endocrinologie et Métabolisme, University of Louvain, Faculty of Medicine, UCL 55.30, Avenue Hippocrate 55, B-1200 Brussels, Belgium

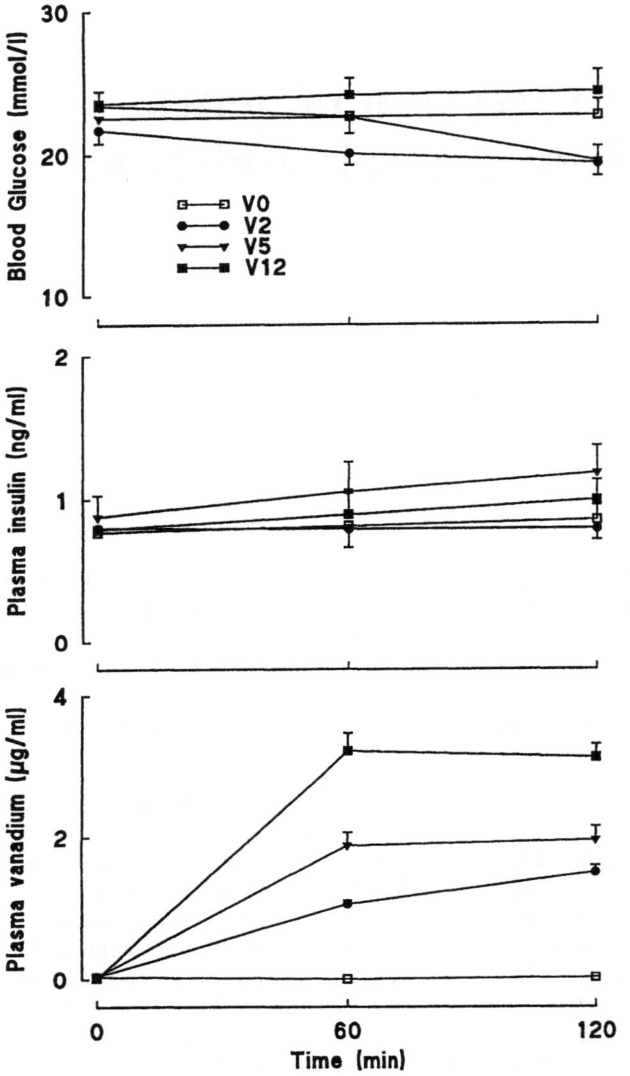

Fig. 1. Time course of the effects of a 2 h perfusion of vanadate on plasma glucose, insulin and vanadium levels in STZ-diabetic rats. The studies were performed in anesthetized animals (pentobarbital, 60 mg/kg, ip; ketamine, 5 mg/kg, im). Catheters were inserted in the right jugular vein for infusion and in the left carotid artery for blood samplings. A tracheotomy was performed to avoid respiratory difficulties. Body temperature was monitored and maintained at 37°C. Thirty minutes after completion of surgery, two successive blood samples were taken for measurement of basal glycemia and insulinemia. Saline (V0) or Na_3VO_4 (adjusted to pH 7.4) was then infused at a total dose of 2 mg/kg (V2), 5 mg/kg (V5), or 12 mg/kg (V12) over a 2 h period. Values are the mean ± SEM for 8–11 rats in each group. (Experiment performed by C Van Tilt).

(PEPCK), and of the liver glucose transporter (GLUT2), which are elevated and contribute to hepatic glucose overproduction in diabetic rats, were decreased toward normal values after vanadate administration (Fig. 2) [13, 14]. All these changes in mRNA levels were accompanied by simi-

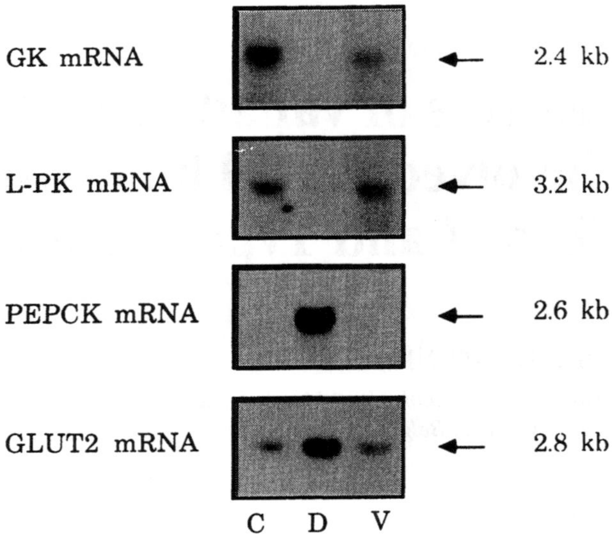

Fig. 2. Northern blot analysis of glucokinase (GK), L-type pyruvate kinase (L-PK), phosphoenolpyruvate carboxykinase (PEPCK) and GLUT2 transporter mRNA in liver from non-diabetic control (C), untreated diabetic (D) and vanadate-treated diabetic rats (V). This figure is representative of six separate experiments. (From ref. [13])

lar changes in enzyme activities or protein amount [13]. Hence, vanadate given to diabetic rats induces a shift of the predominating gluconeogenic flux, with subsequent hepatic glucose overproduction, into a glycolytic flux by correction of impaired pretranslational regulatory mechanisms. These effects were only partly reproduced by treatment with phlorizin, a drug which decreases glycemia merely by inducing a renal leak of glucose [16]. This indicates that the *in vivo* correction of gene expression is not simply explained by relief from glucose toxicity. A direct insulin-like action of vanadate, similar to that reported *in vitro*, is likely to play a major role in the *in vivo* effects of the element on liver enzyme genes. Vanadate has indeed been found to induce L-PK mRNA in primary cultures of hepatocytes [17] and to inhibit PEPCK gene transcription in hepatoma cells [18]. The latter effect required, however, concentrations of the element (0.5–2 mM) [18] one or two orders of magnitude higher than those reached in the plasma (~20 µM) [3, 8] or in the liver (~20 µmol/kg) [4, 8] of treated rats.

Recently, we examined whether vanadate administration to STZ-diabetic rats could also reverse the impaired expression of genes involved in key steps of *lipogenesis* in the two major sites implicated in this metabolic pathway: liver and white adipose tissue [19]. To this end, mRNA levels of acetyl-CoA carboxylase (ACC) and fatty acid synthase (FAS) were measured in rats treated with vanadate. Vanadate totally restored ACC mRNA and partially restored FAS mRNA (to 70 % of non-diabetic levels) in the liver. The activities of both hepatic lipogenic enzymes were also increased. By contrast in white adipose tissue, vanadate modified neither expression nor activity of either lipogenic enzyme, which remained strongly blunted (< 10 % of non-diabetic values). Thus, vanadate treatment of diabetic rats produces a tissue-specific correction of lipogenic enzyme gene expression. This correction results from a reversal of impaired pretranslational regulatory mechanisms. Since glucose 6-phosphate is thought to be the inducer [20] and T_3 an amplifier of lipogenic genes [21], these effects could be mediated by a selective restoration of liver glycolytic flux, and by an improvement of thyroid function, which is usually observed in vanadate-treated diabetic rats.

Finally, vanadate treatment of diabetic rats has also been shown to inhibit the increased expression of the mitochondrial 3-hydroxy-3-methylglutaryl-CoA synthase, the key regulatory enzyme of liver *ketogenesis* [14].

Besides liver, muscle is another site of vanadate action in the insulin-deficient diabetic rat. Vanadate treatment corrected the reduced glycogen synthase activity in skeletal muscle of diabetic rats [22]. It has also been reported to reverse the decreased amount of the insulin-sensitive glucose transporter, GLUT4, by increasing the corresponding levels of mRNA [23].

Insulin-resistant animals

Besides insulin deficiency, resistance of target tissues to the action of insulin is the other major cause of glucose intolerance. Vanadate also improves glucose homeostasis in hyperinsulinemic, insulin-resistant rats and mice (models of type II diabetes) [24–27]. Oral administration of vanadate for more than three months to genetically obese and mildly glucose-intolerant *fa/fa* rats markedly decreased plasma insulin levels and improved tolerance to glucose loads [24]. This improvement could not simply be explained by the lower body weight gain induced by the anorexigenic effect of the element, nor by a decrease in insulin counter-regulatory hormones [28]. Changes in FFA levels with subsequent attenuation of the glucose-fatty acid cycle were also excluded [29]. The beneficial effects of vanadate can rather be ascribed to a direct correction of impaired tissue sensitivity to insulin. Euglycemic-hyperinsulinemic clamps have shown that the increment of whole-body glucose disposal brought about by insulin was larger in *fa/fa* rats treated with vanadate, and that the increase in glucose uptake occurred essentially in muscle [30]. This is of particular interest because muscle is the major site of insulin-mediated glucose disposal [31]. Since we found no change in insulin receptor number and affinity [29], and others did not observe any alteration of receptor tyrosine kinase activity [26], vanadate action is likely to involve, in this model, step(s) distal to the receptor.

As glucose transport is the rate-limiting step for glucose metabolism [32], we investigated whether vanadate treatment affects the expression of the insulin-responsive glucose transporter, GLUT4. Endurance exercise training, which is also an effective means of reducing muscle insulin-resistance in the *fa/fa* rat has indeed been found to induce a significant rise in muscle GLUT4 peptide levels [33]. This was not the case for vanadate which did not modify either GLUT4 mRNA or protein (Fig. 3) [29]. Different molecular mechanisms may, thus, ultimately lead to a reversal of the impaired muscle insulin sensitivity. Yet unknown, those mechanisms involved in V action could be related to a *functional* improvement of glucose transporters due to more efficient translocation to the plasma membrane and/or increased intrinsic activity.

In conclusion, vanadate treatment restores fuel homeostasis in liver and muscle of insulin-deficient diabetic rats. In the liver, this improvement is in part due to a correction of impaired expression of key genes involved in glucose, lipid or ketone bodies metabolism.

On the other hand, vanadate also proves efficient in insulin-resistant diabetes. In the genetically obese fa/fa rat, the muscle is the major site of action of the element. Vanadate treatment markedly reduces muscle insulin resistance probably through a functional improvement of GLUT4.

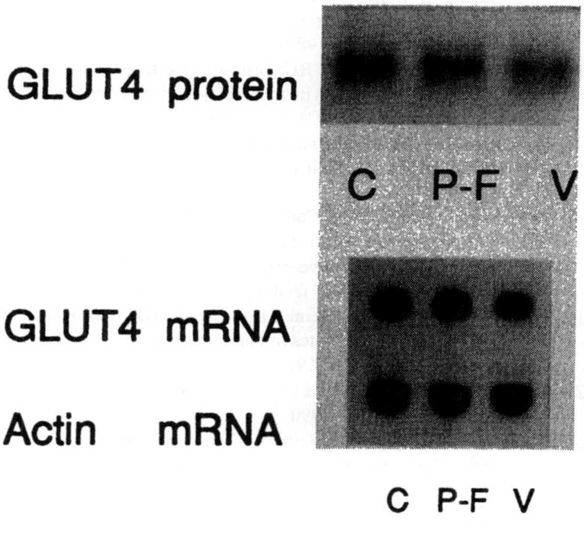

Fig. 3. Western blot of GLUT4 protein and Northern blots of GLUT4 and actin mRNA in tibialis anterior of fa/fa rats from three experimental groups. Control rats (C) were untreated and fed ad libitum; vanadate-treated rats (V) received ad libitum Na_3VO_4 in food and drinking solutions; a third group of rats received a restricted amount of food and were pair-fed (P-F) with vanadate-treated rats in order to ensure a similar slowing of body weight gain. This figure is representative of five separate experiments. (From ref. [29])

Acknowledgments

Author's work was supported by grants of the FNRS, FRSM and FDS, Brussels, Belgium. S.M.B. is Chercheur Qualifié of the Fonds National de la Recherche Scientifique.

References

1. Brichard SM, Lederer J, Henquin JC: The insulin-like properties of vanadium: a curiosity or a perspective for the treatment of diabetes? Diabete Metab 17: 435 440, 1991
2. Heyliger CE, Tahiliani AG, McNeill JH: Effect of vanadate on elevated blood glucose and depressed cardiac performance of diabetic rats. Science 227: 1474–1477, 1985
3. Meyerovitch J, Farfel Z, Sack J, Shechter Y: Oral administration of vanadate normalizes blood glucose levels in streptozotocin-treated rats. J Biol Chem 262: 6658–6662, 1987
4. Gil J, Miralpeix M, Carreras J, Bartrons R: Insulin-like effects of vanadate on glucokinase activity and fructose 2, 6-bisphosphate levels in the liver of diabetic rats. J Biol Chem 263: 1868–1871, 1988
5. Brichard SM, Okitolonda W, Henquin JC: Long term improvement of glucose homeostasis by vanadate treatment in diabetic rats. Endocrinology 123: 2048–2053, 1988
6. Blondel O, Bailbe D, Portha B: *In vivo* insulin resistance in streptozotocin-diabetic rats – evidence for reversal following oral vanadate treatment. Diabetologia 32: 185–190, 1989
7. Ramanadham S, Mongold JJ, Brownsey RW, Cros GH, McNeill JH: Oral vanadyl sulfate in treatment of diabetes mellitus in rats. Am J Physiol 257: H904–H911, 1989
8. Pugazhenthi S, Khandelwal RL: Insulinlike effects of vanadate on hepatic glycogen metabolism in nondiabetic and streptozocin-induced diabetic rats. Diabetes 39: 821–827, 1990
9. Venkatesan N, Avidan A, Davidson MB: Antidiabetic action of vanadyl in rats independent of *in vivo* insulin-receptor kinase activity. Diabetes 40: 492–498, 1991
10. Malabu UH, Dryden S, McCarthy HD, Kilpatrick A, Williams G: Effects of chronic vanadate administration in the STZ-induced diabetic rat. The antihyperglycemic action of vanadate is attribuable entirely to its suppression of feeding. Diabetes 43: 9–15, 1994
11. Henquin JC, Carton F, Ongemba LN, Becker DJ: Improvement of mild hypoinsulinaemic diabetes in the rat by low doses of vanadium. J Endocrinol 142: 555–561, 1994
12. Granner D, Pilkis S: The genes of hepatic glucose metabolism. J Biol Chem 265: 10173–10176, 1990
13. Brichard SM, Desbuquois B, Girard J: Vanadate treatment of diabetic rats reverses the impaired expression of genes involved in hepatic glucose metabolism: effects on glycolytic and gluconeogenic enzymes, and on glucose transporter GLUT2. Mol Cell Endocrinol 91: 91–97, 1993
14. Valera A, Rodriguez-Gil JE, Bosch F: Vanadate treatment restores the expression of genes for key enzymes in the glucose and ketone bodies metabolism in the liver of diabetic rats. J Clin Invest 92: 4-11, 1993
15. Miralpeix M, Carballo E, Bartrons RI, Crepin K, Hue L, Rousseau GG: Oral administration of vanadate to diabetic rats restores liver 6-phosphofructo-2-kinase content and mRNA. Diabetologia 35: 243–248, 1992
16. Brichard SM, Henquin JC, Girard J: Phlorizin treatment of diabetic rats partially reverses the abnormal expression of genes involved in hepatic glucose metabolism. Diabetologia 36: 292–298, 1993
17. Miralpeix M, Decaux JF, Kahn A, Bartrons R: Vanadate induction of L-type pyruvate kinase mRNA in adult rat hepatocytes in primary cultures. Diabetes 40: 462–464, 1991
18. Bosch F, Hatzoglou M, Park EA, Hanson RW: Vanadate inhibits expression of the gene for phosphoenolpyruvate carboxykinase (GTP) in rat hepatoma cells. J Biol Chem 265: 13677–13682, 1990
19. Brichard SM, Ongemba LN, Girard J, Henquin JC: Tissue-specific correction of lipogenic enzyme gene expression in diabetic rats given vanadate. Diabetologia 37: 1065–1072, 1994
20. Foufelle F, Gouhot B, Pegorier JP, Perdereau D, Girard J, Ferre P: Glucose stimulation of lipogenic enzyme gene expression in cultured white adipose tissue. A role for glucose 6-phosphate. J Biol Chem 267: 20543–20546, 1992
21. Iritani N: Nutritional and hormonal regulation of lipogenic-enzyme gene expression in rat liver. Eur J Biochem 205: 433–442, 1992
22. Rossetti L, Laughlin MR: Correction of chronic hyperglycemia with vanadate, but not with phlorizin, normalizes *in vivo* glycogen repletion and *in vitro* glycogen synthase activity in diabetic skeletal muscle. J Clin Invest 84: 892–899, 1989
23. Strout HV, Vicario PP, Biswas C, Saperstein R, Brady EJ, Pilch PF, Berger J: Vanadate treatment of streptozotocin diabetic rats restores expression of the insulin responsive glucose transporter in skeletal muscle. Endocrinology 126: 2728–2732, 1990
24. Brichard SM, Pottier AM, Henquin JC: Long term improvement of glucose homeostasis by vanadate in obese hyperinsulinemic fa/fa rats. Endocrinology 125: 2510–2516, 1989
25. Brichard SM, Bailey CJ, Henquin JC: Marked improvement of glucose homeostasis in diabetic ob/ob mice given oral vanadate. Diabetes 39: 1326–1332, 1990
26. Meyerovitch J, Rothenberg P, Shechter Y, Bonner-Weir S, Kahn CR: Vanadate normalizes hyperglycemia in two mouse models of non-insulin-dependent diabetes mellitus. J Clin Invest 87: 1286–1294, 1991
27. Pugazhenthi S, Angel JF, Khandelwal RL: Long-term effects of vanadate on glycogen metabolizing and lipogenic enzymes of liver in genetically diabetic (db/db) mice. Metabolism 40: 941–946, 1991
28. Brichard SM, Ongemba LN, Kolanowski J, Henquin JC: The influence of vanadate on insulin counter-regulatory hormones in obese fa/fa rats. J Endocrinol 131: 185–191, 1991
29. Brichard SM, Assimacopoulos-Jeannet F, Jeanrenaud B: Vanadate treatment markedly increases glucose utilization in muscle of insulin-resistant fa/fa rats without modifying glucose transporter expression. Endocrinology 131: 311–317, 1992
30. Brichard SM, Ongemba LN, Henquin JC: Oral vanadate decreases muscle insulin resistance in obese fa/fa rats. Diabetologia 35: 522–527, 1992
31. Kraegen EW, James DE, Jenkins AB, Chisholm DJ: Dose response curves for *in vivo* insulin sensitivity in individual tissues in rats. Am J Physiol 248: E353-E362, 1985
32. Morgan HE, Cadenas E, Regen DM, Park CR: Regulation of glucose uptake in muscle. II. Rate limiting steps and effect. J Biol Chem 236: 262–268, 1961
33. Friedman JE, Sherman WN, Reed MJ, Elton CW, Dohm GL: Exercise training increases glucose transporter protein GLUT4 in skeletal muscle of obese Zucker (fa/fa) rats. FEBS Lett 268: 13–16, 1990

Molecular and Cellular Biochemistry **153**: 125–129, 1995.

Decrease in protein tyrosine phosphatase activities in vanadate-treated obese Zucker (fa/fa) rat liver

Subbiah Pugazhenthi, Feridoon Tanha, Bruce Dahl and Ramji L. Khandelwal

Department of Biochemistry, University of Saskatchewan, Saskatoon, Saskatchewan, Canada, S7N 5E5

Abstract

The inhibitory action of vanadate towards protein tyrosine phosphatase (PTPase) has been considered as a probable mechanism by which it exerts insulin-like effects. In this study, we have examined the *in vivo* effects of vanadate on PTPases in the liver of obese Zucker rats, a genetic animal model for obesity and type II diabetes. These animals were characterized by hyperinsulinemia and mild hyperglycemia. The number of insulin receptors were significantly ($p < 0.01$) decreased in liver. After chronic administration of vanadate in obese rats, 80% decrease in the plasma levels of insulin was observed. The insulin receptor numbers were significantly ($p < 0.01$) higher in vanadate-treated obese rats as compared to the untreated ones. The hepatic PTPase activities in cytosolic and particulate fractions, with phosphorylated poly glu:tyr (4:1) and the insulin receptor peptide (residues 1142–1153) as substrates, increased in obese rats. In vanadate-treated obese rat livers, the PTPase activities in both subcellular fractions with these substrates decreased significantly ($p < 0.001$). The decreases in PTPase activities from these groups of rats were further supported by chromatography on a Mono Q column. These data support the view that inhibition of PTPases plays a role in the insulin-mimetic action of vanadate. (Mol Cell Biochem **153**: 125–129, 1995)

Key words: vanadate, obesity, diabetes, protein tyrosine phosphatase

Introduction

Protein tyrosine phosphatases (PTPase), a diverse family of enzymes are known to play a role in signal transduction and cell cycle regulation [1]. The autophosphorylation of insulin receptor and the resultant tyrosine kinase activity are essential for the signal transduction by insulin [2]. PTPases which can reverse these processes have been implicated to play a role in the regulation of insulin receptor tyrosine kinase activity [3]. Some of these PTPases expressed in liver tissue have been identified [4].

Salts of vanadium have been shown to exert insulin-like effects in various tissues such as liver, muscle and adipocytes [5]. The mechanism of their action is not clearly understood. However, there are studies showing that activation of insulin receptor tyrosine kinase might be involved [6-8]. Orthovanadate being a potent inhibitor of PTPases could prevent the inactivation of insulin receptor tyrosine kinase and prolong the action of the hormone [7]. Vanadate has been shown to inhibit hepatic PTPases *in vivo* in streptozotocin-induced diabetic rats [9,10]. The aim of this study is to examine whether the liver PTPases are altered in obese Zucker rats, animal model for type II diabetes and, if so, whether vanadate can normalize them *in vivo*.

Materials and methods

Materials

[γ-^{32}P]ATP, and [^{125}I]insulin were obtained from New England Nuclear (Lachine, Canada). Sodium orthovanadate, poly glutamate: tyrosine (4:1), aprotinin were purchased from Sigma Chemical Co. (St. Louis, Missouri). Insulin receptor peptide (1142–1153) was synthesized at Afdeling Biochemie, Katholieke Universiteit (Leuven, Belgium). AG-1X2 column was obtained from Bio-Rad (Missisauga, Canada).

Address for offprints: R.L. Khandelwal, Department of Biochemistry, University of Saskatchewan, 107 Wiggins Road, Saskatoon, SK, S7N 5E5, Canada

Experimental animals

Five week-old Zucker rats were divided into three groups of 6 rats each: lean (Fa/fa) control, obese (fa/fa) control and obese (fa/fa)-vanadate treated. The treatment and the collection of tissue were by the method described by us earlier [11]. The rats in treated group received sodium orthovanadate through drinking water (80 mg%) for four months. Plasma glucose was assayed enzymatically using a diagnostic kit from Sigma. Plasma insulin was measured by the double-antibody method [12]. Partial purification of insulin receptor from liver and the binding studies with [^{125}I]insulin were carried out by the procedure described earlier [8].

Preparation of cytosolic and particulate fractions from liver

The liver samples were homogenized in 50 mM HEPES buffer (pH 7.6) containing 250 mM sucrose, 1 mM EDTA, 1 mM DTE, 0.1 mM PMSF, 1 mM benzamidine, 0.1 mg/ml aprotinin, 20 μg/ml leupeptin. The homogenate was centrifuged at 10,000 × g for 15 min and the supernatant obtained was centrifuged further at 100,000 × g for 1 h. After separating the soluble cytosolic fraction, the pellet was rehomogenized in the same buffer, but without sucrose and with 1% Triton X-100. The sample was shaken at 4°C for 1 h, centrifuged at 100,000 × g for 90 min and the supernatant was collected as particulate fraction.

Preparation of ^{32}P-labeled substrates

PGT or IR peptide (1142–1153) were phosphorylated using the tyrosine kinase (TK-1) isolated from spleen [13] in 25 mM HEPES buffer (pH 7.5) containing 0.5 mM [γ-^{32}P]ATP (600 CPM/pmol), 10 mM $MgCl_2$, 100 μM sodium orthovanadate at room temperature for two days. Phosphorylated PGT was precipitated with 15% TCA, dissolved in 25 mM HEPES buffer (pH 7.5) and dialysed against the same. Phosphorylated IR peptide was purified in the AG-1X2 column (acetate form, 200–400 mesh) with 30% acetic acid, lyophilised, and passed through Sepadex G-10 to remove vanadate.

PTPase assays

^{32}P-labeled IR peptide and PGT were incubated with 10 μg protein from cytosolic or particulate samples at 30°C for 10 min in 50 mM HEPES buffer (pH 7.5), 1 mM EDTA, 1 mM DTE and 1 mg/ml BSA in a final volume of 50 μl. The final concentration of carryover Triton X-100 in the reaction mixture was 0.1% in case of particulate samples. The assay was terminated with the addition of 200 μl of cold TCA (10%) and 20 μl of 10 mg/ml BSA. After centrifugation the radioactivity of the supernatant was measured in the PGT phosphatase assay. In case of IR peptide, [^{32}P]Pi in the supernatant was extracted with isopropyl alcohol in presence of sodium phosphate and ammonium molybdate and counted [13].

Column chromatography

Solubilized particulate fractions (3 mg protein) from lean control, obese control and obese-vanadate treated rat liver samples were subjected to fast protein liquid chromatography in a Mono Q column (HR5/5, Pharmacia). The column was equilibrated with the 20 mM Tris buffer (pH 7.0) containing 1 mM EDTA, 1 mM DTE, and 0.1 mg/ml aprotinin and developed at the flow rate of 0.5 ml/min with a 20 ml NaCl gradient (0–500 mM). The particulate samples were dialysed against the equilibration buffer before loading on the column. Fifty fractions (0.5 ml each) were collected and PTPase assays were carried out using labeled PGT and IR peptide with 10 μl samples.

Results

Body weight, the plasma levels of glucose and insulin and the hepatic insulin receptor binding capacities in different groups of animals are shown in Table 1. Obese control rats had significantly ($p < 0.001$) higher body weight as compared to the lean ones. However, the body weight of vanadate-treated obese was comparable to that of lean rats. The elevated plasma levels of glucose and insulin in obese rats were decreased by 36 and 80%, respectively in vanadate-treated obese rats. The insulin receptor binding capacity in obese liver was decreased by 60% which was restored to near lean control levels by vanadate.

PTPase activities in cytosolic and particulate fractions from the liver samples of different groups of rats are shown in Table 2. In obese control rats, the hepatic PTPase activities of cytosolic and particulate fractions were significantly ($p < 0.001$) elevated over the lean control values with both PGT and IR peptide as substrates. The PTPase activities of the liver samples from vanadate-treated obese rats were decreased by 35–60%. In vanadate-treated lean rats also the hepatic PTPase activities decreased in cytosolic and particulate fractions (results not shown).

The activities of PTPase in particulate fractions were further characterized by their separation on a Mono Q column by FPLC. The results on PTPase activities with IR peptide as substrate in liver samples from lean control, obese control and obese-vanadate treated rats are shown in Fig. 1. Three distinct peaks of PTPase activities were obtained.

Table 1. Effect of vanadate treatment on body weight, plasma levels of glucose and insulin, and hepatic insulin binding capacity in obese Zucker rats

Animals	Body weight (g)	Plasma glucose (mM)	Plasma insulin (pM)	Total insulin binding capacity (fmol/mg prot.)
Lean control	462 ± 7	6.4 ± 0.9	505 ± 52	254.3 ± 4.0
Obese control	712 ± 31***	10.1 ± 0.9*	2633 ± 407***	104.0 ± 5.3***
Obese vanadate-treated	407 ± 40[c]	6.5 ± 0.3[b]	514 ± 69[c]	222.8 ± 8.0[c]

Values are mean ± S.E.M. of 6 rats. Comparisons were made by student's *t* test. *p < 0.05 and ***p < 0.001 vs lean control. [b]p < 0.01 and [c]p < 0.001 vs obese control.

Table 2. Effect of vanadate treatment on hepatic phosphotyrosine phosphatase activities with poly Glu:Tyr (PGT) and insulin receptor peptide 1142-1153 (IRP) as substrates in obese Zucker rats.

Animals	PGT		IRP	
	Particulate	Cytosol	Particulate	Cytosol
Lean control	601 ± 25	547 ± 23	309 ± 16	625 ± 50
Obese control	817 ± 47***	793 ± 21***	471 ± 13***	921 ± 29***
Obese vanadate-treated	528 ± 44[c]	500 ± 38[c]	278 ± 17[c]	464 ± 44[c]

Values are mean ± S.E.M. of 6 rats. Comparisons were made by student's *t* test. ***p < 0.001 vs lean control. [c]p < 0.001 vs obese control.

Peak I was the major one followed by III and II. The assay of pooled peaks indicated that in obese liver, the level of I increased by 18% and that of III by 32% as compared to the lean ones. Both of these peaks in vanadate-treated obese rats decreased and were comparable to lean rats. Similar changes were observed with PGT as substrate (results not shown).

Discussion

Vanadate is known to have a potent inhibitory action towards purified PTPases *in vitro*. In this study, vanadate-induced normalization of plasma glucose and insulin levels in obese Zucker rats was accompanied by inhibition of PTPase activities in liver. Similar *in vivo* inhibition of hepatic PTPase has been demonstrated by others in STZ-diabetic rats [9, 10]. The decrease in PTPase activity in fa/fa obese Zucker rats observed by us was mainly associated with peak III after separation on Mono Q column of FPLC. The increased activity in obese Zucker rat liver and the inhibition by vanadate were relatively lower in Mono Q column purified peaks as compared to the crude fractions. It is possible, therefore, that other factors may also be involved in the regulation of PTPase activity.

In STZ-diabetic rats, a model for insulin deficient diabetes, PTPase activities have been found to to be altered in a time-dependent and tissue specific manner [9, 10]. Hepatic particulate PTPase activities were elevated during the first two weeks after STZ injection which returned to normal control levels by 30 days [9, 10]. In chronic diabetes, the metabolic aberrations induced by this insulin deficiency could have an effect on PTPase activities. Begum *et al* have reported differences in diabetes-induced changes in adipocyte and liver PTPases of STZ-treated rats. Obese Zucker rats are characterized by insulin resistance, hyperinsulinemia, mild hyperglycemia and hypertriglyceridemia [14]. There are no studies to show the changes in PTPase activities in this insulin resistant diabetic animal model. We have observed significant increases in PTPase activities in cytosolic and particulate fractions. These changes could play a role in the development of insulin resistance.

Chronic administration of vanadate to obese Zucker rats leads to improvement in glucose homeostasis, prevention of hyperinsulinemia, and increased glucose utilization in muscle [15, 16]. We had earlier reported the normalization of reduced membrane associated N-myristoyltransferase activity in Zucker rat liver after vanadate treatment [11]. In other animal models for type II diabetes also, vanadate administration results in the correction of hyperinsulinemia and the elevated lipogenic activities in liver are lowered to the lean control levels [17, 18].

Though the insulin-mimetic action of vanadate has been demonstrated by numerous *in vitro* and *in vivo* studies, its mechanism of action is not clearly understood. Vanadate stimulates insulin receptor kinase activity in various tissues [6–8]. Inhibition of PTPase by this agent is likely to enhance the insulin receptor kinase activity. Vanadate has been shown to prolong the action of insulin in adipocytes probably through inhibition of PTPase [7]. Hyperinsulinemia of type II diabetic animals is the result of insulin resistance. Hence inhibition of PTPase can lead to improved insulin sensitivity and decrease of plasma insulin levels. Some studies have shown that vanadate could act at the post-receptor level [19, 20]. Insulin receptor kinase-induced tyrosine phosphorylation of cellular proteins is essential in the signal transduction of insulin. Hence vanadate action at the post-receptor level could also involve inhibition of PTPases.

Vanadate has been shown to increase the tyrosine phosphorylation in many tissues. Vanadate-induced contraction of smooth muscle is accompanied by increased tyrosine phosphorylation [21]. Vanadate in combination with hydrogen peroxide increased the formation of inositol phosphate in various cell lines and this effect correlated with enhanced tyrosine phosphorylation [22]. However, in some studies vanadate has been shown to exert insulin-like effects

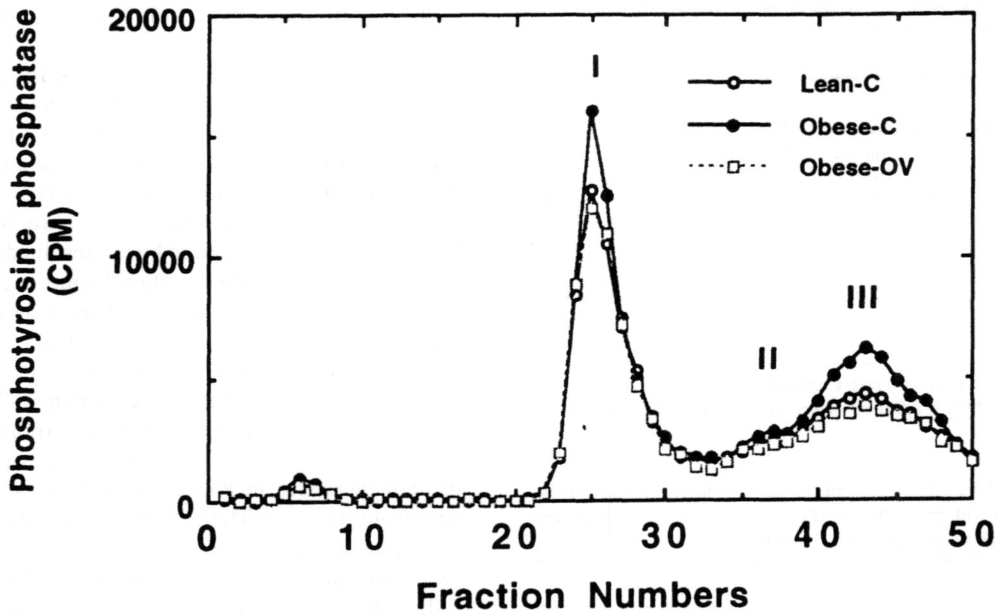

Fig. 1. Mono Q column chromatography of liver pariculate fractions. Three mg protein samples of solubilized particulate fractions from lean (--O--), obese control (--●--) and obese-vanadate treated (--□--) rats were separated on a Mono Q column as described in methods. The fractions were assayed for PTPase with labeled insulin receptor peptide (1142–1153 residues) as substrate.

which did not correlate tyrosine phosphorylation patterns [23, 24].

Vanadium has been shown to accumulate in liver at a concentration of 20 μM in treated rats [25]. Hence this trace element will be present in the cytosolic fraction prepared for PTPase assay. However, the final expected concentration of vanadate in the reaction mixture is around 50 nM at which it does not cause inhibition of PTPase *in vitro*. Further, the crude as well as the purified peaks of particulate fractions are unlikely to contain vanadate. Therefore, the observed changes are due to decreases in PTPase activities and not due to inhibition by vanadate during the assay. Further studies are needed to examine whether PTPases are inactivated or downregulated.

Acknowledgement

This research was supported by the Medical Research Council of Canada and the Canadian Diabetes Association.

References

1. Fischer EH, Charbonneau H, Tonks NK: Protein tyrosine phosphatases: A diverse family of intracellular and transmembrane enzymes. Science 253: 401–406, 1991
2. Rosen OM: After insulin binds. Science 237: 1452–1458, 1987
3. Goldstein BJ: Protein-tyrosine phosphatases and the regulation of insulin action. J Cell Biochem 48: 33–42, 1992
4. Hashimoto N, Zhang W-R, Goldstein BJ: Insulin receptor and epidermal growth factor receptor dephosphorylation by three major rat liver protein-tyrosine phosphatases expressed in a recombinant bacterial system. Biochem J 284: 569–576, 1992
5. Shechter Y: Insulin-mimetic effects of vanadate. Possible implications for future treatment of diabetes. Diabetes 39: 1–5, 1990
6. Tamura S, Brown TA, Whipple JH, Fujita-Yamaguchi Y, Dubler RE, Cheng K, Larner J: A novel mechanism for the insulin-like effect of vanadate on glycogen synthase in rat adipocytes. J Bio Chem 259: 6650–6658, 1984
7. Fantus IG, Ahmad F, Deragon G: Vanadate augments insulin-stimulated insulin receptor kinase activity and prolongs insulin action in rat adipocytes. Diabetes 43: 375–383, 1994
8. Pugazhenthi S, Khandelwal RL: Does the insulin-mimetic action of vanadate involve insulin receptor kinase? Mol Cell Biochem 127/128: 211–218, 1993
9. Meyerovitch J, Backer JM,, Kahn CR: Hepatic phosphotyrosine phosphatase activity and its alterations in diabetic rats. J Clin Invest 84: 976–983, 1989
10. Begum N, Sussman KE, Draznin B: Differential effects of diabetes on adipocyte and liver phosphotyrosine and phosphoserine phosphatase activities. Diabetes 40: 1620–1629, 1991
11. King MJ, Pugazhenthi S, Khandelwal RL, Sharma RK: Membrane-associated N-myristoyltransferase activity is reduced in obese (fa/fa) Zucker rat liver. Biochem Biophys Res Commun 196: 665–670, 1993
12. Morgan CR, Lazarow A: Immunoassay of insulin: two antibody system: plasma insulin levels of normal, subdiabetic and diabetic rats. Diabetes 12: 115-126, 1963
13. Swarup G, Subrahmanyam G: Purification and characterization of a protein-tyrosine phosphatase from rat spleen which dephosphorylates

and inactivates a tyrosine-specific protein kinase. J Biol Chem 264: 7801–7808, 1989

14. Bray GA: The Zucker-fatty rat: a review. Federation Proc 36: 148–153, 1977
15. Brichard SM, Pottier AM, Henquin JC: Long term improvement of glucose homeostasis by vanadate in obese hyperinsulinemic fa/fa rats. Endocrinology 125: 2510–2516, 1989
16. Brichard SM, Assimacopoulos-Jeannet F, Jeanrenaud B: Vanadate treatment markedly increases glucose utilization in muscle of insulin-resistant fa/fa rats without modifying glucose transporter expression. Endocrinology 131: 311–317, 1992
17. Pugazhenthi S, Angel JF, Khandelwal RL: Long-term effects of vanadate treatment on glycogen metabolizing and lipogenic enzymes of liver in genetically diabetic (db/db) mice. Metabolism 40: 941–946, 1991
18. Khandelwal RL, Pugazhenthi S: *In vivo* effects of vanadate on hepatic glycogen metabolizing and lipogenic enzymes in insulin-dependent and insulin-resistant diabetic animals. Mol Cell Biochem 153: 87–94, 1995
19. Green A: The insulin-like effect of sodium vanadate on adipocyte glucose transport is mediated at a post-insulin receptor level. Biochem J 238: 663–669, 1986
20. D'Onofrio F, Le MQU, Chiasson J-L, Srivastava AK: Activation of mitogen activated protein (MAP) kinases by vanadate is independent of insulin receptor autophosphorylation. FEBS Lett 340: 269–275, 1994
21. Di Salvo J, Semenchuk LA, Lauer J: Vanadate-induced contraction of smooth muscle and enhanced protein tyrosine phosphorylation. Arch Biochem Biophys 304: 386-391, 1993
22. Zick Y, Sagi-Eisenberg R: A combination of H_2O_2 and vanadate concomitantly stimulates protein tyrosine phosphorylation and polyphosphoinositide breakdown in different cell lines. Biochemistry 29: 10240–10245, 1990
23. Strout HV, Vicario PP, Saperstein R, Slater EE. The insulin-mimetic effect of vanadate is not correlated with insulin receptor tyrosine kinase activity nor phosphorylation in mouse diaphragm *in vivo*. Endocrinology 124: 1918–1924, 1989
24. Mooney RA, Bordwell KL, Luhowskyj S, Casnellie JE: The insulin-like effect of vanadate on lipolysis in rat adipocytes is not accompanied by an insulin-like effect on tyrosine phosphorylation. Endocrinology 124: 422–429, 1989
25. Pugazhenthi S, Khandelwal RL: Insulinlike effects of vanadate on hepatic glycogen metabolism in nondiabetic and streptozotocin-induced diabetic rats. Diabetes 39: 821–827, 1990

Molecular and Cellular Biochemistry **153:** 131–137, 1995.

Evidence for selective effects of vanadium on adipose cell metabolism involving actions on cAMP-dependent protein kinase

Roger W. Brownsey and Gordon W. Dong
Department of Biochemistry and Molecular Biology, Medical Block 'A'/ Copp Building, University of British Columbia, 2146 Health Sciences Mall, Vancouver, British Columbia, Canada V6T 1Z3

Abstract

The insulin-like effects of vanadium *in vivo* are likely to be achieved at micromolar concentrations. Demonstrated effects of vanadium on adipose tissue of streptozotocin-diabetic rats include inhibition of basal and stimulated rates of lipolysis and effects on fat cell protein phosphorylation. The studies described below examined the effects of vanadium (to a maximum concentration of 0.5 mM) on adipose cells or tissue *in vitro*. Vanadium, added as a vanadyl-albumin complex or as sodium orthovanadate, produced a marked (greater than 50%) inhibition of isoproterenol-stimulated lipolysis. Inhibition of lipolysis equivalent to that seen with insulin, was achieved with ~ 100 μM vanadium. In contrast, no insulin-like stimulation of *de novo* fatty acid biosynthesis was observed with vanadium below 0.5 mM. Surprisingly, the antilipolytic effects of vanadium persisted in the presence of cilostamide, an inhibitor of the insulin-sensitive isoform of cyclic nucleotide phosphodiesterase. Studies with purified preparations of the catalytic subunit of cyclic AMP-dependent protein kinase revealed dose-dependent inhibition with vanadyl-glutathione (to a maximum of ~40% inhibition). Equivalent inhibition of cyclic AMP-dependent phosphorylation of Kemptide (~50%) was observed upon incubation of freshly-prepared fat-pad supernatant fractions with vanadyl-glutathione. These results suggest that effects of low concentrations of vanadium may be mediated, at least in part, by actions on the catalytic subunit of cyclic AMP-dependent protein kinase. (Mol Cell Biochem: **153:** 131–137, 1995)

Key words: vanadium, insulin action, lipid metabolism, adipose tissue, lipolysis, cyclic AMP-dependent protein kinase

Introduction

Insulin resistance and insulin depletion (separately or in combination) can lead to the development of a markedly catabolic state such as that observed in experimental or clinical diabetes. Complex changes in intermediary metabolism in diabetes include not only the impairment of glucose homeostasis but also effects on protein and lipid metabolism (for reviews see [1–3]). Among observed changes in lipid metabolism, the loss of insulin regulation is reflected by marked elevations in rates of adipose tissue lipolysis and consequently elevated levels of circulating non-esterified long-chain fatty acids which play a significant role in promoting hepatic synthesis and export of very low density lipoproteins [4–7].

In previous studies, we have reported that the elevated basal and epinephrine-stimulated rates of lipolysis in white adipose tissue and elevated circulating levels of triacylglycerols in streptozotocin (STZ)-diabetic rats are corrected by chronic treatment of animals with dietary vanadium [8, 9]. Studies *in vitro* have also confirmed that vanadium compounds antagonize the lipolytic responses to catecholamines of rat adipose cells [10]. The rate of adipose tissue lipolysis is critically dependent upon the extent of phosphorylation of the rate-limiting enzyme hormone-sensitive lipase, which is directly phosphorylated and activated by cyclic AMP-dependent protein kinase [11]. It seems most likely that the counter-regulatory effects of insulin involve suppression of the phosphorylation of hormone-sensitive lipase or stimulation of enzyme dephosphorylation (or both). We have previously noted evidence for the persistent activation of cyclic

Address for offprints: R.W. Brownsey, Department of Biochemistry and Molecular Biology, Medical Block 'A'/Copp Building, University of British Columbia, 2146 Health Sciences Mall, Vancouver, British Columbia, Canada V6T 1Z3

AMP-dependent protein kinase in extracts of adipose tissue from STZ-diabetic rats and that this effect of the diabetic state was also corrected by chronic treatment with vanadium *in vivo* [8].

The studies presented here have addressed the possible mechanisms by which treatment of adipose cells with vanadium may lead to suppression of hormone-stimulated lipolysis *in vitro*. We considered it especially important to consider the possible action of vanadium on cyclic nucleotide phosphodiesterase activity because compelling evidence has accumulated to support the initial report of Loten & Sneyd [12] indicating that insulin induces activation of a specific isoform (known as the 'particulate', low Km, cGMP-inhibitable isoform) of cyclic nucleotide phosphodiesterase (cGI-PDE). Activation of cGI-PDE appears to involve phosphorylation by an insulin-stimulated protein seryl/threonyl kinase and may contribute importantly, if not exclusively, to the antilipolytic effects of the hormone [13–15]. In addition, there is evidence that some effects of vanadium may be prevented by the PDE inhibitor 3-isobutyl-1-methylxanthine [16]. Moreover, recent reports have demonstrated direct activation of the cGI-PDE isoform *in vitro* by vanadium [17, 18]. Surprisingly, we observed that the antilipolytic effects of vanadium were still apparent following PDE inhibition. Further, it appears that direct effects of vanadium on cyclic AMP-dependent protein kinase itself may explain, at least in part, the antilipolytic actions of this transition metal.

Materials and methods

Materials

Most laboratory chemical reagents and solvents were obtained from Fisher Scientific or BDH Chemicals Canada Ltd. (Vancouver, B.C., Canada). Biochemicals were mostly obtained from Sigma (St Louis, MO, USA), including proteinase inhibitors, catalytic subunit of cyclic AMP-dependent protein kinase (and the corresponding inhibitor peptide), Kemptide and defatted bovine serum albumin. Reagents required for the assay of glycerol were obtained from Boehringer Mannheim (Laval, Quebec, Canada) and radioisotopes were from Amersham International (Oakville, Ontario, Canada). P81 phosphocellulose paper was obtained from Whatman International (Maidstone, Kent, U.K.).

Preparation of tissues, cells and cell extracts

Male Wistar rats weighing 160–200 g were maintained on 12 h light/dark cycles (with the dark cycle beginning at 8:00 pm) and allowed free access to food (Purina laboratory chow) and water until time of killing between 9:00 and 10:00 am. Epididymal adipose tissue was removed and dissected into small (~50–100 mg) pieces for direct incubations in Krebs-Henseleit bicarbonate-buffered medium as described previously [19]. Alternatively, isolated adipose cells were prepared as described initially [20] and subsequently modified [21]. Cell extracts were obtained by Polytron homogenization as described [22] and supernatant fractions prepared by centrifugation at 4°C (30 min at 25,000 × g).

Metabolic measurements and enzyme assays

To estimate rates of lipolysis, glycerol output from adipose tissue or cells was determined as described [23]. *De novo* fatty acid biosynthesis was determined as previously [24] and the activity of cyclic AMP-dependent protein kinase assayed by filter paper method as recently reviewed [25].

Other Techniques

Protein determinations employed the dye-binding method [26]. Statistical analyses involved Students' *t*-test for simple comparisons or ANOVA for multiple comparisons. Curve-fitting and associated linear or other regression analyses were carried out with Enzfitter or SigmaPlot/SigmaStat packages. Differences were considered significant with $p = 0.05$ or less.

Results

The initial premise of these studies was to focus on effects of vanadium *in vitro* on isolated rat adipose tissue which might be important physiologically and which could therefore be detected at concentrations (in the micromolar range) which might be observed *in vivo*. The studies involved incubation of adipose tissue pieces or adipose cells (isolated by collagenase digestion) in the presence of sodium orthovanadate or vanadyl sulfate. The use of vanadium (V) as orthovanadate might be considered preferable *in vitro* (under oxidizing conditions at neutral pH), especially at low concentrations (below 200 μM) where the monomeric species predominates [27]. In contrast, the effects of vanadium (IV) might be more difficult to interpret because of the possibility of oxidation in aerated physiological buffers. We were therefore surprised to find that effects of vanadyl sulfate on adipose tissue were at least as marked and generally more reliable than effects observed with sodium orthovanadate. Vanadium (IV) was therefore used in most studies reported below. It is important to note that solutions containing vanadyl sulfate were prepared by first mixing with bovine serum albumin to prevent oxidation or precipitation prior to addition to the physiological saline (Krebs-Henseleit buffer). Analysis of absorption spectra indicated the formation of stable albumin-vanadyl complex (with a major absorption maximum close to 750 nm), in agreement with previous studies based upon electron paramagnetic resonance spectroscopy [28].

We first examined effects of vanadium on fatty acid biosynthesis by following the incorporation of [^{3}H] from tritiated water into saponifiable tissue lipids. In these experiments, we obtained the anticipated stimulation of fatty acid synthesis in the presence of insulin, from a basal value of 6.8$^+/_-$1.7 μg atoms [H] per g wet weight of tissue/h (mean $^+/_-$ S.E.M. for 9 independent experiments) to a maximum of 21.2$^+/_-$5.0 μg atoms [H]/g/h in the presence of insulin (stimulation by insulin 3.7$^+/_-$0.7-fold, $p < 0.01$). Maximal insulin effects were observed at hormone concentrations in the range 100–200 μU/ml and so the concentration used routinely was 500 μU/ml. Under identical conditions, we saw no significant stimulation of fatty acid synthesis with either sodium orthovanadate or vanadyl sulfate when used up to 0.25 mM. The highest value observed (at 0.5 mM vanadyl sulfate) was 12.7$^+/_-$2.4 ug atoms [H]/g/h (a stimulation of 1.88$^+/_-$0.36-fold). Similar results were obtained in a series of 6 independent experiments with isolated fat cells. In this case, the stimulation with insulin was smaller than with adipose tissue (170$^+/_-$9% control) and no significant stimulation was observed with vanadium up to 0.25 mM (no values differed from control by more than 20 %).

Antilipolytic effects of vanadium

Previous studies indicated that treatment of STZ-diabetic rats with orally-administered vanadium led to a reduction in circulating levels of triacylglycerols. In addition, when adipose tissue from these experimental animals was subsequently incubated *in vitro*, it was found that the markedly elevated basal and epinephrine-stimulated lipolysis observed in STZ-diabetic animals was no longer apparent in those animals which had been fed diets supplemented with vanadium [8, 9]. We therefore examined the effects of vanadium on lipolysis assessed by the rate of glycerol release from either adipose tissue pieces or from isolated adipose cells (Fig. 1). In preliminary experiments, a dose-dependent stimulation of lipolysis by isoproterenol was observed. For subsequent experiments, an isoproterenol concentration of 5×10^{-7} M was used since this gave a convenient near-maximal stimulation of glycerol output (measured over a 30 min incubation period). On average, the activation of triacylglycerol breakdown was in the range 5–7 fold. The results in Fig. 1 illustrate that the addition of vanadyl sulfate to adipocytes for 30 min prior to the addition of isoproterenol, led to a dose-dependent inhibition of lipolysis. Inhibition was apparent with vanadium at 10 μM (approximately 20% inhibition) and progressed to approximately 60% inhibition with extracellular vanadyl sulfate at 300 μM. In experiments with intact adipose tissue, increasing the vanadium concentration to 0.5 mM had no further effect on lipolysis, although with isolated adipose cells, the degree of inhibition was further enhanced (to ~ 80%). In these experiments, the maximal inhibition seen with insulin (1 mU/ml) was equivalent in cells and adipose tissue pieces (~25–35%). These experiments therefore confirmed earlier studies [10] that vanadium induces rapid and potent antilipolytic actions *in vitro*. Further experiments revealed that pre-incubation with vanadium prior to addition of isoproterenol was not essential to observe the antilipolytic effects. Substantial inhibition of lipolysis could be observed with simultaneous addition of vanadium together with isoproterenol (data not shown).

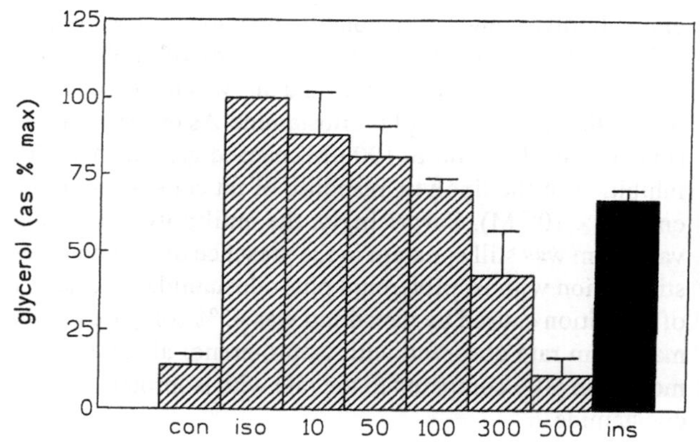

Fig. 1. Antilipolytic effects of vanadium. Isolated rat adipose cells were incubated in Krebs-Henseleit buffer and glycerol release into the medium determined following incubation (30 min) in the absence of hormones ('con'), in the presence of isoproterenol ('iso', 5×10^{-7} M), isoproterenol plus vanadium (numerical values indicate μM concentration of vanadyl sulfate added as a complex with serum albumin) or isoproterenol plus insulin ('ins', 1 mU/ml – filled bar). Values for glycerol output are expressed relative to isoproterenol alone (100 %). Absolute rates of glycerol release in control and isoproterenol-treated conditions were respectively 0.85 +/– 0.15 and 6.23 +/– 1.73 μmol/h/gm dry wt of cells. All results (mean plus bars to indicate S.E.M.) were derived from 7 independent experiments with different batches of cells. Treatments with insulin and vanadium (50 μM or higher) all gave significant inhibition of the effects of isoproterenol ($p < 0.02$).

Vanadium and cyclic AMP phosphodiesterase

In view of evidence discussed in the introduction, we anticipated that incubation of adipose tissue with cilostamide (a potent and selective inhibitor of cGI-PDE) would abolish the antilipolytic effects of vanadium. As anticipated, cilostamide alone produced a small but consistent activation of endogenous lipolysis and more notably, potentiated the lipolytic effect of isoproterenol (Fig. 2). Stimulation of lipolysis with isoproterenol alone was ~5-fold and this was increased to ~7-fold with isoproterenol plus cilostamide. As observed above (Fig. 1) vanadyl sulfate (100 μM) caused greater than 60% inhibition of the lipolytic response of fat cells to isoproterenol (5×10^{-7} M). Surprisingly, the antilipolytic effect of vanadium was still apparent in the presence of the combined stimulation with isoproterenol plus cilostamide. The degree of inhibition caused by vanadium was 45% compared to the maximum rate observed with isoproterenol alone and was more than 50% compared to the rate with isoproterenol plus cilostamide.

Vanadium and cyclic AMP-dependent protein kinase

We next examined the possibility that vanadium might inhibit lipolysis through a direct effect on cyclic AMP-dependent protein kinase. For these experiments, fresh extracts of rat adipose tissue were used as a source of protein kinase activity as well as purified preparations of the catalytic subunit of the enzyme. In both cases, vanadium (IV) was used as a preformed equimolar mixture of vanadyl sulfate plus glutathione (designated 'vanadyl-glutathione'). Based upon several spectroscopic studies, it appears that vanadyl (IV) is likely to represent the most significant intracellular form of this metal owing to reduction by intracellular glutathione [29] and formation of complexes with proteins and organic phosphates – most likely including nucleotides [30–32]. It may be noted that glutathione alone caused no inhibition of protein kinase activity. Indeed SH-reagents are essential for maintenance of activity of the catalytic subunit of cyclic AMP-dependent protein kinase (which is suspended in 100 mM DTT for storage) and the amount of glutathione added in the mixture with vanadyl is very small compared to the total SH-reagent concentration in the assay mixture (~ 2 mM).

(a) Studies with purified preparations of protein kinase
Protein kinase activity was determined (in 10 min assays) by following incorporation of [^{32}p] from [γ-^{32}P]ATP into Kemptide using a filter-paper assay method. Addition of vanadyl-glutathione over the concentration range 10–100 μM (for 5 min prior to assay) led to a dose-dependent inhibition of Kemptide phosphorylation to a maximum of ~40% inhibition (Fig. 3). Increasing the vanadium concentration up to 0.2 or 0.5 mM did not produce any further kinase inhibition.

(b) Studies with fat-pad supernatant fractions
Rat fat-pad pieces were first incubated in Krebs-Henseleit buffer to minimize any activation of cyclic AMP-dependent protein kinase caused at the time of killing. Tissue homogenates and supernatant fractions were then prepared and Kemptide phosphorylation determined following incubation with 8-bromo cyclic AMP (0.5 μM) in the absence or presence of vanadyl-glutathione (100 μM). Significant Kemptide phosphorylation was observed in the presence of the specific inhibitor of cyclic AMP-dependent protein kinase. This ba-

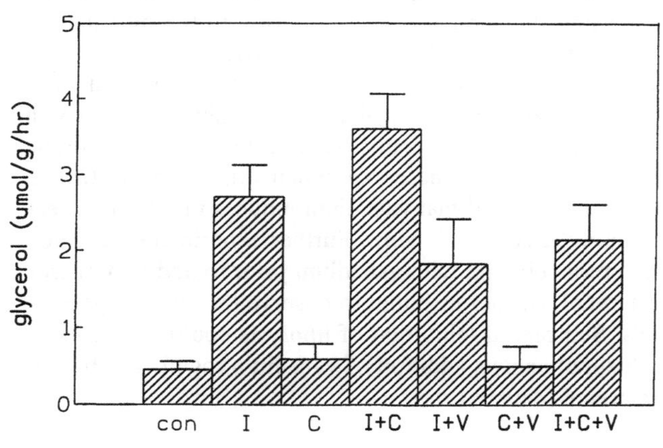

Fig. 2. Vanadium and cyclic AMP phosphodiesterase. Isolated fat cells were incubated and glycerol output determined as described (Fig. 1 legend). Cells were incubated in the absence of hormones ('con') or with single or multiple additions indicated by the following letter codes: 'I' = isoproterenol (5×10^{-7}M), 'C' = cilostamide (1 μM), 'V' = vanadyl sulfate (100 μM, presented as a complex with serum albumin). Values (mean plus bar to indicate SEM) are from 6 independent experiments with different batches of cells and are expressed as μmol/h/g dry weight of cells.

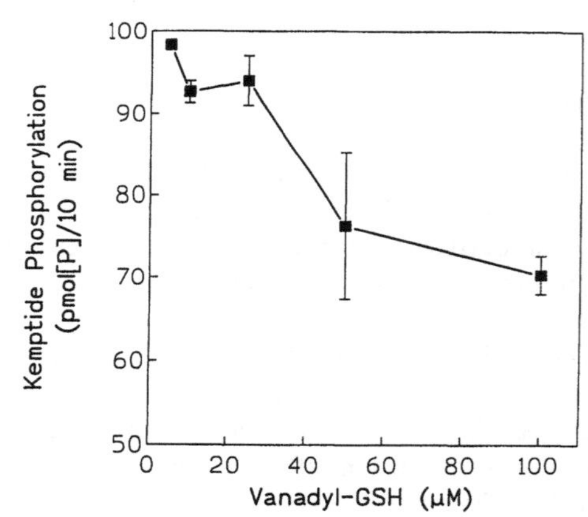

Fig. 3. Effects of vanadium on the activity of the catalytic subunit of cyclic AMP-dependent protein kinase. The phosphorylation of Kemptide by purified preparations of protein kinase was determined in the absence or presence of the indicated concentrations of vanadyl sulfate (added in the presence of glutathione as described in the text). Results represent the mean and S.E.M. of values determined in three independent experiments with different kinase preparations and are expressed relative to maximum activity observed in the absence of vanadium (normalized to 100 pmol [P]/10 min assay. The absolute incorporation of [^{32}P] in the absence of vanadium was 5.9 pmol [P]/10 min. Significant kinase inhibition was observed at vanadium concentrations of 50 and 100 μM ($p < 0.05$).

sal Kemptide phosphorylation ($7.1^{+}/_{-}1.1$ pmol[P]/10 min assay) was not significantly altered by inclusion of vanadyl-glutathione. Addition of 8-bromo cyclic AMP led to a 2–3 fold activation of Kemptide phosphorylation ($7.1^{+}/_{-}2.4$ pmol[P]/10 min above basal) and this effect was diminished (to $3.7^{+}/_{-}1.8$ pmol[P]/10 min above basal) in the presence of vanadyl-glutathione (Fig. 4A). All values are means $^{+}/_{-}$ S.E.M. from 9 independent experiments. Examination of the dose-dependent activation of Kemptide phosphorylation by 8-bromo cyclic AMP revealed that inhibitory effects of vanadyl-glutathione (100 μM) were apparent over the entire concentration range of nucleotide (Fig. 4B), reflecting a reduction in V_{max} of cyclic AMP-dependent protein kinase rather a change in sensitivity to the allosteric activator (and was confirmed by curve-fitting analysis). As with the other analyses reported above, the effects of vanadium on Kemptide phosphorylation were rapid, with a t½ of approximately 2.5 min in these enzyme assays.

Discussion

The studies described here have confirmed earlier reports that vanadium is able to exert a potent and rapid antilipolytic effect in rat adipose tissue. We observed similar effects when vanadium was added to extracellular buffers as sodium orthovanadate – V(V) – or as vanadyl sulfate – V(IV). In contrast to the marked effects of vanadium on fat cell lipolysis, we could find no evidence for a similarly potent effect on fatty acid biosynthesis. This is in contrast to the findings of one report [33] but in agreement with another report [10]. In this latter study, antilipolytic effects of vanadate were apparent at 10 μM and maximal at 100 μM, whereas activation of fatty acid biosynthesis was not apparent below 1 mM. Our results suggest that effects of vanadium which have been observed *in vitro* only at high concentrations (in the mM range) may not necessarily be expressed *in vivo*, since circulating concentrations achieved by administration of vanadium to experimental animals remain in the micromolar range [34, 35]. On the basis of the observations described above, it is possible that vanadium may act in a counterregulatory or 'anti-catabolic' manner *in vivo*, while its ability to act in a positively anabolic role, at least in terms of adipose tissue metabolism, may be questioned. It is intriguing in this regard, that treatment of STZ-diabetic rats with vanadium can overcome the associated hyperglycemia and hyperlipidemia but is not apparently able to restore normal growth rates of the affected animals.

The form of vanadium which is effective in different experimental situations requires some comment. We wondered if the presentation of cells with extracellular vanadium (IV) as vanadyl sulfate might be futile if oxidation to vanadium

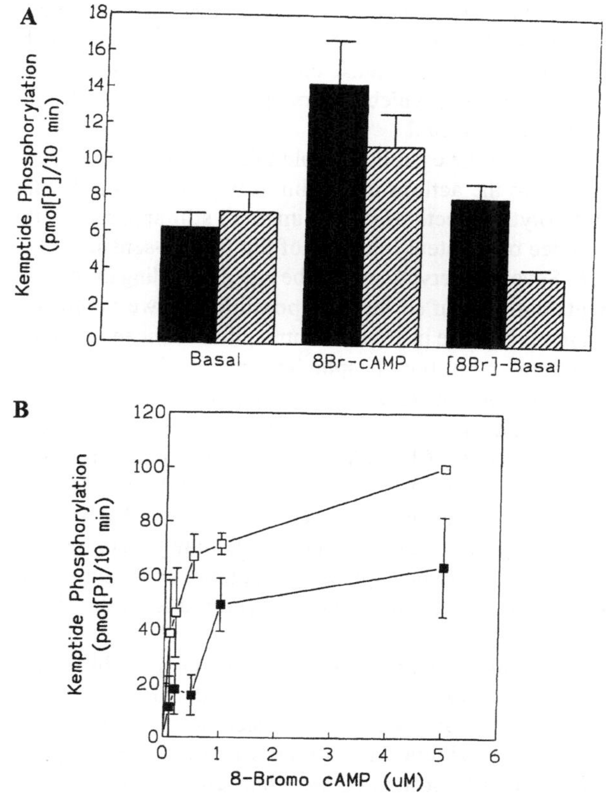

Fig. 4. Effects of vanadium on cyclic AMP-dependent protein kinase activity in fat-pad supernatant fractions. Rat epididymal fat pads were incubated and homogenized prior to preparation of supernatant (25,000 × g) fractions as described (see methods). Phosphorylation of Kemptide was determined (Panel A) in the presence of inhibitor peptide specific for cAMP-dependent protein kinase ('Basal') or in the presence of 8-bromo cyclic AMP ('8Br-cAMP', 0.5 μM). Results are given as mean and SEM values from 9 independent experiments and are expressed as pmol [P] incorporated per assay (10 min) in the absence (solid bars) or presence (hatched bars) of vanadium (100 μM vanadyl sulfate, in the presence of glutathione). Values at the right side of panel A illustrate the cyclic nucleotide-stimulated Kemptide phosphorylation following subtraction of 'basal' values. Kemptide phosphorylation (Panel B) was determined with the indicated concentrations of 8-bromo cyclic AMP, in the absence (open symbols) or presence (closed symbols) of vanadyl sulfate (100 μM, in the presence of glutathione). Results are the mean and SEM of values obtained in five independent experiments. Inhibition by vanadium was significant ($p < 0.05$) at all tested concentrations of 8-Bromo cAMP.

(V) occurred rapidly. However, we found that the presence of bovine serum albumin served to stabilize vanadyl sulfate even under oxidizing conditions at neutral pH.

Spectroscopic evidence for a stable complex between vanadyl and albumin was obtained and this complex clearly facilitated presentation of vanadium to the cell surface for uptake. It seems possible that despite the possible hazards of employing vanadyl in extracellular buffers, there may even be some selective advantages. For example the uptake of vanadyl may be more rapid or saturated at lower concentrations than that of orthovanadate or alternatively there may be a kinetic advantage because the reduction of orthovanadate to vanadium (IV) which occurs upon entry to the cell interior will be obviated.

In view of the established role of cyclic AMP phosphodiesterase in the actions of insulin, it was surprising that the antilipolytic effects of vanadium were still apparent in the presence of a potent inhibitor of the insulin-sensitive cGI-PDE. Several reservations may be noted regarding the experiments carried out so far. It is possible that we obtained a fortuitous balance in effective intracellular concentrations of vanadium and of the phosphodiesterase inhibitor such that vanadium was able to prevent phosphodiesterase inhibition. The fact that cilostamide can block the antilipolytic effects of insulin itself [36] would argue against this possibility. It is also possible that vanadium could lead to the activation of an isoform of phosphodiesterase distinct from that affected by insulin. Indeed several cyclic nucleotide phosphodiesterase are known to be expressed in adipose cells and so further detailed experiments will be required to evaluate this possibility.

Despite the reservations noted above, the possibility that vanadium may exert antilipolytic effects distal to the synthesis of cyclic AMP is worthy of consideration. We have provided direct evidence for effects of vanadium on cyclic AMP-dependent protein kinase which could account for diminished phosphorylation and activation of hormone-sensitive lipase in the presence of vanadium. We have recently discovered that similar observations have also been noted by others (Crans *et al.* personal communication). The evidence so far available would suggest an effect of vanadium on the protein kinase catalytic subunit, rather than interference with the binding of cyclic AMP. It is clearly important to establish the mechanism by which vanadium may affect cyclic AMP-dependent protein kinase. Many studies have been based on the contention that vanadium acts as an analogue for phosphate. It is also possible that vanadium in its intracellular form (VO^{2+}) might mimic a divalent metal cation (Crans *et al.* in preparation). Comparison of the effects of vanadium on different protein kinases, and how these effects relate to the well-established effects of vanadium on various phosphatase activities should be most revealing.

To conclude, at least two significant implications may be drawn from the experiments described above. First, the actions of low concentrations of vanadium *in vivo* may be considerably more selective than one might conclude from pleiotropic effects observed with mM and higher concentrations of vanadium used in studies *in vitro*. For example it is not yet entirely clear that vanadium can reproduce the general anabolic (and perhaps growth-supporting) function lost by insulin depletion in STZ-diabetes. Second, the role of the 'glucose-fatty acid cycle', proposed originally by Randle and colleagues [3, 4], in the actions of vanadium should be carefully evaluated. Specifically, it is possible that through a primary effect on lipolysis, vanadium may be able to bring about normalization of diverse branches of metabolism. The diminished supply of non-esterified fatty acids would be expected to relieve inhibition of the pyruvate dehydrogenase complex (PDH) and thereby encourage pyruvate oxidation at the expense of *beta*-oxidation. The decline in cellular acetyl-CoA/CoA ratios would not only lead to deactivation of PDH kinase (and hence dephosphorylation and activation of PDH) but would also diminish the allosteric activation of pyruvate carboxylase and therefore suppress gluconeogenesis and hepatic glucose output.

Acknowledgements

Work in this laboratory is supported by grants (to RWB) from the Medical Research Council of Canada and from the Canadian Diabetes Association. In addition, exchange/travel was sponsored by a grant from NATO jointly to RWB and Professor R.M. Denton (Bristol, U.K.). We are grateful to Drs Nick Redpath and Chris Proud and Professor Richard Denton (all of the Department of Biochemistry, University of Bristol, U.K.) for samples of highly-purified cyclic AMP-dependent protein kinase and Dr Kenneth Murray (SmithKline Beecham, U.K.) for Cilostamide.

References

1. Cahill GF: Physiology of insulin in man. Diabetes 20: 785–789, 1971
2. Moller DE, Flier JS: Insulin resistance – mechanisms, syndromes and implications. N Engl J Med 325: 938–948, 1991
3. Randle PJ, Priestman DA, Mistry SC, Halsall A: Glucose fatty acid interactions and the regulation of glucose disposal. J Cell Biochem 55: 1–11 (suppl), 1994
4. Randle PJ, Hales CN, Garland PB, Newsholme EA: The glucose fatty-acid cycle. Its role in insulin sensitivity and the metabolic disturbances of diabetes mellitus. Lancet 1: 785–789, 1963
5. Nikkila EA: Control of plasma and liver triglyceride kinetics by carbohydrate metabolism and insulin. Adv Lipid Res 7: 63–134, 1969
6. Saudek CD, Eder HA: Lipid metabolism in diabetes mellitus. Am J Med 66: 843–852, 1979
7. Howard BV, Howard WJ: Dyslipidemia in non-insulin-dependent diabetes mellitus. Endocri Rev 15: 263–274, 1994
8. Ramanadham S, Mongold JJ, Brownsey RW, Cros GH, McNeill JH: Oral vanadyl sulfate in treatment of diabetes mellitus in rats. Am J Physiol 257: H904–H911, 1989
9. Ramanadham S, Brownsey RW, Cros GH, Mongold JJ, McNeill JH: Sustained prevention of myocardial and metabolic abnormalities in diabetic rats following withdrawal from oral vanadyl treatment. Metabolism, 38: 1022–1028, 1989
10. Duckworth WC, Solomon SS, Liepnieks J, Hamel FG, Hand S, Peavy DE: Insulin-like effects of vanadate in isolated rat adipocytes. Endocrinol 122: 2285–2289, 1988

11. Nilsson NO, Stralfors P, Fredrickson G, Belfrage P: Regulation of adipose tissue lipolysis: effects of noradrenaline and insulin on phosphorylation of hormone-sensitive lipase and lipolysis in intact rat adipocytes. FEBS Lett 111: 125–130, 1980
12. Loten EG, Sneyd JGT: An effect of insulin on adipose-tissue adenosine 3',5'-cyclic monophosphate phosphodiesterase. Biochem J 120: 187–193, 1970
13. Degerman E, Smith CJ, Tornqvist H, Vasta V, Belfrage P, Manganiello VC: Evidence that insulin and isoprenaline activate the cGMP-inhibited low-Km cAMP phosphodiesterase in rat fat cells by phosphorylation. Proc Natl Acad Sci USA 87: 533–537, 1990
14. Shibata H, Robinson FW, Soderling TR, Kono T: Effects of okadaic acid on insulin-sensitive cAMP phosphodiesterase in rat adipocytes. Evidence that insulin may stimulate the enzyme by phosphorylation. J. Biol Chem 266: 17948–17953, 1991
15. Vasta V, Smith CJ, Calvo J, Belfrage P, Manganiello VC: Insulin and isoproterenol induce phosphorylation of the particulate cyclic GMP-inhibited, low Km cyclic AMP phosphodiesterase (cGI PDE) in 3T3-L1 adipocytes. Biochem Biophys Res Commun 183: 1070–1075, 1992
16. Sera M, Tanaka K, Morita T, Ueki H: Increasing effect of vanadate on lipoprotein lipase activity in isolated rat fat pads. Arch. Biochem. Biophys. 279: 291–297, 1990
17. Souness JE, Maslen C, Scott LC: Effects of solubilization and vanadate/glutathione complex on inhibitor potencies against eosinophil cyclic AMP-specific phosphodiesterase. FEBS Lett 302: 181–184, 1992
18. Ueki H, Okuhama R, Sera M, Inoue T, Tominaga N, Morita T: Stimulatory effect of vanadate on 3',5'-cyclic guanosine monophosphate-inhibited low Michaelis-Menten constant 3',5'-cyclic adenosine monophosphate phosphodiesterase activity in isolated rat fat pads. Endocrinol 131: 441–446, 1992
19. Brownsey RW, Dong GW, Lam V, McGreer W: Studies on protein phosphorylation using subcellular fractions from insulin-treated white adipose tissue of rats. Biochem. Cell Biol 66: 296–308, 1988
20. Rodbell M: Metabolism of isolated fat cells. I. Effects of hormones on glucose metabolism and lipolysis. J Biol Chem 239: 375–380, 1964
21. Honnor RC, Dhillon GS, Londos C: cAMP-dependent protein kinase and lipolysis in rat adipocytes. 1. Cell preparation, manipulation and predictability in behaviour. J Biol Chem 260: 15122–15129, 1985
22. Winz R, Hess D, Aebersold R, Brownsey RW: Unique structural features and differential phosphorylation of the 280-kDa component (isozyme) of rat liver acetyl-CoA carboxylase. J Biol Chem 269: 14438–14445, 1994
23. Garland PB, Randle PJ: A rapid enzymatic assay for glycerol. Nature 196: 987–988, 1962
24. Stansbie D, Brownsey RW, Crettaz M, Denton RM: Acute effects *in vivo* of anti-insulin serum on rates of fatty acid synthesis and activities of acetyl-coenzyme A carboxylase and pyruvate dehydrogenase in liver and epididymal adipose tissue of fed rats. Biochem J 160: 413–416, 1976
25. Casnellie JE: Assay of protein kinases using peptides with basic residues for phosphocellulose binding. Methods Enzymol. 200: 115–120, 1991
26. Bradford MM: A rapid and sensitive method for the quantitation of microgram quantities of protein utilizing the principle of protein-dye binding. Anal Biochem 72: 248–254, 1976
27. Crans DC: Aqueous chemistry of labile oxovanadates: relevance to biological studies. Comments Inorg Chem 16: 1–33, 1994
28. Chasteen ND, Francavilla J: An electron paramagnetic resonance study of vanadyl(IV)-serum albumin complexes. J Phys Chem 80: 867–871, 1976
29. Macara IG, Kustin K, Cantley LC: Glutathione reduces cytoplasmic vanadate – mechanism and physiological implications. Biochim Biophys Acta 629: 95–106, 1980
30. Sakurai H, Shimomura S, Fukuzawa K, Ishizu K: Detection of oxovanadium (IV) and characterization of its ligand environment in subcellular fractions of the liver of rats treated with pentav alent vanadium (V). Biochem Biophys Res Commun 96: 293–298, 1980
31. Nechay BR, Nanninga LB, Nechay PSE: Vanadyl (IV) and vanadate (V) binding to selected endogenous phosphate, carboxyl and amino ligands: calculations of cellular vanadium species distribution. Arch Biochem Biophys 251: 128–138, 1986
32. Chasteen ND: Vanadyl(IV) electron nuclear double resonance/electron spin echo envelope modulation spin probes. Methods Enzymol 237: 232–244, 1993
33. Shechter Y, Karlish SJD: Insulin-like stimulation of glucose oxidation in rat adipocytes by vanadyl (IV) ions. Nature 284: 556–558, 1980
34. Ramanadham S, Heyliger C, Gresser MJ, Tracey AS, McNeill JH: The distribution and half-life for retention of vanadium in the organs of normal and diabetic rats orally fed vanadium (IV) and vanadium (V). Biol Trace Elem Res 30: 119–124, 1990
35. Hamel FG, Solomon SS, Jespersen AS, Blotcky A, Rack E, Duckworth WC: Alteration of tissue vanadium content in diabetes. Metabolism 42: 1503–1505, 1993
36. Schmitz-Peiffer C, Reeves ML, Denton RM: Characterization of the cyclic nucleotide phosphodiesterase isoenzymes present in rat epididymal fat cells. Cell Signalling 4: 37–49, 1992

Molecular and Cellular Biochemistry **153**: 139–144, 1995.

The enhancement by pervanadate of tyrosine phosphorylation on prostatic proteins occurs through the inhibition of membrane-associated tyrosine phosphatases

Martin Boissonneault[1], Alcide Chapdelaine[2] and Simone Chevalier[3]

The Montreal General Hospital Research Institute[1,3], Department of Surgery (Urology)[2,3], McGill University; Maisonneuve-Rosemont Research Centre[1,2,3], Departments of Biochemistry[1,3] and Medicine[2,3], University of Montreal, Canada

Abstract

The relapse of prostate cancer during endocrine therapies is attributed to the proliferation of growth factor (GF)-dependent epithelial cells. Such cells are present but in a quiescent state in the normal adult human and dog (experimental model) prostates. GF-signaling pathways involve the activation of protein tyrosine kinases (PTK) whose action is also modulated by phosphotyrosine protein phosphatases (PTPs). To that effect, we have previously reported that dividing canine prostatic epithelial cells exhibited high levels of phosphotyrosyl-(pY)-proteins which were greatly enhanced when incubated in the presence of vanadate. The aim of this study, performed with pervanadate (pV), was to determine whether pV acts either directly by stimulating prostatic PTKs or indirectly by inhibiting PTPs. Upon fractionation, most of the PTK activity was found in membranes of dividing cells and pV selectively increased its activity. This was due to an inhibition of intrinsic PTPs, as demonstrated by dephosphorylation of endogenous pY-proteins which was abolished by pV. This activity was very sensitive to pV (IC_{50}: 150 nM) and was due to non-secreted forms of prostatic acid phosphatase (PAP), a pV inhibited-enzyme, as well as to PTP-1 B, as demonstrated by gel filtration, isoelectric focusing and probing with antibodies. These enzymes were also detected in membranes from human hyperplastic/neoplastic prostates but only PTP-1 B was present in those of prostatic carcinoma PC3 cells. These PTPs, bound to membranes of dividing cells (normal vs neoplastic) where activated PTKs are also located, may be of importance in the development and progression of prostatic proliferative diseases. (Mol Cell Biochem **153**: 139–144, 1995)

Key words: pervanadate, prostatic acid phosphatase, phosphotyrosine protein phosphatase, PTP-1 B, protein tyrosine kinase, prostate

Introduction

Most growth factors and some peptide hormones transmit their intracellular signal through pathways involving both protein tyrosine kinases (PTKs) and phosphotyrosine protein phosphatases (PTPs). Up to now, PTKs have been studied more extensively than PTPs but the identification of their natural substrates remains incomplete. The respective roles of these enzymes in the control of cell proliferation in the prostate still remain unclear. With this respect, it has recently been demonstrated that prostatic acid phosphatase (PAP) regulates cell growth by phosphotyrosine (pY) dephosphorylation. Indeed, when the c-DNA encoding for the secretory form of PAP was transfected into PAP-deficient prostatic carcinoma PC3 cells, the latter exhibited a diminished growth rate paralleled by a decrease in pY content [1]. In addition, it has been reported that in prostatic epithelial cells, there is an inverse correlation between PTK and PAP activities, the proliferating cells showing high PTK and low PAP while differentiated cells exhibited high PAP and low PTK [2, 3]. This is consistent with the facts that PAP hydrolyses pY-containing substrates [4–8], that the major PTP in prostatic epi-

Address for offprints: S. Chevalier, The Montreal General Hospital Research Institute, 1650 Cedar Avenue, R1-137, Montreal, Canada, H3G 1A4

thelial cells co-purifies with cellular forms of PAP [2] and that its activity is strongly reduced by vanadate [2, 3, 5–8], an inhibitor of PTPs [9]. However, the effect of vanadate on the level of pY-proteins is highest in dividing prostatic epithelial cells even though their PAP activity is low [3, 10]. This puzzling result could indicate that vanadate directly stimulates PTKs or else that it inhibits intracellular PTPs, related or not to non-secretory forms of PAP. In the present investigation, we have tested these hypotheses using pervanadate (peroxy-vanadate, pV), a more potent inhibitor of PTPs than vanadate [11].

Materials and methods

Materials

Vanadate (Na_3VO_4) was obtained from Aldrich (Milwaukee, WI). The synthetic PTK substrate poly(Glu,Tyr), a copolymer of sodium L-glutamate:L-tyrosine (4:1), p-nitrophenyl phosphate (p-NPP), benzamidine, phenylmethylsulfonyl fluoride (PMSF) and mouse monoclonal human PAP antibody were purchased from Sigma Chemical Co. (St. Louis, MO), and mouse monoclonal antibody against PTP-1B (Ab-1) from Oncogene Science (Manhasset, NY). Ampholines, goat anti-rabbit and goat anti-mouse peroxidase-conjugated antibodies were from Bio-Rad (Mississauga, Ont.) and rabbit polyclonal antibody against phosphotyrosine from UBI (Lake Placid, NY). Leupeptin, aprotinin and pepstatin were from Boehringer-Mannheim (Laval, Que.). Reagents for enhanced chemiluminescence (ECL) Western blot detection were obtained from Amersham (Oakville, Ont.), Superose 12 HR from Pharmacia (Dorval, Que.) and γ-^{32}P]ATP (7000 Ci/mmol) from Dupont (Dorval, Qué.).

Subcellular fractionation of prostatic tissue and cells
The methods used for the culture of normal canine prostatic epithelial (CPE) cells have been described previously [12]. Briefly, the cells were allowed to attach over a 3-day period at 37°C in a humidified chamber under 5% CO_2-95% air. All subsequent procedures were then carried out at 4°C. The adhering cells were first scraped into buffer A [25 mM HEPES-NaOH, pH 7.4, 0.25 M sucrose, 2 mM ethylene diamine tetraacetic acid (EDTA), 5 mM β-mercaptoethanol, and protease inhibitors (1 mM PMSF, 40 mM benzamidine, 20 µg/ml leupeptin, 10 µg/ml aprotinin, 4 µg/ml pepstatin)], and then pelleted by a 500 g centrifugation. The cells were homogenized with a Potter (50 strokes of a tight-fitting pestle) and sonicated 3 × 15 sec at 200 W (Branson). Nuclei were removed by a 10 min centrifugation at 1000 g. The supernatant was centrifuged for 20 min at 12000 g to remove other organelles and debris. Following a subsequent centrifugation of the supernatant at 105000 g for 60 min, the soluble fraction (S105) was saved and the pellet was extracted by resuspension in buffer A containing 2% Triton X-100, 0.25 M KCI and incubation on ice for 60 min prior to a second centrifugation for 60 min at 105000 g. The supernatant of this last centrifugation was considered as the particulate fraction (P105).

Human prostatic carcinoma PC3 cells, obtained from the American Type Culture Collection (Rockville, MD), were cultured at confluency in RPMI medium supplemented with 10% fetal calf serum. Cells were scraped into buffer A and submitted to the same procedures as described above for CPE cells.

Human hyperplastic/neoplastic prostates were obtained either at autopsy (within 8 h *post mortem*) of men over sixty-five years of age or at surgery for prostatic carcinoma. They were rapidly dissected and frozen at –80°C until use. After thawing and mincing, the tissue was first homogenized with a Polytron in 4 vol. of ice-cold buffer A and then processed as described for cultured cells. Preparations, referred to as BPH in figures, were aliquoted and kept frozen at –80°C until use.

PTK assay
The PTK activity of prostatic fractions (10–25 µg proteins) was determined by phosphorylation of 50 µg poly(Glu,Tyr) at room T for 20 min in 20 mM HEPES buffer, pH 7.4, containing 20 mM $MgCl_2$, 5 mM $MnCl_2$, 5 mM β-mercaptoethanol, 0.1% (v/v) Triton X-100, 10 µM ATP (2 µCi [γ-^{32}P] ATP/assay) and, when indicated, 50 µM pV (freshly prepared by mixing equimolar amounts of H_2O_2 and Na_3VO_4), in a final volume of 100 µl. Samples were then processed as described previously [13]. Parallel assays, omitting enzyme preparations or poly(Glu,Tyr), were used as blanks. Results of PTK activity were expressed either in cpm or in pmol of Pi transferred from ATP to poly(Glu,Tyr)/min/mg protein.

Phosphatase assay
Phosphatase activity, which could antagonize that of PTKs, was determined in 20 mM HEPES at pH 7.4 using 10 mM p-NPP as substrate and 50 µM pV when indicated under conditions yielding linear relationships between activity and time as well as enzyme concentration (not shown). The cleavage of p-NPP was monitored by measuring the absorbance of p-NP liberated at 410 nm following incubation at 37°C for 30 min and the subsequent addition of NaOH at a final concentration of 0.2 M.

In vitro protein phosphorylation-dephosphorylation and Western blot analysis
For protein phosphorylation, each fraction (50 µg proteins) was incubated for 20 min at room T under conditions described

above for PTK assays, omitting poly(Glu,Tyr), in absence or presence of 50 μM pV, starting the reaction with 100 μM unlabelled ATP and stopping it with TCA (10% final).

To study the effect of pV on the dephosphorylation of prostatic proteins, the latter were first allowed to phosphorylate *in vitro* by the action of endogenous PTKs in presence of ATP, as described above. The reaction was stopped with 20 mM EDTA instead of TCA. Dephosphorylation was then allowed to proceed for 30 min in the presence or absence of 50 μM pV. Proteins were finally precipitated with TCA (10% total).

In both instances, after standing at 4°C for 30 min, proteins were pelleted by centrifugation, washed with acetone, solubilized at 100°C for 3 min in SDS-PAGE sample solution [13] and separated by electrophoresis on 10% polyacrylamide gels in presence of sodium dodecyl sulfate (SDS) at 25 mA/gel. Proteins were then transferred to nitrocellulose membranes and immunoblotted with a rabbit polyclonal anti-phosphotyrosine antibody (1/1000 in 10 mM Tris and 150 mM NaCl, pH 7.4 (TBS) containing 1% milk powder). Goat anti-rabbit IgGs coupled to peroxidase (1/2000 in TBS containing 1% milk powder) were then used as second antibodies and the detection was achieved using the ECL kit.

Isoelectric focusing
Isoelectric focusing under native conditions was carried out at 500 V and 4°C for 4 h in analytical polyacrylamide gels containing 2% Ampholines (pH 4–6) using three 16 cm tubes. pI values were determined on one gel and phosphatase activity was visualized by incubating the others for 30 min at 37°C in 0.1 M sodium acetate buffer, pH 5.0 containing 4 mM α-naphthyl phosphate, 2.5 mM p-p'-p'' pararosaniline and 0.1% Fast Garnet GBC salt.

Gel filtration of particulate fractions and immunodetection of specific phosphatases
An aliquot (1 mg proteins) of P105 fractions from each extract was submitted to size-exclusion chromatography on a Superose 12 HR column (1.5 × 50 cm) equilibrated in 25 mM imidazole-iminodiacetic acid buffer, pH 7.4, containing 10% glycerol, 100 mM NaCl, 0.1% Triton X-100 and 0.5 mM β-mercaptoethanol. The column was connected to a FPLC apparatus (Pharmacia). Fractions of 200 μl were collected and either assayed for their PTK and phosphatase activities or frozen at –80°C until use. The detection of PTP-1B and PAP in peaks of phosphatase activity following chromatography was performed by Western blot analysis after the addition of 2X SDS-PAGE sample solution, under the conditions described above, using mouse monoclonal anti-human PTP-1B or PAP, a goat anti-mouse IgG coupled to peroxidase and the ECL kit.

Results and discussion

Pervanadate enhances the phosphorylation of the synthetic substrate poly(Glu, Tyr) and that of endogenous proteins in the particulate fraction
We have previously reported that 88% of the PTK activity in dividing prostatic cells and less than 40% of the activity in BPH tissue was associated with their particulate fraction [10, 13]. Fig. 1 (upper panel) shows that the specific PTK activity on the synthetic substrate poly(Glu,Tyr) in the absence of pV was 3–10 fold higher in P105 fractions from CPE and PC3 cells than in their corresponding S105 ones while both fractions exhibited similar activities in BPH tissues. Interestingly, the activity in P105 fractions from all prostatic preparations was greatly enhanced when 50 μM pV was present in the assays (147%, 95%, and 152%, respectively). This concentration of pV was selected since, in control experiments with CPE cells, it had been observed that 50 and 100 μM pV yielded identical activation of PTK activity. In contrast, the

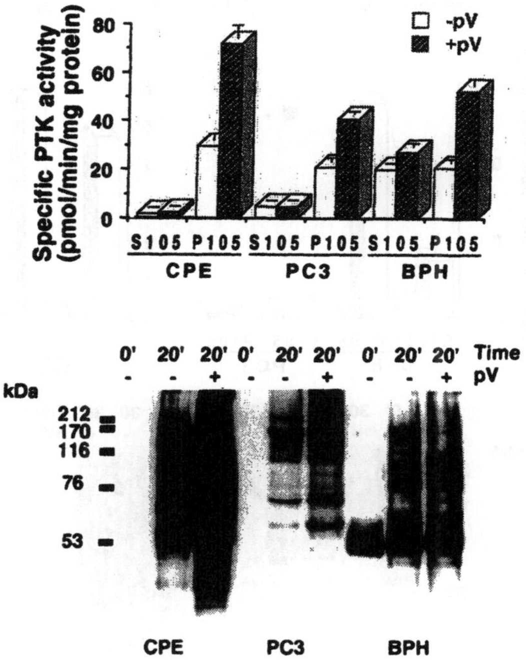

Fig. 1. Effect of pervanadate on the PTK activity of fractions from CPE and PC3 cells, and BPH tissue. Upper panel: specific activity on subcellular fractions using poly(Glu,Tyr) as a substrate. Assays were performed in absence (closed bars) or presence (hatched bars) of 50 μM pV. Lower panel: PTK activity of P105 fractions evidenced by the level of phosphotyrosylation of endogenous proteins after a 20 min incubation with ATP in presence (+) or absence (–) of 50 μM pV. The position of molecular weight standards is indicated on the left.

PTK activity in S105 fraction was almost unaffected by this treatment. For comparison, tartrate, a classical PAP inhibitor [14], had no effect on the specific PTK activity of both S105 and P105 fractions for all preparations, when tested at concentrations up to 5 mM (data not shown).

When the PTK activity of these P105 fractions was assayed by probing the level of pY-proteins (Fig. 1, lower panel), pV was shown to strongly enhance the phosphorylation of almost all proteins detected. As subcellular fractionations were performed in absence of phosphatase inhibitors, it was not surprising to observe low levels of pY-proteins prior to the addition of ATP (Fig. 1, lower panel, lanes 0′).

Pervanadate inhibits the phosphatase activity of all fractions

To examine whether the stimulatory effect of pV on the PTK activity could be due to an inhibition of phosphatases, this hydrolytic activity was assayed on p-NPP at pH 7.4 in all fractions. Fig. 2, upper panel, shows that the specific phosphatase activities were more elevated in P105 fractions, the highest being the P105 of BPH tissue, and that in all fractions, the activity was almost abolished by 50 µM pV. IC_{50}'s (in µM) for P105 fractions were 0.15 (CPE), 0.07 (PC3) and 0.30 (BPH); those for S105 fractions were 3.3 (CPE), 17 (PC3) and 0.50 (BPH). Tartrate, to a lesser degree than pV, could also inhibit the phosphatase activity of these fractions: P105 and S105 fractions from BPH tissue and the S105 fraction from CPE cells had IC_{50}'s of 100–160 µM while all other fractions exhibited IC_{50}'s in the mM range.

Even though the majority of phosphatase activity (65–80%, not shown) was found in S105 fractions, we have focused our subsequent studies on enzymes present in P105 fractions since pV, but not tartrate, increased the specific PTK activity in these preparations preferentially.

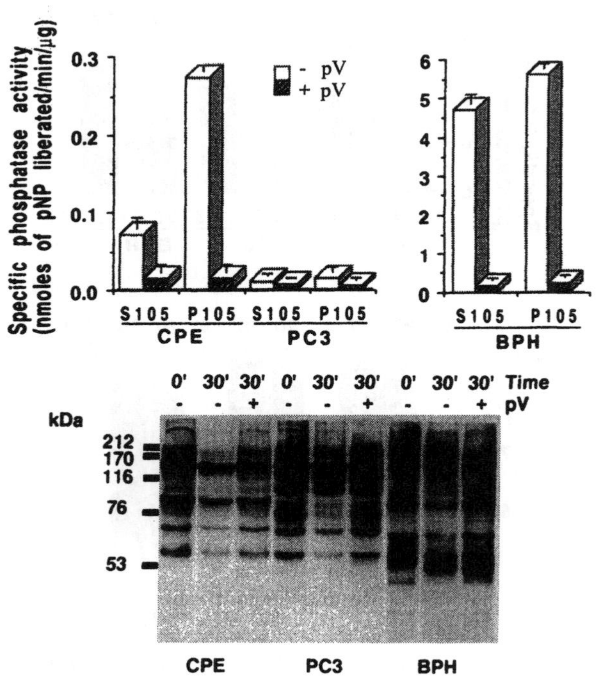

Fig. 2. Effect of pervanadate on the phosphatase activity of fractions from CPE and PC3 cells, and BPH tissue. Upper panel: specific activity using p-NPP as a substrate. Lower panel: levels of phosphotyrosylated endogenous proteins after a 30 min incubation in absence (–) or presence (+) of 50 µM pV. The position of molecular weight standards is indicated on the left.

Presence of PTPs in particulate fractions

To verify whether pV would prevent the dephosphorylation of pY-proteins in P105 fractions, we first allowed the *in vitro* tyrosylphosphorylation on endogenous proteins to occur through the action of PTKs, in presence of ATP for 20 min; this reaction was then blocked by the addition of EDTA and dephosphorylation was allowed to proceed for 30 min in the presence or absence of pV. Results (Fig. 2, lower panel) show that pV indeed prevented dephosphorylation of pY-proteins, thereby indicating the presence of PTPs active on natural substrates.

Characterization of phosphatases

Isoelectric focusing under native conditions was performed to compare the phosphatases present in the P105 fraction from CPE cells to those secreted in dog seminal plasma (considered as PAP and strongly inhibited by pV [6]). Fig. 3 shows that intracellular phosphatases, bound to membranes of CPE cells, have more basic pIs (5.45–5.80) than those species found in seminal plasma (4.75–5.0).

Upon gel filtration, the P105 fraction from CPE cells resulted in the resolution of PTK and phosphatase activities into two peaks (Fig. 4, upper panel). Similar patterns were obtained with the corresponding fractions from PC3 cells and BPH tissue (not shown). In contrast with the results obtained with crude P105 fractions, no stimulation by pV was observed when the PTK activity of peak fractions was tested on poly(Glu,Tyr). This indicates that under those conditions, pV has no direct effect on PTKs. However, phosphatase peaks remained highly sensitive to pV inhibition.

As PTP-1 B is a ubiquitously expressed enzyme bound to endoplasmic reticulum membranes [15], its presence was sought for and indeed demonstrated but only in the high-Mr peaks of phosphatase activity (A) of the three P105 fractions analyzed (Fig. 4, lower panel). Since PTP-1 B, a 50 kDa protein [16], was only found in high-Mr peaks, it might be bound to other proteins.

When PAP was similarly immunodetected, it was found in both phosphatase peaks (A and B) from P105 fraction of BPH tissue (Fig. 4, lower panel) but not in those from CPE and PC3 cells (not shown). It is noteworthy that PC3 cells do

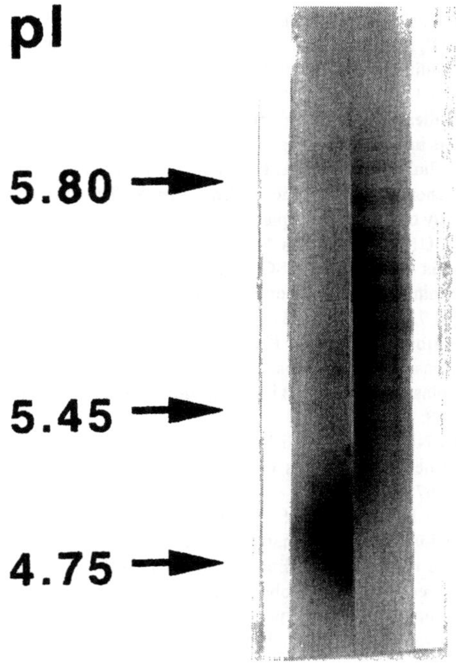

Fig. 3. Isoelectric focusing under native conditions. Phosphatase activity of dog seminal plasma (left) and P105 fraction from CPE cells (right). pI values are also indicated.

not express PAP [1]. On the other hand, the antibody to human PAP has already been shown to cross-react with canine PAP, for instance in cultured CPE cells [5]; moreover, their acid phosphatase activity increased with the duration of the culture period [17] when the rate of division is low [18]. It is thus possible that PAP was not detected in CPE cells because, after three days of culture, adherent cells have entered into the cell cycle and are in a state of rapid division [12, 17, 18]. Their low PAP levels at this stage thus support the hypothesis that this enzyme acts as a negative growth regulator. The finding that PAP was present in membranes from BPH tissue and was resolved into two peaks is also of interest since most studies concerning this enzyme were undertaken on the secretory isoforms found in human seminal plasma or else, on the soluble enzyme found in secretory granules and/or cytosol of prostate tissue mainly composed of differentiated cells [4]. This study, focusing on membrane-associated phosphotyrosine phosphatases, has shown the presence of membrane-bound PAP as well as of PTP-1B; these enzymes have similar sizes on SDS-PAGE, they are both sensitive to pV, are active at acid and neutral pH and, altogether, have more basic pIs than secreted PAP.

In conclusion, from our previous study on protein phosphorylation in prostate epithelial cells, it was established that vanadate influenced the PTK-PTP enzyme systems [3, 10]. In the present investigation with pV, this effect was specifically attributed to the inhibition of PTPs present in membrane fractions of dividing cells and BPH tissue. Since PTK levels in such cells are elevated and that the highest specific activities are found in membranes, it is suggested that membrane-bound PTPs regulate the degree of protein phosphotyrosylation which can be important for a mitogenic signal transduction pathway occurring in androgen-independent prostatic epithelial cells (CPE and PC3). Further studies are required to understand the specific roles of these membrane-bound PTPs in the prostate.

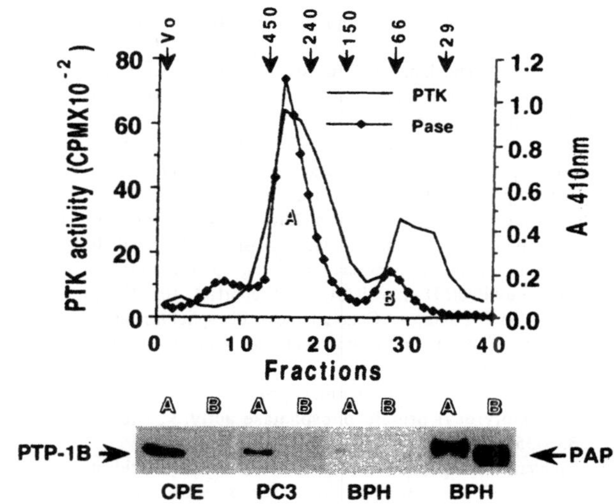

Fig. 4. Gel filtration of the P105 fraction from CPE cells and immunodetection of PTP-1B and PAP. Upper panel: the P105 fraction from CPE cells was chromatographed on a Superose 12 HR column and fractions were assayed on p-NPP for determination of phosphatase activity and on poly(Glu,Tyr) for PTK activity. Arrows and numbers above them indicate, respectively, the elution position of molecular weight standards and their mass in kDa. Fractions 16 and 30 were also assayed on poly(Glu,Tyr) in the presence or absence of 50 μM pV. Lower panel: immunoblot analyses of phosphatases: the three P105 fractions yielding similar chromatographic patterns, fractions 15 (peak A) and 27 (peak B) were probed with antibodies against PTP-1 B and PAP, the latter being shown only for BPH tissue.

Acknowledgments

This research was supported by the Medical Research Council of Canada. S. Chevalier is the recipient of a senior scholarship from 'Le Fonds de la Recherche en Sante du Quebec'.

References

1. Lin MF, DaVolio J, Garcia-Arenas R: Expression of prostatic acid phosphatase activity and the growth of prostate carcinoma cells. Cancer Res 52: 4600–4607, 1992
2. Lin MF, Clinton GM: Human prostatic acid phosphatase and its

phosphotyrosyl-protein phosphatase activity. Adv Prot Phosphatases 4: 199–228, 1987
3. Tessier S, Chapdelaine A, Chevalier S: Effect of vanadate on protein phosphorylation and on acid phosphatase activity in the canine prostate. Mol Cell Endocrinol 64: 87–94, 1989
4. Li HC, Chernoff J, Chen LK, Krischonbaum A: A phosphotyrosyl-protein phosphatase activity associated with acid phosphatase from human prostate gland. Eur J Biochem 138: 45–51, 1984
5. Chevalier S, Bourassa C, Bleau G, Roberts KD, Chapdelaine A: Acid phosphatases as local modulators of prostatic cell growth. In Advances in gene technology: molecular biology of the endocrine system. Edited by Puett D *et al.* ICSU short reports 4: 106–107,1986
6. Chevalier S, Landry D, Chapdelaine A: Phosphotyrosine phosphatase activity of human and canine acid phosphatases of prostatic origin. Prostate 12: 209–219, 1988
7. Lin MF, Clinton GM: Human prostatic acid phosphatase has phosphotyrosyl protein phosphatase activity. Biochem J 235: 351–357, 1986
8. Nguyen L, Chapdelaine A, Chevalier S: Prostatic acid phosphatase in serum of patients with prostatic cancer is a specific phosphotyrosine acid phosphatase. Clin Chem 36: 1450–1455, 1990
9. Swarup G, Cohen S, Garbers DL: Inhibition of membrane phosphotyrosyl-protein phosphatase activity by vanadate. Biochem Biophys Res Commun 107: 1104–1109, 1982
10. Bourassa C, Nguyen LT, Durocher Y, Roberts KD, Chevalier S: Prostate epithelial cells in culture: phosphorylation of protein tyrosyl residues and tyrosine protein kinase activity. J Cell Biochem 46: 291–301, 1991
11. Faure R, Baquiran G, Bergeron JJM, Posner Bl: The dephos-phorylation of insulin and epidermal growth factor receptors. J Biol Chem 267: 11215–11221, 1992
12. Chevalier S, Bleau G, Roberts KD, Chapdelaine A: Nonsteroidal serum factors involved in the regulation of the proliferation of canine prostatic epithelial cells in culture. Prostate 5: 503–512, 1984
13. Durocher Y, Chapdelaine A, Chevalier S: Tyrosine protein kinase activity of human hyperplastic prostate and carcinoma cell lines PC3 and DU145. Cancer Res 49: 4818–4823, 1989
14. Mathes G, Richmond SG, Sprunt DH: The use of L-tartrate in determining 'prostatic' serum acid phosphatase: a report of 514 cases. J Urol 75: 143–150, 1959
15. Frangioni JV, Beahm PH, Shifrin V, Jost CA, Nell BG: The nontransmembrane tyrosine phosphatase PTP-1 B localizes to the endoplasmic reticulum via its 35 amino acid C-terminal sequence. Cell 68: 545–560, 1992
16. Tonks NK, Diltz CD, Fisher EH: Purification of the major protein tyrosine phosphatases of human placenta. J Biol Chem 263: 6722–6728, 1988
17. Dionne FT, Chevalier S, Bleau G, Roberts KD, Chapdelaine A: Induction of acid phosphatase synthesis in canine prostatic epithelial cells *in vitro*. Mol Cell Endocrinol 33: 113–123, 1983
18. Chevalier S, Bleau G, Roberts KD, Chapdelaine A: Proliferation and differentiation of canine prostatic epithelial cells in culture. Mol Cell Endocrinol 24: 195–208, 1981

Molecular and Cellular Biochemistry **153:** 145–150, 1995.

Contractile effects of vanadate on aorta rings from virgin and pregnant rats

Jean St-Louis[1,2], Benoit Sicotte[1], Eric Breton[1] , Ashok K. Srivastava[3]

Centres de recherche, [1]Hôpital Ste-Justine and [3]Hôpital Hotel-Dieu de Montréal, [2]Department of Obstetrics and Gynecology, and [3]Department of Medicine, University of Montréal, Montréal, Qc, Canada

Abstract

The present study was undertaken to characterize the contractile effects of vanadate on thoracic aorta rings from virgin and term-pregnant rats. Vanadate caused concentration-dependent contraction in rat aortic rings with an EC_{50} (concentration producing 50% maximum response) of 0.10 mM. Contractions in response to vanadate were equivalent to the ones measured with 1 μM phenylephrine. The effects of vanadate were not affected by indomethacin (up to 10 μM), an inhibitor of prostanoid cyclooxygenase, but were blocked in a concentration-dependent manner by staurosporine (0.1–1.0 μM), an inhibitor of protein kinase C. Vanadate exhibited a significant decrease of contractile responses in aorta of pregnant as compared to virgin rats. When aortic rings were bathed in presence of different concentrations of vanadate, the concentration-response curve to phenylephrine was shifted to the left, but maximum response was not affected. The potentiation of the contractions to phenylephrine by vanadate was significantly more prominent in aorta of virgin than of pregnant rats. These results suggest that the contractile effect of vanadate on rat aorta is independent of endogenous prostanoids and may be mediated by protein kinase C-dependent pathway. These results also show that the contractile response to vanadate on the rat aorta is impaired during pregnancy. (Mol Cell Biochem: **153:** 145–150, 1995)

Key words: vanadate, pregnancy, vascular smooth muscle, staurosporine

Introduction

Vanadate, the commonly occurring form of the trace element vanadium, has been shown to exert a myriad of biological effects on several systems, both *in vivo* and *in vitro* [1, 2]. Vanadate has also been shown to induce contraction in various types of smooth muscle, [3–11]. It has been suggested that vanadate mediates increase in intracellular Ca^{2+} levels. Voltage-operated calcium channels might play a prominent role in mediating the contractile responses to vanadate [12–17].

Pregnancy is associated with decreased blood pressure and peripheral resistance and marked increases in plasma volume and cardiac output [18, 19]. Simultaneously, the contractile effects of vasoconstrictors, such as angiotensin II, vasopressin and α-adrenomimetics, decrease in blood vessels of pregnant compared to virgin rats [20]. Recent studies have shown that the mechanism responsible for these blunted responses to vasoconstrictors might involved an alteration in calcium mobilization and/or storage by vascular smooth muscle during pregnancy [21].

Since the effects of vanadate on smooth muscle contractility might involve Ca^{2+} mobilization and since the responses to vasoconstrictors are blunted during pregnancy, the present investigation was undertaken to examine if vanadate-mediated contractile response is also altered in aortic rings of pregnant rats. In addition, a possible role of protein kinase in mediating this response was also evaluated using a selective protein kinase C inhibitor.

Materials and methods

Female Sprague-Dawley rats (Charles River Canada, St-Constant, QC) aged 10–11 weeks were mated with age-matched males. The morning on which vaginal smears were found to contain spermatozoa was labeled day 1 of pregnancy. The pregnant females were then placed in individual

Address for offprints: Centre de Recherche, Hôpital Ste-Justine, 3175 Chemin Côte Ste-Catherine, Montréal, Q.C. H3T 1C5, Canada cages until used on the 22nd day of gestation. Virgin rats of the same age served as controls without considering the stage of the estrous cycle. The animals were housed in facilities of the Research Center of Hospital Ste-Justine, which is accredited by the Canadian Council on Animal Care. The protocol was approved by the local animal care committee.

Rats were sacrificed by decapitation. The thoracic aorta was rapidly removed and placed in cold Krebs bicarbonate solution (KBS) of the following composition, in mM: NaCl, 118; KCl, 4.65; $NaHCO_3$, 25; $CaCl_2$, 2.5; $MgSO_4$, 1.18; KH_2PO_4, 1.18; and dextrose, 5.5. The solution was bubbled with a mixture Of 95% O_2: 5% CO_2; pH was 7.4. The aorta was cleaned of fat and extraneous tissues and cut into 4 consecutive rings (2–3 mm) which were placed in individual jacketed tissue baths (10 ml, Radnoti Glass, Monrovia, CA) maintained at 37°C. The lumen of each ring was rubbed with a needle to remove the endothelium. In each experiment, 4 rings from 2 different rats were used. Aortic rings were equilibrated for 60 min in KBS under an optimal passive tension of 2.0 g [22] with frequent washing and tension adjustment. After equilibration, the tissues were challenged with 1.0 μM phenylephrine (PE). At plateau response, carbachol (0.1 mM) was added to verify endothelial removal. Tension was measured by force-displacement transducers connected to a personal computer through the Work Bench® program (Kent Scientific, Litchfield, CT).

Cumulative concentration-response curves to vanadate (0.01–1 mM) were measured. In the experiments involving tissues from pregnant rats, one of the 4 rings of both virgin and pregnant rats served as control, while each of the other three were preincubated with a different concentration of indomethacin (1, 3 and 10 μM), an inhibitor of prostanoids synthesis, or of staurosporine (30, 100 and 100 nM), an inhibitor of protein kinase C, added 10 min before measuring the curve. In another set of experiments, cumulative concentration-response curves to phenylephrine (PE, 10^{-11}–10^{-5} M) were measured in the absence and presence of 0.01, 0.1 and 0.3 mM vanadate.

Each concentration-response curve was analysed by computer fitting to a 4-parameters logistic equation (Inplot, Graph Pad™, San Diego, CA) to evaluate the concentration producing 50% of maximum response (EC_{50}) and the maximum asymptote of the curve (E_{max}, maximum response). With the responses to PE, the minimum asymptote of the curve was also evaluated. Comparison of the curves was performed by factorial analysis of variance (GB-STAT, Dynamic Microsystems Inc, Silver Spring, MD) on mean pD_2 (negative log of the EC_{50}), on mean E_{max} and on the lower asymptote of the curve (in the experiments with PE). In figures, data are expressed as mean experimental points with their standard error along with the best fitted curve to these points.

All salts employed in these experiments were of analytical grade obtained from Fisher Scientific (Montréal, QC). Sodium orthovanadate, phenylephrine hydrochloride (PE) and carbachol hydrochloride were purchased from Sigma Chemical (St. Louis, MO) and indomethacin and staurosporine from Research Biochemical Inc. (Natick, MA).

Results

The left panel of Fig. 1 shows a representative tracing of cumulative concentration-response curve to vanadate along with the control response to 1 μM PE in aortic rings of virgin rats. Addition of carbachol (Car, 0.1 mM) shows that the endothelium was indeed not functional. A small response to vanadate was recorded at 30 μM vanadate, while maximum response was reached at 1 mM of the stimulant. This is confirmed by the mean concentration-response curve to vanadate (Fig. 1, right panel). In this set of experiments, we obtained a maximum response to vanadate of 1.68 ± 0.39 g with an EC_{50} of 0.10 mM. The tracing also shows that the maximum response to vanadate was higher than the control response to PE (1 μM), indicating that, in aortic rings of virgin rats, the response to vanadate is at least as effective as that of the α-adrenoceptor agonist.

In order to verify if the contractile response to vanadate was indirectly mediated by endogenous liberation of derivatives of arachidonic acid, aorta were incubated in the absence and presence of different concentrations of indomethacin. Figure 2A indicates that indomethacin (up to 10 μM) did not significantly alter the smooth muscle response to vanadate. On the other hand, staurosporine (0.03–1 μM), a selective inhibitor of protein kinase C, decreased the contractile response of vanadate in a concentration-dependent manner, without affecting sensitivity to the stimulant (Fig. 2B). Response to vanadate was almost completely abolished with 1 μM staurosporine.

Responses to vanadate were markedly blunted in aortic rings of pregnant (Fig. 3). In rings from virgin rats, 30 μM of vanadate induced a significant contractile response, whereas 0.1 mM vanadate was required to induce similar degree of response in tissues of pregnant animals. Assuming that identical maximum response would have been reached in both groups (Fig. 3B), the IC_{50} of vanadate would have increased from 0.123–0.658 mM, suggesting a marked rightward shift of the vanadate concentration-response curve in aorta rings of pregnant as compared to virgin rats.

To further assess if vanadate exposure could alter the contractile responses of phenylephrine, aortic rings of virgin and pregnant rats were bathed in the absence and the presence of 0.01, 0.1 and 0.3 mM vanadate. This manoeuvre induced a concentration-dependent increase in tone in tissues of virgin but not of pregnant rats (lower asymptote of the concentration-response curves in Fig. 4). On top of these responses, a

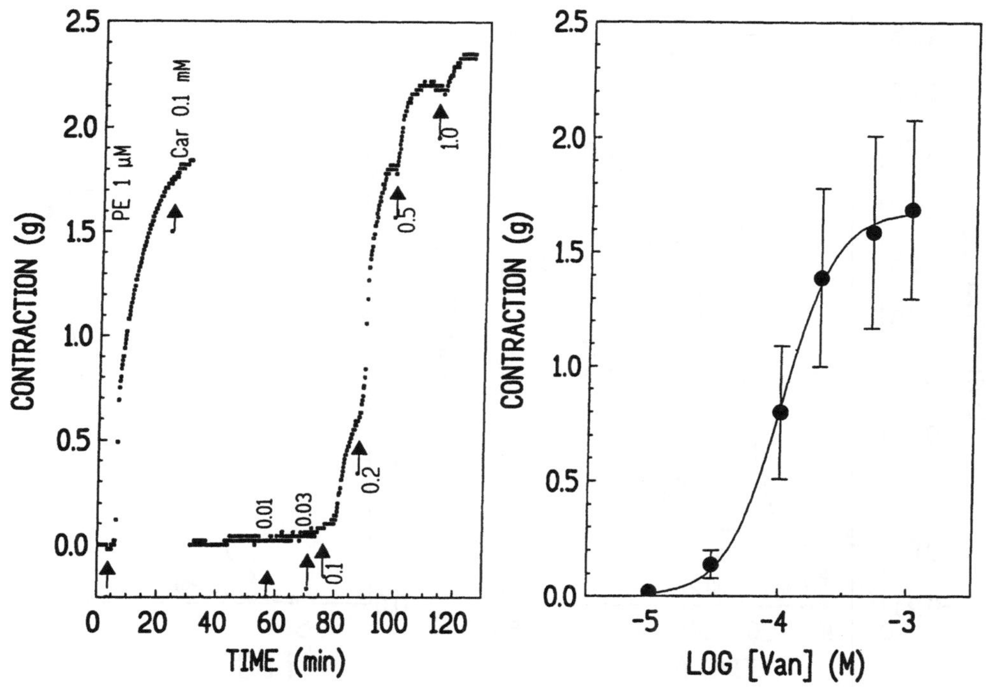

Fig. 1. Concentration-response curve of vanadate in aortic rings of virgin rats. A) dot graph of computer recording of a typical experiment showing the effects of phenylephrine (PE, 1 µM) and of carbachol (Car, 0.1 mM) on the top of PE response and concentration-response curve to vanadate. Arrows indicate the time of addition of the substance, concentration of vanadate is given in mM. B) mean concentration-response curve (n = 10) of vanadate.

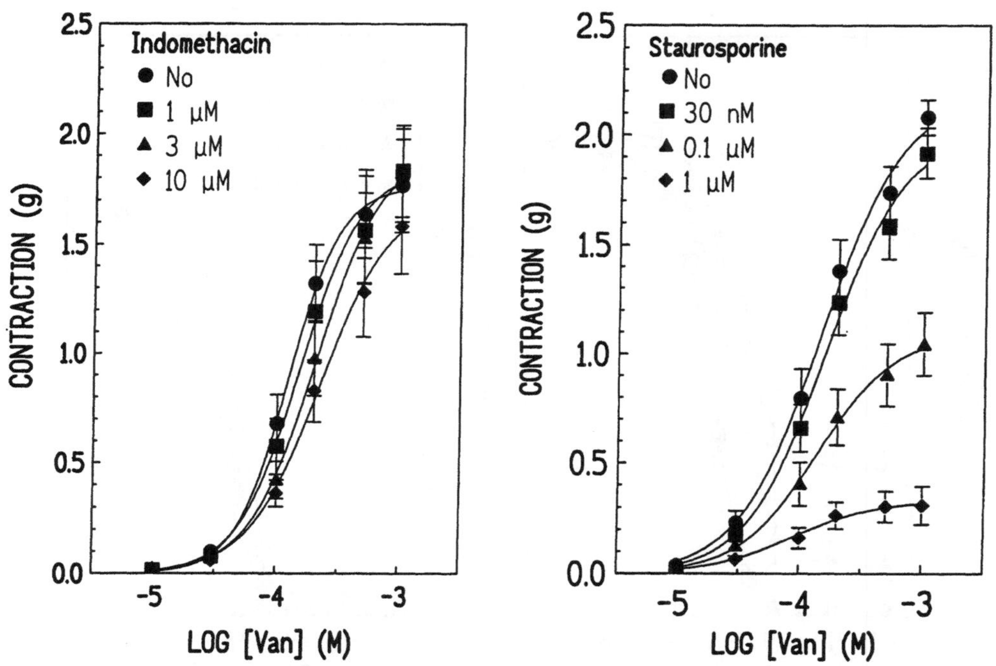

Fig. 2. Effects of indomethacin (left panel) and staurosporine (right panel) on concentration-response curve of vanadate in aortic rings of virgin rats. Abscissa, logarithm concentration of vanadate in molar; ordinate contractile response in gram.

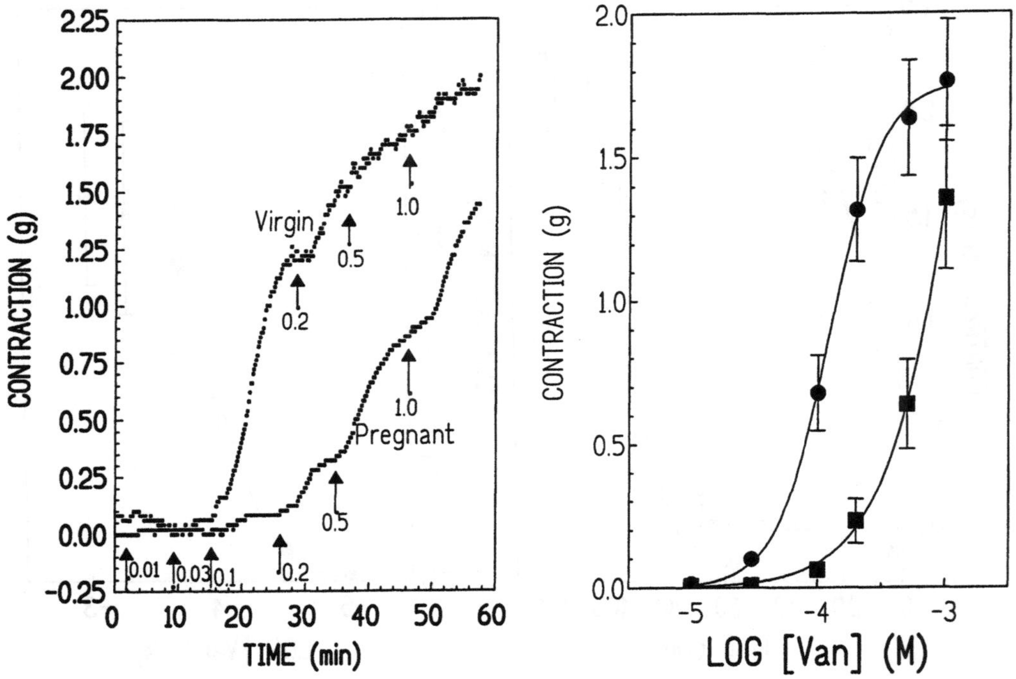

Fig. 3. Experimental computer tracing (left panel) and mean (right panel) concentration-response curves of vanadate in aortic rings of virgin (circles) and pregnant (squares) rats. In the left panel, numbers indicate the concentration of vanadate in mM given at the arrow.

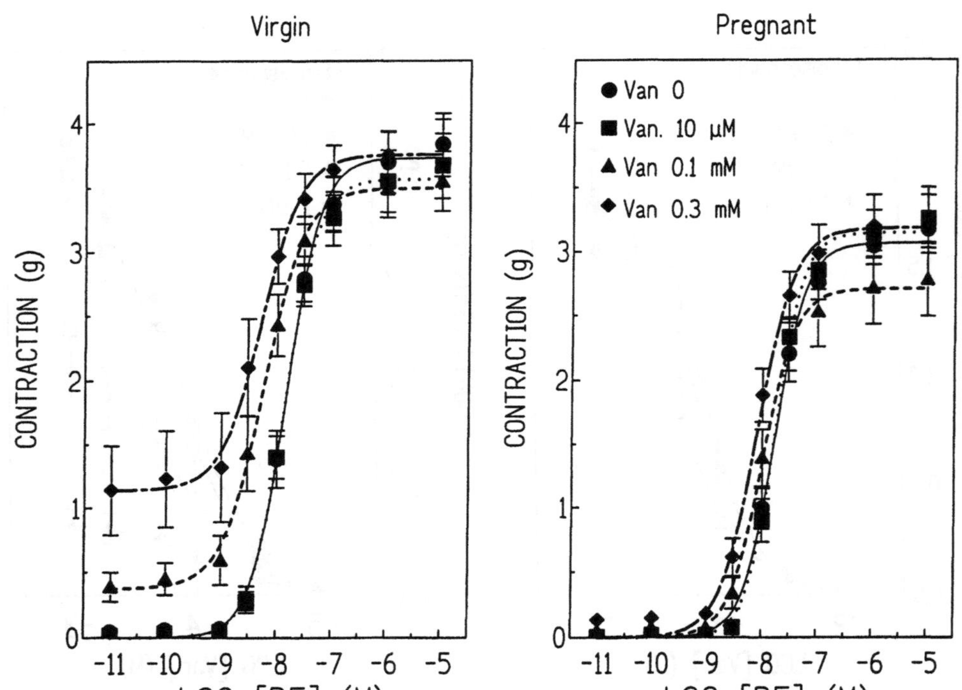

Fig. 4. Concentration-response curves of phenylephrine in aortic rings of virgin (left panel) and pregnant (right panel) rats. Rings were preincubated (10 min) in the absence (circles) and presence of the different concentrations of vanadate.

concentration-response curve to PE was measured in each tissues. In aortic rings of virgin rats, the prior activation by vanadate resulted in an increased potency to PE. Indeed, EC_{50} to PE was decreased ($p < 0.01$) from 15 nM in the absence of vanadate to 5 nM in the presence of 0.3 mM of the salt. Although vanadate did not induce an increase in tone in tissues of pregnant rats, EC_{50} of PE was also decreased from 17 to 8 nM in the same conditions. Maximum responses to PE was not affected by vanadate in both groups of aortic ring.

Discussion

The data presented in this paper indicate that the contractile effects of vanadate on rat aorta is independent of endogenous prostanoids and that it may be modulated by protein kinase C-dependent pathway. Furthermore, these results also show that the effect of vanadate on smooth muscle contractility is blunted during pregnancy. The mechanisms of this resistance to vanadate are not clear but appear to be similar to the ones responsible for the resistance to other vasoconstrictors such as phenylephrine.

The contractions induced by vanadate in rat aortic rings were unaffected by indomethacin, at concentrations known to fully inhibit synthesis of prostanoids by cyclooxygenase in rat aorta [22, 23]. Similar findings were reported previously [5]. This last report also indicates that contractile responses to vanadate are not mediated by the release of norepinephrine [5, 6]. All together these studies indicate that the effect of vanadate on vascular smooth muscle is not indirectly mediated by the liberation of endogenous contractile substances such as prostanoids and α-adrenergic agents.

Staurosporine, a highly selective inhibitor of protein kinase C, which at higher concentrations inhibits other protein kinases including protein tyrosine kinases [24], was able to antagonize contractions stimulated by vanadate at concentrations as low as 100 nM. This suggests an involvement of protein kinase C in vanadate-induced smooth muscle contractility. Although the precise mechanism by which protein kinase C modulates contractile activity is not known, it has been suggested that protein kinase C-mediated phosphorylation of a thin filament associated protein, caldesmon, may play some role [25]. Vanadate is a potent inhibitor of protein tyrosine phosphatase [26]. It was recently demonstrated [10] that vanadate-induced contraction of the guinea-pig taenia coli was accompanied by tyrosine phosphorylation of three proteins. This phosphorylation as well as the contractile response induced by vanadate were inhibited by 2.5–5 μM genistein, a potent inhibitor of tyrosine kinase [24]. They also showed that chelation of extracellular calcium with 5 mM EGTA totally reversed the contractions induced by vanadate, without effect on tyrosine phosphorylation. This suggests that protein tyrosine phosphorylation may be an early step in the cascade of cellular events leading to smooth muscle contraction after stimulation with vanadate. Recent data suggest that pervanadate (vanadyl hydroperoxide) contracts rat aorta rings via an inhibition of tyrosine phosphatase pathways, resulting in an enhancement of tissue phospho-tyrosilated proteins [27]. This increased tyrosine phosphorylation was proposed to result in increased Ca^{2+} influx through voltage-sensitive calcium channels [11]. Indeed, the effects of pervanadate were markedly decreased by 1 μM nifedipine. These results are in contrast to the study of Sanchez-Ferrer *et al.* [6] who reported that the contractile effects of vanadate in cat cerebral and femoral arteries were not affected by verapamil, a calcium entry blocker, but slightly decreased in Ca^{2+}-free (+EGTA) medium. Similar results were reported in guinea-pig aorta showing that D-600 (Methoxy-verapamil) slightly inhibited the contractile responses to vanadate and norepinephrine [4]. When using higher concentrations of verapamil (10 μM) in rat portal vein, or performing the experiment in the absence of Ca^{2+}, a complete block of the stimulating effect of vanadate was observed [17]. It thus appears that the dependence of the contractile effects of vanadate on Ca^{2+} influx varies with vessel type or the type or dosage of calcium channel blocker used. Further experiments are needed to ascertain this aspect.

Our results also demonstrate that the contractile response of vanadate was significantly decreased in aorta rings of pregnant as compared to virgin rats. This appears to results in a much lower sensitivity of the aortic smooth muscle to vanadate. Blunted responses to vanadate in rings of pregnant rats is similar, although of greater magnitude, to the reported decreased responses to vasoconstrictors (α-adrenoceptor agonists, angiotensin II, etc.) whose receptors are coupled to G-proteins [20, 22]. Recent work from this laboratory [28] indicates that the contractile effects of Bay 10 K 8644, a dihydropyridine calcium channel opener, is blunted in aortic rings of pregnant rats. Candura *et al.* [11] suggested that both vanadate and Bay K 8644 share a similar mode of action in the rabbit isolated ileum in facilitating calcium influx through voltage-dependent calcium channels. Furthermore, we have reported that the inhibitory effects of nifedipine, a dihydropyridine calcium channel blocker, is also decreased in isolated vascular tissues of pregnant rats [28]. These observations support the concept of a decreased gain of post-receptor coupling mechanisms in vascular tissues of pregnant rats. This is further substantiated by the observation that isolated mesenteric microvessels of pregnant rats show an hyperpolarization (up to 6 mV) in comparison to similar tissues of virgin rats [29]. The exact location of this proposed partial uncoupling of stimulation-contraction mechanisms remains to be identified. However, the present study reports evidence that protein kinase can be implicated.

We have also observed that vanadate potentiated the response of rat aortic rings to lower than maximal concentrations of phenylephrine. This suggest that vanadate and phenylephrine share a common pathway to induce contractions of vascular smooth muscle. Similar conclusion was reached by Ozaki and Urakawa [4] in guinea-pig aorta. The smaller increased responses to phenylephrine by the presence of vanadate in rings of pregnant rats, without changes in maximum response, indicates that this pathway may be down-regulated in pregnancy.

In conclusion, we have shown that the contractile effects of vanadate in aorta rings of virgin rats is independent of tissue prostanoid release. The concentration-dependent inhibition of vanadate-induced contractions by staurosporine suggests a possible role of protein kinase C in the action of vanadate to contract vascular smooth muscle. We have also observed that the response of aorta rings to vanadate is markedly blunted in pregnancy. This supports previous reports indicating that the blunted responses to vasoconstrictors observed during pregnancy are linked to partial uncoupling of stimulation-contraction mechanisms in vascular smooth muscle.

Acknowledgements

This work was supported by a grant from BIOPEDIA Inc (Fondation de l'Hôpital Ste-Justine) and by a senior scholarship from the Fonds de la Recharche en Santé du Québec to J.S.L and a grant from Québec Heart and Stroke Foundation to A.K.S.

References

1. Nechay BR, Nanninga LB, Nechay PSE *et al.*: Role of Vanadium in biology. Fed Proc 45: 123–132, 1986
2. Shechter Y: Insulin-mimetic effects of vanadate: possible implications for future treatment of diabetes. Diabetes 39: 1–5, 1990
3. Huot S, Muldoon S, Panmani M, Clough D, and Haddy FJ: Effects of sodium vanadate (Na_3VO_4) on wall tension and Na^+ - K^+ pump activity in isolated canine saphenous vein. Fed Proc 38: 1036-a, 1979
4. Ozaki H, Urakawa N: Effects of vanadate on mechanical responses and Na-K pump in vascular smooth muscle. Eur J Pharmacol 68: 339–347, 1980
5. Shimada T, Shimamura K, Sunano S: Effects of sodium vanadate on various types of vascular smooth muscles. Blood Vessels 23: 113–124, 1986
6. Sanchez-Ferrer CF, Marin J, Lluch M, Valverde A, Salaices M: Actions of vanadate on vascular tension and sodium pump activity in cat isolated cerebral and femoral arteries. Br J Pharmacol 93: 53–60, 1988
7. Hudgins PM, Bond GH: Alteration by vanadate of contractility in vascular and intestinal smooth muscle preparations. Pharmacology 23: 156–164, 1981
8. Nayler RA, Sparrow MP: Mechanism of vanadate-induced contraction of airways smooth muscle of the guinea-pig. Br J Pharmacol 80: 163–172, 1983
9. Sunano S, Kato S, Moriyama K, Shimamura K: Influences of sodium on the contractile action of vanadate in depolarized vas deferens and ureter of the guinea-pig. Arch Int Pharmacodyn Therap 293: 196–208, 1988
10. Di Salvo J, Semenchuck LA, Lauer J: Vanadate-induced contraction of smooth muscle and enhanced protein tyrosine phosphorylation. Arch Biochem Biophys 304: 386–391, 1993
11. Candura SM, Manzo L, Marraccini P, Coccini T, Tonini M: Investigation into vanadate-induced potentiation of smooth muscle contractility in the rabbit isolated ileum. Life Sci 54: 237–244, 1993
12. Jandhyala BS, Hom GJ: Physiological and pharmacological properties of vanadium. Life Sci 33: 1325–1340, 1983
13. Erdmann E, Werdan K, Krawietz W, Schmitz W, Scholz H: Vanadate and its significance in biochemistry and pharmacology. Biochem. Pharmacol 33: 945–950, 1984
14. Cole WC, McPherson CD, Sontag D: ATP-regulated K^+ channels protect the myocardium against ischemia/reperfusion damage. Circ Res 69: 571–581, 1991
15. Dupont Y and Bennett N.Vanadate inhibition of the Ca2+ dependent conformational change of the sarcoplasmic reticulum Ca2+-ATPase. Febs letters.139.237–240. 1982.
16. Richelmi P, Mirabelli F, Salis A, Finardi G, Berte F, Bellomo G.On the role of mitochondria in cell injury caused by vanadate-induced Ca^{2+} overload. Toxicology 57: 29–44, 1989
17. Shimamura K, Sunano S: Effects of sodium vanadate on the smooth muscle of the rat portal vein. Naunyn-Schmiedeberg's Arch Pharmacol 337: 347–353, 1988
18. Barron WM, Stamoutsos BA, Lindheimer MD: Role of volume in the regulation of vasopressin secretion during pregnancy in the rat. J Clin Invest 73: 923–932, 1984
19. Lundgren Y, Karlsson K, Ljungblad U: Circulatory changes during pregnancy in spontaneously and renal hypertensive rats. Clin Sci 57: 337s–339s, 1979
20. St-Louis J, Massicotte G, Parent A: Effet anti-hypertenseur de la grossesse: Influence de la réactivité vasculaire. Médecine/Science 6: 358–365, 1988
21. St-Louis J, Sicotte B: The blunted responses to phenylephrine in aortic rings of pregnant rats is linked to a decrease in extracellular calcium mobilization. Hypertens Pregn 12: 256 (Abstr.), 1993
22. St-Louis J, Sicotte B: Prostaglandin- or endothelium-mediated vasodilation is not involved in the blunted responses of blood vessels to vasoconstrictors in pregnant rats. Am J Obstet Gynecol 166: 684–692, 1992
23. Rees DD, Palmer RMT, Schulz R, Hodson HF, Moncada S: Characterization of three inhibitors of endothelium nitric oxide synthase *in vitro* and *in vivo*. Br J Pharmacol 101: 746–752, 1990
24. Hidaka H, Kobayashi R: Pharmacology of protein kinase inhibitors. Ann Rev Pharmacol Toxicol 32: 377–397, 1992
25. Adam LP, Gapinski CJ, Hathaway DR: Phosphorylation sequences in h-caldesmon from phorbol ester stimulated canine aortas. FEBS Lett 302: 223–226, 1992
26. Swarup G, Cohen S, Garbers DL: Inhibition of membrane phosphotyrosyl-protein phosphatase activity by vanadate. Biochem Biophys Res Comm 107: 1104–1109, 1982
27. Laniyonu A, Saifeddine M, Ahmad S, Hollenberg MD: Regulation of vascular and gastric smooth muscle contractility by pervanadate. Br J Pharmacol 113: 403–410, 1994
28. Roy B, Sicotte B, Brochu M, St-Louis J: Effects of nifedipine and Bay K 8644 on myotropic responses in aortic rings of pregnant rats. Eur J Pharmacol 280: 1–9, 1995
29. Meyer MC, Brayden JE, McLaughlin MK: Characteristics of vascular smooth muscle in the maternal resistance circulation during pregnancy in the rat. Am J Obstet Gynecol 169: 1510–1516, 1993

Molecular and Cellular Biochemistry **153:** 151–155, 1995.

In vivo modulation of N-myristoyltransferase activity by orthovanadate

Martin J. King[1], Subbiah Pugazhenthi[2], Ramji L. Khandelwal[2] and Rajendra K. Sharma[3]

[1]Saskatoon Cancer Centre, 20 Campus Drive, Saskatoon; [3]Department of Pathology and Saskatoon Cancer Centre, Royal University Hospital, University of Saskatchewan, Saskatoon; [2]Department of Biochemistry, University of Saskatchewan, Saskatoon, Canada

Abstract

N-Myristoyltransferase (NMT) catalyses the transfer of myristate from myristoyl-CoA to the NH_2-terminal glycine residue of several proteins and are important in signal transduction. STZ-induced diabetes (an animal model for insulin-dependent diabetes mellitus, IDDM) resulted in a 2-fold increase in rat liver NMT activity as compared with control animals. In obese Zucker (fa/fa) rats (an animal model for non-insulin dependent diabetes mellitus, NIDDM) there was a ~4.7-fold lower liver particulate NMT activity as compared with the control lean rat livers. Administration of sodium orthovanadate to the diabetic rats normalised liver NMT activity. These results would indicate that the rat liver particulate N-myristoyltransferase activity appears to be inversely proportional to the level of plasma insulin, implicating insulin in the control of N-myristoylation. (Mol Cell Biochem **153**: 151–155, 1995)

Key words: sodium orthovanadate, diabetes, N-myristoyltransferase, liver, membrane-associated, vanadate, obese Zucker rat

Abbreviations: NMT – N-myristoyl-CoA:protein N-myristoyltransferase; IDDM – insulin-dependent diabetes mellitus; NIDDM – non-insulin-dependent diabetes mellitus; NIP_{71} – 71 kDa N-myristoyltransferase inhibitor protein; NAF_{45} – 45 kDa N-myristoyltransferase activating factor

Introduction

N-terminal fatty acid modification of cellular, viral and onco-proteins, by myristate is catalysed by the enzyme(s) N-myristoyl:protein N-myristoyltransferase (NMT; EC 2.3.1.97). NMT activity, and thus myristoylation, is essential for eukaryote cell growth. However, the mechanism, function and regulation of myristoylation are at present poorly understood. Many of the known myristoylated proteins appear to have important roles in cellular regulation and signal transduction, such as; the catalytic subunit of cAMP-dependent protein kinase [1], $pp60^{src}$ [2] and Ca^{2+}- and calmodulin-dependent protein phosphatase [3]. The importance of the myristate moiety, and thus NMT itself, has been demonstrated with mutant proteins lacking the myristate modification i.e. non-myristoylated $pp60^{v\text{-}src}$ possesses full tyrosine kinase activity but is no longer transforming [4]. Additionally, a lack of NMT activity is lethal to an eukaryote cell [5]. Perturbations caused by diabetes mellitus will frequently impair signal transduction pathways and thus a cell's ability to respond to a given stimulus.

Diabetes mellitus may be divided into two broad categories; insulin-dependent diabetes mellitus (IDDM), or juvenile-onset diabetes, and the more common non-insulin-dependent diabetes mellitus (NIDDM). Perturbations caused by different forms of diabetes will frequently impair signal transduction pathways and thus ability of cells to respond to a given stimulus. In many cases sodium orthovanadate (Na_3VO_4; an insulinlike agent) has been observed to alleviate the effects of juvenile-onset as well as non-insulin-dependent diabetes

Address for offprints: R.K. Sharma, Department of Pathology and Saskatoon Cancer Centre, Royal University Hospital, University of Saskatchewan, 20 Campus Drive, Saskatoon S7N 4H4, Canada both *in vivo* and *in vitro* [6–8]. Vanadium is group V trace element essential for the normal functioning and growth in higher eukaryotes [9]. As with many of the transitional metals vanadium exists in several valance, or oxidation, states (+II–+V). Different physiological effects have been attributed to different valence states of vanadium [10, 11]. However, it is not always clear which valance state is active *in vivo*. For example, vanadate (+V) has been known to inhibit several enzymes *in vitro* which deal with the transfer of phosphate, i.e., phosphotyrosyl protein phosphatases, alkaline phosphatase and certain phosphoseryl/threonyl protein phosphatase, ATPases and phosphofructokinase [12–14]. The insulin-like effects observed with vanadate were thought to be due to the action of vanadate on either the tyrosine kinase activity of the insulin receptor (stimulation) and/or of phosphotyrosyl protein phosphatase activity (inhibition) [12, 15, 16]. Another valance state of vanadium, vanadyl (+IV), has also been observed to exhibits insulin-like effects [10, 17]. Administration of either sodium orthovanadate or vanadyl sulphate to streptozotocin-induced diabetic rats has resulted in normalisation of plasma glucose, inhibition of lipolysis, and activation of glycogen synthase [7, 8, 10, 18]. Vanadate (+V) is reduced to vanadyl (+IV), via a nonenzymatic glutathione catalysed process, in the cytoplasm of red blood cells and adipocytes [19] which has led to the hypothesis that the valance state of vanadium which is important in mimicking the effects of insulin is vanadyl (+IV). Vanadate incubated with cell suspensions of either red blood cells or adipocytes led to a biphasic uptake of vanadate; a rapid initial equilibrium phase followed by a slower prolonged reduction coupled uptake [19]. However as these studies do not allow for the continuous replenishing of vanadate in animals, the contribution of vanadate in mimicking insulin can not be eliminated. Quantitative reduction of vanadate to vanadyl was achieved only after about 23 h. *In vivo* studies, vanadate is usually administered in drinking water and so consequently the supply of vanadate is being constantly replenished in the animal [17].

Due to the diverse actions and mechanisms of insulin, affecting cellular phosphorylation and dephosphorylation, as well as a variety of other secondary messengers, i.e., DAG, it is likely that both valance states of vanadium can mimic certain actions of insulin. However, while vanadate may act by inhibiting protein dephosphorylation vanadyl is a poor inhibitor of phosphotyrosyl protein phosphatases *in vitro*. Consequently, the insulin-like actions of the two valance states of vanadium are likely to work via different mechanisms.

Some of the studies on NMT in diabetes mellitus which has been carried out in our laboratory are reviewed in this article.

Rat N-myristoyltransferase activity

NMT activity was present in both rat brain and liver (brain> liver, with respect to NMT activity per mg protein). The subcellular distribution of liver NMT activity was inverse to that observed in rat brain (Table 1) [20–23]. Rat brain possessed 8.3-fold higher NMT activity in the soluble phase per mg of protein as compared per mg of particulate protein. Whereas, with the rat liver, greater NMT activity was observed in the particulate phase per mg protein; 4.7-fold in the Sprague-Dawley and 10.3-fold in the obese Zucker rats, as compared with the soluble fraction (Table 1) [21, 22]. Adjusting for the total yields of soluble and particulate protein from rat brain and liver resulted in only 2% of the total rat brain NMT activity compared with 47–63% of the total rat liver NMT activity being membrane-associated. Bovine brain in known to contain NIP_{71}, a particulate inhibitor of NMT [23], and NAF_{45}, a cytosolic activator of NMT [24]. Inhibitor activities have been observed in rat liver and brain and these may account in part for the subcellular differences observed [25]. Vanadate-treatment did not change the proportion of enzyme activity in the soluble or particulate fractions (Fig. 1) [22].

Insulin-dependent diabetes mellitus and vanadate modulation of N-myristoyltransferase activity

STZ treatment increased rat liver NMT activity 1.5–2.1 fold in all subcellular fractions of the cell (Fig. 1A). Identical increases were observed with either cAMP-dependent protein

Table 1. Subcellular distribution of rat liver NMT activity

Cellular fraction	NMT Activity		
	Brain	Liver	
		Sprague-Dawley	Zucker
Homogenate	3.10 ± 0.05 (100)	0.354 ± 0.050 (100)	0.128 ± 0.029 (100)
Soluble	4.36 ± 0.19 (141)	0.154 ± 0.054 (43)	0.085 ± 0.067 (66)
Particulate	0.32 ± 0.25 (10)	0.810 ± 0.064 (289)	0.920 ± 0.140 (719)

Liver and brain NMT activity was assayed as described using cAMP-dependent protein kinase derived peptide (463 μM) as substrate. For details of experimental conditions see King and Sharma (1991). Results are expressed as unit per assay or (in parentheses) as percentage of activity compared with the homogenates (mean ± S.E.M., n = 4–18).

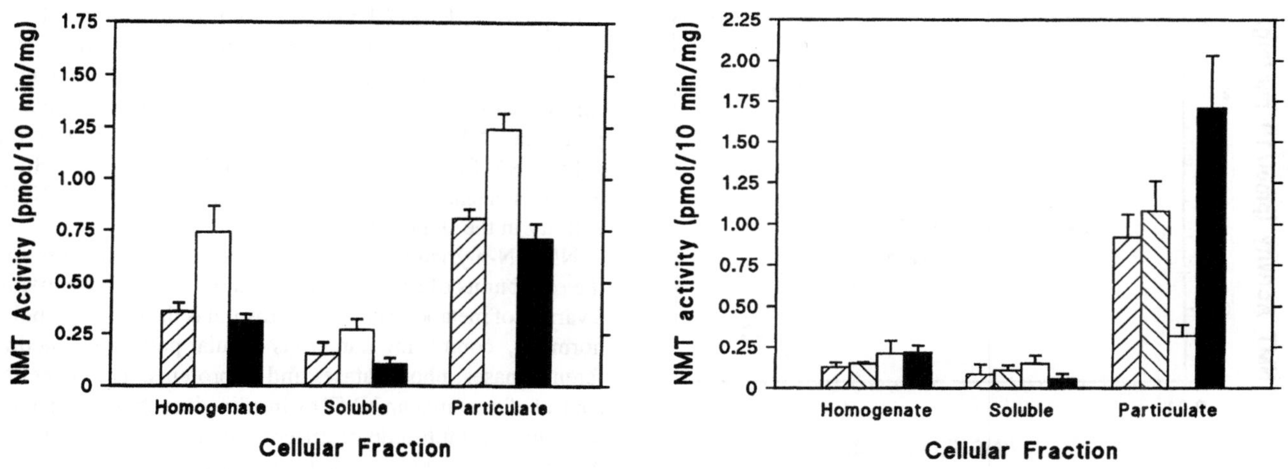

Fig. 1. Effect of diabetes and vanadate treatment on NMT activity in rat liver. (A) Rat liver NMT from STZ-induced diabetic rats was assayed using cAMP-dependent protein kinase derived peptide (463 μM) as substrate. For details of experimental conditions see King *et al.* [21]. Control rat (hatched, //), STZ-induced diabetic rat (open), VO_4^{3-}-treated STZ-induced diabetic (solid). (B) Rat liver NMT from obese Zucker rats was assayed using cAMF-dependent protein kinase derived peptide (463 μM) as substrate. For details of experimental conditions see King *et al.* [22]. Lean control rat (hatched, / /), VO_4^{3-}-treated lean (hatched, \\\\), obese (fa/fa) rat (open), VO_4^{3-}-treated obese (fa/fa) rat (solid).

kinase derived peptide or $pp60^{src}$ derived peptide as substrate for NMT assay. Vanadate treatment of the STZ-diabetic rats resulted in the normalisation of elevated NMT (75–120% of the control activities; Fig. 1A). Changes in NMT activity were not due to any corresponding change in the apparent K_m of peptide substrate; control, 50 μM vs. STZ-diabetic, 54 μM vs. vanadate-treated STZ-diabetic, 56 μM. Vanadate (100 μM) had no significant effect on liver NMT activity *in vitro*, i.e., 1.42 ± 0.38 pmol/10 min/mg vs. 1.51 ± 0.51 pmol/10 min/mg with respect to particulate NMT activity from STZ-diabetic rats in the absence and presence of vanadate, respectively [21].

Non-insulin-dependent diabetes mellitus and vanadate modulation of N-myristoyltransferase activity

In contrast with IDDM, obese or NIDDM resulted in a 4.7-fold decrease in liver NMT activity. However, this phenomena was observed only in the particulate phase (Fig. 1B). This loss in particulate NMT activity was not mirrored by any corresponding increase in the soluble NMT activity, indicating that there was no simple translocation of active NMT from the membrane fraction to the soluble phase (Fig. 1B). Vanadate-treatment of these obese Zucker rats lead to a 5.3-fold increase in the particulate NMT activity, resulting in a 58% increase over the vanadate-treated lean Zucker rat particulate NMT activity (Fig. 1B). No subsequent increase in NMT activity was observed with vanadate-treated lean rats. Nor was the subcellular distribution of liver NMT activity affected by vanadate-treatment (Fig. 1B). Interestingly, NMT activity from rat brain showed no significant variation between the control, vanadate-treated control, diabetic and vanadate-treated diabetic animals [22]. However, as the brain is an insulin-insensitive organ this is not an uncommon phenomena.

Discussion

Diabetes results in a modulation of NMT activity in the liver, an important insulin target tissue. The particulate NMT activity (the major liver activity) was inversely proportional to the plasma insulin concentrations (Fig. 2). Administration of sodium orthovanadate to the diabetic rats resulted in normalisation of these NMT activities. In addition, the vanadate administration resulted in the particulate activity becoming independent of the plasma insulin concentration [comparison of vanadate-treated control (lean) rats, vanadate-treated STZ-diabetic and vanadate-treated obese (fa/fa) rats vs. plasma insulin concentration; r = –0.031]. The mechanism of action of insulin and the effects of diabetes on NMT are at present unknown. Modulation of the liver NMT activity could be due to direct effects on NMT, or indirectly by modifying a corresponding rat NIP_{71} and/or NAF_{45} activity; affecting the protein concentrations; protein expression or mRNA stability; or the specific activities of NMT. The differences observed between liver and brain may be due to target tissue or inherent differences in NMT forms between tissues [20–23]. NMT has been observed in multiple forms

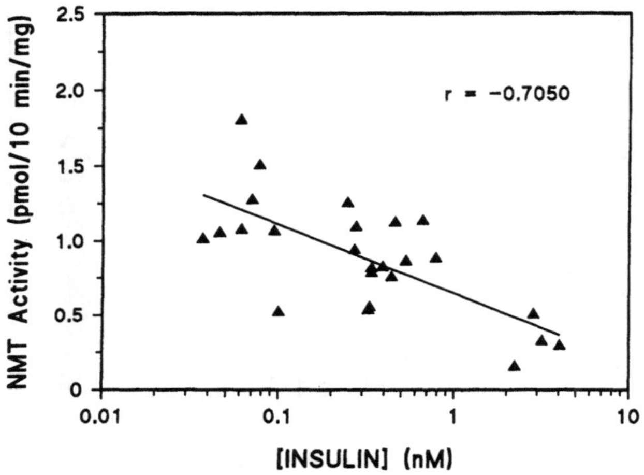

Fig. 2. Correlation of plasma insulin concentration with NMT activity in rat liver. Rat liver particulate NMT activity and plasma insulin levels were determined as described in King *et al.* [22]. Results are expressed as pmol myristoylpeptide formed/10 min/mg rat extract vs. plasma insulin (nM).

in bovine tissue [26] and mouse cell lines [27]. Changes in enzyme concentration vs. specific activity or subunit composition will require suitable tools to answer, i.e., anti-NMT antibodies. The lack of any direct effect of vanadate on NMT activity *in vitro* is not necessarily an indication that vanadyl is the active vanadium oxidation state involved with NMT modulation. Vanadate may regulate NMT indirectly, by modulating NIP_{71} or NAF_{45} activity. Vanadate treatment of cells is known to increase cellular phosphotyrosyl protein phosphorylation [28], consequently modulation of the specific activity of NMT could be controlled by a phosphorylation/dephosphorylation reaction. That the insulin receptor (a tyrosine kinase) itself is directly phosphorylating NMT, or any NMT effector, is unlikely as no common precautions against phosphotyrosyl protein phosphatases were employed [12, 13]. Rat liver is known to contain high activity of phosphotyrosyl protein phosphatases fully active in the homogenisation buffers employed [12, 13, 29]. However, the possibility of a metallophosphatase (like Ca^{2+}- and calmodulin dependent protein phosphatase) being involved can not be eliminated.

NMT is inhibited by palmitoyl-CoA. Any increase in free fatty acid concentration will inevitably lead to an increase in the palmitoyl-CoA concentration. The obese Zucker rats had elevated triglyceride concentrations, this would result in an increase in the free fatty acid and consequently in palmitoyl-CoA concentrations. However, it is unlikely that this is the mechanism of inhibition observed with the NMT as NMT isolated from the vanadate-treated obese Zucker rats, whose triglyceride levels were still 15-fold elevated over the lean rats, possessed NMT levels elevated over the control values (Fig. 1B). A potential mechanism to explain the variation of NMT activity could be enzyme translocation. Under this scenario the translocation would be from an activated particulate NMT (possibly NAF_{45} associated) to an inactive soluble NMT (possibly NIP_{71} associated) for NIDDM. However, this mechanism would not explain the increase in NMT activity in the IDDM model.

NMT N-myristoylates several kinases, phosphatases and the α-subunits of several G-proteins (α_i and α_o) essential in a variety of metabolic pathways. Insulin is the major anabolic hormone, controlling numerous cellular pathways, via different kinases, phosphatases and G-proteins. The potential control of mammalian NMT by insulin gives the cell a handle to regulate a multitude of different metabolic pathways by the control of a single enzyme, or group of isozymes [23, 24, 26]. Therefore, the obvious importance of NMT in cellular regulation and signal transduction demands more research, such that a complete understanding of the regulation and expression of NMT may be achieved. This in itself may provide valuable insights into the mechanisms of action of insulin and diabetes mellitus.

Acknowledgements

This work was partially supported by the Medical Research Council of Canada. Dr. Martin J. King is a recipient of a Research Fellowship from the Health Services Utilization and Research Commission of Saskatchewan.

References

1. Aitken A, Cohen P, Sandkarn S, Williams DH, Calder AG, Smith A, Klee CB: Identification of the NH_2-terminal blocking group of calcineurin B as myristic acid. FEBS Lett 150: 314–318, 1982
2. Buss JE, Kamps MP, Sefton BM: Myristic acid is attached to the transforming protein of Rous sarcoma virus during or immediately after synthesis and is present in both soluble and membrane-bound forms of the protein. Mol Cell Biol 4: 2697–2704, 1984
3. Carr SA, Biemann K, Shoji S, Parmellee DC, Titani K: n-Tetradecanoyl is the NH2-terminal blocking group of the catalytic subunit of cyclic AMP-dependent protein kinase from bovine cardiac muscle. Proc Natl Acad Sci USA 79: 6128–6131, 1982
4. Duronio RJ, Towler DA, Heuckeroth RO, *et al*: Disruption of the yeast N-myristoyl transferase gene causes recessive lethality. Science 243: 796–800, 1989
5. Kamps MP, Buss JE, Sefton BM: Mutation of NH_2-terminal glycine of $p60^{src}$ prevents both myristoylation and morphological transformation. Proc Natl Acad Sci USA 82: 4625–4628, 1985
6. Pugazhenthi S, Khandelwal RL, Angel JF: Insulin-like effect of vanadate on malic enzyme and glucose-6-phosphate dehydrogenase activities in streptozotocin-induced diabetic rat liver. Biochim Biophys Acta 1083: 310–312, 1991

7. Pugazhenthi S, Angel JF, Khandelwal RL: Long-term effects of vanadate treatment on glycogen metabolizing and lipogenic enzymes of liver in genetically diabetic (db/db) mice. Metabolism 40: 941–946, 1991
8. Khandelwal RL, Pugazhenthi S: *In vivo* effects of vanadate on hepatic glycogen metabolism and lipogenic enzymes in insulin-dependent and insulin-resistant diabetic animals. Mol Cell Biochem, this issue: 000–000, 1995
9. Hamel FG, Solomon SS, Stranik A, Blotcky A, Rack E, Duckworth WC: Alteration of tissue vanadium content in diabetes. Metabolism 42: 1503–1505, 1993
10. Cros GH, Mongold JJ, Serrano JJ, McNeill JH: Enhanced *in vivo* sensitivity of vanadyl-treated diabetic rats to insulin. Can J Physiol Pharmacol 68: 486–491, 1990
11. Duckworth WC, Solomon SS, Liepnieks J, Hamel FG, Hand S, Peavy DE: Insulin-like effects of vanadate in isolated rat adipocytes. Endocrinology 122: 2285–2289, 1988
12. King MJ, Sale GJ: Insulin-receptor phosphotyrosyl-protein phosphatases. Biochem J 256: 893–902, 1988
13. King MJ, Sale GJ: Assay of phosphotyrosyl protein phosphatase using synthetic peptide 1142–1153 of the insulin receptor. FEBS Lett 237: 137–140, 1988
14. Degani H, Gochin M, Karlish SDJ, Shechter Y: Electron paramagnetic resonance studies and insulin-like effects of vanadium in rat adipocytes. Biochemistry 20: 5795–5799, 1981
15. Smith DM, Sale GJ: Evidence that a novel serine kinase catalyses phosphorylation of the insulin receptor in an insulin-dependent and tyrosine kinase-dependent manner. Biochem J 256: 903–909, 1988
16. Sale GJ: Recent progress in our understanding of the mechanism of action of insulin. Int J Biochem 20: 897–908, 1988
17. Ramanadham SA, Mongold JJ, Brownsey RW, Cros GH, McNeill JH: Oral vanadyl sulphate in treatment of diabetes mellitus in rats. Am J Physiol 257: H904–H911, 1989
18. Pugazhenthi S, Khandelwal RL: Insulinlike effects of vanadate on hepatic glycogen metabolism in nondiabetic and streptozotocin-induced diabetic rats. Diabetes 39: 821–827, 1990
19. Macara IG, Kustin K, Cantley LC: Glutathione reduces cytoplasmic vanadate mechanism and physiological implications. Biochim Biophys Acta 629: 95–106, 1980
20. King MJ, Sharma RK: N-myristoyl transferase assay using phosphocellulose paper binding. Anal Biochem 199: 149–153, 1991
21. King MJ, Pugazhenthi S, Khandelwal RL, Sharma RK: Elevated N-myristoyl transferase activity is reversed by sodium orthovanadate in streptozotocin-induced diabetic rat. Bioch Biophys Acta 1165: 259–262, 1993
22. King MJ, Pugazhenthi S, Khandelwal RL, Sharma RK: Membrane-associated N-myristoyltransferase activity is reduced in obese (fa/fa) Zucker rat liver. Biochem Biophys Res Commun 196: 665–670, 1993
23. King MJ, Sharma RK: Identification, purification and characteriztion of a membrane-associated N-myristoyltransferase inhibitor protein from bovine brain. Biochem J 291: 635–639, 1993
24. King MJ, Sharma RK: Differential activation of bovine brain N-myristoyltransferase(s) by a cytosolic activator. Biochem Biophys Res Commun 212: 580–588, 1995
25. McIlhinney RAJ, McGlone K: Characterisation of a myristoyl CoA:glycylpeptide N-myristoyl transferase activity in rat brain: Subcellular and regional distribution. J Neurochem 54: 110–117, 1990
26. King MJ, Sharma RK: Demonstration of multiple forms of bovine brain myristoyl CoA:protein N-myristoyl transferase. Mol Cell Biochem 113: 77–81, 1992
27. Boutin JA, Ferry G, Ernould A-P, Maes P, Remond G, Vincent M: Myristoyl-CoA:protein N-myristoyltransferase activity in cancer cells. Purification and characterisation of a cytosolic isoform from the murine leukemia cell line L1210. Eur J Biochem 214: 853–867, 1993
28. Klarland JK: Transformation of cells by an inhibitor of phosphatases acting on phosphotyrosine in proteins. Cell 41: 707–717, 1985
29. King MY, Sale GJ: Dephosphorylation of insulin-receptor autophosphorylation sites by particulate and soluble phosphotyrosyl-protein phosphatases. Biochem J 266:251–259, 1990

Molecular and Cellular Biochemistry **153**: 157–160, 1995.

Regulation and control of glucose overutilization in erythrocytes by vanadate

Najma Zaheer Baquer, Amit Kumar Saxena and Poonam Srivastava
School of Life Sciences, Jawaharlal Nehru University, New Delhi 110 067, India

Abstract

The insulin mimetic effect of vanadate in *in vitro* incubation of erythrocytes with high glucose concentrations showed an increase in sorbitol accumulation and glucose utilization using U-^{14}C-glucose. Aldose reductase inhibitors and vanadate addition reversed the sorbitol accumulation, whereas insulin could not reverse it. Increased glucose utilization was also normalized with vanadium compounds. Increased activity of aldose reductase and sorbitol levels in diabetic animals were also normalized with vanadate treatment. (Mol Cell Biochem **153**: 157–160, 1995)

Key words: insulin mimetic, vanadate, RBC, glucose utilization, aldose reductase

Introduction

Measurement of sorbitol content in erythrocytes suggested the presence of aldose reductase [1]. Aldose reductase has been purified from erythrocytes by Das & Srivastava and Hanada *et al.* [2, 3]. The polyol pathway in erythrocytes is activated during persistent hyperglyceaemia resulting in an intracellular accumulation of sorbitol in diabetes [4, 5]. Earlier work from our laboratory showed that oral administration of vanadate to diabetic rat normalized the increased aldose reductase activity and sorbitol levels in kidney [6]. Strong correlation has been reported between polyol pathway and small blood vessel complications of diabetes mellitus [7]. During diabetes this pathway is responsible for sorbitol accumulation in erythrocytes thus resulting in membrane deformation in red blood cells leading to pathogenic consequences [5, 8].

Materials and methods

Animals

Albino male rats of Wistar strain weighing between 210–230 g (2 months old) were used in all the experiments.

Induction of diabetes

Alloxan diabetes was produced essentially by the method of Sochor *et al.* [9].

Treatment of animals and preparation of erythrocytes

Treatment of animals with vanadate was essentially carried out as described by Saxena *et al.* [6].

Preparation of RBC

Rats were anaesthetized and blood was collected in heparin by puncturing the heart , blood was suspended in 3 volumes of phosphate buffer saline (100 mM potassium phosphate pH 7.4/0.15 M NaCl) and centrifuged at 1000 g for 10 min. Red blood cells were washed twice with the same buffer. For enzyme estimation packed RBC were hemolyzed with 3 volumes of 10 mM potassium phosphate buffer pH 7.0/5 mM 2 mercaptoethanol. Hemolyzate was centrifuged at 10,000 g for 30 min. The supernatant was passed through a DEAE cellulose (DE-52) column and the enzyme eluted with a linear gradient of 0–200 mM NaCl.

Address for offprints: N.Z. Baquer, School of Life Sciences, Jawaharlal Nehru University, New Delhi 110 067, India

Determination of aldose reductase activity

The aldose reductase activity was determined in the partially purified erythrocyte extracts described above by the method of Das & Srivastava and Hamada *et al.* [2, 3]. One unit of enzyme activity is defined as the amount required to oxidize on μmole of NADPH per gram tissue wet weight per min at 25°C.

Determination of sorbitol content

Packed RBC were deproteinized with 3 volumes of cold 6% PCA. The precipitate was centrifuged and supernatant was neutralized at 4°C with 2 M K_2CO_3 and centrifuged at 1000 g for 5 min to obtain the protein free supernatant for measurement of sorbitol. Sorbitol was measured by the modified method of Malone *et al.* [10] using purified sorbitol dehydrogenase. NADPH produced was measured fluorimeterically using a Schimadzu RE-540 spectrofluorometer.

Measurement of total glucose oxidation

Total glucose oxidation of RBC was measured using $U^{14}C$ glucose. All incubations were performed in a shaking water bath at 37°C according to the method of Malone *et al.* and Hothersall [10, 11]. Details of incubation and additions are given in the tables. Details of all procedures for determination of $^{14}CO_2$ are as described by Hothersall *et al.* [11].

Haemoglobin estimation

Haemoglobin was determined by the method of Van Kampen and Zÿlstra [12].

Chemicals

All the substrates, enzymes, coenzymes, vanadium compounds and other chemicals were purchased form Sigma Chemicals Co. (St. Louis, MO). Radioactive glucose was from Bhabha Atomic Research Centre (BARC), Bombay, India.

Results

When compared to the age matched control rats a significant increase of 65% was observed in the aldose reductase activity of red cells obtained from diabetic animals with a five fold increase in the sorbitol level. Percentage changes in aldose reductase activity and sorbitol levels are shown in Fig. 1 Vanadate administration normalized the aldose reductase activity and the sorbitol levels.

The effect of the vanadiun compounds and aldose reductase inhibitors on sorbitol accumulation in *in vitro* incubated erythrocytes is shown in Table 1. When RBCs were incubated with high glucose, an increase in sorbitol accumulation was found, a condition resembling hyperglycaemia. Insulin addition did not reverse the measured accumulation of sorbitol, whereas sodium orthovanadate, in a final concentration of 4 mM effectively lowered the sorbitol content significantly. Aldose reductase inhibitors also significantly lowered the levels of sorbitol .

When packed RBCs were incubated with normal glucose (5 mM) and high glucose (20 mM) concentration for two hours using $U^{14}C$ glucose to study glucose oxidation, a three fold increase in glucose oxidation was found with high glucose concentration. Results showed that vanadium compounds above 1 mM caused a marked lowering in the elevated glucose oxidation in the erythrocytes incubated with 20 mM glucose. Vanadyl sulphate was the most effective of the vanadium compounds in this respect. The results are shown in Table 2.

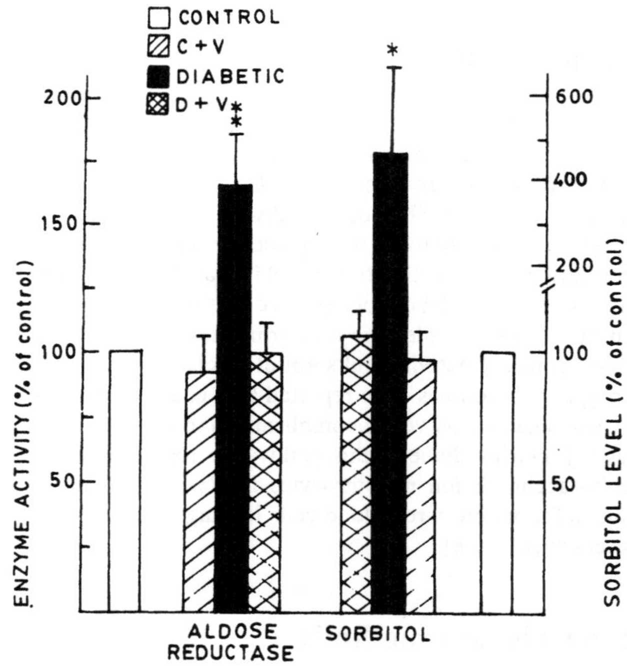

Fig. 1. Percent change in aldose reductase activity and sorbitol levels in erythrocytes of diabetic and vanadate treated control and diabetic rats (taking control as 100%). Details of the experimental procedure are given in Materials and methods. Abbreviations C+V, control rats treated with vanadate; D+V, diabetic rat treated with vanadate; Fishers P values are *p < 0.001 and **p < 0.01.

Table 1. In vitro effect of vanadium and aldose reductase inhibitors on sorbitol accumulation in erythrocytes incubated in high glucose medium

Experimental Conditions	Sorbitol nm/gHb	Fold increase from control
Control RBC (80 mg/dl glucose)	2.87 ± 0.32	1.0
RBC + glucose (500 mg/dl)	15.10 ± 0.98[a]	5.7
RBC + glucose (500 mg/dl)+ Insulin (16 ng/ml)	15.02 ± 0.64[a]	5.6
Vanadium compounds (4 mM final)		
Sodium orthovanadate	4.16 ± 0.39[b]	1.4
Vanadyl sulphate	4.10 ± 0.37[c]	1.4
Sodium metavanadate	7.11 ± 0.48[a]	2.4
Vanadium pentoxide	6.90 ± 0.53[a]	2.4
Aldose Reductase Inhibitors		
Diphenyl hydantoin (0.4 mM)	3.57 ± 0.52	1.3
Quercetin (0.4 mM)	3.33 ± 0.30	1.2
Tetramethylene glutaric acid (1.2 M)	3.79 ± 0.47	1.4

Each value is a mean ± S.E.M. of five separate experiments. P values are shown as a = $p < 0.001$, b = $p < 0.01$ and c = $p < 0.05$ vs control. Packed RBC were incubated in KRB for 6 h. Control glucose (low concentration 80 mg/dl) and with 500 mg/dl glucose. Effect of vanadium compounds and aldose reductase inhibitors on sorbitol accumulation was measured in the presence of high glucose (500 mg/dl). Sorbitol was assayed as described in Materials and Methods. RBC is Red Blood Cells.

Table 2. Effect of Vanadium compounds and insulin on total glucose oxidation in rat erythrocytes

Experimental Conditions	μmol glucose oxidized per hour per gm Hb	Percentage increase from control
RBC + Glucose		
Glucose (5 mM)	2.33 ± 0.12	100
Glucose (20 mM)	6.49 ± 0.32[a]	278
RBC + Glucose (20 mM)		
Sodium orthovanadate	2.87 ± 0.25	123
Vanadyl sulphate	2.00 ± 0.16	86
Sodium metavanadate	3.37 ± 0.31[c]	145
Vanadium pentoxide	3.28 ± 0.28	141
Insulin (16 ng/ml)	6.21 ± 0.40[a]	267

Each value is a mean ± S.E.M. of five separate experiments. P values are shown as a = $p < 0.001$, b = $p < 0.01$ and c = $p < 0.05$ vs. control. Packed RBC were preincubated with Krebs Ringer buffer at 37°C for 20 min in rubber capped vials with centre wells. Uniformly labelled ^{14}C glucose was then added in the reaction mixture and erythrocytes were then incubated for 2 h with the additions given above, vanadium compounds used were all 4 mM (final). Reaction was stopped with 5N HC1 and $^{14}CO_2$ was collected in hyamine hydroxide and counted in scintillation counter.

Discussion

The activity of aldose reductase in erythrocytes is very low and diabetes caused a significant increase in aldose reductase activity as well as sorbitol accumulation. The action of vanadate, in reversing the sorbitol accumulation, seems primarily due the glucose lowering ability of the trace metal, thereby normalizing the increased glucose flux and activation of aldose reductase in insulin independent tissues [6].

In vitro incubation of erythrocytes in high glucose concentration increased the sorbitol content significantly, the erythrocyte intracellular sorbitol levels reflects the severity of diabetes. *In vitro* insulin addition to these erythrocytes could not prevent the increased sorbitol accumulation, aldose reductase inhibitors and vanadium salts were however effective in reversing the sorbitol accumulation

Further to substantiate that vanadiun compounds could control glucose overutilization in erythrocytes total glucose oxidation using $U^{14}C$ glucose was studied. A three fold increase in total glucose oxidation was effectively reversed by vanadium compounds (Table 2). Such a reversal effect was not observed when insulin was added in the incubations.

Earlier studies form our laboratory and others have shown the effect of vanadium in reversing diabetic effects in kidney liver and heart [6, 13, 17]. Vanadate seems to effectively control the metabolic pathways in insulin independent tissues both in *in vitro* and *in vivo* conditions, possibly by regulating the expression of glucose transporters in a tissue specific manner. Recent studies have also demonstrated that vanadate treatment restores the expression of genes for glucose transporters and for key enzymes of glucose metabolism in diabetic rat liver [18, 19]. Vanadate probably acts at a site distant to insulin receptor, possibly by utilizing an insulin-independent alternative pathway, which has also been suggested earlier [20, 21, 22) enabling it to perform its antidiabetic effect on insulin-insensitive tissues also. Vanadate, therefore, has a future potential as an insulin-insensitive drug for the treatment of diabetes. It can be taken orally and maintains a stable long lasting normoglycaemia.

Acknowledgements

AKS and PS are grateful to the UGC for Senior Research Fellowships. Authors thank M.A. Askar and Mr. Saini for the preparation of manuscripts and figures.

References

1. Travis SF, Morrison AD, Clements RS, Winegred AL, Oski FA: Metabolic alterations in human erythrocyte sorbitol produced by increase in glucose concentration. J Clin Invest 50: 2104–2111, 1971
2. Das B, Srivastava SK: Purification and properties of aldose reductase and aldehyde reductase II from human erythrocyte. Arch Biochem Biophys 238: 670–679, 1985
3. Hamada Y, Kito R, Raskin P: Increased erythrocyte aldose reductase activity in Type I diabetic patients. Diabetic Med 226–231, 1991
4. Srivastava SK, Ansari NH, Hair GA, Jaspan J, Rao MB, Das B: Hyperglyeaemia induced activation of human erythrocyte aldose reductase and alteration in kinetic properties. Biochem Biophys Acta 870: 302–331, 1986
5. Robey C, Mahapatra AD, Cohen MP, Suarez S: Sorbinil partially prevents decreased erythrocyte deformibility in experimental diabetes mellitus. Diabetes 36: 1010–1013, 1987
6. Saxena AR, Srivastava P, Kale RK, Baquer NZ: Effect of vanadate administration on polyol pathway in diabetic rat kidney. Biochem Int 26: 59–68, 1992
7. Brownlee M, Cerami A: The Biochemistry of the Complications of diabetes. Ann Rev Biochem 50: 385 432, 1981
8. Gabay RH: The sorbitol pathway and complications of diabetes. N Engl J Med 288: 831–836, 1973
9. Sochor M, Baquer NZ, McLean P: Glucose over and underutilization in diabetes: Comparative studies on the changes in activities of enzymes of glucose metabolism in rat kidney and liver. Mol Physiol 7: 51–68, 1985
10. Malone JI, Knox G, Benford S, Tedasco TA: Red cell sorbitol: An indicator of diabetic control. Diabetes 29: 861–864, 1980
11. Hothersall JS, Baquer NZ, McLean P: Pathways of carbohydrate metabolism in peripheral nervous tissue. The contribution of alternative route of glucose utilization in peripheral nerve and brain. Enzyme 27: 259–267, 1982
12. Van Kampen EJ, Zijlstra WG: Standardization of haemalobinometery II. The Hemiglobinicyanide method. Clin Chem Acta 6: 538–544, 1961
13. Saxena AK, Srivastava P, Baquer NZ: Effect of vanadate on glycolytic enzymes and malic enzyme in insulin dependent and independent tissues of diabetic rats. Eur J Pharmacol 216: 123–126, 1992
14. Saxena AK, Srivastava P, Kale RK, Baquer NZ: Impaired antioxidant status in diabetic rat liver: Effect of vanadate. Biochem Pharmacol 40: 539–542, 1993
15. Srivastava P, Saxena AK, Kale RK, Baquer NZ: Insulin like effects of lithium and vanadate on the altered antioxidant status of diabetic rats. Res Commun Chem Pathol Pharmacol 80(3): 283–293, 1993
16. Heyliger CE, Tahiliani AG, McNeil JH: Effect of vanadate on elevated blood glucose and depressed cardiac performance of diabetic rats. Science 227: 1474–1476, 1985
17. Pugazenthi S, Khandelwal RE: Insulin like effects of vanadate on hepatic glycogen metabolism in non-diabetic and streptozotocin-induced diabetic rats. Diabetes 39: 821–827, 1990
18. Brichard SM, Desguois B, Girard J: Vanadate of diabetic rats reverses the impaired expression of genes involved in hepatic glucose metabolism: Effect of glycolytic and gluconeogenic enzyme and on glucose transporter GLUT-2. Mol Cell Endocrinol 8: 81–97, 1993
19. Valera A, Rodriguez-Gil JE, Bosoh F: Vanadate treatment restores the expression of genes for the key enzymes in the glucose and ketone bodies metabolism in the liver of diabetic rats. J Clin Invest 29: 4–11, 1993
20. Guan A: The insulin like effect of sodium vanadate on adipocyte glucose transport mediated port insulin receptor level. Biochem J 238: 665–669, 1986
21. Blondel O, Sison J, Chevalier B, Bernard P: Impaired insulin action but normal insulin receptor activity in diabetic rat liver: Effect of vanadate. Endocrinol Med 258: E459–E467; 1990
22. Strout HV, Vicario PP, Biswas C, Saperstein R, Brady EJ, Plich PF, Berger J: Vanadate treatment of streptozotocin diabetic rats restores expression of the insulin responsive glucose transporter in skeletal muscle. Endocrinol 126: 2728–2732, 1990

Molecular and Cellular Biochemistry **153**: 161–166, 1995.

In vitro and *in vivo* antineoplastic effects of ortrovanadate

Tony F. Cruz, Ariela Morgan and Weixian Min

Connective Tissue Research Group, Samuel Lunenfeld Research Institute, Department of Pathology, Mount Sinai Hospital and University of Toronto, Toronto, Canada

Abstract

In the present study we have demonstrated that orthovanadate at concentrations of 5–10 uM is cytotoxic to proliferating cells including primary cultures and tumour cell lines. However, concentrations of up to 50 uM did not affect the viability of non-proliferating cells. The cytotoxicity appears to be dependent on the vanadium concentration rather than on the oxidation state of vanadium or the vanadium compound. Furthermore, tumour cell lines with different proliferative rates were equally sensitive to orthovanadate cytotoxicity. Although the mechanisms responsible for the cytotoxicity are not known, addition of H_2O_2 potentiated orthovanadate cytotoxicity suggesting that hydroxyl or vanadium radicals may be involved. *In vivo* subcutaneous injections of orthovanadate into mice containing MDAY-D2 tumours resulted in the inhibition of tumour growth by 85–100%. These data indicated that orthovanadate at concentrations greater than 5 uM has antineoplastic properties and may be useful as a chemotherapeutic agent. (Mol Cell Biochem **153**: 161–166, 1995)

Key words: orthovanadate, antineoplastic, cytotoxicity, proliferation

Introduction

Although vanadium is most well known for its insulin mimetic effects and for ameliorating experimental diabetes mellitus [1, 2], it also affects signal transduction mechanisms, protooncogene expression and intracellular enzyme activities [3–5]. Vanadate's inhibition of tyrosine phosphatases was thought to provide these compounds with growth factor mimetic activities [6]. However, other studies have demonstrated that micromolar concentrations of orthovanadate (5–10 uM) inhibit cellular growth and even result in cytotoxicity in highly proliferating conditions [7–8]. Rijksen *et al.* (1993) [9] demonstrated that orthovanadate enhanced cell proliferation when added in the presence of a single growth factor such PDGF at suboptimal concentrations but inhibited cell proliferation when added with multiple growth factors. Thus, vanadate appears to be inhibitory or cytotoxic to cells in actively proliferating conditions. Furthermore, Thompson *et al.* (1984) [10] showed that vanadium has antineoplastic effects *in vivo*.

Although the mechanisms responsible for the cytotoxicity are not understood, orthovanadate is capable of interacting with intracellular superoxide and hydrogen peroxide to produce hydroxyl radicals and vanadium oxygen intermediates which may also be highly toxic to cells [11, 12]. Orthovanadate also affects intracellular pH and calcium, inhibits ATPases, oxidizes NADH, modulates several second messengers and protooncogene expression [5] which may be responsible for growth inhibition and cytotoxicity. In chondrocytes, we have previously found that orthovanadate inhibits growth factor and interleukin 1 induction of fos expression [4]. Since fos expression is required for cell proliferation, it seems likely that inhibition of this and other transcription factors required for cell proliferation may be partially responsible. In the present study we investigated the cytotoxic effects of orthovanadate on proliferating and non-proliferating cells, various tumour cell lines and tumour growth in mice.

Address for offprints: T. Cruz, Samuel Lunenfeld Research Institute, Mount Sinai Hospital, 600 University Avenue, Toronto M5G 1X5, Canada

Materials and methods

Cytotoxicity assays

Primary chondrocyte cultures were freshly isolated from calf cartilage as previously described [13]. The cells were either plated at high (2.0×10^6 cells per well) or low (5×10^5 cells per well) cell density on a six well plate. The cells were incubated in Ham's F12 medium containing 5% FBS overnight. Fresh medium with or without 5% FBS was added to the cells and incubated for an additional 24 h. Following the incubation, fresh medium of the same composition is added to the cells containing different concentrations of orthovanadate ranging from 0–50 uM and incubated for 48 h. The medium was removed, the cells washed and then trypsinized. The cells were stained with Trypan blue and counted. All cytotoxicity experiments were performed in quadruplicate and repeated at least three times.

The astrocytoma cell line (HTB14) was obtained from the American Type Culture Collection. The drug resistant cell lines KB3-1, KB8, KB8-5 and KB8-5-11 were obtained from Dr. Irene Andrulis, Mount Sinai Hospital, Toronto. The MDAY-D2 haematopoetic cell line was obtained from Dr. Jim Dennis, Mount Sinai Hospital, Toronto. These cells were plated or grown in suspension in DMEM containing 5% FBS. After incubation for 24 h, orthovanadate or other vanadium compounds was added. After 48 h, the percentage of cell survival was determined as described above.

Effect of orthovanadate administration on MDAY-D2 tumour growth

Ten week old DBA/2j female mice were acclimatized for 3–4 days. Each mouse was injected subcutaneously in the posterior lateral side with 4×10^5 cells in 100 ul of PBS. On day 5, the mice were divided into two groups: one group received subcutaneous injections of 100 ul of PBS and another group received 100 ul of PBS containing 500 ug of orthovanadate daily. The orthovanadate was administered subcutaneously on the opposite tumour-free posterior lateral side. On day 14, the mice were sacrificed, weighed and tumours dissected and weighed.

Results

Cytotoxicity of orthovanadate on non-proliferating and proliferating cells

The cytotoxic effect of orthovanadate on chondrocyte primary cultures plated at either high cell density (non-proliferating), or at low cell density in the presence (proliferating) or absence (non-proliferating) of FBS was determined (Fig. 1). Cells grown at high cell density or at low cell density in serum free conditions do not proliferate. Under these conditions, concentrations of up to 50 uM orthovanadate did not affect cell viability of non-proliferating chondrocytes. However, concentrations of 5–10 uM orthovanadate were cytotoxic to proliferating chondrocytes plated at low cell density in the presence of serum. The cells undergo morphological changes such as cell shrinkage and rounding, and chromatin condensation. The morphological changes and cytotoxic effects were concentration and time dependent. These changes are observed by 16 h at the higher orthovanadate concentrations, and by 48 h at all concentrations higher than 5 uM. These data suggest that proliferating chondrocytes are sensitive to orthovanadate, whereas non-proliferating cells are resistant to orthovanadate cytotoxicity.

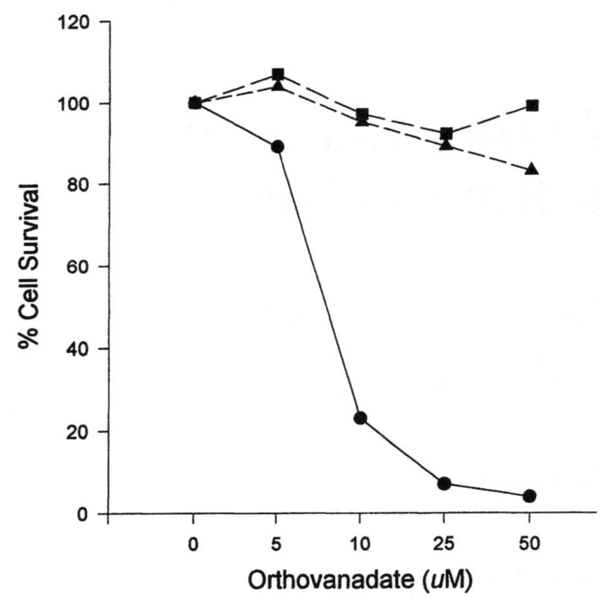

Fig. 1. Orthovanadate cytotoxicity on proliferating and non-proliferating chondrocyte cultures. Chondrocytes plated at high cell density (2×10^6 cells per well) (■) with 5% FBS or at low cell density (5×10^5 cells per well) with (●) and without (▼) 5% FBS were treated with varying concentrations of orthovanadate in Ham's F12 medium for 48 h. The percentage of cell survival was determined as described under Methods. Each value represents a mean of four samples and the variability was less than 10%.

Orthovanadate cytotoxicity on tumour cell lines and drug resistant cell lines

Since orthovanadate appeared to be selectively cytotoxic to proliferating cells, we investigated the effect of orthovanadate on several tumour cell lines. The cytotoxic effects

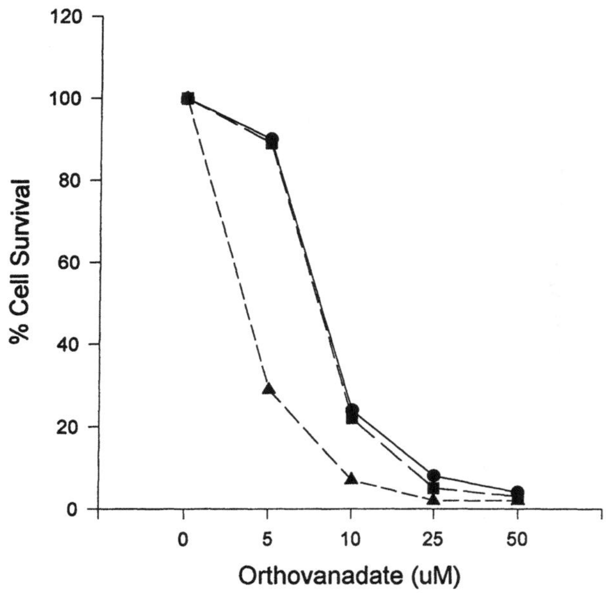

Fig. 2. Comparison of orthovanadate cytotoxicity among different tumour cell lines. Human astrocytoma (HTB14) (■), mouse endothelioma (EOMA) (▼) and haematopoietic (MDAY-D2) (●) cells were incubated with varying concentrations of orthovanadate for 48 h. The percentage of cell survival was determined as described under Methods. Each value represents the mean of four samples and the variability was less than 10%.

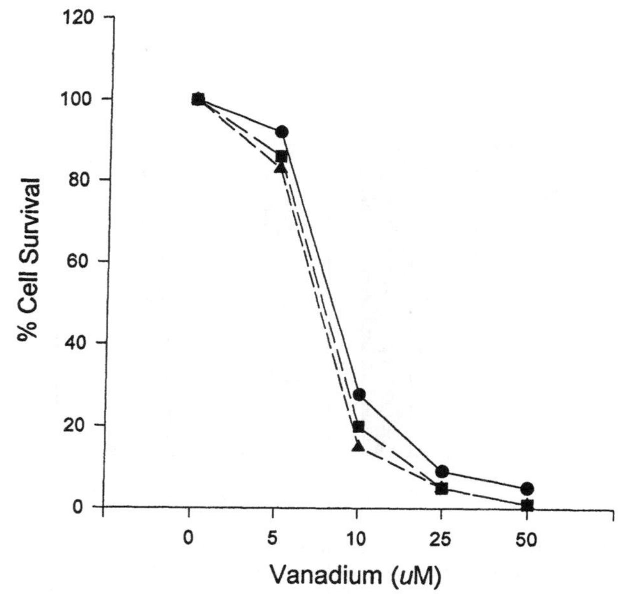

Fig. 3. Cytotoxic effects of different forms of vanadium on MDAY-D2 cells. MDAY-D2 were incubated with varying concentrations of vanadium in the forms of orthovanadate (●), vanadyl sulphate (■) and peroxovanadate (▼) for 48 h. The percentage of cell survival was determined as described under Methods. Each value represents the mean of four samples and the variability was less than 10%.

of orthovanadate on an adherent human astrocytoma cell line (HTB14), a mouse haematopoietic cell line (MDAY-D2) grown in suspension and an adherent mouse endothelial cell line (EOMA) are demonstrated in Fig. 2. Orthovanadate was cytotoxic to all of the cell lines examined. Between 5–10 uM of orthovanadate was required for cell toxicity, and concentrations of 25 uM or greater reduced cell density by over 98% in 48 h.

To determine whether orthovanadate is toxic to drug resistant cell lines, we compared the effect of orthovanadate on three cell lines, KB8, KB8-5, and KB85-11, which have increasing drug resistance respectively, relative to the parent cell line, KB3-1. These drug resistant cell lines are not killed by several classes of chemotherapeutic agents such as colchicine, vinblastine and doxorubicin [14]. As demonstrated in Fig. 4, orthovanadate was equally cytotoxic to all of the drug resistant cell lines.

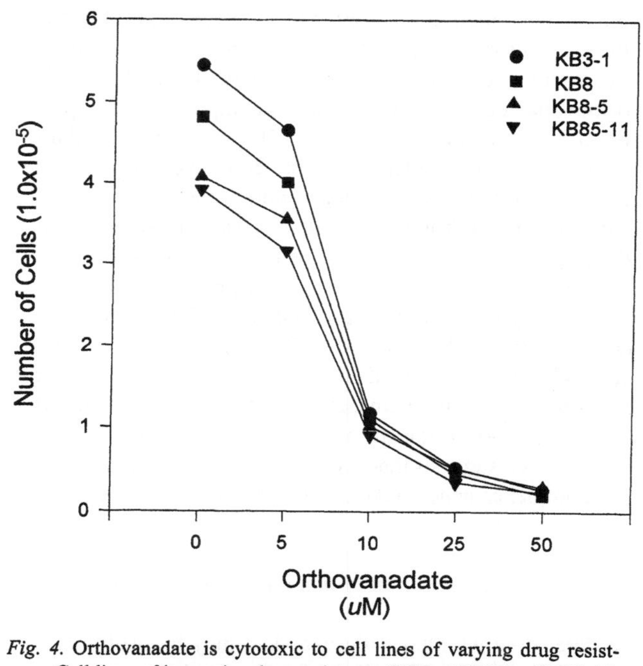

Fig. 4. Orthovanadate is cytotoxic to cell lines of varying drug resistance. Cell lines of increasing drug resistance (KB8 < KB8-5 < KB85-11) and the parent cell line, KB3-1, were incubated in media containing 0–50 uM orthovanadate for 48 h. The cells were harvested and the number of viable cells determined as described under Methods. Each value represents the mean of four samples and the variability was less than 10%.

Efficacy of different forms of vanadium

Figure 3 compares the cytotoxic effect of orthovanadate, vanadyl sulphate and peroxovanadate on MDAY D2 cell cultures. The cytotoxic effects of all three vanadium compounds showed a similar concentration dependence. These

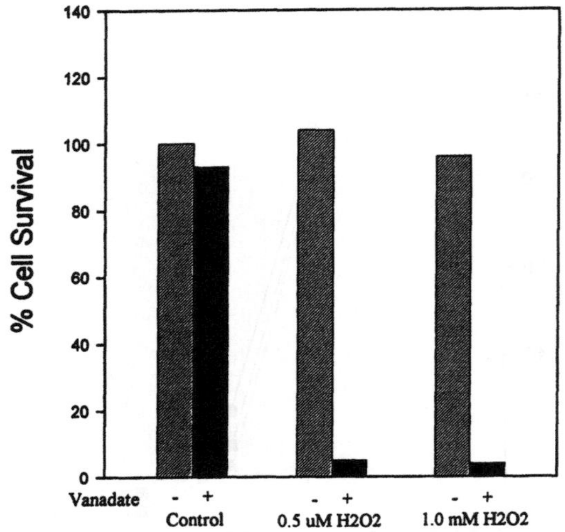

Fig. 5. Hydrogen peroxide potentiates orthovanadate cytotoxicity. MDAY-D2 cells were incubated with either medium alone, 0.5 mM H_2O_2 or 1.0 mM H_2O_2 in the absence (lined bars) and presence (black bars) of 5 uM orthovanadate for 24 h. The percentage of cell survival was determined under Methods. Each value represents the mean of four experiments and the variability was less than 10%.

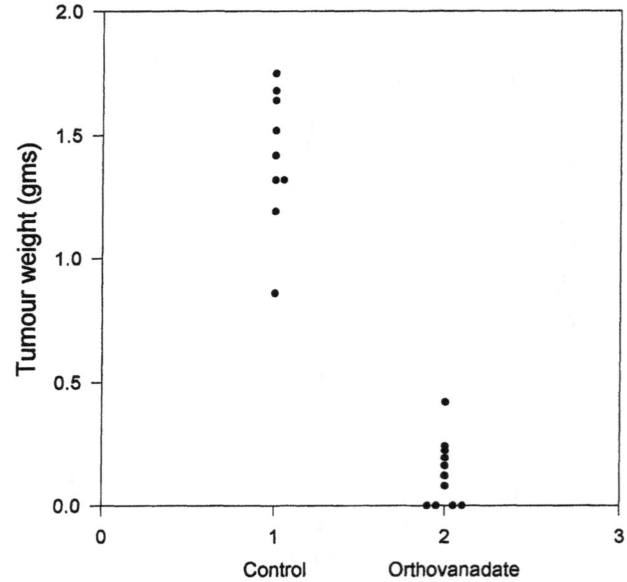

Fig. 6. Effect of *in vivo* orthovanadate treatment on tumour growth. Mice were injected with MDAY-D2 cells subcutaneously. After five days, the mice received daily injections of PBS with and without 500 ug of orthovanadate on the opposite tumour-free posterior lateral side. On day 14, the mice were sacrificed, the tumour removed and weighed. The results represent the tumour weight from each mouse.

data would suggest that the vanadium cytotoxicity on proliferating cells is dependent on vanadium concentration and not on the oxidation state or the form of vanadium. Although the mechanism by which vanadium compounds generate their cytotoxic effects is not known, it is well established that orthovanadate and other vanadium compounds are capable of reacting with superoxide and H_2O_2 to generate vanadium-oxygen molecules and hydroxyl radicals which may promote cellular cytotoxicity. In order to determine whether orthovanadate and hydrogen peroxide acted synergistically in inducing cytotoxicity, cells were treated with suboptimal concentrations of orthovanadate in the presence and absence of H_2O_2. Figure 5 demonstrates that incubation of MDAY-D2 cells with 0.5 mM and 1 mM H_2O_2 for 24 h had no effect on cell viability. However, addition of H_2O_2 in combination with 5 uM orthovanadate increased cell cytotoxicity significantly compared to orthovanadate alone. The potentiation of cell toxicity by H_2O_2 suggests that hydroxyl radicals or vanadate-oxygen complexes generated may be responsible for cell death.

Orthovanadate inhibition of tumour growth in vivo

To determine whether orthovanadate may be useful as an antineoplastic agent, the effect of orthovanadate treatment on a subcutaneous MDAY-D2 tumour mouse model was examined. The weight of tumours from individual mice is demonstrated in Fig. 6. Mice with small tumours (Day 5) treated daily with 500 ug of orthovanadate on the opposite tumour-free posterior side for 9 days showed a significant decrease in tumour growth compared to mice treated with vehicle alone. In control animals, the tumour weights varied from 0.86–1.74 g, whereas in orthovanadate treated mice, four mice did not have detectable tumours and 11 mice had tumours varying from 0.08–0.47 g. Orthovanadate treatment either completely inhibited tumour formation or reduced tumour growth by over 85% when compared to controls. In comparison, the average body weight of the control group was 22.5 ± 1.7 g and the orthovanadate treated group was 19.4 ± 2.1 g.

Discussion

In the present study we have demonstrated that orthovanadate and other vanadium compounds are cytotoxic to proliferating cells but not to non-proliferating cells in culture. Concentrations of 5–10 uM orthovanadate were highly toxic to proliferating chondrocytes plated at low cell density in the presence of serum. However, orthovanadate concentrations of 50 uM did not affect chondrocyte viability when cells were plated under non-proliferating conditions. Using non-tumorigenic cells, Rijkesen *et al.* (1993) [9] also demonstrated that the inhibitory effects of orthovanadate on cell proliferation are dependent on the level of growth factors.

Similarly, Afshari *et al.* (1993) [8] reported that 5–10 uM orthovanadate is toxic to Syrian hamster embryonic cells incubated in the presence of serum. These data suggest that orthovanadate cytotoxicity is dependent on cellular proliferation. Orthovanadate treatment induced several morphological changes such as cell shrinkage and rounding, and chromatin condensation which is characteristic of apoptosis induced by many chemotherapeutic agents and gamma- or X-radiation [15, 16]. The selective cytototoxic effects of orthovanadate on proliferating cells may be related to the requirement of cell proliferation for the induction of apoptosis. We are currently investigating this possibility.

It has been suggested that orthovanadate inhibition of tyrosine phosphatase activity may be directly involved in cytotoxicity [9] or potentiates the cytotoxicity of other chemotherapeutic agents such as fluorouracil [17]. However, this seems unlikely since peroxovanadate and vanadyl sulphate are more potent inhibitors of tyrosine phosphatases than orthovanadate (results not shown), yet all three vanadium compounds were equally cytotoxic to proliferating cells. Cytotoxicity appeared to be dependent on vanadium concentrations rather than on the oxidation state of the vanadium compound or its ability to inhibit tyrosine phosphatases. However, the concentrations required for cytotoxicity are similar or slightly lower than the concentrations required to increase intracellular levels of tyrosine phosphate, activate MAP kinases, and inhibit fos and collagenase expression [3, 4, 13, 18]. Although orthovanadate stimulation of MAP kinase or inhibition of fos expression do not appear to be involved in cytotoxicity, these data suggest that orthovanadate at these concentrations alters a variety of intracellular events which are associated with cellular proliferation. If orthovanadate also stimulates MAP kinases in chondrocytes, then one might have expected a stimulation of fos expression rather than an inhibition. Thus, orthovanadate is also acting down stream of MAP kinase activation or through a separate mechanism in the inhibition of fos expression. In this respect, we have recently shown that orthovanadate stimulated AP-1 binding to AP-1 containing oligonucleotides but inhibits interleukin 1 and fibroblast growth factor induction of CAT expression in chondrocytes transfected with an AP-1-CAT construct (results not shown).

Orthovanadate was equally cytotoxic to a variety of primary tumour and drug resistant cell lines which possessed different proliferative rates. The cell type and proliferative rate of tumour cells do not determine the sensitivity to orthovanadate. However, the mechanisms by which orthovanadate induced cytotoxicity in proliferating cells remains unknown. It is possible that orthovanadate inhibition of other enzyme activities (ATPases, ribonuclease) or the production of hydroxyl radicals through its reactivity with superoxides and hydrogen peroxide may be involved [5, 9, 13]. Since non toxic levels of H_2O_2 potentiated the cytotoxicity of orthovanadate, the involvement of hydroxyl or vanadium radical intermediates produced is suggested. In support of this possibility, vanadate induction of lipid peroxidation and glutathione depletion in hepatocytes is also thought to be responsible for vanadate-induced hepatotoxicity [19, 20]. In this respect, it is interesting that a phenotypic characteristic of tumour cells is altered levels of antioxidant enzymes such as manganese and copper-zinc superoxide dismutase and catalase which would result in increased levels of superoxides and hydrogen peroxide [21, 22]. Thus, the higher levels of intracellular reactive oxygen species in proliferating tumour cells may provide a mechanism to explain the increased sensitivity of tumour cells to orthovanadate cytotoxicity.

The present study demonstrates that orthovanadate reduced tumour growth in mice. Subcutaneous injections of orthovanadate daily for nine days reduced MDAY-D2 tumour growth in mice by 85–100%. Although the animals showed some stress immediately following each injection, their body weight over the treatment period was approximately 85–90% that of the untreated animals. These data indicate that orthovanadate may be a useful antineoplastic agent for the treatment of tumours, particularly drug resistant tumours. In addition, the gastritis induced by oral administration of orthovanadate or vanadyl sulphate [23] may be explained by the cytotoxic effects of these vanadium compounds on the rapidly proliferating surface epithelial cells in the gastrointestinal tract. Whether these general toxic effects can be abolished by conjugating vanadium with organic molecules or by supplementing its administration with antioxidants remains to be determined.

Acknowledgments

We would like to acknowledge Angiogenesis Technologies Inc. for financial support.

References

1. Shechter Y: Perspective in diabetes: insulin-mimetic effects of vanadate, possible implications for future treatment of diabetes. Diabetes 39: 1–5, 1990
2. Meyerovitch J, Farfel Z, Sack J, Schechter Y: Oral administration of vanadate normalizes blood glucose levels in streptozotocin-treated rats: characterization and model of action. J Biol Chem 262: 6658–6662, 1987
3. D'Onorfio F, Le MQU, Chiasson J-L, Srivastava AK: Activation of mitogenm. FEBS Lett 340: 269–275, 1994
4. Conquer JA, Grima DT, Cruz TF: Orthovanadate inhibits interleukin-1 and phorbol ester induced collagenase production by chondrocytes. Ann New York Acad Sci 732: 447–450, 1994
5. Stern A, Yin X, Tsang SS, Davison A, Moon J: Vanadium as a modulator of cellular regulatory cascades and oncogene expression. Biochem Cell Biol 71: 103–112, 1993

6. Swarup G, Cohen S, Garbers DL: Inhibition of membrane phosphotyrosyl-protein phosphatase activity by vanadate. Biochem Biophys Res Commun 107: 1104–1109, 1982
7. Gordon JA: Use of vanadate as protein-phosphotyrosine phosphatase inhibitor. Enzymol 201: 477–482, 1990
8. Afshari CA, Kodama S, Bivins HM, Willard TB, Fujiki H, Barrett JC: Induction of Neoplastic Progression in syrian hamster embryo cells treated with protein phosphatase inhibitors. Cancer Res 53: 1777–1782, 1993
9. Rijksen G, Voller MCW, Van Zoelen EJJ: Orthovanadate both mimics and antagonizes the transforming Growth *B* action on normal rat kidney cell. J Cell Physiol 514: 393–401, 1993
10. Thompson H, Chasteen ND, Mroker LD: Dietary vanadyl(IV) sulfate inhibits chemically-induced mammary carcinogenesis. Carcinogenesis 5: 849–851, 1984
11. Ravi Shankar HN, Ramasarma T: Multiple reaction in vanadyl-V (IV) oxidation by H_2O_2. Mol Cell Biochem 129: 9–29, 1993
12. Shi X, Dalal NS: Hydroxyl radical generation in the DADH/ microsomal reduction of vanadate. Free Rad Res Comms 17: 369–376, 1992
13. Cruz TF, Mills G, Prikker PH, Kandel AK: Inverse correlation between tyrosine phosphorylation and collagenase production in chondrocytes. Biochem J 269:717–721, 1990
14. Akiyama S, Fojo A, Hanover JA, Pastan I, Gottesman MM: Isolation and genetic characterization of human KB cell lines resistant to multiple drugs. Somatic Cell Mol Genetics 11: 117–126, 1985
15. Barry MA, Beehnke CA, Eastman A: Activation of programmed cell death (Apoptosis) by cisplatin, other anticancer drugs, toxins and hyperthermia. Biochem Pharmacol 40: 2353–2362, 1990
16. Dive C, Hickman JA: Drug target interactions: only the first step in the commitment to a programmed cell death? Br J Cancer 64: 192–196, 1991
17. Kovach JS, Svingen PA, Schaid DJ: Levamisole potentiation of fluorouracil antiproliferative activity mimicked by orthovanadate, an inhibitor of tyrosine phosphatase. J National Cancer Inst 84: 515–519, 1992
18. Grima DT, Kandel RA, Pepinsky B, Cruz TF: Lipocortin 2 (Annexin 2) is a major substrate for constitutive tyrosine kinase activity in chondrocytes. Biochemistry 33: 2921–2926, 1994
19. Younes M, Strubelt O: Vanadate-induced toxicitv towards isolated perfused rat livers: the role of lipid peroxidation. Toxicology 66: 63–74, 1991
20. Stacey NH, Kappus H: Comparison of methods of assessment of metal-induced lipid peroxidation in isolated rat hepatocytes. J Toxicol Environ Health 9: 277–284, 1982
21. Oberley LW, Buettner GR: Role of superoxide dismutase in cancer: a review. Cancer Res 39: 1141–1149, 1979
22. Sun Y: Free radicals, antioxidant enzymes and carcinogenesis. Free Rad Biol Med 8: 583–599, 1990
23. Zaporowska H, Wasilewski W, Stotwinska M: Effect of chronic vanadium administration in drinking water to rats. Biometals 6: 3–10, 1992

Molecular and Cellular Biochemistry **153:** 167–171, 1995.

Membrane – vanadium interaction: A toxicokinetic evaluation

Raj K. Upreti
Biomembrane Division, Industrial Toxicology Research Centre, M.G. Marg, Post Box No.80, Lucknow-226 001, India

Abstract

Vanadium is an important trace metal widely distributed in environment. Interaction of vanadate with skeletal muscle sarcolemma and basement membrane has been focussed. Scatchard analysis indicated the presence of more than one binding site for vanadate. Vanadate inhibits sarcolemmal and intestinal brush border membrane enzymes in a non-competitive manner. Membrane-vanadium interaction may lead to several structural and functional changes. The binding of vanadium to basement membrane may have some protective role. (Mol Cell Biochem **153:** 167–171, 1995)

Key words: vanadium, sarcolemma, brush border, membrane enzymes

Introduction

Unlike organic pollutants, the trace metals are not biodegradable and they tend to build up the ecosystem to levels which may be toxic. Vanadium is an important trace metal that is widely distributed in environment. It is considered as an essential element but it is highly toxic when introduced in excessive doses to animals and humans [1]. It is used extensively in various types of industry, and exposure to high vanadium levels is not uncommon. Due to its wide industrial use, the biological actions of vanadium are of interest [2]. Vanadium compounds enter the body primarily through the lungs where they are absorbed slowly and excreted mainly in the urine [1]. Vanadium fumes induce inflamatory changes in the mucous membranes of the respiratory tract in exposed humans and animals. Inhalation exposure to vanadium can cause conjuctivitis, pharyngitis, rhinitis, chronic productive cough and tightness of the chest [3]. In case of severe exposure, vanadium may cause cellular necrosis to liver and kidney. Vanadium salts have been used medicinally as antiseptic, spirochetocide , antituberculotic and antianemic agents [3]. Further, vanadate has been reported to counteract glucagon effects in isolated rat hepatocytes [4]. The medicinal use may also result in gastrointestinal disorders and nervous system effects [5]. The absorption of vanadium through the gastrointestinal tract is low [6, 7]. However, vanadium has been reported in tissues and urine with in hours after a single [8] and repeated oral exposure in rats [9]. Much higher absorption of vanadium in young rats have been observed due to a greater nonselective permeability of the undeveloped intestinal barrier [10]. Vanadium is rapidly distributed in tissues after inhalation or oral exposures. There is an initial accumulation in the lungs, kidneys, liver and muscles. However, retention of vanadium occurs primarily in the bone [8, 10].

Membrane as a model for toxicity evaluation

Chemicals encountered in occupation or in environment undergo physico-chemical interactions with reactive biological entities to manifest their pathological, physiological or pharamacological effects. The prime target sites for xenobiotic toxicity are surface membranes possibly due to their exposed location and chemical reactivity. The interaction of a chemical with either protein or lipid components of the cell membrane may substantially alter membrane structure and function.

In addition to the well defined and widely worked out erythrocyte plasma membrane, the other important surface membranes which come directly in contact with xenobiotics are skeletal muscle cells sarcolemma and intestinal brush border membrane (BBM). Sarcolemma is made up of a

Address for offprints: R.K. Upreti, Biomembrane Division, Industrial Toxicology Research Centre, M.G. Marg, Post Box No. 80, Lucknow-226 001, India plasma membrane, a felt-like electron dense basal lamina, and an overlying reticular lamina containing collagen and reticular fibrils embedded in an amorphos matrix [11,12]. Basal lamina and reticular lamina together constitute basement membrane (B.M.). Sarcolemma plays an important role in the excitation and contraction phenomenon of muscles [13]. In general, basement membrane is known to play significant role in filteration, structural organization of tissues and cell-cell adhesion [14]. The intestinal BBM is highly specialized plasma membrane responsible for digestive and absorptive functions and their closed vesicles retain the original orientation of the membrane [15].

Sarcolemmal and basement membrane interaction of vanadium

While there are many reports which describe the inhibition of membrane bound ATPase and other enzymes by vanadate, studies on skeletal muscle sarcolemmal enzymes are scanty. The preceding text is mainly focussed on frog skeletal muscle sarcolemma. Owing to the fact that bulk of the literature is available on the mechanism of action of vanadate on Na^{+}+K^{+}-ATPase from a wide range of sources and the functions of this enzyme in skeletal muscle are little appreciated as compared to Ca^{2+}-ATPase. The present article, therefore, does not deal much with $Na^{+}K^{+}$-ATPase. We have demonstrated the presence of Ca^{2+}-ATPase, Mg^{2+}-ATPase, Ca^{2+}+Mg-ATPase, 5′-nucleotidase, alkaline phosphatase and acetyl cholinesterase in purified frog skeletal muscle sarcolemma. Most of the enzymes were inhibited by vanadate, with the prominent exception of acetyl cholinesterase (Figs 1 and 2). In frog sarcolemma, the ion transporting ATPases, 5′-nucleotidase and alkaline phosphatase behaved similarly to vanadate and can be considered in the same class due to their common feature distinguished by the formation of a phosphoenzyme intermediate during the reaction cycle. Acetyl cholinesterase is structurally and functionally different from the above enzymes and therefore its response against vanadate is different. It has been shown that vanadate stimulates rather to inhibit acetyl cholinesterase in rat ventricular strips, electric eel and also in erythrocytes [16]. However, frog sarcolemmal esterase revealed no significant effect of vanadate.

Vanadate inhibition of frog skeletal muscle sarcolemmal ATPases was non-competitive type. Inhibitor constant of Ca^{2+}-ATPase at two different substrate concentrations revealed Ki value of 4.6×10^{-5} M. Additional enzyme kinetic studies indicated that vanadate did not effect the Km (1.10 mM) of the enzyme but it reduced the Vmax (0.80 umole Pi/mg/h to 0.55 umole). It has also been shown that vanadate acts as a non-competitive inhibitor to plasma membrane ATPases of yeast and corn root [17, 18]. The mechanistic view of the action of vanadate on phosphoenzyme ion transport ATPases and other enzymes of this catagory can be understood by considering the structure of these enzyme molecules. All members of this class are inhibited by vanadate. Vanadate enters the reaction sequence by the back door and forms a stable inactive complex in a reversible reaction with the E_2 conformation of the enzyme in a stoicheometry of one vanadate bound per active site for phosphorylation [19]. Vanadate is also known to induce the formation of two-dimensional crystalline arrays of Ca^{2+}-ATPase molecule in sarcoplasmic reticulum membrane vesicles [20].

Not much is available in literature on the mechanism of action of vanadate on 5′-nucleotidase. The presence of 5′-nucleotidase is controversial in all membranes. However, we observed good activity of this enzyme in frog skeletal sarco-

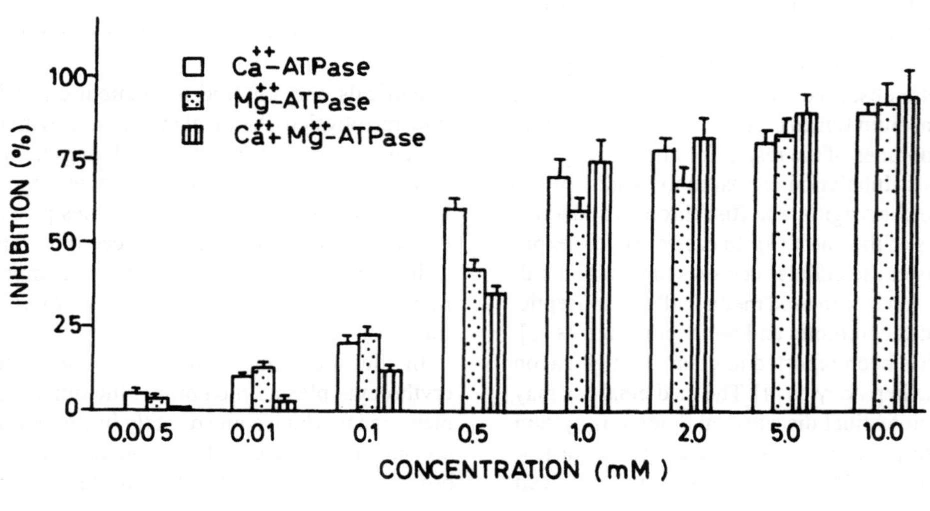

Fig. 1. Inhibition pattern of frog skeletal muscle sarcolemmal ATPases by vanadate.

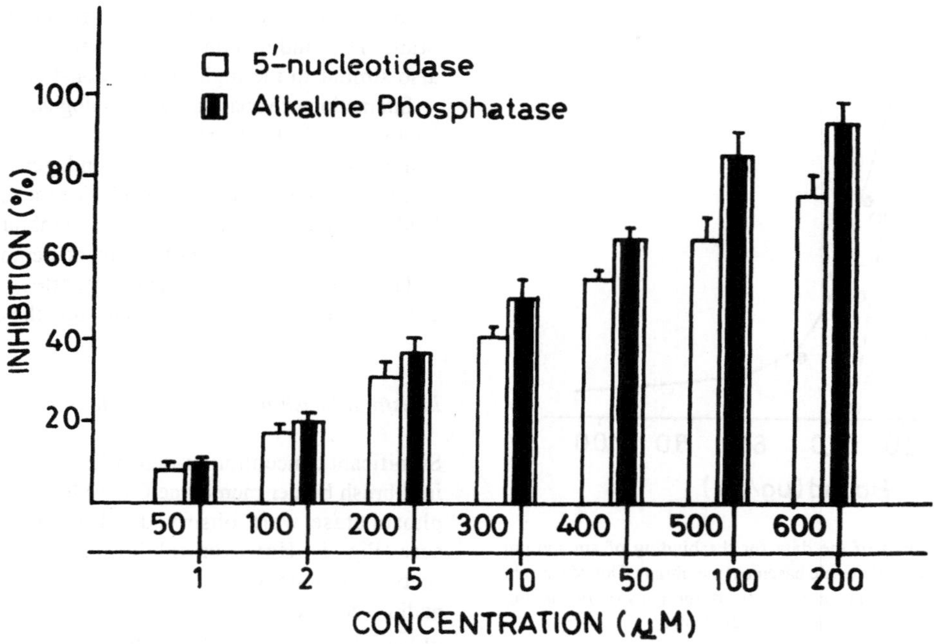

Fig. 2. Inhibition pattern of frog skeletal muscle sarcolemmal 5'-nucleotidase and alkaline phosphatase by vanadate. The upper line vanadate concentrations refer to 5'-nucleotidase and the lower line to alkaline phosphatase.

lemma. 5'-nucleotidase was also inhibited non-competitively by vanadate. Inhibitor constant (Ki) value was 6.2 × 10^{-5} M. Further, vanadate did not effect the Km (0.75 mM) of the enzyme but reduced the Vmax (0.066–0.038 umole). Similar inhibitory response of vanadate on 5'-nucleotidase could be due to its resemblence with ATPases in at least carrying out the reaction cycles in similar manner.

The effect of vanadate on the inhibition of alkaline phosphatase have been studied in few cases [21–23]. However, the mode of vanadate action on alkaline phosphatase in most of the studies are not well discussed. Kinetic studies on alkaline phosphatase of frog sarcolemma demonstrated that vanadate inhibits in a non-competitive manner. The inhibitor constant (Ki) was found to be 4.2 uM which is comparable with the findings on rat mesentric artery and human liver alkaline phosphatase [21, 22]. The higher sensitivity of alkaline phosphatase to vanadate may be, in part, due to that it is closely related to Na^{+}+K^{+}-ATPase as is the case with K^{+}-activated p-nitrophenyl phosphatase. Since vanadate is a potent inhibitor of Na^{+}+K^{+}-ATPase and it also inhibits K^{+}-dependent phosphatase approximately with the similar potential, it is likely for the vanadate to inhibit alkaline phosphatase in a similar fashion.

Vanadate binding studies with isolated frog skeletal muscle sarcolemma and basement membrane revealed a concentration-dependent binding. Scatchard analysis indicated biphasic curves with a high and low affinity components for

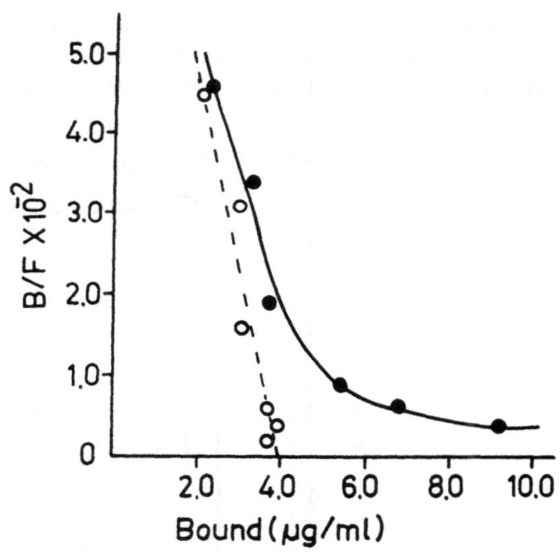

Fig. 3. Scatchard analysis of the data for the binding of ammonium metavanadate to frog skeletal muscle sarcolemma. B is the concentration of bound and F is the concentration of free vanadium. Dotted straight line was obtained after applying correction factor for non specific binding sites showing the presence of one major binding site.

both membranes (Figs 3 and 4). This suggests the existence of more than one binding site responsible for binding of vanadium. The binding constants for high affinity binding site of sarcolemma were n_1 = 3.8 n mole/mg protein, k_1 = 1.57 ×

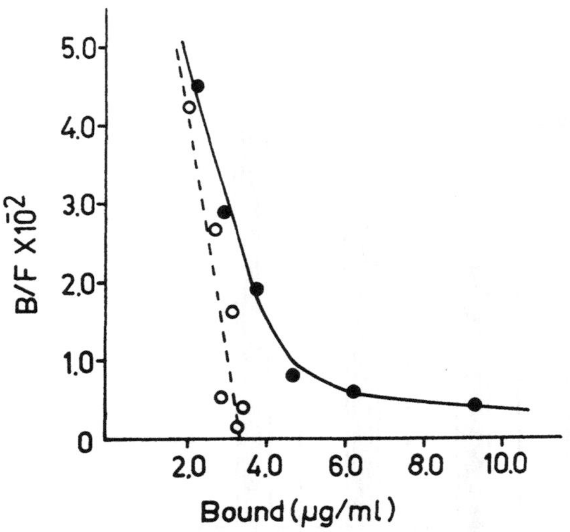

Fig. 4. Scatchard analysis of the data for the binding of ammonium metavanadate to frog skeletal muscle basement membrane. Dotted straight line was obtained after applying correction factor for non specific binding sites showing the presence of one major binding site.

10^7 M^{-1} and that of basement membrane were $n_1 = 4.1$ n mole/mg protein, $k_1 = 1.31 \times 10^7$ M^{-1}. Their respective low affinity binding constants were $n_2 = 11.4$ n mole/mg protein, $k_2 = 6.55 \times 10^5$ M^{-1} and $n_2 = 12.8$ n mole/mg protein, $k_2 = 4.69 \times 10^5$ M^{-1}. Applying correction factor to minimize the non-specific binding sites, the biphasic curve was converted into a straight line suggesting that there exist at least one major binding site in sarcolemma as well as in basement membrane. The binding constants obtained after correction were also similar to the high affinity constants.

Interaction of vanadium with biological entities are of great interest. Besides binding to membranous components, vanadium compounds can also bind to some endogenous phosphate, carboxyl and amino-ligands present inside the cell [24]. However, binding of vanadium with basement membrane suggest that the cells having an outer layer of basement membrane such as skeletal muscle, lens capsule and kidney cortex, could play significant protective role against its entry into the cell [25].

Intestinal brush border membrane interaction of vanadium

Significant concentration-dependent inhibition of rat intestinal brush border membrane Ca^{2+}+Mg^{2+}-ATPase and alkaline phosphatase were observed following vanadium exposure (Fig. 5). However, inhibition of disaccharidases and acetyl cholinesterase activities were not observed over the vanadium concentrations upto 40 mM. The toxicokinetic study of Ca^{2+}-Mg^{2+}-ATPase and alkaline phosphatase further indicated that the inhibition caused by vanadium was non-competitive type [26]. This is in agreement with earlier findings on sarcolemma membrane ATPase and alkaline phosphatase and further suggest a generalized mode of vanadium action. These enzymes are involved in the movement of important intermediates across the intestinal brush border membrane. Therefore, inhibition of ATPase and phophatase by vana-

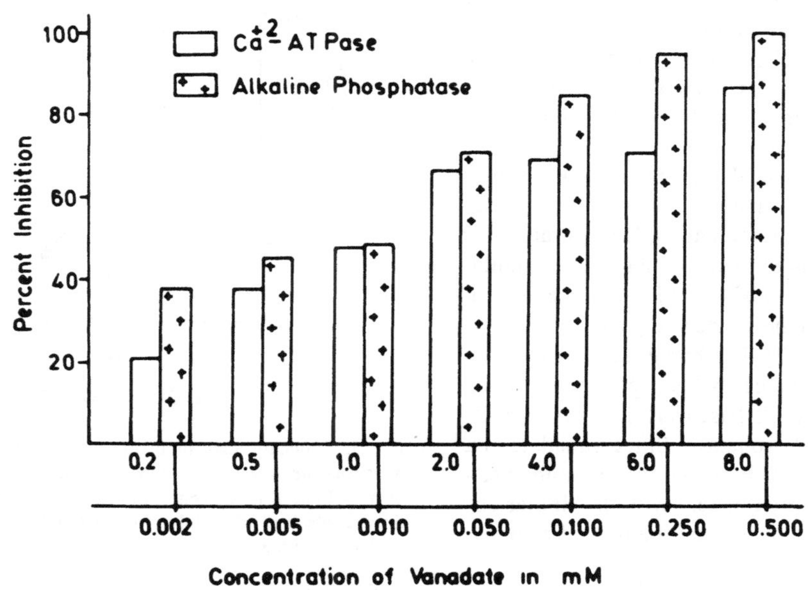

Fig. 5. Inhibition pattern of rat intestinal brush border membrane Ca^{2+}-ATPase and alkaline phosphatase by vanadate. The upper line vanadate concentrations refer to Ca^{2+}-ATPase and the lower line to alkaline phosphatase. [from Gupta K, Upreti RK, Kidwai AM: Bull Environ Contam Toxicol 52: 919–926, 1994].

dium may bring about deleterious effects at least in higher concentrations.

Conclusions

This article has focused on vanadium interaction with skeletal muscle sarcolemmal membrane and intestinal brush border membrane enzymes. Membrane enzymes inhibited by vanadate in a non-competitive manner suggests that it binds the enzyme molecules either at the active site or any other available site. Furthermore, vanadate binding with surface membranes also indicates the existance of some structural constituents (vis-a-vis membrane enzymes) responsible for its binding. The binding of vanadate to enzyme(s) could cause the conformational changes resulting into the enzymatically inactive state.

Acknowledgement

Author is grateful to Dr. A.M. Kidwai for his keen interest and valuable suggestions.

References

1. Jandhyala BS, Hom GJ: Physiological and pharmacological properties of vanadium. Life Sci 33: 1325–1340, 1983
2. Nechay BR: Mechanisms of action of vanadium. An Rev Pharmacol Toxicol 24: 501–524, 1984
3. Domingo JL, Llobet JM, Tomas JM, Corbella J: Short term toxicity studies of vanadium in rats. J Appl Toxicol 5: 418–421, 1985
4. Miralpeic M, Gil J, Rosa JL, Carreras J, Bartrons R: Vanadate counteracts glucagon effects in isolated rat hepatocytes. Life Sci 44: 1491–1497, 1989
5. Hammond PB, Beliles RP: Metals In: J. Doull, C.D. Klaassen, M.O. Amdur (eds). Toxicology, the Basic Science of Poisons, MacMillan Publishing, NY, 1980 pp 460–470
6. Roshchin AV, Ordzhonikidze EK, Shalganova IV: Vanadium-toxicity, metabolism, carrier state. J Hyg Epidemiol Microbiol Immunol 24: 377–383, 1980
7. Conklin AW, Skinner CS, Felten TL, Sanders CL: Clearance and distribution of intratracheally instilled vanadium-48 compounds in the rat. Toxicol Lett 11: 199–203, 1982
8. Edel J, Sabbioni E: Retention of intratracheally instilled and ingested tetravalent and pentavalent vanadium in the rat. J Trace Elem Electrolytes Health Dis 2: 23–30, 1988
9. Parker RD, Sharma RP: Accumulation and dipletion of vanadium in selected tissues of rats treated with vanadyl sulfate and sodium orthovanadate. J Environ Pathol Toxicol. 2 : 235–245, 1978
10. Edel J, Pietra R, Sabbioni E, Marfante E, Springer A, Ubertalli L: Disposition of vanadium in rat tissues at different age. Chemosphere 13: 87–93, 1984
11. Sanes JR, Marshall LM, McMahan UJ: Reinnervation of muscle fiber basal lamina after removal of myofibrils. J cell Biol 78: 176–198, 1978
12. Borg TK, Caulfield JB: Morphology of connective tissue in muscle. Tissue & Cell 12: 197–207, 1980
13. Weber A, Murray JM: Molecular control mechanism in muscle contraction. Physiol Rev 53: 612–673, 1980
14. Grant ME, Codfrey HJ, Orkin RW: Current concept of basement membrane structure and function. Biosci Rep 1: 819–842, 1981
15. Klip A, Grinstein S, Semenza G: Transmembrane disposition of the phlorizine binding protein of intestinal brush border. FEBS Lett 99: 91–96, 1979
16. Catalan RE, Martinez AM, Aragones MD, Godoy JE: Activation of acetylcholinesterase by vanadate. Neuropharmac 28: 1119–1122, 1985
17. Dufour JP, Boutry M, Goffeau A: Plasma membrane ATPase of yeast. Comparative inhibition studies of the purified and membrane bound enzymes. J Biol Chem 255: 5735–5741, 1980
18. Tu S-I, Sliwinski BJ: Mechanistic investigation of corn root plasma membrane ATPase. Arch Biochem Biophys 241: 348–355, 1985
19. Huang WH, Askari A: Simultaneous binding of ATP and vanadate to $(Na^{+}+K^{+})$-ATPase. J Biol Chem 259: 13287–13291, 1981
20. Dux L, Martonosi A: Ca^{2+}-ATPase crystals in sarcoplasmic reticulum; Effect of trypsin digestion. J Biol Chem 258: 10111–10115, 1983
21. Kwan CY: Characteristics of plasmalemma alkaline phosphatase of rat mesentric artery. Blood Vessels 20: 109–121, 1983
22. Chakraborttey A, Stinson RA: Properties of membrane bound and solubilized forms of alkaline phosphatase from human liver. Biochim Biophys Acta 839: 174–180, 1985
23. Farley JR, Baylink DJ: Skeletal alkaline phosphatase activity as a bone formation index *in vitro*. Metabol Clin Exp 35: 563–571, 1986
24. Nechay BR, Nanniuga LB, Nechay PSE: Vanadyl and vanadate binding to selected endogenous phosphate, carboxyl and amino ligands: Calculations of cellular vanadium species distribution. Arch Biochem Biophys 251: 128–138, 1986
25. Ali N, Upreti RK, Kidwai AM: Sarcolemma as model for testing toxicity of chemicals. Ind J Biochem Biophys 25: 209–214, 1988
26. Gupta K, Upreti RK, Kidwai AM : Toxicokinetic study of rat intestinal brush border membrane enzymes following in vitro exposure to lead and vanadium. Bull Environ Contam Toxicol 52: 919–926, 1994

PART III

POTENTIAL USE IN THERAPY AND TOXICOLOGICAL STUDIES

Molecular and Cellular Biochemistry **153**:175–180, 1995.

Increased potency of vanadium using organic ligands

John H. McNeill, Violet G. Yuen, Soter Dai and Chris Orvig[1]

Faculty of Pharmaceutical Sciences and [1]Department of Chemistry, University of British Columbia, Vancouver, B.C., V6T 1Z3, Canada

Abstract

The *in vivo* glucose lowering effect of orally administered inorganic vanadium compounds in diabetes was first reported in our laboratory in 1985. While both vanadate and vanadyl forms of vanadium are orally active, they are still not well absorbed. We have synthesized several organic vanadium compounds and one compound, bis(maltolato)oxovanadium(IV) or BMOV, has been extensively investigated. BMOV proved effective in lowering plasma glucose and lipids in STZ-diabetic rats when administered in drinking water over a 25 week period. The maintenance dose (0.18 mmol/kg/day) was approximately 50% of that required for vanadyl sulfate (VS). Secondary complications of diabetes were prevented by BMOV and no marked toxicity was noted. Oral gavage of STZ-diabetic rats with BMOV also reduced blood glucose levels. The ED_{50} for BMOV was 0.5 mmol/kg, while for VS the estimated ED_{50} was 0.9 mmol/kg. BMOV was also effective by the intraperitoneal route in STZ-diabetic rats. The ED_{50} was 0.08 mmol/kg compared to 0.22 mmol/kg for VS. Some animals treated p.o. or i.p. remained euglycemic for up to 14 weeks. An i.v. infusion of BMOV of 0.05 mmol/kg over a 30 min period reduced plasma glucose levels by 50% while VS was not effective. (Supported by the CDA). (Mol Cell Biochem **153**: 175–180, 1995)

Key words: Vanadium, diabetes, insulin-mimetic, hypertension

Introduction

Vanadium is a Group V transition element that exists in many oxidation states and is ubiquitous in nature. First discovered by Nils Sefstrom in 1831, vanadium was named after the Norse goddess Vanadis because of its crystalline beauty. Vanadium was used at the turn of the century by a French physician to treat a number of disease states including diabetes [1]. However, it was not until the 1970's that the current interest in vanadium was revived with the discovery that vanadate was a potent phosphatase inhibitor when used *in vitro* [2, 3].

A large body of *in vitro* experimental work exists that demonstrates the insulin-mimetic actions of vanadium. In both the vanadate (+5 oxidation state) and vanadyl (+4 oxidation state) forms, vanadium mimics the actions of insulin on carbohydrate and fat metabolism. In carbohydrate metabolism vanadium stimulates hexose transport, glucose transporter expression, glucose oxidation and glycogen synthesis in a number of tissues [4–7]. At the cellular level, vanadium has been shown not to bind directly to the insulin receptor but it does increase the autophosphorylation of the β subunit of the receptor, tyrosine kinase activity and phosphorylation of substrates [8–12]. Given the reported insulin-mimetic properties of vanadium *in vitro*. the possible *in vivo* insulin-like effects of vanadium were tested in animal models of diabetes mellitus.

There are a number of animal models of diabetes available for the examination of drug effects on the diabetic condition. Alloxan and streptozotocin (STZ) injection chemically induce a model of Type I diabetes mellitus in which the animal is hyperglycemic, hypoinsulinemic, hyperlipidemic, hyperphagic and polydipsic; however, insulin treatment is not always required to maintain these animals since the severity of the diabetic state depends on the dose of the drug used [13]. Another method of producing a hypoinsulinemic model of diabetes mellitus is to partially pancreatectomize rats by the removal of approximately 90% of the pancreas [14]. BB Wistar rats are a commercially

Address for offprints: J.H. McNeill, Faculty of Pharmaceutical Sciences, 2146 East Mall, Vancouver, British Columbia, V6T 1 Z3, Canada available strain of rats that are genetically prone to the development of insulin-dependent Type I diabetes [15, 16]. In these models vanadium has been shown to be effective in reducing plasma glucose levels to normal or near normal, in restoring altered plasma lipid levels and improving the glucose response to an oral or intravenous glucose tolerance test [17–20].

Type I diabetes mellitus is not the only diabetic state in which vanadium has been shown to be effective. There are several animal models of Type II diabetes characterized by mild to marked hyperglycemia, hyperinsulinemia, obesity and glucose intolerance. These models include the STZ-injected neonatal rat, ob/ob mice and the fa/fa Zucker rat. Vanadium has improved peripheral glucose utilization, hepatic glycogen content and tolerance to oral glucose [21–23].

Inorganic vanadium *in vivo*

The first evidence for the *in vivo* insulin-mimetic actions of vanadium were demonstrated by Heyliger *et al* in 1985. Sodium metavanadate was administered at a maximal concentration of 0.8 mg/mL in 0.9% NaCI drinking water to both control and STZ-diabetic rats. Vanadate treatment decreased food and fluid intake in the diabetic treated group and fluid intake in control treated animals. Plasma glucose levels in diabetic treated rats were restored to within normal parameters without an increase in circulating plasma insulin levels. It was also observed that vanadium treatment decreased plasma insulin levels in control treated animals without altering plasma glucose values. In addition, the isolated working heart preparation, done at the termination of the experiment to evaluate the effect of vanadium on the impaired heart function observed in diabetes, showed that vanadium restored heart function to normal. Vanadate treatment did not affect the body weight gain in the diabetic treated animals but did significantly reduce weight gain in the control treated animals and in both groups of animals problems with diarrhea and dehydration were noted [24].

Since a review of the literature suggested that vanadyl was less toxic than vanadate [25], the insulin-mimetic effects of vanadyl were subsequently examined. Vanadyl was administered in the drinking water at concentrations ranging from 0.25–1 mg/mL. There was a dose dependent increase in the number of animals that responded with improved plasma glucose levels following treatment. Ramanadham *et al* showed that 63 days of treatment with vanadyl normalized plasma glucose, triglyceride, T_4 and creatinine levels, and restored heart function in STZ-diabetic animals. However, the slowed body weight gain observed in control animals treated with vanadate was not improved with vanadyl treatment [17].

One question that was raised was whether vanadium treatment would be effective if the start of therapy was delayed. Cam *et al* administered vanadyl sulfate in the drinking water starting 3, 10 and 17 days after STZ injection. The time of onset of treatment did not alter the glucose lowering properties of vanadyl [20].

Removal of vanadyl treatment for 13 weeks after an initial 3 week period of treatment showed that plasma glucose, lipid and thyroid hormone levels and heart function in STZ-diabetic treated animals were still normalized [26]. An oral glucose tolerance test showed an improved glucose response without a concomitant increase in circulating plasma insulin levels. The maintenance effects following withdrawal of vanadium therapy have subsequently been repeated in a number of experiments [27, 28]. The mechanism of action for the sustained euglycemia observed following withdrawal of vanadium therapy is not known.

A one year toxicity study has shown that vanadyl sulfate treatment improved or prevented the tissue damage seen in the kidney of diabetic animals. Vanadyl treatment normalized serum transaminases and urea nitrogen and creatinine levels in diabetic treated animals. Cataract formation, a common secondary characteristic in diabetes, was prevented with vanadium therapy [29, 30].

Because the STZ model of diabetes is not completely insulin deficient, a question arose concerning the effectiveness of vanadium in controlling elevated glucose levels in the BB Wistar rat model. BB rats were stabilized on exogenous insulin prior to the start of vanadyl treatment. Vanadyl sulfate was administered in the drinking water at a maximal concentration of 0.75 mg/mL for 6 months. It was found that the dose of exogenous insulin required could be significantly reduced with vanadium treatment. However, when exogenous insulin levels were reduced by greater than 50% there was a crossover point at which the beneficial effects of vanadium and insulin in combination were diminished and there was a detrimental effect on body weight gain. This suggests that vanadyl was not effective in mimicking the anabolic effects of insulin [19].

A current hypothesis in hypertension research is that insulin resistance and hyperinsulinemia may indirectly play a role in the development of elevated blood pressure (Fig. 1; [31]) thus a drug intervention which can reduce plasma insulin levels may be effective in lowering blood pressure.

Since treatment with either vanadate and vanadyl results in a significant reduction in circulating plasma insulin levels in control treated animals, a number of experiments were done to examine the hypertension/insulin resistance hypothesis. Spontaneously hypertensive rats (SHR) are a genetic strain of animals that develop essential hypertension starting at 8 weeks of age. While this model is hyperinsulinemic, there is some controversy as to whether SHR are insulin resistant [32–35]. Bhanot *et al* showed that treatment of SHR with vanadyl sulfate in the drinking water could prevent the

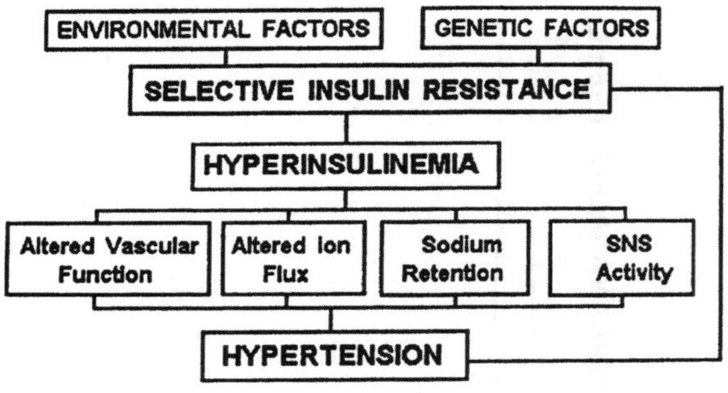

Fig. 1. Schematic representation of the possible inter-relationship between hypertension and insulin resistance and hyperinsulinemia. (Modified from Rocchini, 1992).

development of hypertension and reduce plasma insulin levels to normal. Secondly, a group of SHR were allowed to develop hypertension and then vanadium treatment was initiated; it was found that plasma insulin levels were restored to normal and high blood pressure was reduced by approximately 50% of untreated animals. Hyperinsulinemia and insulin resistance may contribute to the etiology of hypertension [35].

The fructose animal model represents an acquired form of systolic hypertension where the increased blood pressure is diet induced [36]. Similar effects of vanadium treatment on elevated blood pressure and plasma insulin levels have been shown in the fructose-induced hypertensive rat [37].

Organic vanadium *in vivo*

While a reduced food intake and a slowed body weight gain in control treated animals have been attributed to vanadium treatment, pair fed studies conducted in this laboratory and others have shown that the insulin-mimetic actions of vanadium cannot be attributed to the redudion in food intake [21, 35]. Therefore since inorganic vanadium is poorly absorbed from the gastrointestinal (Gl) tract and [38, 39] some Gl difficulties have been reported with both vanadyl and vanadate, our laboratory and others have synthesized a number of organic vanadium compounds.

The first compound manufactured for our laboratory was a cysteine/vanadyl complex, naglivan (Fig. 2). Naglivan was synthesized in France and was investigated in collaboration with Dr. G. Cros, Montpellier, France. Naglivan is water insoluble and therefore it was administered as a suspension by oral gavage. Treatment with naglivan alone was not initially effective in reducing plasma glucose to normal and thus it was administered in combination with insulin. The combination therapy was effective in reducing plasma glucose levels, restoring elevated plasma lipid levels and heart function in STZ-diabetic rats. The reduction in body weight gain observed in non-diabetic animals with inorganic vanadium was not seen with naglivan over an 8 week treatment period. After a 3 week combination therapy regime insulin was withdrawn and diabetic animals that continued to receive naglivan alone demonstrated an improved glucose control. A subsequent experiment showed that diabetic animals could be euglycemic with naglivan alone when the dose of naglivan was increased [40].

Bis(maltolato)oxovanadium(IV) (BMOV) (Fig. 3), a maltol/vanadyl compound, was developed in a collaboration with Dr. C. Orvig in the Department of Chemistry at the University of British Columbia [41]. BMOV is a potent example of a series of compounds designed to be orally absorbed by passive diffusion as a result of their properties of water solubility, electrical neutrality and low molecular weight. BMOV was administered to STZ-diabetic rats in the drinking water at a maximal concentration of 0.75 mg/mL for 6 months. BMOV restored plasma glucose levels to normal in 8/12 animals and restored elevated plasma lipid parameters and heart function in all diabetic treated rats. There was a strong correlation between improved heart function and long term glucose control. However unlike vanadyl, BMOV treatment did not result in the sustained euglycemic response observed following withdrawal of treatment [42].

There was no effect of BMOV therapy on body weight gain in control treated animals for the first 10 weeks of treatment as compared to inorganic vanadium therapy which reduced body weight gain within 1–2 weeks of treatment initiation.

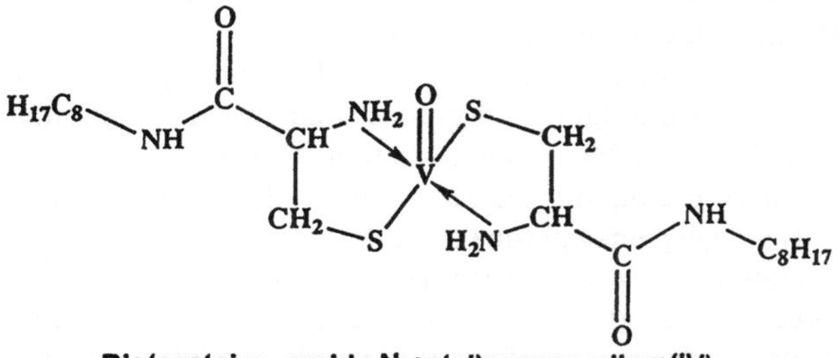

Bis(cysteine, amide N-octyl)oxovanadium(IV)

Fig. 2. Bis(N-octylcysteinamide)oxovanadium(IV) – Naglivan.

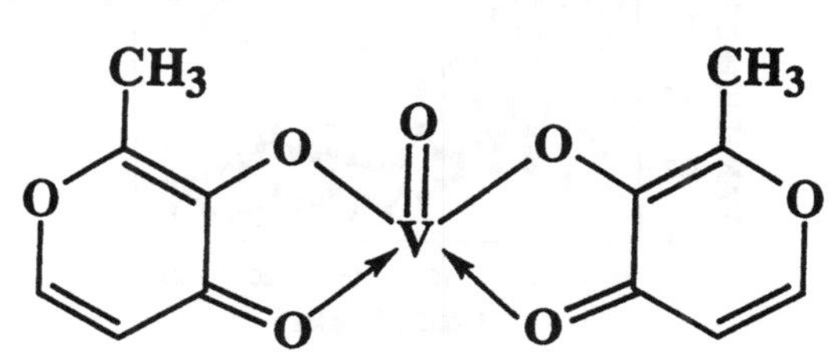

Bis(maltolato)oxovanadium(IV)

Fig. 3. Bis(maltolato)oxovanadium(IV) – BMOV.

However, the reduction in circulating plasma insulin levels in control treated animals was similar to that seen with vanadyl. Chronic BMOV treatment was effective in ameliorating the pathological abnormalities associated with STZ-induced diabetes mellitus [43].

It was observed that the diabetic treated group could be subdivided based on the glucose response to BMOV with 4/12 diabetic treated animals never demonstrating stable glucose control. There was no difference in the dose of vanadium administered between control and euglycemic diabetic treated animals whereas the uncontrolled diabetic animals had significantly elevated and highly variable levels of vanadium intake. Figure 4 shows the dose of BMOV required to maintain the euglycemic effect. This dose was 2–3 times lower than that required with either vanadate or vanadyl; 0.18 mmol/kg/day for BMOV as compared to 0.4–0.6 mmol/kg/day for vanadate or vanadyl [44].

A number of acute experiment were conducted to compare the glucose lowering properties of BMOV and vanadyl sulfate. BMOV and vanadyl sulfate were administered by oral gavage at a concentration of 175 mg/kg (0.55 and 0.82 mmol/kg, respectively) or by intraperitoneal injection at a concentration of 20 mg/kg (0.063 and 0.091 mmol/kg, respectively). BMOV was found to be 2–3 times as potent as vanadyl sulfate by either route of administration. An euglycemic response to both organic and inorganic vanadium could be observed within hours of administration. The lower plasma glucose levels were sustained with BMOV in 15–20% of animals for up to 14 weeks following a single administration. Intravenous administration of vanadium produced a plasma glucose lowering effect only when given by infusion over an extended period of time. A bolus or slow bolus injection of vanadium into the tail vein did not lower plasma glucose levels. A determination of plasma vanadium levels following a slow bolus injection (Fig. 5) suggested that the

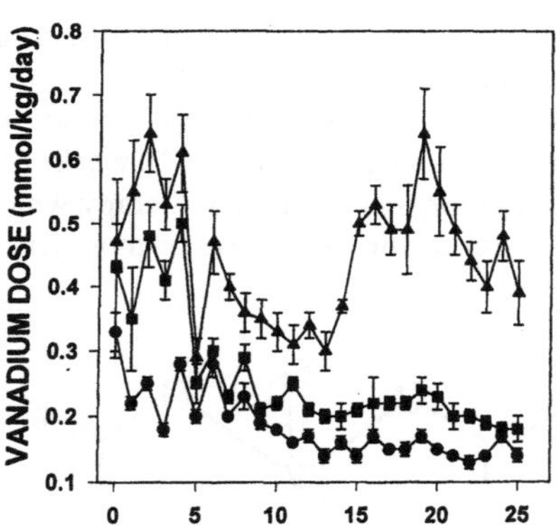

Fig. 4. The vanadium dose for control-treated (●, n = 8), diabetic-treated responders (■, n = 8) and diabetic-treated partial responders (▲, n = 4) for 25 weeks of treatment with BMOV.

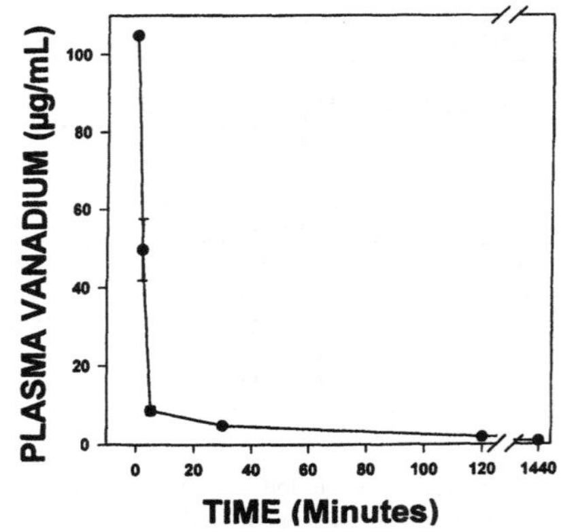

Fig. 5. Plasma vanadium levels following slow i.v. bolus administration of BMOV 32 μmol/kg over 7 to 10 minutes.

vanadium was cleared from the circulation, either by redistribution to tissues or by excretion from the body, before it could affect glucose levels [45].

Both oral dose response and intraperitoneal dose response curves have been done to compare the effectiveness of BMOV and vanadyl sulfate following administration of a single dose. At the highest doses administered BMOV produced an euglycemic response in 100% of animals treated as compared to 80% to 90% with vanadyl sulfate. The ED_{50} following oral administration indicated that BMOV was 2 times as potent as vanadyl sulfate; 0.5 mmol/kg for BMOV as compared to 0.92 mmol/kg for vanadyl sulfate. Similarly, BMOV was 3 times as potent by intraperitoneal injection [45].

BMOV administration to SHR animals had similar effects on elevated blood pressure and hyperinsulinemia as was seen with inorganic vanadium. The dose required to produce this reduction in blood pressure was 0.35–0.45 mmol/kg/day of BMOV as compared to 0.4–0.6 mmol/kg/day of vanadyl sulfate [46].

BMOV has also been used in fa/fa (fatty) Zucker rats to examine the effectiveness of organic vanadium in Type II diabetes. BMOV at a maximal concentration of 0.5 mg/mL for 14 weeks of treatment reduced plasma insulin levels from 180 μU/mL to normal (50 μU/mL) by week 4. At these concentrations BMOV did not affect body weight gain in lean controls but did significantly reduce body weight in the fatty treated group. BMOV administration at a maximal concentration of 0.2 mg/mL did not effect food and fluid intake, body weight gain or plasma cholesterol levels in fatty treated animals. At the lower concentration BMOV did significantly reduce plasma glucose and triglyceride levels. Fed plasma insulin levels were significantly reduced from 260 to140 μU/mL. An oral glucose tolerance test showed an improved glucose tolerance in fatty treated animals regardless of the con-

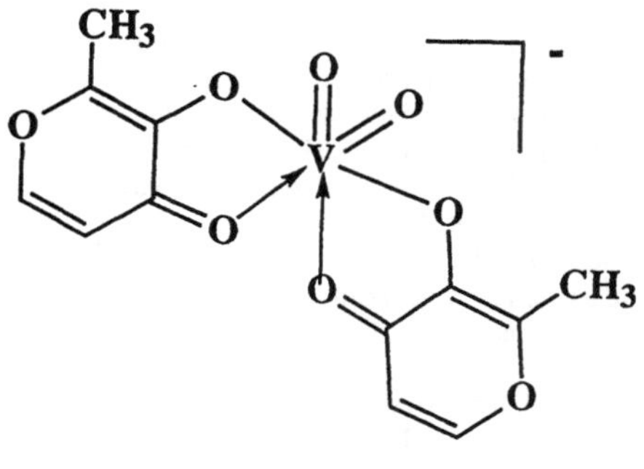

Fig. 6. Bis(maltolato)dioxovanadate(V) – BMO2V.

centration of BMOV.(unpublished observations)

Bis(maltolato)dioxovanadate(V) (BMO2V) is an anionic vanadium (V) analog of BMOV prepared in Dr. Orvig's laboratory (Fig. 6). Chronic oral administration of BMO2V in the drinking water was found to be less effective than BMOV and required higher doses of vanadium to obtain the plasma glucose lowering effects. Acute administration of BMO2V was as effective as BMOV in reducing plasma glucose levels but was not as well tolerated. All animals treated acutely with BMO2V had severe diarrhea that lasted for at least 12 h.(unpublished observations)

Bis(kojato)oxovanadium(IV) (BKOV) is a kojic acid/vanadyl complex directly analogous to BMOV that was more soluble in water than BMOV (Fig. 7). BKOV was less effective than BMOV in lowering plasma glucose levels both

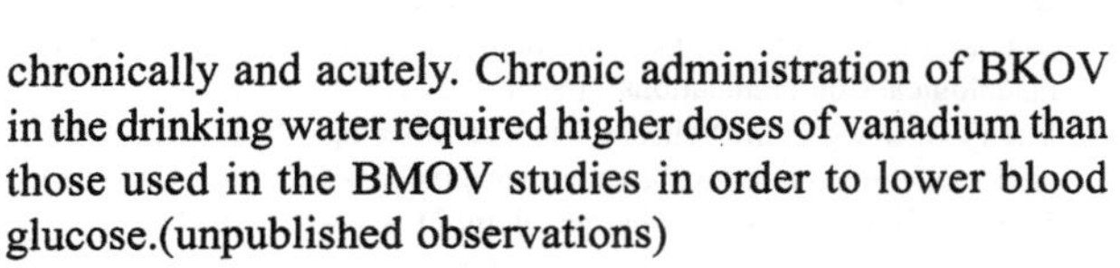

Fig. 7. Bis(kojato)oxovanadium(IV) – BKOV.

chronically and acutely. Chronic administration of BKOV in the drinking water required higher doses of vanadium than those used in the BMOV studies in order to lower blood glucose.(unpublished observations)

Summary

Both organic and inorganic vanadium are effective compounds in the treatment of hyperglycemia in diabetes mellitus and the prevention of the secondary complications associated with this disease state. Organically chelated vanadium complexes appear to be as effective insulin-mimetic agents at significantly lower doses. These compounds reduced the gastrointestinal side effects of vanadium treatment and did not effect body weight gain and food and fluid intake in control treated animals. As reported elsewhere in this same volume, early trials with vanadyl in diabetic human volunteers have shown positive findings which are consistent with the results obtained in animal studies discussed in this and other papers. There is a possible role for the use of these compounds in the treatment of diabetes mellitus, either alone or as an adjunct therapy along with insulin.

Acknowledgements

We wish to thank Sylvia Chan for secretarial assistance, Dr. Lucio Gelmini for the preparation of bis(maltolato) oxovanadium(IV), bis(maltolato)dioxovanadate(V) and bis (kojato)oxovanadium(IV) and Mary Battell for her advice and support. Studies quoted in this paper from our laboratory were supported by the Canadian Diabetes Association, the Medical Research Council of Canada, the Heart and Stroke Foundation of British Columbia and Yukon and the Natural Sciences and Engineering Research Council of Canada.

References

1. Willsky GR: Vanadium in the biosphere. In: N. Dennis Chaasteen (ed.). Vanadium in biological systems. Kluwer Academic Publishers, Dordrecht, 1990, pp 124
2. Cantley LC Jr, Josephson L, Warner R, Yanagisawa M, Lechene C, Guidotti G: Vanadate is a potent (Na,K)-ATPase inhibitor found in ATP derived from muscle. Journal of Biological Chemistry 252 (21): 7421–7423, 1977
3. Nechay BR: Mechanisma of action of vanadium. Annual Review Pharmacology and Toxicology 24: 501–524, 1984
4. Jackson T, Salhanick AI, Sparks JD, Sparks CE, Bolognino M, Amatrude JM: Insulin-mimetic effects of vanadate in primary cultures of rat hepatocytes. Diabetes 37: 1234–1240, 1988
5. Duckworth WC, Solomon SS, Liepnieks J, Hamel FG, Hand S, Peavy DE: Insulin-like effects of vanadate in isolated rat adipocytes. Endocrinology 122: 2285–2289, 1988
6. Shechter Y, Karlish SJD: Insulin-like stimulation of glucose oxidation in rat adipocytes by vanadyl (IV) ions. Nature 284: 556–558, 1980
7. Tamura S, Brown TA, Whipple JH, Fujita-Yamaguchi Y, Dubler RE, Chemg K, Larner J: A novel mechanism for the insulin-like effect of vanadate on glycogen synthase in rat adipocytes. Journal of Biological Chemistry 259 (10): 6650–6658, 1984.
8. Tamura S, Brown TA, Dubler RE, Larner J: Insulin-like effect of vanadate on adipocyte glycogen synthase and on phosphorylation of 95,000 dalton subunit of insulin receptor. Biochem Biophys Research Communication 113: 8042, 1983

9. Klarlund JK: Transformation of cells by an inhibitor of phosphatases acting on phosphotyrosineinproteins. Cell 4: 707–711, 1985
10. Mooney RA, Bordwell KL, Luhowsky S, Casnellie JE: The insulin-like effect of vanadate on lipolysis in rat adiposytes is not accompanied by an insulin-like effect on tyrosine phosphorylation. Endocrinology 124: 422–429, 1989
11. Bernier M, Laird DM, Lane MD: Effect of vanadate on the cellular accumulation of pp15, an apparent product of insulin receptor tyrosine kinase activity. Journal of Biological Chemistry 263 (27): 13626–13634, 1988
12. Fantus EG, Kadota S, Deragon G, Foster B, Posner B: Pervanadate [peroxide(s) of vanadate] mimics insulin action in rat adipbocytes via activation of the insulin receptor tyrosine kinase. Biochemistry 28: 8864–8871, 1989
13. Junod A, Lambert AE, Stauffacher W, Renold AE: Diabetogenic action of streptozotocin: relationship of dose to metabolic response. Jounal of Clinical Investigation 48: 2129–2139, 1969
14. Kaufmann F, Rodriguez RR: Subtotal pancreatectomy in five different rat strains: incidence and course of development of diabetes. Diabetologia 27: 38–43, 1984
15. Nakhooda AF, Like AA, Chappel CI, Murray FT, Marilss EB: The spontaneously diabetic Wistar rat; metabolic and morphologic studies. Diabetes 26: 100–112, 1977
16. Nakhooda AF, Like AA, Marliss EB: Diabetes mellitus in the 'BB' Wistar rat. In: E.J. Andrews, B.C. Ward and N.H. Altman (eds). Spontaneous Animal Models of Human Disease. Volume I, Academic Press, Toronto, 1979, pp 131–136
17. Ramanadham S, Mongold JJ, Brownsey RW, Cros GH, McNeill JH: Oral vanadyl sulfate in the treatment of diabetes mellitus in rats. American Joumal of Physiology 257: H904–H911, 1989a
18. Rossetti L, Giaccari A, Klein-Robbenhaar E, Vogel LR: Insulinomimetic properties of trace elements and characterization of their *in vivo* mode of action. Diabetes 39: 1243–1250, 1990
19. Battell ML, Yuen VG, McNeill JH: Treatment of BB rats with vanadyl sulfate. Pharmacological Communications 1 (4): 291–301, 1992
20. Cam MC, Pederson RA, Brownsey RW, McNeill JH: Long-term effectiveness of oral vanadyl sulfate in streptozotocin-diabetic rats. Diabetologia 36: 218–224, 1993a
21. Brichard SM, Pattier AM, Henquin JC: Long term improvement of glucose homeostatis by vanadate in obese hyperinsulinemic fa/fa rats. Endocrinology 126: 2510–2516, 1989
22. Brichard SM, Bailey CJ, Henquin JC: Marked improvement of glucose homeostasis in diabetic ob/ob mice given oral vanadate. Diabetes 39: 1326–1332, 1990
23. Blondel O, Simon J, Shevalier B, Portha B: Impaired insulin action but normal insulin receptor activity in diabetic rat liver: effect of vanadate. American Joumal of Physiology 258 (Endocrinol Metab 21): E459–E467, 1990
24. Heyliger CE, Tahiliani AG, McNeill JH: Effect of vanadate on elevated blood glucose and depressed cardiac performance of diabetic rats. Science 227: 1474–1477, 1985
25. Roschin AV, Ordizhonidze EK, Shalganova IV: Vanadium-toxicity, metabolism, carrier state. Jour Hyg Epid Microbiol Immunol 24: 377–383, 1980
26. Ramanadham S, Brownsey RW, Cros GH, Mongold JJ, McNeill JH: Sustained prevention of myocardial abnormalities in diabetic rats following withdrawal from oral vanadyl treatment. Metabolism 38: 1022–1028, 1989b
27. Pederson RA, Ramanadham S, Buchan AMJ, McNeill JH: Long-term effects of vanadyl treatment on streptozocin-induced diabetes in rats. Diabetes 38: 1390–1395, 1989
28. Cam MC, Faun J, McNeill JH: Concentration-dependent glucose-lowering effects of oral vanadyl are maintained following treatment withdrawal in streptozotocin diabetic rats. Metabolism 1994 (in press)
29. Dai S, Thompson KH, McNeill JH: One-year treatment of streptozotocin-induced diabetic rats with vanadyl sulphate. Pharmacology and Toxicology 74: 101–109, 1994a
30. Dai S, McNeill JH: One-year treatment of non-diabetic and streptozotocin-diabetic-rats with vanadyl sulphate did not alter blood pressure or haematological indices. Pharmacology and Toxicology 74: 110–115, 1994b
31. Rocchini AP: Is insulin resistance in hypertension a generalized phenomena? In: U. Smith, N.E. Bruun, T. Hedner and B. Hokfelt (eds). Hypertension is an insulin resistant disorder: genetic factors and cellular mechanisms. Excerpta Medica, 1992, pp 213–225
32. Reaven GM, Chang H: Relationship between blood pressure, plasma insulin and triglyceride concentration and insulin action in spontaneously hypertensive and Wistar Kyoto rats. American Journal of Hypertension 4: 34–38, 1991
33. Reaven GM, Chang H, Hoffman BB, Azhar S: Resistance to insulin stimulated glucose uptake in adipocytes isolated from spontaneously hypertensive rats. Diabetes 38: 1155–1160, 1989
34. Buchanan TA, Youn JH, Campese VM, Sipos GF: Enhanced glucose tolerance in spontaneously hypertensive rats: pancreatic β-cell hyperfunction with normal insulin sensitivity. Diabetes 41: 872–878, 1992
35. Bhanot S, McNeill JH: Vanadyl sulfate lowers plasma insulin and blood pressure in spontaneously hypertensive rats. Hypertension 24: 625–632, 1994a
36. Hwang IS, Ho H, Hoffman BB, Reaven GM: Fructose induced insulin resistance and hypertension in rats. Hypertension 10: 512–516, 1987
37. Bhanot S, McNeill JH, Bryer-Ash M: Vanadyl sulfate prevents fuctose-induced hyperinsulinemia and hypertension in rats. Hypertension 23: 308–312, 1994b
38. Conklin AW, Skinner CS, Feltin TL, Sanders CL: Clearance and distribution of intratracheally instilled vanadium-48 compounds in the rat. Toxicological Letters 11: 199–203, 1982
39. Underwood EJ: Vanadium In: Trace elements in human and animal nutrition. Academic Press, New York, 1977, pp 388–397
40. Cam MC, Cros GH, Serrano JJ, Lazaro R, McNeill JH: *In vivo* antidiabetic actions of naglivan, an organic vanadyl compound in streptozotocin-induced diabetes. Diabetes Research and Clinical Practice 20: 111–121, 1993b
41. McNeill JH, Yuen VG, Hoveyda HR, Orvig C: Bis(maltolato) oxovanadium(IV) is a potent insulin mimic. Journal of Medicinal Chemistry 35 (8): 1489–1491, 1992
42. Yuen VG, Orvig C, Thompson KH, McNeill JH: Improvement in cardiac dysfunction in streptozotocin-induced diabetic rats following chronic oral administration of bis(maltolato)oxovanadium(IV). Canadian Journal of Physiology and Pharmacology 71: 270–276, 1993a
43. Dai S, Yuen VG, Orvig C, McNeill JH: Prevention of diabetes-induced pathology in STZ-diabetic rats by bis(maltolato)oxovanadium(JV). Pharmacological Communications 3 (4): 311–321, 1993
44. Yuen VG, Orvig C, McNeill JH: Glucose-lowering effects of a new organic vanadium complex, bis(maltolato)oxovanadium(IV). Canadian Journal of Physiology and Pharmacology 71: 263–269, 1993b
45. Yuen VG, Orvig C, McNeill JH: Comparison of the glucose lowering properties of vanadyl sulfate and bis(maltolato)oxovanadium(IV) following acute and chronic administration. Canadian Journal of Physiology and Pharmacology 73: 55–64, 1995
46. Bhanot S, Bryer-Ash M, Cheung A, McNeill JH: Bis(maltolato)oxovanadium(IV) attenuates hyperinsulinemia and hypertension in spontaneously hypertensive rats. Diabetes 43: 857–861, 1994c

Molecular and Cellular Biochemistry **153**: 181–190, 1995.

In vivo effects of peroxovanadium compounds in BB rats

J.F. Yale, C. Vigeant, C. Nardolillo, Q. Chu, J-Z. Yu, A. Shaver[1] and B.I. Posner[2]

From The McGill Nutrition and Food Science Centre, the [1]Department of Chemistry and the [2]Polypeptide Hormone Laboratory, McGill University, Montreal, Qc, Canada, H3A 1A1

Abstract

Peroxovanadium compounds, each containing an oxo ligand, one or two peroxo anions, and an ancillary ligand in the inner coordination sphere of vanadium, were synthesized, crystallized and characterized by ^{51}V NMR as > 95% pure. They markedly decreased plasma glucose in insulin-deprived diabetic BB rats, with a nadir occurring between 60 and 100 min after intravenous, intraperitoneal or subcutaneous administration. Plasma glucose was reduced after oral administration in insulin-treated and in insulin-deprived BB rats. When compared to sodium orthovanadate, peroxovanadium compounds exhibited a markedly greater potency on a molar basis, and in relation to their toxicity. The *in vivo* potency can be predicted by the degree of phosphotyrosine phosphatase inhibition observed *in vitro*. These are the first agents other than insulin that can acutely and markedly reduce plasma glucose in hypoinsulinemic diabetic BB rats. (Mol Cell Biochem **153**: 181–190, 1995)

Key words: vanadate, peroxovanadium compounds, BB rats, hypoglycemic agents

Abbreviations: V – Vanadium; pV – peroxovanadium; IRK – insulin receptor kinase; PTP – phosphotyrosine phosphatase; BW – body weight; IV – intravenous; IP – intraperitoneal; SC – subcutaneous

Introduction

Vanadium compounds (V) have been shown to possess numerous insulin-like effects. Thus, vanadate was shown to stimulate glucose transport and oxidation in isolated rat adipocytes and skeletal muscle [1–3], and glycogen synthesis in rat liver and diaphragm [1]; and to inhibit gluconeogenesis in hepatocytes [1]. Further *in vitro* studies demonstrated that, like insulin, vanadate induced lipogenesis and inhibited lipolysis in rat adipocytes [4–6], but unlike insulin did not affect protein synthesis or degradation in isolated skeletal muscle [7]. *In vivo*, numerous studies have shown that the oral administration of vanadate and vanadyl sulphate improved hyperglycemia in insulinopenic (streptozotocin-diabetic rats [8–32], partially pancreatectomized rats [33, 34], and BB rats [35, 36]) and insulin-resistant (Zucker fa/fa rats [37], db/db mice [38], and ob/ob mice [38,39]) animal models of diabetes.

In 1987, we observed that the combination of vanadate and H_2O_2 was markedly synergistic in mimicking insulin and activating the insulin receptor kinase (IRK) in rat adipocytes [40]. In further studies we showed that pV promoted tyrosine phosphorylation and activation of the IRK [41], and correspondingly inhibited an IRK-associated phosphotyrosine phosphatase (PTP) [42] with a potency 10^3 times that of vanadate. More recently, we synthesized a number of highly pure, stable pV compounds each containing an oxo ligand, one or two peroxo anions, and an ancillary ligand in the inner coordination sphere of vanadium. The various ancillary ligands confer unique properties on each compound and markedly influence the potency of these compounds as IRK activators and PTP inhibitors in cultured cells and *in vitro* [43].

Address for offprints: J-F. Yale, McGill Nutrition and Food Science Centre, Royal Victorial Hospital, room H685, 687 Pine Avenue West, Montreal, Qc, H3A 1A1, Canada

In this paper, we describe some of the *in vivo* effects of these peroxovanadium compounds in diabetic BB rats. Our initial oral experiments revealed potent effects of peroxovanadium compounds but may have been affected by alterations in food intake. In acute parenteral studies we showed that the pV compounds had potent *in vivo* effects on blood glucose unrelated to issues of food intake.

Materials and methods

Animals

Female Wistar rats (180–210 g) were purchased from Charles River Canada Inc., St. Constant, QC. Diabetic BB rats were obtained from Dr. Pierre Thibert, Animal Resources Division, Health and Welfare Canada, Ottawa, ON. The BB rats were kept in metabolic cages, in temperature (20°C) and humidity (70%) controlled rooms, in laminar flow hoods, with 12 h lightdark cycles. They were fed *ad libitum* (Purina Rat Chow, Ralston Purina, St. Louis, MO, USA), except for periods of fasting prior to testing. The diabetic BB rats were treated with beef/pork Ultralente insulin injections. The insulin dosages were adjusted to maintain a mild glycosuria by decreasing 0.5 unit/day if negative, increasing 0.5 unit/day if > 5 g/dl and/or ketonuria, and keeping the same dose otherwise. Individual rats were only used in one experiment each.

Vanadate and peroxovanadium compounds

Sodium orthovanadate was acquired from Sigma (St. Louis, MO, USA). The following pV compounds were synthesized by Dr Jesse Ng of the Department of Chemistry, and were crystallized and characterized by ^{51}V NMR as >95% pure as previously described [43]: $K[VO(O_2)_2(phen)]\cdot 3H_2O$ or bpV(phen); $K_2[VO(O_2)_2(pic)]\cdot H_2O$ or bpV(pic), $K[VO(O_2)_2(4,7\text{-}Me_2phen)]$ or bpV(Me_2phen); $NH_4[VO(O_2)(2,6\text{-}pdc)\text{-}H_2O]\cdot H_2O$ or mpV (2,6-pdc), where phen = 1,10-phenanthroline, pic = picolinic acid anion,4,7-Me_2phen = 4,7-dimethyl-1,10-phenanthroline and 2,6-pdc = pyridine-2,6-dicarboxylic acid dianion.

The compounds were dissolved in phosphate buffered saline (PBS) and the pH was adjusted to 7.4 before administration.

Effects of oral peroxovanadium compounds on the oral glucose tolerance in wistar rats

After a 16 hour fast, Wistar rats received by gavage either bpV(phen) [20 or 40 μmol/100 g BW] or the phosphate buffer alone (n = 8), simultaneously with glucose (2.5 g/kg, 50 g/dL)). Blood samples (200 μL) were taken from the tip of the tail without anesthesia for plasma glucose measurements at 0, 30, 60, 120, 180 and 240 min.

Effects of oral peroxovanadium compounds in insulin-treated BB rats

In the days preceding the experiment, the subcutaneous (SC) insulin dosage of the BB rats was adjusted to achieve glucose levels between 10 and 20 mmol/L in the absence of ketonuria. While continuing the daily insulin injections [at 16 h], bpV(phen) [20 μmol/100 g BW BID; 8 h and 16 h] (n = 8) or the phosphate buffer alone (n = 8) were initiated on day 0. Blood samples (200 μL) were taken from the tip of the tail without anesthesia for plasma glucose measurements before and 8 and 24 h after the initiation of bpV(phen). Food was *ad libitum*.

Effects of oral peroxovanadium compounds in insulin-deprived BB rats

In the days preceding the experiments, the subcutaneous (SC) insulin dosage of the diabetic BB rats was adjusted to achieve glucose levels between 15 and 25 mmol/L in the absence of ketonuria. Insulin treatment was then stopped, followed 20 h later by the initiation of bpV(phen) by gavage at dosages of 20, 10, 5 or 2 μmol/100 g BW BID (n = 4 per group), or the phosphate buffer alone (n = 8). Blood samples (200 μL) were taken from the tip of the tail without anesthesia for plasma glucose measurements at times 0, 8, 24 and 32 h. Food was *ad libitum*.

Acute hypoglycemic effects of intravenous pV compounds in diabetic BB rats

In the days preceding the experiments, the subcutaneous (SC) insulin dosage of the BB rats was adjusted to achieve glucose levels between 15 and 25 mmol/L in the absence of ketonuria. After a 4 h fast and 20 h after their last insulin injection, the rats were lightly anesthesized with ether and received a tail-vein injection of one of the following compound: bpV(pic) [0.15 μmol/100 g BW], bpV(Me_2phen) [0.6 μmol/100 g BW], vanadate [1.2 μmol/100 g BW], or the phosphate buffered saline vehicle alone (PBS). Blood samples (200 μL) were taken from the tip of the tail at 1 and 15 min before the injection (baseline plasma glucose level), as well as every 20 min after the injection, for 180 min.

Acute effects of intraperitoneal and subcutaneous injections of peroxovanadium compounds in diabetic BB rats

After a 4 h fast and 20 h after their last insulin injection, the rats received, in the absence of anesthesia, an injection of bpV(pic) 0.4 μmol/100 g BW IP, bpV(pic) 3.6 μmol/100 g BW SC or the phosphate buffer alone. Blood samples (200 μL) were taken from the tip of the tail at 1 and 5 min before the injection (baseline plasma glucose level), as well as every 20 min after the injection, for 180 min.

Assays

The blood samples were centrifuged for 5 minutes in a Beckman Microfuge B. Plasma glucose determinations were performed on a Glucose Analyzer II (Beckman Instruments, Brea, CA).

Statistical analysis

For the time-course data, the effects of time and drug were analyzed by analysis of variance for repeated measurements. The BMDP Statistical Software (Los Angeles, CA) was used for all calculations.

Results

Effects of oral peroxovanadium compounds on the oral glucose tolerance in Wistar rats

The effects of bpV(phen) on the plasma glucose response to an oral glucose load are depicted in Fig. 1. Compared to the control animals, the rats treated with bpV(phen) 40 μmol/100 g BW had a significantly lower increase in plasma glucose over the 3 hours of the test ($p < 0.0001$).

Effects of oral peroxovanadium compounds in insulin-treated BB rats

When bpV(phen) [20 μmol/100 g BID] was administered to insulin-treated BB rats (Fig. 2), the first dose, administered in the morning 16 h after the last insulin dose, had no major effect over the next 8 h. The second dose, administered concomitantly with the daily insulin injection, produced dramatic decreases in plasma glucose over the next 16 h, causing hypoglycemia in all treated rats (ranging from 1.2–3.0 mmol/l) with death occurring in 3/8 bpV(phen)-treated rats.

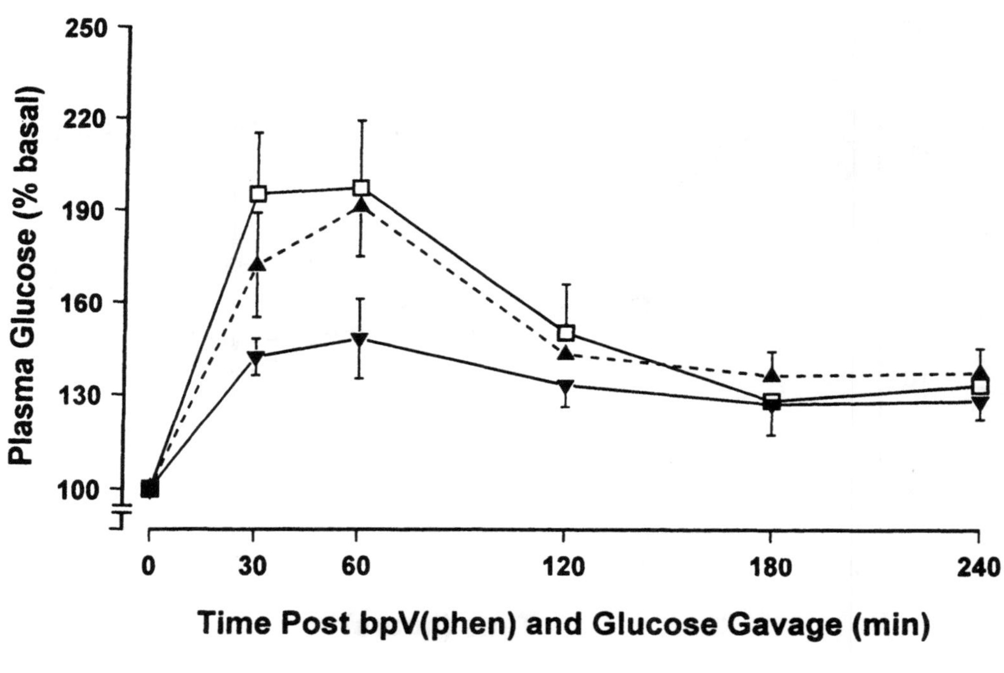

Fig. 1. Effects of oral bpV(phen) on oral glucose tolerance test. bpV(phen) 20 μmol/100 g BW (▲---▲, n = 8), 40 μmol/100 g BW (▼——▼, n = 8) or the phosphate buffer alone (□——□, n = 8) were administered by gavage to Wistar rats, simultaneously with oral glucose (2.5 g/kg). Data presented as mean ± SEM. Plasma glucose concentrations at time 0 were bpV(phen) 20 μmol/100 g: 5.3 ± 0.3 mmol/L, bpV(phen) 40 μmol/100 g: 5.3 ± 0.2 mmol/L and phosphate buffer alone: 5.7 ± 0.4 mmol/L.

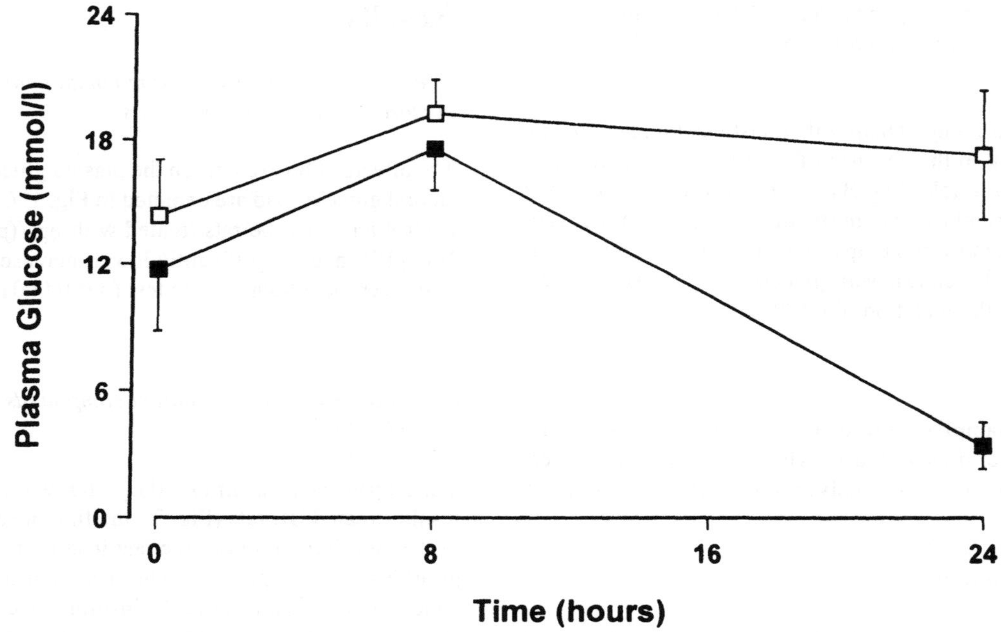

Fig. 2. Effects of oral bpV(phen) on plasma glucose in insulin-treated diabetic BB rats. bpV(phen) 20 µmol /100 g BID (■——■, n = 8) or the phosphate buffer alone (□——□, n = 8) were administered by gavage to diabetic BB rats at times 0 (8 h am) and 8 h (16 h). The BB rats received their usual dose of sub-cutaneous insulin at 16 h. Data presented as mean ± SEM.

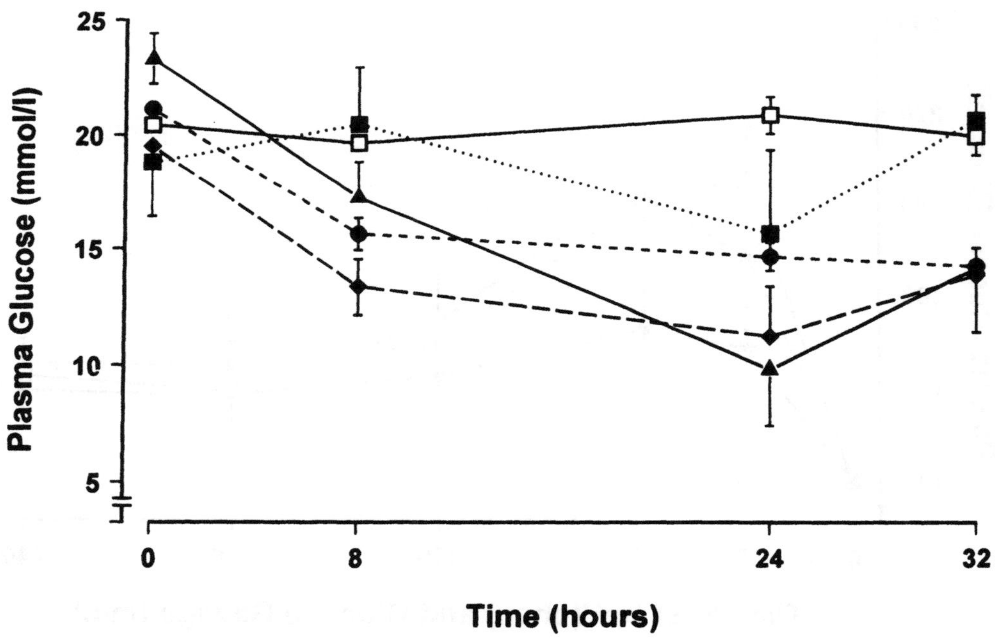

Fig. 3. Effects of oral bpV(phen) on plasma glucose in insulin-deprived diabetic BB rats. bpV(phen) 2 (■ . . . ■, n = 4), 5 (●---●, n = 4), 10 (♦---♦, n = 4) or 20 (▲——▲, n = 4) µmol/100 g BW BID or the phosphate buffer alone (□——□, n = 12) were administered by gavage at times 0 (8 h am) and 8 h (16 h) in diabetic BB rats. The last insulin injection was administered 16 h before time 0. Data presented as mean ± SEM.

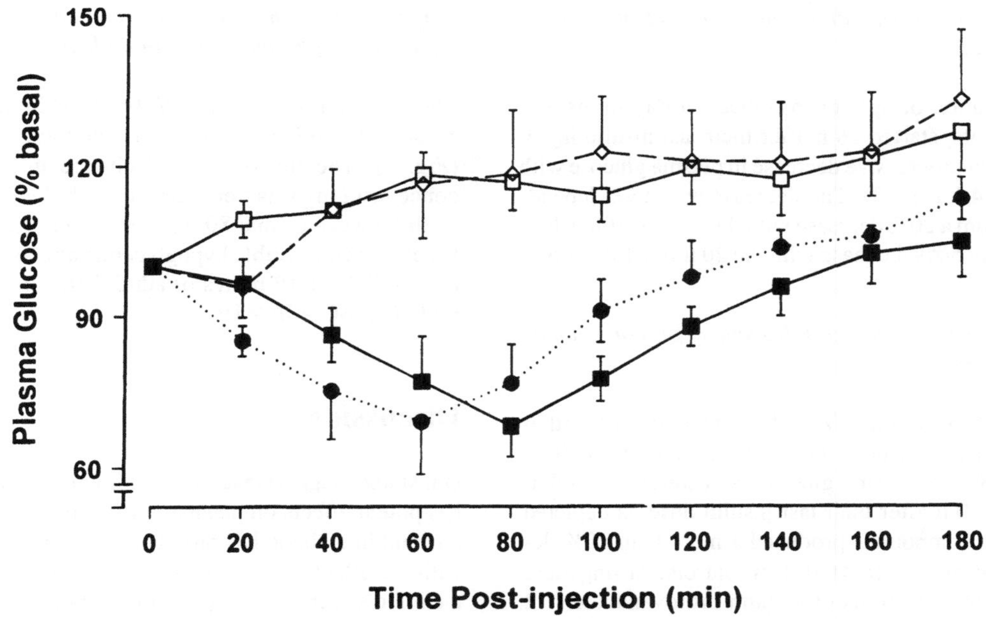

Fig. 4. Time-course of the effects of intravenous peroxovanadium compounds in diabetic BB rats. Acute effect of a bolus IV injection of pV compounds, vanadate or the buffer alone on plasma glucose concentrations in diabetic BB rats 20 h after their last SC insulin dose. PBS alone as controls (□——□, n = 10), vanadate 1.2 μmol/100 g BW (◊---◊, n = 6), bpV(pic) 0.15 μmol/100 g BW (■——■, n = 8), bpV(Me_2phen) 0.6 μmol/100 g BW (●····●, n = 3). Plasma glucose concentrations at the time of injection were: PBS: 16.7 ± 1.5; Vanadate: 19.2 ± 1.7; bpV(pic): 16.6 ± 1.4 and bpV (Me_2phen): 17.8 ± 2.1 mmol/L.

Data shown as mean (± SEM) percent of plasma glucose concentrations at the time of injection for each rat.

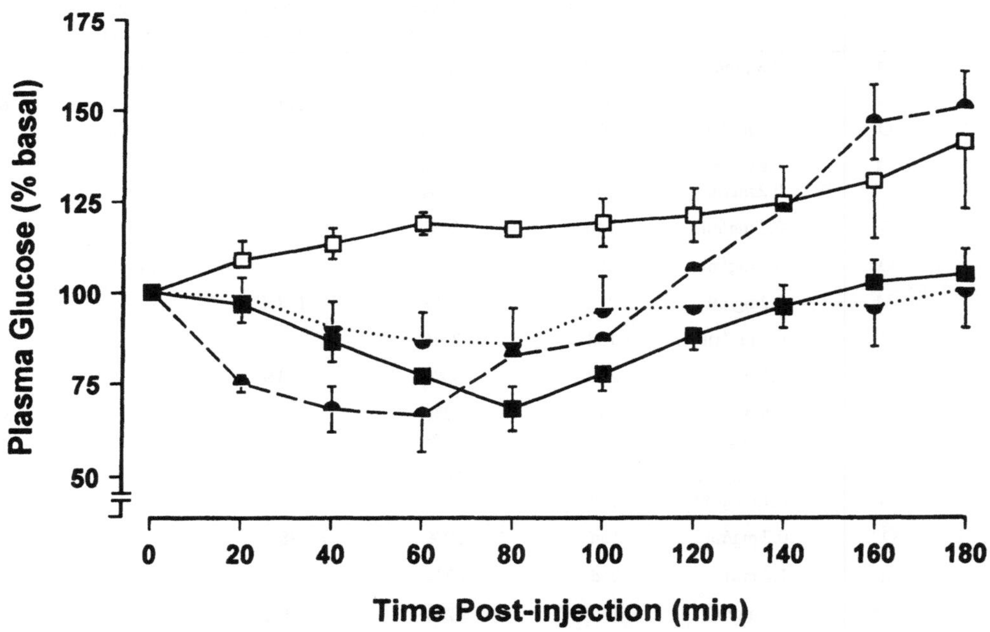

Fig. 5. Effects of different parenteral routes of administration of bpV(pic) on plasma glucose levels in BB rats. Acute effect of a single injection of bpV(pic) on plasma glucose concentrations in diabetic BB rats 20 h after their last SC insulin dose. The plasma glucose concentrations at the time of injection are indicated for each dose (mmol/L, ± SEM), followed by the number of rats studied at each dose. PBS IP as controls (□——□, 13.4 ± 1.7, n =5). bpV (pic) 0.15 μmol/100 g BW IV (■——■, 16.6 ± 1.4 , n = 7), 0.4 μmol/100 g BW IP (◓---◓, 17.7 ± 0.9, n = 6), and 3.6 μmol/100 g BW SC (◒...◒, 20.0 ± 1.0, n = 6). Data shown as mean (± SEM) percent of plasma glucose concentrations at the time of injection for each rat.

Effects of oral peroxovanadium compounds in insulin-deprived BB rats

The administration of bpV(phen) twice per day to insulin-deprived BB rats, started 16 h after their last insulin injection, produced a progressive decrease in plasma glucose with a nadir after 24 h (Fig. 3). The decrease was dose-dependent, ranging from a 20% decrease with the 2 μmol/100 g dose to a greater than 50% decrease with the 20 μmol/100 g dose.

Acute hypoglycemic effects of intravenous pV compounds in diabetic BB rats

The effect of a single IV bolus injection of either bpV(Me_2phen) [0.6 μmol/100 g BW] or bpV(pic) [0.15 μmol/100 g BW] on plasma glucose concentrations in diabetic BB rats, 20 h after their last insulin dose, is depicted in Fig. 4. Both compounds produced a more than 20% decrease in plasma glucose ($p < 0.001$ vs controls), lasting more than 2 h. A single IV bolus of sodium orthovanadate (1.2 μmol/100 g BW) did not produce a hypoglycemic effect in diabetic BB rats.

Acute effects of intraperitoneal and subcutaneous injections of pV compounds in diabetic BB rats

The injection of bpV(pic) IV, IP or SC in diabetic BB rats induced a significant ($p < 0.01$) decrease in plasma glucose (Fig. 5). The time-course of change in plasma glucose concentrations was very similar with the 3 routes, with a nadir between 60 and 80 min. However, the doses required to achieve comparable hypoglycemic effects were higher using the SC and IP routes of administration than via the IV route (i.e. SC > IP > IV).

Discussion

Our studies suggest that the oral administration of bpV(phen) has potent effects on plasma glucose in both normal Wistar rats and in diabetic BB rats. These results are in agreement with the effects of vanadium previously reported, although the effects seem more rapid with the peroxovanadium compounds. Studies on Vanadate and Vanadyl sulphate, mostly conducted in streptozotocin-diabetic rats (Tables 1 and 2)

Table 1. Effects of vanadate in stroptozotocin-diabetic rats

		Dose	Duration	Plasma glucose*	Plasma insulin*	Body weight*	Food intake*
Heylinger *et al*[8]	O	0.8 mg/ml	42d	43%	141%	110%	
Brichard *et al*[15]	O	0.2 mg/ml	60d	57%	200%	89%	61%
	O	0.5 mg/ml	60d	37%	280%	89%	44%
Meyerovitch *et al*[14]	M	0.8 mg/ml	4d	18%			
		0.2 mg/ml	25d	35%			
Gil *et al*[9]	O	0.7 mg/ml	14d	26%		103%	42%
Sekar *et al*[30]	O	0.3 mg/ml	60d	37%		148%	
Blondel *et al*[10]	M	0.2mg/ml	20d	31%	118%		55%
Bollen *et al*[28]	O	0.7 mg/ml	18d	25%			
Pugazhenthi *et al*[13]	O	0.5 mg/ml	35d	43%	183%		53%
Domingo *et al*[21]	O	0.5 mg/ml	14d	62%			54%
	M	0.2 mg/ml	14d	56%			
Domingo *et al*[22]	O	0.23 mg/ml	28d	50%			56%
	M	0.15 mg/ml	28d	37%			44%
Valera *et al*[23]	O	0.7 mg/ml	15d	39%	88%		
Oster *et al*[24]	M	1.2 mM	26d	63%		96%	35%
Madsen *et al*[25]	VP	0.1 mg/ml	14d	72%	86%	132%	67%
Malabu *et al*[29]	M	1.0 mg/ml	21d	43%	131%	95%	60%

M : Metavanadate
O : Orthovanadate
VP : Vanadium Pentoxide
* : Expressed in % of that observed in the untreated control diabetic rats

Table 2. Effects of vanadyl sulphate in stroptozotocin-diabetic rats

	Dose	Duration	Plasma glucose*	Plasma insulin*	Body weight*	Food intake*
Ramanadham *et al*[12]	1.0 mg/ml	70d	38%	133%	114%	
Ramanadham *et al*[35]	0.25 mg/ml	14d	63%		92%	
	0.50 mg/ml	14d	47%		88%	
	0.75 mg/ml	14d	33%		92%	
	1.00 mg/ml	14d	24%		81%	
Mongold *et al*[31]	0.25 mg/ml	39d	75%		87%	89%
	0.50 mg/ml	39d	70%		78%	70%
	0.75 mg/ml	39d	45%		79%	62%
	1.00 mg/ml	39d	41%		93%	66%
Domingo *et al*[21]	1.10 mg/ml	14d	30%			51%
Domingo *et al*[22]	0.31 mg/ml	28d	64%			80%
Venkatesan *et al*[26]	0.5 mg/ml	12d	57%	116%		65%
Thompson *et al*[27]	0.5–1.0 mg/ml	28d	49%		89%	54%
Cam *et al*[32]	0.75 mg/ml	150d	54%	123%	100%	54%

* : Expressed in % of that observed in the untreated control diabetic rats

have previously shown impressive decreases of plasma glucose to values between 25 and 75% of the starting glucose levels, with minor side effects such as weight loss and decreased food intake. Recent reports have raised doubts about the significance of the results of these *in vivo* studies by revealing potential confounding factors as detailed below.

It was first shown that euglycemia persisted for more than 13 weeks after the end of a 3 week course of vanadyl sulphate in streptozotocin-diabetic rats [16, 38, 44], suggesting that the long term effects seen with chronic vanadyl sulphate treatment might be related to a regeneration of beta cells. Brichard had previously reported significantly greater pancreatic insulin levels in streptozotocin-diabetic rats after 60 days of vanadate treatment [15], and most studies have revealed higher plasma insulin levels with Vanadate treatment (Tables 1 and 2). Chicheportiche *et al* [45] and others had previously observed that euglycemia (for example through islet transplantation) could lead to regeneration of beta cells in streptozotocin-diabetic rats. Obih *et al* showed that vanadyl sulphate, started before streptozotocin, could prevent its diabetogenic effects [46]. To avoid these confounding factors we performed our studies on diabetic BB rats, whose diabetes is felt to be due to a permanent autoimmune destruction of the β cells [47].

Malabu *et al* [36] have suggested that most of the effects of vanadate might be related to the decreased food intake caused by vanadate. In untreated diabetic rats pair-fed to match the vanadate treated animals there was identical improvement in blood glucose values. Madsen *et al* [32] observed that oral vanadate reduces Na^+ dependent glucose transport in rat small intestine, raising the possibility that some of the effects might be related to slower glucose absorption. These effects could also explain the effects we have seen in our oral studies. A slower absorption of glucose through the gut could explain the lower glucose increase seen in bpV(phen)-treated Wistar rats after the ingestion of oral glucose. Both a slower absorption of glucose and a decrease in food intake could also explain the dramatic decrease in plasma glucose seen in bpV(phen)treated diabetic BB rats receiving their usual insulin dosages, and the decrease observed in the insulin-deprived bpV(phen)-treated diabetic BB rats compared to untreated animals. In order to circumvent these factors, we studied the acute effects of peroxovanadium compounds administered parenterally to fasted diabetic BB rats.

In this paper, we report the potent and acute hypoglycemic effects of two peroxovanadium compounds (bpV(pic) and bpV(Me_2phen) in diabetic BB rats when administered intravenously. Their potency, on a molar basis, is also much greater than that of sodium orthovanadate. This is in agreement with our recent finding that pV compounds are 50–1000 fold more effective than vanadate in reproducing insulin actions *in vitro* [41–43, 48]. These potent effects were obtained in the absence of acute mortality at the dosages used. Our data show a marked and acute reduction in plasma glucose levels following the IV injection of pV compounds to fasted rats. This establishes that these agents can have insulin-mimetic effects *in vivo* independent from effects on food intake or absorption.

Whether vanadate can have hypoglycemic potential in the absence of circulating insulin has not previously been demonstrated. The hypoinsulinemic models of diabetes used have included streptozotocin-diabetic rats, partially pancreatectomized rats and BB rats. Except for the BB rats, these models involve a partial destruction of pancreatic β cell mass and the animals do not develop ketoacidosis even in the absence of insulin treatment. As in human insulindependent diabetes, the spontaneous diabetes of the BB rat is characterized by the acute appearance at a young age (60–140 days) of hyperglycemia, glycosuria, ketonuria and weight loss, leading to a rapid death unless insulin therapy is instituted. Battell *et al.* [24] reported a decrease in the insulin requirements of BB rats treated with oral vanadyl sulphate but could not withdraw insulin treatment in these rats. We studied the hypoglycemic effects of pV compounds in these insulinopenic rats. To ensure their insulin-deprived state, the animals were used 16 h after their last insulin injection, at which time their plasma glucose levels exceeded 15 mmol/L. The pV compounds are the first agents, with the exception of insulin, that can acutely decrease plasma glucose in insulin-deprived diabetic BB rats (Fig. 4). This acute decrease cannot be related to a decrease in food intake or intestinal glucose absorption since both control and peroxovanadium injected animals were fasted during the course of the study.

The compound bpV(pic) was effective when administered IV, IP or SC, as shown in Fig. 5. Compared to the IV route, the IP route required slightly greater doses. In contrast, the dose of SC injections was 24 times greater than that required when the IV route was used. While the IP and SC routes required higher doses than the IV administration to produce the same effects, their time-course of action was remarkably similar with a nadir between 60 and 100 min (Fig. 5). This suggests rapid absorption of these compounds from their sites of injection.

We recently reported that the peroxovanadium compounds activated the insulin receptor kinase (IRK), and that this capacity appeared to derive from an inhibition of the phosphotyrosine phosphatase (PTP). The potency of the peroxovanadium compounds that we studied *in vivo* to inhibit PTP are displayed in Fig. 6 along with the minimally effective dosage of each compound (defined as the smallest dose that produced a mean decrease in plasma glucose ≥ 20% in diabetic BB rats). A significant correlation ($r = 0.97$, $p < 0.01$) was found between the capacity of the compounds to inhibit PTP and their *in vivo* hypoglycemic potency. In contrast, no correlation was found between the IRK activation and the *in vivo* potency (data not shown). The assessment of PTP inhibition, therefore, may be a good predictor of the *in vivo* potency of these compounds.

In summary, we report the potent acute hypoglycemic effects of peroxovanadium compounds administered parenterally in insulin-depleted diabetic BB rats. The *in vivo* potency may be predicted from the degree of phosphotyrosine phosphatase inhibition observed *in vitro*. Potent effects were observed after oral administration, although different mechanisms such as inhibition of food intake or delayed glucose absorption may also have played a role. More studies are

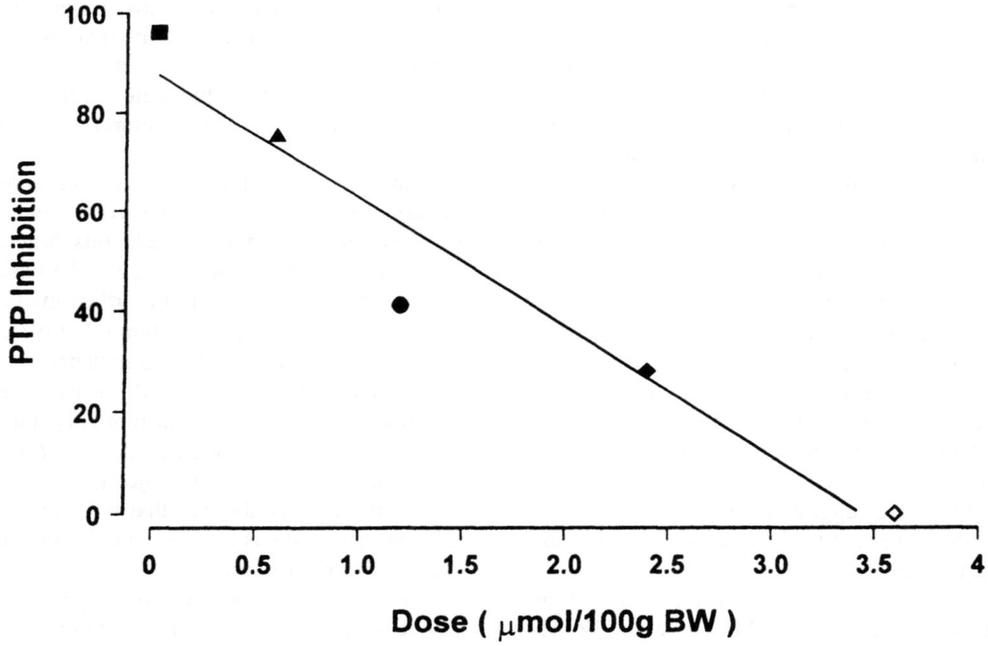

Fig. 6. Correlation between the hypoglycemic activities of peroxovanadium compounds and their effects on phosphotyrosine phosphatase inhibition. Correlation between the peroxovanadium compound dose required to induce a 20% decrease in plasma glucose levels in diabetic BB rats and the phosphotyrosine phosphatase inhibition as described elsewhere. ■: bpV(pic); ▲: bpV(phen); ●: bpV(Me_2phen); ♦: mpV(2,6-pdc); ◊: Vanadate.

required on the effects of these agents after oral administration and on their long term toxicity.

Acknowledgements

The authors express their profound appreciation to Dr Jesse Ng for preparing the pV compounds.

This study was supported by grants from the Medical Research Council of Canada (MA9275, J.-F.Y.; and MT-4182, B.I.P.) and from Marion Merrell Dow Canada, Laval, QC.

References

1. Tolman EL, Barris E, Burns M, Pansini A, Partridge R: Effects of vanadium on glucose metabolism *in vitro*. Life Sciences 25: 1159–1164, 1979
2. Dubyak GR, Kleinzeller A: The insulin-mimetic effects of vanadate in isolated rat adipocytes. J Biol Chem 255: 5306–5312, 1980
3. Schechter Y, Karlish SJD: Insulin-like stimulation of glucose oxidation in rat adipocytes by vanadyl (iv) ions. Nature (Lond) 284: 556–558, 1980
4. Degani H, Gochin M, Karlish SJD, Shechter Y: Electron paramagnetic resonance studies and insulin-like effects of vanadium in rat adipocytes. Biochemistry 20: 5795–5799, 1989
5. Duckworth WC, Solomon SS, Liepnieks J, Hamel FG, Hand S, Peavy DE: Insulin-like effects of vanadate in isolated rat adipocytes. Endocrinology 122: 2285–2289, 1988
6. Mooney RA, Bordwell KL, Luhowskyj S, Casnellie JE: The insulin-like effect of vanadate on lipolysis in rat adipocytes is not accompanied by an insulin-like effect on tyrosine phosphorylation. Endocrinology 124: 422–429, 1989
7. Clark AS, Fagan JM, Mitch WE: Selectivity of the insulin-like actions of vanadate on glucose and protein metabolism in skeletal muscle. Biochem J: 232: 273–276, 1975
8. Heylinger CE, Tahiliani AG, McNeill JH: Effect of vanadate on elevated blood glucose and depressed cardiac performance of diabetic rats. Science 277: 1474–1477, 1985
9. Gil J, Miralprix M, Carreras J, Bartrons R: Insulin-like effects of vanadate on glucokinase activity and fructose 2,6-biphosphate levels in the liver of diabetic rats. J Biol Chem 263: 1868–1871, 1988
10. Blondel O, Bailbe D, Portha B: *In vivo* insulin resistance in streptozotocin-diabetic rats-evidence for reversal following oral vanadate treatment. Diabetologia 32: 185–190, 1989
11. Bendayan M, Gingras D: Effects of vanadate administration on blood glucose and insulin levels as well as on the exocrine pancreatic function in streptozotocin-diabetic rats. Diabetologia 32: 561–567, 1989
12. Ramanadham S, Mongold JJ, Brownsey RW, Cros GH, McNeill JH: Oral vanadyl sulphate in treatment of diabetes mellitus in rats. Am J Physiol 257: H904–H911, 1989
13. Pugazhenti S, Khandelwal RL: Insulinlike effects of vanadate on hepatic glycogen metabolism in nondiabetic and streptozotocin-induced diabetic rats. Diabetes 39: 821–827, 1990
14. Meyerovitch J, Farfel Z, Sack J, Shechter Y: Oral administration of vanadate normalizes blood glucose levels in streptozotocin-treated rats. J Biol Chem 262: 6658–6662, 1987
15. Brichard SM, Okitolonda W, Henquin JC: Long term improvement of glucose homeostasis by vanadate treatment in diabetic rats. Endocrinology 123: 2048–2053, 1988
16. Pederson RA, Ramanadham S, Buchan AMJ, McNeill J: Long-term effects of vanadyl treatment on streptozotocin-induced diabetes in rats. Diabetes 38: 1390–1395, 1989
17. Bollen M, Miralpeix M, Ventura F, Toth B, Bartrons R, Stalmans W: Oral administration of vanadate to streptozotocin-diabetic rats restores the glucose-induced activation of liver glycogen synthase . Biochem J 267: 269–271, 1990
18. Sakurai H, Tsuchiya K, Nukatsuka M, Sofue M, Kawada J: Insulin-like effects of vanadyl ion on streptozotocin-induced diabetic rats. J of Endocrinol 126: 451–459, 1990
19. Serradas P, Bailbe D, Blondel O, Portha B: Abnormal b-cell function in rats with non-insulin-dependent diabetes induced by neonatal streptozotocin: effect of *in vivo* insulin, phlorizin, or vanadate treatments. Pancreas 6: 54–62, 1991
20. Ramanadham S, Brownsey RW, Cros GH, Mongold JJ, McNeill JH: Sustained prevention of myocardial and metabolic abnormalities in diabetic rats following withdrawal from oral vanadyl treatment. Metabolism 38: 1022–1028, 1989
21. Domingo JL, Gomez M, Llobet JM, Corbella J, Keen CL: Improvement of glucose homeostasis by oral vanadyl or vanadate treatment in diabetic rats is accompanied by negative side effects. Pharmacology and Toxicology 68: 249–253, 1991
22. Domingo JL, Gomez M, Llobet JM, Corbella J, Keen CL: Oral vanadium administration to streptozotocin-diabetic rats has marked negative side-effects which are independent of the form of vanadium used. Toxicology 66: 279–287, 1991
23. Valera A, Rodriguez-Gil JE, Bosch F: Vanadate Treatment Restores the Expression of Genes for Key Enzymes in the Glucose and Ketone Bodies Metabolism in the Liver of Diabetic Rats. J Clin Invest 92: 4–11, 1993
24. Oster MH, Llobet JM, Domingo JL, Gedrman JB, Keen CL: Vanadium treatment of diabetic Sprague-Dawley rats results in tissue vanadium accumulation and pro-oxidant effects. Toxicology 83: 115–130, 1993
25. Madsen KL, Porter VM, Fedorak RN: Oral vanadate reduces Na(+)-dependent glucose transport in rat small intestine. Diabetes 42: 1126–32, 1993
26. Venkatesan N, Avidan A, Davidson MB: Antidiabetic Action of Vanadyl in rats independent of *in vivo* insulin-receptor kinase activity. Diabetes 40: 492–498, 1991
27. Thompson KH, Leichter J, McNeill JH: Studies of Vanadyl Sulfate as a Glucose-Lowering Agent in STZ-Diabetic rats. Biochem and Biophys Research Comm 197: 1549–1555, 1993
28. Bollen M, Miralpeix M, Ventura F, Toth B, Bartrons R, Stalmans W: Oral administration of vanadate to streptozotocin-diabetic rats restores the glucose-induced activation of liver glycogen synthase. Biochem J 267: 269–271, 1990
29. Malabu UH, Dryden S, McCarthy HD, Kilpatrick A, Williams G: Effects of chronic vanadate administration in the STZ-induced diabetic rat: the antihyperglycemic action of vanadate is attributable entirely to its suppression of feeding. Diabetes 43: 9–15, 1994
30. Sekar N, Kanthasany A, William S, Subramanian S, Govindasamy S: Insulin actions of vanadate in diabetic rats. Pharmacological Research 22: 207–217, 1989
31. Mongold JJ, Cros GH, Vian L, Tep A, Ramanadham S, Siou G, Diaz J, McNeill JH, Serrano JJ: Toxicological Aspects of Vanadyl Sulphate on Diabetic Rats: Effects on Vanadium Levels and Pancreatic b-Cell Morphology. Pharmacology and Toxicology 67: 192–198, 1990
32. Cam MC, Pederson RA, Brownsey RW, McNeill JH: Long-term effectiveness of oral vanadyl sulphate in streptozotocin-diabetic rats. Diabetologia 36: 218–224, 1993
33. Rossetti L, Laughlin MR: Correction of chronic hyperglycemia with vanadate, but not with phlorizin, normalizes *in vivo* glycogen repletion and *in vitro* glycogen synthase activity in diabetic skeletal muscle. J Clin Invest 84: 892–899, 1989

34. Rossetti L, Giaccari A, Klein-Robbenhaar E, Vogel LR: Insulinomimetic properties of trace elements and characterization of their *in vivo* mode of action. Diabetes 39: 1243–1250, 1990
35. Ramanadham S, Cros GH, Mongold JJ, Serrano JJ, McNeill JH: Enhanced *in vivo* sensitivity of vanadyl-treated diabetic rats to insulin. Can J Physiol Pharmacol 68: 486–491, 1990
36. Battell ML, Yuen VG, McNeill JH: Treatment of BB rats with vanadyl sulphate. Pharmacology Communications 1: 291–301, 1992
37. Brichard SM, Pottier AM, Henquin JC: Long term improvement of glucose homeostasis by vanadate in obese hyperinsulinemic *fa/fa* rats. Endocrinology 125: 2510–2516, 1989
38. Meyerovitch J, Rothenberg P, Shechter Y, Bonner-Weir S, Kahn CR: Vanadate normalizes hyperglycemia in two mouse models of non-insulin-dependent diabetes mellitus. J Clin Inv 87: 1286–1294, 1991
39. Brichard SM, Bailey CJ, Henquin JC: Marked improvement of glucose homeostasis in diabetic *ob/ob* mice given oral vanadate. Diabetes 39: 1326–1332, 1990
40. Kadota S, Fantus IG, Deragon G, Guyda HJ, Posner BI: Stimulation of insulin-like growth factor II receptor binding and insulin receptor kinase activity in rat adipocytes. J Biol Chem 262: 8252–8256, 1987
41. Fantus IG, Kadota S, Deragon G, Foster B, Posner BI: Pervanadate (peroxide(s) of vanadate) mimics insulin action in rat adipocytes via activation of the insulin-receptor tyrosine kinase. Biochemistry 28: 8864–8871, 1989
42. Faure R, Baquiran G, Bergeron JJM, Posner BI: The dephosphorylation of insulin and epidermal growth factor receptors. J Biol Chem 267: 11215–11221, 1992
43. Posner BI, Faure R, Burgess JW, Bevan AP, Lachance D, Zhang-Sun G, Fantus IG, Ng JB, Hall DA, Soo Lum B, Shaver A: Peroxovanadium compounds: A new class of potent phosphotyrosine phosphatase inhibitors which are insulin mimetics. J Biol Chem 269: 4596–4604, 1994
44. Cros G, Mongold JJ, Serrano JJ, Ramanadham S, McNeill JH: Effects of vanadyl derivatives on animal models of diabetes. Molecular and Cellular Biochemistry 109: 163–166, 1992
45. Chicheportiche D, Darquy S, Lepeintre J, Khosrovani C, Rouchette J, Capron F, Halban P, Reach G: Implantation d'îlots de porc microencapsulés chez le rat diabétique. Diabete et Metabolisme 15: IV, 1989 (abstract)
46. Obih P, Wingard M, Olubadewo J. Effectiveness of alzet minipump administration of vanadyl sulphate in prevention of STZ-induced diabetes. FASEB J 6: A1591, 1992 (abstract)
47. Yale JF, Grose M, Marliss EB: Pathogenesis of the Type I Diabetes Mellitus of the BB Rat. In: F. Labrie and L. Proulx (eds). Endocrinology, Exerpta Medical International Congress Series #655, Amsterdam, pp 89–91, 1984
48. Kadota S, Fantus IG, Deragon G, Guyda HJ, Hersh B, Posner BI: Peroxide(s) of vanadium: a novel and potent insulin-mimetic agent which activates the insulin receptor kinase. Biochem. Biophys Res Comm 147: 259–266, 1987

Molecular and Cellular Biochemistry **153**: 191–195, 1995.

Long-term antidiabetic activity of vanadyl after treatment withdrawal: Restoration of insulin secretion?

Gerard H. Cros[1], Margaret C. Cam[2], Jean-Jacques Serrano[1], Gerard Ribes[3] and John H. McNeill[2]

[1]*Laboratoire de Pharmacologie, Faculte de Pharmacie, F-34060 Montpellier France;* [2]*Department of Pharmacology and Toxicology, Faculty of Pharmaceutical Sciences, The University of British Columbia, Vancouver BC, V6T 1Z3, Canada;* [3]*Laboratoire de Pharmacologie, Faculté de Médecine and UMR CNRS 9921, F-34060 Montpellier, France*

Abstract

In its vanadate (V^{5+}) or vanadyl (V^{4+}) forms, vanadium has been demonstrated to possess antidiabetic activity. Oral treatment of streptozotocin (STZ)-diabetic animals with either form is associated with correction of hyperglycemia, and prevention of diabetes-induced complications, although weight gain is unaffected. Vanadium treatment of non-diabetic animals lowers plasma insulin levels by reducing insulin demand, as these animals remain normoglycemic. These results suggest that vanadium has *in vivo* insulin-mimetic or insulin-enhancing effects, in agreement with several *in vitro* observations.

Chronic treatment with vanadium has also been shown to result in sustained antidiabetic effects in STZ-diabetic animals long after treatment has ceased. Thus, at 13 weeks after withdrawal from treatment, corrected animals had normalized glucose and weight gain, and improved basal insulin levels. In addition, near-normal glucose tolerance was found despite an insignificant insulin response. Since vanadium accumulates in several tissue sites (e.g. bone, kidney) when pharmacological doses are administered, it is possible that stored vanadium may be important in maintaining near-normal glucose tolerance at least in the short-term following withdrawal from treatment. Recently, following withdrawal of vanadyl treatment up to 30 weeks, diabetic animals which had remained normoglycemic and had normalized glucose tolerance showed improvements in plasma insulin levels both in the basal state and in response to oral glucose, as compared to those which had reverted to hyperglycemia. The observed significant improvements in insulin capacity over the long-term (>3 months) suggests that a restored and/or preserved insulin secretion may be essential for maintained reversal of the diabetic state over a prolonged period after treatment is withdrawn. (Mol Cell Biochem **153**: 191–195, 1995)

Key words: vanadyl, glucose, glucose-tolerance, vanadium stores, pancreatic insulin secretion

Introduction

Oral administration of vanadium as either vanadate or vanadyl has been demonstrated to induce normoglycemia in the STZ-diabetic rat [1, 2], and the genetically obese db/db mouse [3], as well as to reduce the insulin requirement in the spontaneously diabetic BB rat [4, 5]. In addition, long-term correction of diabetes has been observed after treatment is withdrawn, thus raising the possibility that vanadyl treatment may also induce the preservation or restoration of endogenous insulin secreting capacity. The aim of the present review article is to discuss the possible mechanism(s) of this apparent diabetes-reversing property, and to suggest directions for future research in order to assess its potential therapeutic relevance.

Address for offprints: G. Cros, Laboratoire de Pharmacologie, Faculté de Pharmacie, F-34060 Montpellier, France

Sustained post-withdrawal antidiabetic effects of vanadyl

Effects after short-term (3-week) vanadyl treatment

The sustained 'post-withdrawal' effects of vanadyl sulfate ($VOSO_4$) was first demonstrated by Ramanadham *et al.* [6]. In this study, 3-day STZ-diabetic animals were treated with $VOSO_4$ at various concentrations (0.25, 0.50, 0.75 and 1 mg/ml) in the drinking water for three weeks. The percentage of animals that responded to euglycemia was lowest in the 0.25 mg/ml group (38%), while all diabetic animals treated with higher concentrations became euglycemic. Surprisingly, at 13 weeks following withdrawal from treatment, it was found that while only 24% of the animals which had responded to euglycemia had subsequently reverted to hyperglycemia, some diabetic animals remained normoglycemic and glucose tolerant (25, 43, 83 and 60% for 0.25, 0.50, 0.75 and 1 mgiml, respectively). At this time, various parameters classically altered in STZ-diabetes were normalized or prevented: weight gain, and plasma insulin, T_4, triglyceride, cholesterol and phospholipid levels were restored. As well, isolated working heart function and adipose lipolytic rates were normal.

However, there were some discrepancies in the characterization of the pancreatic β-cells and insulin secretory function in these animals [7]. For instance, plasma insulin levels in the basal state were normalized, although insulin release remained deficient (35% of control) during an oral glucose tolerance test (OGTT). As well, immunohistochemical studies indicated that the pancreatic islet area comprised of β cells was only slightly diminished (68% of control), suggesting that vanadyl treatment somehow activated the regeneration of beta cells or prevented a complete destruction possibly via the elimination of glucose toxicity. However, this finding was in contrast to pancreatic function measured *in situ* which showed that glucose-induced insulin response in these animals remained marginal and was equivalent to only 12% of control. At this time, vanadium was undetectable in blood, but was present in kidney at ~0.05 μg/g.

Effects after chronic (52-week) vanadyl treatment

In a recent study by Dai *et al* [8], STZ-diabetic and non-diabetic rats were treated with $VOSO_4$ in the drinking water for 52 weeks. Following one year of treatment in control animals, despite no change in food intake, there was a significant reduction in both weight gain and plasma insulin. In diabetic animals, treatment significantly reduced food and fluid intake, glucose and lipid levels, and the occurrence of cataracts. Some animals were then maintained after withdrawal from treatment for an additional 16-week period. After treatment was withdrawn, there was an immediate rise in food and fluid intake in the diabetic animals followed by a progressive return to control levels from 6–9 weeks after withdrawal. In addition, fed glycemia continued to be maintained at <13 mM, while fasting plasma glucose and insulin levels, as well as triglyceride and cholesterol levels, did not differ significantly from non-diabetic controls. Blood vanadium levels were similar between both treated control and diabetic animals at the end of treatment (~0.4 μg/ml), but were below detection levels (<0.02 μg/ml) by 6 weeks after treatment was withdrawn.

Effects of severity of diabetes on post-withdrawal euglycemia

More recently, in a study by Cam *et al.* [9], vanadyl treatment was initiated 3, 10 or 17 days after the induction of STZ-diabetes at an initial concentration of 0.75 mg/ml, and was increased by 0.25 mg/ml increments in those rats which failed to respond to the point of euglycemia. There was no significant effect of delayed times of initiation on the response rate to various concentrations. Retrospective examination, however, revealed that the more severely diabetic animals (more pronounced hyperglycemia and hypoinsulinemia at the start of treatment) had required higher concentrations of $VOSO_4$ to achieve euglycemia, thus allowing for subsequent separation into low-concentration (LC, 0.75–1.0 mg/ml) and high-concentration (HC, 1.25 and 1.5 mg/ml) responder categories. After a 10-week treatment period, all animals had responded to normoglycemia, and there were no significant differences in either plasma glucose or insulin between LC and HC. Whereas treatment reduced plasma insulin in LC from pretreatment levels, there was no further reduction in HC.

Following 10 weeks of treatment, vanadyl was withdrawn and animals were monitored for 20 weeks. Mean plasma glucose levels in 12/13 animals in LC were normal up to 5 months after treatment, whereas 5/6 animals in HC reverted to hyperglycemia at various times after withdrawal. Plasma insulin levels gradually increased over the 20 weeks in treated non-diabetic controls (CT) and LC, whereas in HC, plasma insulin was found to rise initially, and subsequently drop to pre-withdrawal levels. The downward trend following a peak level in plasma insulin paralleled the recurrence of hyperglycemia in HC, suggesting an exhaustion of already limited insulin stores in the more severely diabetic group of animals. An OGTT was performed following short (3-week) and long (20-week) term withdrawal periods (see Fig. 1). At 3 weeks, all animals were still euglycemic, and while there was a slight impairment in glucose response in HC and LC relative to CT (Fig. 1A), insulin release was lowest in HC, intermediate in LC and highest in CT (Fig. 1B). However,

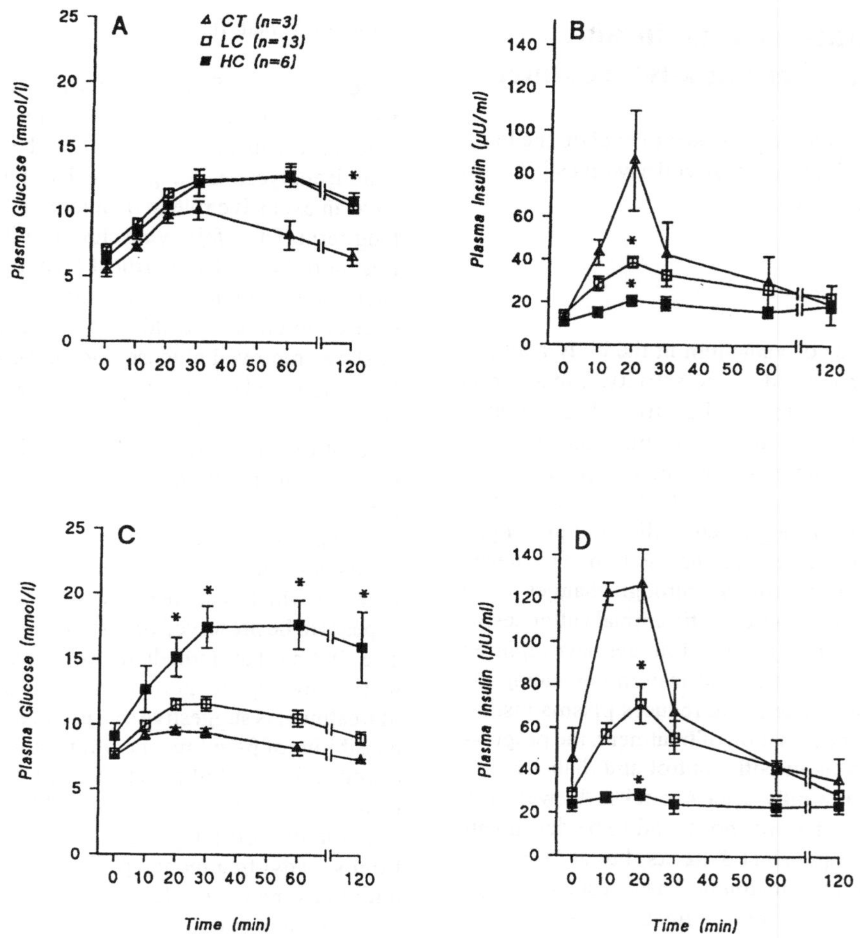

Fig. 1. Plasma glucose (A,C) and insulin (B,D) response over 120 minutes following a 1 g/kg oral glucose dose in overnight-fasted animals at 3 (A,B) and 20 (C,D) weeks after withdrawal from 10-week vanadyl treatment period.

glucose tolerance was not found to be correlated to insulin release at this time; hence, LC and HC showed identical glucose tolerance despite a comparatively lower insulin response in HC (Fig. 1B). At 20 weeks after withdrawal, only HC was found to have marked glucose intolerance (Fig. 1C). Unlike at 3 weeks, the degree of glucose tolerance was closely associated with the level of insulin response. Thus, while insulin release remained unsubstantial in HC, it was improved markedly in LC (Fig. 1D), coinciding with the maintained near-normal glucose tolerance in this group.

In the same paper [9], the immediate and post-withdrawal responses to the more orally potent organo-mineral vanadyl derivative, naglivan (bis(cysteine, amide N-octyl)oxovanadium IV), were found to be similarly dependent on plasma insulin levels at the start of treatment. After a 10-week treatment period, it was found that normoglycemia in 4/7 animals given naglivan (50–150 mg/kg/day), was maintained for 30 weeks after withdrawal from treatment, which was accompanied by a sustained improvement in plasma insulin. An OGTT performed at 30 weeks showed that integrated insulin release was improved to ~50% in responders while that of non-responders remained at ~20% of normal animals.

In summary, the prolonged antidiabetic activity of $VOSO_4$ and its organic derivative, naglivan, has been confirmed and extended to 30 weeks after chronic treatment is withdrawn. It appears that the initial diabetic state (degree of β-cell loss) of the animal administered STZ determines the eventual response to treatment and the ability of the animal to remain normoglycemic for an indefinite period after cessation of treatment. This suggests that over the long term, prevention of hyperglycemia and diabetes-related complications following withdrawal from treatment may be dependent on the level of residual insulin secretory capacity in these animals which is preserved by vanadyl treatment.

Possible mechanisms of maintained normoglycemia after vanadyl treatment

Several mechanisms could be proposed to explain the long-term posttreatment activity of vanadyl derivatives [9]. Two hypotheses will be discussed:

Vanadium stores

The prolonged presence of vanadium in the body may potentially maintain some antidiabetic activity. Vanadium is stored in various organs with long half-lives. For example, blood, muscle, liver, kidney, and bone vanadium half-lives in normal rats after a single subsutaneous injection were determined to be 11.4, 27.2, 29.3, and 376 h, respectively [10]. Although a more thorough study following prolonged post-treatment periods is necessary, the duration of influence of stored vanadium can be estimated through changes in the levels of plasma insulin of non-diabetic animals after cessation of vanadium treatment [9]. Several studies have reported that vanadium, due to its peripheral insulin-enhancing effects which lowers insulin demand, reduces plasma insulin in control animals. After cessation of treatment, the progressive rise in insulin levels in both control and diabetic (LC and HC) animals occurs only after 4 weeks. As well, the degree of glucose tolerance was not found to be dependent on insulin secretory response at 3 weeks, but at 20 weeks following withdrawal from treatment. These results are suggestive of a period of antidiabetic effectiveness of stored vanadium in the presence of reduced levels of circulating insulin.

Another scenario which can be proposed involves the possibility that vanadium stores may continue to negatively influence insulin release after treatment is withdrawn. Voss *et al.* [11] reported the inhibition of insulin secretion and biosynthesis (but not of insulin storage) in normal rat islets by concentrations of vanadate as low as 1–50 μM. Thus, there exists a potential for small amounts of stored vanadium, especially in the pancreas, to continue to influence insulin secretion after vanadyl treatment is ceased. A two-week study determined that the half-life of pancreatic vanadium following a subcutaneous dose was 49 h with a rate constant for the slow phase of a two-compartment model averaging 7 days [10]. Thus, pancreatic vanadium stores, following chronic treatment, can potentially reduce insulin secretion over a prolonged period. In support, treated control animals continue to display a lower OGTT insulin response relative to untreated controls at 13 weeks following one-year treatment [8].

Pancreatic insulin

Albeit a poor reflection of tissue vanadium stores, blood vanadium levels were undetectable after withdrawal from chronic treatment at a time when the animals continued to exhibit euglycemia, suggesting that stored vanadium may not have an exclusive role in maintaining normoglycemia in the long-term. Thus, following chronic withdrawal (>3 months), it is more likely that a critical level of insulin secretory capacity is important in maintaining euglycemia and normal glucose tolerance. At this time, in animals which continue to show improved glucose homeostasis, basal plasma insulin is increased relative to pretreatment levels, and glucose-stimulated insulin secretion during an OGTT can reach 50% of control. In contrast, animals which revert to hyperglycemia are markedly insulinopenic and fail to show significant insulin response.

Thus, insulin secretion in levels sufficient to maintain euglycemia and positive metabolic balance either alone, or in levels which 'enhance' the presence of stored vanadium, appears to be preserved or restored in STZ-diabetic animals. The finding that immediate, as well as post-treatment, correction are dependent on plasma insulin levels at initiation of treatment is suggestive of a protective effect, although the possibility of β-cell regeneration cannot be ruled out. Interestingly, the level of pancreatic function found in STZ-diabetic animals which either respond to treatment, or continue to maintain normal glucose homeostasis after withdrwal from vanadyl treatment, appears to be similar. Thus, in the initial posttreatment (13-week) study, insulin release from *in situ* pancreas remained at 12% control despite a much improved morphology and insulin content of the islets of Langerhans [7]. Similarly, *in situ* insulin release in the subset of STZ-diabetic rats which responded to the point of normoglycemia was found to be improved over the non-responding treated animals, although it also remained at 12% of control [12]. More recently, it was found that the preservation of ~12% of pancreatic insulin content was sufficient in both untreated and vanadyl-treated diabetic animals to maintain normal glucose homeostasis for an extended period [13].

Future directions

In summary, the long-term (up to 30 weeks) correction of STZ-induced diabetes obtained with orally administered vanadyl or its organo-mineral derivative, naglivan can be hypothesized to result from combined effects of the continued activity of tissue vanadium stores, along with the improved pancreatic insulin secretory capacity in the long-term. However, several issues need to be considered prior to establishing the therapeutic relevance of this particular property of vanadyl. These include (a) the possible occurrence of long-term correction in insulin-resistant models of diabetes, (b) the minimal duration of treatment and dose necessary to obtain chronic post-withdrawal antidiabetic effects, (c) the relative effects of (i) vanadium *per se* (vs. insulin, food restriction, or other treatments), (ii) the route of administration (oral *vs.* injected vanadium), (iii) the oxidation state of the metal (vanadate *vs.* vanadyl) in inducing long-term diabetes correction, (d) the relative roles of stored vanadium *vs.* preserved insulin secretion on the observed correction, (e) the direct effects of administered vanadium on insulin secretion and β-cell proliferation and differentiation.

Some of these issues are explored in the next article of the same journal (see Poucheret *et al.*) using various treatment protocols associated with isolated pancreas studies, determination of tissue vanadium levels and histological studies.

References

1. Heyliger CE, Tahiliani AG, McNeill JH: Effect of vanadate on elevated blood glucose and depressed cardiac performance of diabetic rats. Science 227: 1474–1477, 1985
2. Ramanadham S, Mongold JJ, Brownsey R, Cros GH, McNeill JH: OraJ vanadyl sulfate in treatment of diabetes mellitus in rats. Am J Physiol 257 (Heart Circ Physiol 26): H904–H911, 1989
3. Meyerovitch J, Rothenberg P, Shechter T, Bonner-Weir S, Kahn CR: Vanadate normalizes hyperglycemia in two mouse models of non-insulin-dependent diabetes mellitus. J Clin Invest 87: 1286–1294, 1991
4. Battell ML, Yuen VG, McNeill JH: Treatment of BB rats with vanadyl sulphate. Pharmacol Commun 1: 291–301, 1992
5. Ramanadham S, Cros GH, Mongold JJ, Serrano JJ, McNeill JH: Enhanced *in vivo* sensitivity of vanadyl-treated diabetic rats to insulin. Can J Physiol Pharmacol 68: 486–491,1990
6. Ramanadham S, Brownsey RW, Cros GH, Mongold JJ, McNeill JH: Sustained prevention of myocardial and metabolic abnormalities in diabetic rats following withdrawal from oral vanadyl treatment. Metabolism 38: 1022–1028, 1989b
7. Pederson RA, Ramanadham S, Buchan AMJ, McNeill JH: Long-term effects of vanadyl treatment on streptozotocin-induced diabetes in rats. Diabetes 38: 1390–1395, 1989
8. Dai S, Thompson KH, McNeill JH: One-year treatment of streptozotocin-induced diabetic rats with vanadyl sulfate. Pharmacol Toxicol 74: 101–109, 1994
9. Cam MC, Faun J, McNeill JH: Concentration-dependent glucose lowering effects of oral vanadyl are maintained following treatment withdrawal in streptozotocin-diabetic rats. Metab Clin Exp 44: 332–339, 1995
10. Al-Bayati M, Raabe OG, Giri SN, Knaak JB: Distribution of vanadate in the rat following suboutaneous an oral routes of administration. J Am Coll Toxicol 10: 233–241, 1991
11. Voss C, Herrmann I, Hartmann K, Zuhike H: *In vitro* effect of vanadate on content, secretion and biosynthesis of insulin in isolated islets of normal Wistar rats. Exp Clin Endocrinol 99: 159–163, 1992
12. Cam MC, Pederson RA, Brownsey RW, McNeill JH: Long-term effectiveness of oral vanadyl sulphate in streptozotocin-diabetic rats. Diabetologia 36: 218–224, 1993
13. Cam MC, Li WM, Faun J, McNeill JH: Short-term vanadyl treatment partially preserves pancreatic b-cells and ameliorates chronic diabetes induced by streptozotocin (submitted)

Molecular and Cellular Biochemistry **153**: 197–204, 1995.

Long-term correction of STZ-diabetic rats after short-term i.p. $VOSO_4$ treatment: Persistence of insulin secreting capacities assessed by isolated pancreas studies

Patrick Poucheret[1], René Gross[2], Anne Cadène[1], Michelle Mantéguetti[2], Jean-Jacques Serrano[1], Gérard Ribes[2] and Gérard Cros[1]

[1]*Laboratoire de Pharmacologie, Faculte de Pharmacie, F-34060 Montpellier;* [2]*Laboratoire de Pharmacologie, Faculté de Médecine and UMR CNRS 9921, F-34060 Montpellier, France*

Abstract

We have previously shown that 3 week oral $VOSO_4$ treatment of streptozotocin (STZ, 60 mg/kg)-induced diabetic rats was able to correct diabetes for 13 weeks after treatment withdrawal. In the present study, we investigated whether a short-term (8 days) i.p. $VOSO_4$ treatment was similarly able to reverse the diabetic state. Insulin secretory capacities were assessed at distance of treatment using the isolated pancreas preparation. Seven treatment-groups were performed: high dose $VOSO_4$ -treated diabetics (HVD, 1.3 mM/kg/8 days), food-restricted diabetics (FRD, food adjusted to HVD levels), low dose $VOSO_4$-treated diabetes (LVD, 0.06 mM/kg/day), insulin-treated diabetics (ID, dose adjusted to normalize glycaemia) and $VOSO_4$ (0.06 mM/kg/day) + insulin (dose adjusted to normalize glycaemia in the presence of vanadium)-treated diabetics (IVD), in addition to the corresponding untreated non-diabetic controls (C) and diabetics (D). Our results indicate that long-term correction of diabetes (a) can be obtained after an 8 day treatment using i.p. $VOSO_4$ in diabetic animals retaining some degree of pancreatic function, (b) is not obtained with insulin treatment or food restriction although the association of $VOSO_4$ and insulin was found beneficial, (c) can be prolonged in some individuals for at least 4 months, i.e. in conditions such that tissue vanadium concentrations had returned to values close to pre-treatment levels, (d) is associated with improved and in some cases normalized insulin secretion from isolated pancreas. The protective or corrective role of $VOSO_4$ on diabetes-related pancreatic alterations, as well as the potential of the $VOSO_4$-insulin association should be further studied in view of the possible use of vanadium derivatives in the treatment of diabetes. (Mol Cell Biochem **153:** 197–204, 1995)

Key words: diabetes, vanadium, insulin secretion, rat

Introduction

Vanadium derivatives has been shown to possess *in vivo* insulin-mimetic properties and *in vivo* anti-diabetic properties. *In vitro* studies include demonstration of activation of glucose transport in adipocytes [1], stimulation of glycolysis [2], inhibition of lipolysis [2] and stimulation of lipid synthesis [3, 4]. *In vivo* studies have shown that the administration of vanadium under the forms of vanadyl, vanadate or pervanadate, was able to induce normoglycemia in the STZ-induced diabetes of the rat [5–9] or the genetically acquired type II diabetes of the *db/db* mouse [10, 11], and reduce the requirement of insulin in a model of type I diabetes, the BB rat [12, 13]. In these studies, the administration of vanadium compounds was able to normalize glycaemia and prevent most of the secondary complications of diabetes, while body growth was lower than in non-diabetic controls during treatment.

Vanadyl sulfate administered in drinking water was also shown to possess anti-diabetic activities for 13–20 weeks after cessation of treatment [14–16]. In these studies, prolonged diabetes correction after treatment withdrawal was associated with normal growth of the animals, suggesting a ben-

Address for offprints: G. Cros, Laboratoire de Pharmacologie, Faculté de Pharmacie, F-34060 Montpellier, France eficial effect on endocrine pancreas function. However, studies concerning the endocrine pancreas status of long-term corrected animals have produced discrepancies within the various parameters measured. Indeed, while insulin plasma levels were normalized [14], and glycemic response to Oral Glucose Tolerance Test found similar to that of non-diabetic controls [17], insulin was not released in expected amounts [17]. As well, while histological studies indicated that the number of beta cells was found only slightly diminished as compared to that of controls, suggesting that vanadyl somehow activated beta cell regeneration or prevented the full destruction of beta cells, the insulin response of *in situ* pancreas to glucose and the insulin secretagogue GIP was very low [17]. Therefore, the respective roles of a restored endocrine pancreas function and/or a possible peripheral effects of vanadium present in the tissue after cessation of treatment, remain to be further explored.

The present study was undertaken to further characterize $VOSO_4$-induced long-term correction of diabetes. In order to do so, it appeared necessary to obtain a protocol in which the duration of treatment would be minimal and vanadium bioavailability maximal. As oral administration of vanadium is associated with limited digestive absorption (1–3%), and dosage limited by the occurence of digestive side-effects -such as diarrhea- or variable with the degree of correction (variation of fluid intake) when administered in drinking water, $VOSO_4$ was given *i.p.* and its activity compared to those insulin or food restriction. Endocrine pancreas function was explored using the *isolated* pancreas preparation with glucose as physiological insulin secretion stimulant, pancreas submitted to histological studies, and vanadium tissue concentrations measured.

Materials and methods

Animals and induction of diabetes

Male Wistar rats (180–200 g) were injected with a single dose of streptozotocin (60 mg/kg, i.v.). Only animals having glycaemia ≥ 14 mM (80–95% of animals) were selected for further studies. In these conditions, no spontaneous reversal of diabetes was ever observed. Treatments were initiated 8 days after induction of diabetes.

Establishment of models of long-term correction

We attempted to obtain normoglycemia during *i.p.* $VOSO_4$ treatment and prolongation of its effect after treatment.

Preliminary toxicity studies performed on diabetic animals indicated that single high doses of $VOSO_4$ administered to diabetic animals allowed the evaluation of LD100 (0.7 mM/kg), LD50 (0.5 mM/kg) and MTD (0.2 mM/kg). A single dose of 0.2 mM/kg was associated with a drop in blood glucose in 30% of the animals for a period of 24 h, while lower doses were inefficient. Correction of 100% of the animals required the administration of MTD for 4 days. However, if treatment was stopped at that time, hyperglycemia returned within 48 h.

It was therefore decided to perform two types of protocols: in the first protocol, the dose was adjusted to correct the maximum number of animals and treatment continued, although toxicity might be high, for the time previously shown necessary to obtain long-term correction after treatment had been stopped (8 days). In the second protocol, $VOSO_4$ dosage was reduced to non toxic levels for the same duration of treatment (8 days) and associated with insulin to obtain normoglycemia.

Protocol of treatment 1: In high dose $VOSO_4$-treated diabetics (HVD), treatment was initiated at 0.2 mM/kg for the time necessary to normalize glycemia in 100% of the animals (4 days), then gradually decreased to 0.15 mM/kg for 2 days and 0.1 mM/kg for 2 days. A group of food-restricted diabetics (FRD) in which the amount of food was adjusted to the level of HVD animals for the same period of time (8 days), was included in the study. Animals were sacrificed 24–28 days after cessation of treatment.

Protocol of treatment 2: In Insulin/Vanadyl diabetics (IVD), treatment with $VOSO_4$ (at the acutely inefficient dose of 0.06 mM) was associated with insulin. A $VOSO_4$ (0.06 mM/kg/day)-treated group (low-dose $VOSO_4$-treated diabetics, LVD) and an insulin-treated group (ID) were followed in parallel. Daily (5 pm) insulin treatment (Endopancrine® zinc protamine, Organon Laboratories) was initiated at 3 IU *per* rat and the dose adjusted thereafter according to daily glycemia, in order to keep blood glucose levels as close as possible to 5.55 mM. Non-diabetic controls (C) and non-treated diabetics (D) were followed in parallel throughout both protocols.

Animals were sacrificed 24 days to 4 months (see results) after cessation of treatment in order to test endocrine pancreas function using the isolated pancreas preparation, measure plasma insulin and vanadium tissue levels, and perform histological studies.

Isolated pancreas studies

Pancreases were isolated from animals anesthetized with pentobarbital (60 mg/kg *i.p.*). Blood samples were collected for measurement of plasma glucose and insulin levels.

The technique of Loubatières *et al.* [18] was used to isolate the pancreas from the neighbouring tissues. The organ was then transferred into a plastic chamber maintained at 37°5 C. Perfusion medium was Krebs-Ringer bicarbonate buffer containing 2 g/l bovine serum albumin and 5 mM glucose, and continuously bubbled with a mixture of 95% O_2/5% CO_2.

Infusion pressure was selected to provide a pancreatic outflow of 2.5 ml/min.

The first sample was taken 30 min after initiation of perfusion. Two more samples were collected 10 and 15 min later; glucose concentration was increased from 5 to 11 mM to test the insulin response. Samples were then collected every min for 5 min, then every 5 min for 20 min. Glucose concentration was then reduced to 5 mM and two more samples were collected every 5 min. All samples were collected for 1 min allowing determinations of pancreatic effluent output, and were immediately frozen for insulin assay. Insulin output was calculated by multiplying the hormone concentration (ng/ml) in the effluent by the corresponding flow rate (ml/min).

Plasma insulin as well as the hormone concentration in the effluent of isolated pancreas were determined by the radioimmunological method of Herbert *et al.* [19], with rat insulin as standard. The sensitivity of the method was 0.1 ng/ml.

Histological studies and vanadium tissue concentrations were performed as described previously [20].

Expression and statistical analysis of results

Results are indicated as means ± S.E.M.s. Statistical comparisons were performed using the Student's 't' test or the analysis of variance followed by the multiple comparison test of Newman-Keuls, as appropriate.

Results

General feature of the animals and vanadium tissue levels

Tables 1a–c indicate the general feature of the animals from the various treatment-groups. As previously noted, non-treated diabetics (D) remained hyperglycemic for the entire study period, body growth was impaired, food consumption and fluid intake were higher than the normal control rats (C). HVD group was characterized during the 8 day treatment by a decrease in food and fluid intakes as compared to the D group and mortality was significant (7/24, 29%). All surviving animals were normoglycemic at the end of the treatment period. After cessation of treatment, 3 more animals died within 6 days. However, 12/14 (85%) of the surviving animals were still normoglycemic at the date of sacrifice and growth as well as food and fluid intakes were close to C levels. Normoglycemia could be noted in 6/12 (50%) of FRD during the 8 day restriction period. However, glycemia returned to pre-treatment levels as soon as the animals were fed *ad libitum* and remained elevated thereafter.

Vanadium could be detected only in bone of non vanadium-treated animals (1.13 ± 0.12 μg/g); in contrast, it was still present in HVD animals 26 days after treatment in blood (0.03 ± 0.02 μg/g) and liver (1.65 ± 0.32 μg/g), while bone levels were dramatically elevated (24.68 ± 1.891 μg/g). Therefore a possible influence of stored vanadium in the long-term correction of the animals led us to prolong the duration of the post-treatment period for the second set of experiments.

As opposed to the HVD group, all 18 animals from the LVD group survived and were characterized by the absence of glycemic correction during treatment. Insulin treatment allowed euglycemia in 10/12 animals from IVD or ID groups. The dose of insulin was lower in IVD (1.79 ± 0.11 IU/kg/day) than ID (2.35 ± 0.13 IU/kg/day) animals ($p \leq 0.05$). 50% of IVD animals were still corrected two months later at moment of sacrifice, while most of the LVD animals remained hyperglycemic. Surprisingly, four animals from the LVD group (22%) were normoglycemic at that date and these animals were kept for two additional months. During that time, one of them died spontaneously, one reversed to diabetes and two remained corrected.

Vanadium measurements showed that significant amounts of vanadium were still stored in liver (0.77 ± 0.18 μg/g) and bone (15.32 ± 4.60 μg/g) of LVD animals two months after treatment, while after 4 months vanadium could be only detected in bones, although at levels slightly higher (3.66 ± 1.26 μg/g) than those of non vanadium-treated animals (1.98 ± 0.48 μg/g). Vanadium levels of IVD animals (0.28 ± 0.05, 2.20 ± 0.27 and 25.40 ± 2.60 μg/g for blood, liver and bone, respectively) were significantly higher than those of LVD after the same 2 month post-treatment period, and even higher than those of HVD for blood and liver after 26 days of treatment withdrawal.

Isolated pancreas studies

Figures 1–4 illustrates the insulin responses of the pancreas from the various treatment-groups.

The insulin responses of C animals were similar within the various experimental groups (Fig. 1). Basal insulin output (5 mM glucose) was low, and the increment in glucose concentration from 5 to 11 mM resulted in a clear biphasic response in insulin secretion: insulin output peaked at 2 min of high glucose, then a minimum was achieved after 5 min. A secondary increase in insulin secretion occurred after 5 min.

Figure 1 also illustrates the insulin response recorded after various durations of diabetes. Eight days after STZ injection, basal insulin secretion was still present at a level similar to that of C animals, while increasing glucose concentration resulted in a modest but significant insulin response which peaked after 2 min. From 42 days after STZ on, all isolated pancreas from nontreated diabetics were characterized by low basal insulin output and total lack of response to 11 mM glucose.

Table 1a. Effect of the various treatments on glycaemia (mean ± S.E.M, mM)

Days after initiation of treatment	During treatment 4	8	After treatment 34	71
C	5.85 ± 0.19	6.58 ± 0.31	4.18 ± 0.85	5.94 ± 0.56
D	17.19 ± 1.87*	20.71 ± 1.02*	21.11 ± 2.46*	21.77 ± 2.58*
FRD	20.96 ± 3.94*	9.98 ± 3.32*†	22.98 ± 0.71*	ND
HVD	3.87 ± 0.26*†	3.54 ± 0.34*†	9.29 ± 1.07*†	ND
ID	8.27 ± 2.67†	8.20 ± 2.49†	22.23 ± 0.91*	18.64 ± 0.65*
LVD(1)	15.6 ± 0.94	15.71 ± 1.56*†	16.75 ± 0.62*	15.29 ± 0.30*†
IVD	5.41 ± 0.9†	7.39 ± 1.82†	6.71 ± 1.35†	10.61 ± 1.08*†

(1) While no LVD animal was corrected during treatment, 4 animals became euglycemic 6 weeks after cessation of treatment and were still corrected 2 months after treatment. Between 2 and 4 months after treatment, one of them died, one turned diabetic, and two remained euglycemic. (*) Significandy different ($p \leq 0.05$) from C group. (†) significantly different from D group.

Table 1b. Effect of the various treatments on body weight changes (mean ± S.E.M, g/animaUday)

Days after initiation of treatment	During treatment 0–8	After treatment 9–34	35–71
C	+7.2 ± 0.3	+4.3 ± 0.1	+1.9 ± 0.1
D	+5.6 ± 0.6*	+1.2 ± 0.1*	+0.7 ± 0.2*
FRD	–4.1 ± 1.7*†	+3.8 ± 0.1*†	ND
HVD	–3.6 ± 0.5*†	+5.3 ± 0.5†	ND
ID	+8.3 ± 0.3*†	+2.6 ± 0.4*†	+0.1 ± 0.0*†
LVD	+2.4 ± 0.3*†	+0.8 ± 0.1*	+2.4 ± 0.3*†
IVD	+4.2 ± 0.4*	+4.1 ± 0.4†	+1.1 ± 0.1*

(*) Significantly different ($p \leq 0.05$) from C group. (†) significantly different ($p \leq 0.05$) from D group.

Table 1c. Effect of the various treatments on food and fluid intake (mean ± S.E.M, g/animal/day)

Days after initiation of treatment	During treatment 0–8		After treatment 9–34		35–71	
	Food	Fluid	Food	Fluid	Food	Fluid
C	23.6 ± 0.4	31.8 ± 0.9	28.5 ± 0.8	43.8 ± 1.4	29.4 ± 1.4	42.6 ± 1.9
D	38.0 ± 1.1*	137.1 ± 3.4*	40.2 ± 1.7*	143.1 ± 3.3*	40.0 ± 2.0*	153.0 ± 7.2*
FRD	14.7 ± 1.9*†	52.6 ± 9.3'*†	42.6 ± 2.0*	160.0 ± 12.5*	ND	ND
HVD	14.7* ± 1.9*†	31.1 ± 6.6†	25.5 ± 1.9†	74.1 ± 5.8†	ND	ND
ID	31.8 ± 3.1*	57.9 ± 7.4*†	33.2 ± 1.9?†	102.7 ± 9.9*†	43.3 ± 2.8*	151.7 ± 12.4*
LVD	33.7 ± 0.7*†	113.8 ± 4.8*†	39.2 ± 1.2*	138.5 ± 5.6*	39.1 ± 2.4*	142.5 ± 7.8*
IVD	24.5 ± 0.4†	40.9 ± 1.8*†	28.4 ± 1.9†	53.6 ± 5.9†	29.1 ± 2.0†	54.3 ± 5.7†

(*) Significantly different ($p \leq 0.05$) from C group. (†) significantly different ($p \leq 0.05$) from D group.

Figure 2 illustrates the average pancreatic insulin responses of HVD and FRD animals recorded 26 days after treatment withdrawal (or 42 days after induction of diabetes). Pancreas from the HVD group were characterized by a high basal insulin output, the presence of a clear – although lower than controls' – first phase insulin response to high glucose, whereas no significant second phase response occured. Insulin response of FRD was impaired to a degree similar to that of non-treated diabetics.

Figure 3 illustrates the mean insulin responses of IVD, ID and LVD animals, recorded 2 months after treatment withdrawal (or 74 days after STZ injection). Mean IVD insulin secretory pattern was similar to that of HVD animals: high basal insulin output, significant first phase, and lack of any second phase. Insulin responses of LVD (non-corrected animals) and ID groups were impaired as in non-treated diabetics.

Figure 4 shows the individual insulin responses of the three remaining LVD animals recorded 4 months after treatment. Insulin secretion pattern of euglycemic LVD animals was similar to that of non-diabetic controls (low basal insu-

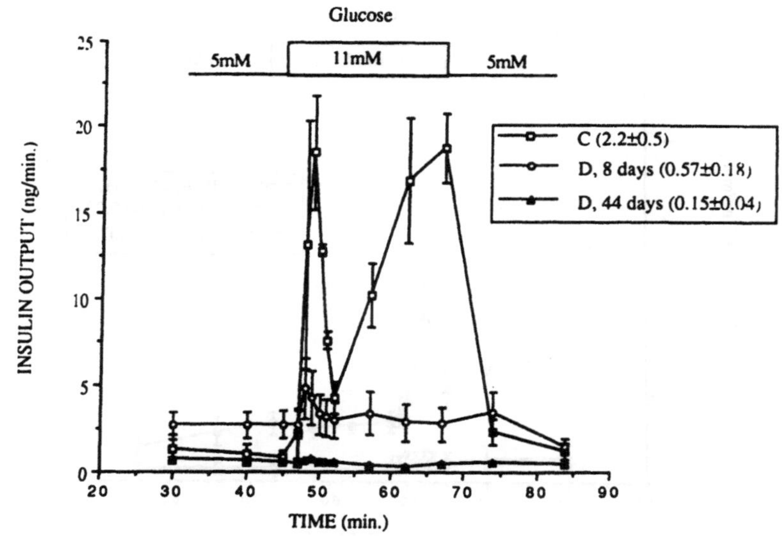

Fig. 1. Effect of STZ injection on the insulin response of isolated pancreas. Pancreas from Non-Diabetic Controls (C, n = 3), and from diabetic animals (D) sacrified 8 days (n = 7) or 42 days (n = 6) after STZ injection, were isolated and perfused according to Materials and methods, and insulin secretion measured under two different conditions of glucose concentration. Values in the legend indicate plasma insulin levels (ng/ml) at sacrifice. Values are means ± S.E.M.

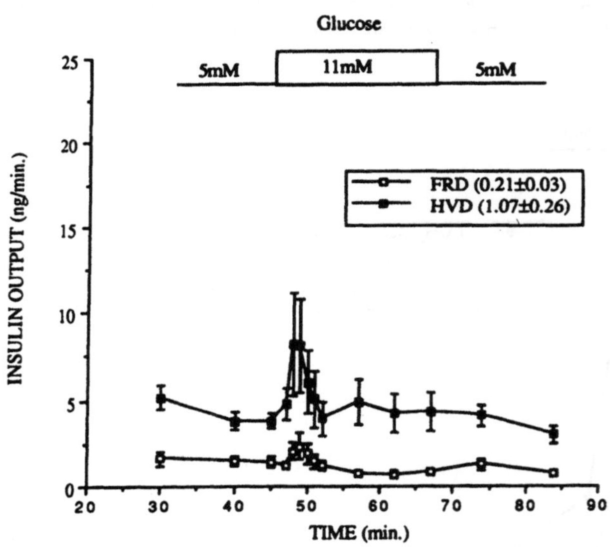

Fig. 2. Effect of treatments of 8-day diabetic rats with high dosage $VOSO_4$ or food restriction on the insulin response of isolated pancreas. Pancreas from 8 day diabetic rats treated with high dosage $VOSO_4$ (HVD, n = 7) or submitted to food restriction (FRD, n = 6) for 8 days were isolated and perfused according to Materials and methods 24–28 days after treatment, and insulin secretion measured under two different conditions of glucose concentration. Values in the legend indicate plasma insulin levels (ng/ml) at sacrifice. Values are means ± S.E.M.

lin secretion, high initial insulin secretory peak, presence of a secondary phase), while that of the non-corrected LVD animal was similar to that of non-treated diabetics.

Mean plasma insulin levels, indicated on Figs 1–4, correlated with the degree of correction of diabetes in the various groups studied.

Histological studies

Histological analysis (not illustrated) of the pancreas from the various treatment-groups showed that pancreatic islets from non-diabetic controls contained a high proportion of well-granulated β-cells. Any insulin staining was hardly visible in non-treated diabetics, including 8 days diabetics. In contrast, some insulin was clearly present in islets from the HVD and IVD groups. The aspect of pancreatic islets from the uncorrected LVD animals was similar to that of corresponding untreated diabetics, while that of the 4-month corrected animals from the same group was similar to that of normal animals.

Discussion

The present study clearly shows that short term (8 day) *i.p.* treatment with $VOSO_4$ alone or in association with insulin is able to induce a long-term period of euglycemia in rats made diabetic by streptozotocin. In addition, pancreatic β-cells from long-term corrected animals retained some insulin-secreting capacities. This study also shows that the capacity of vanadium to induce long-term correction of diabetes was specific of the metal itself, as it could not be reproduced by

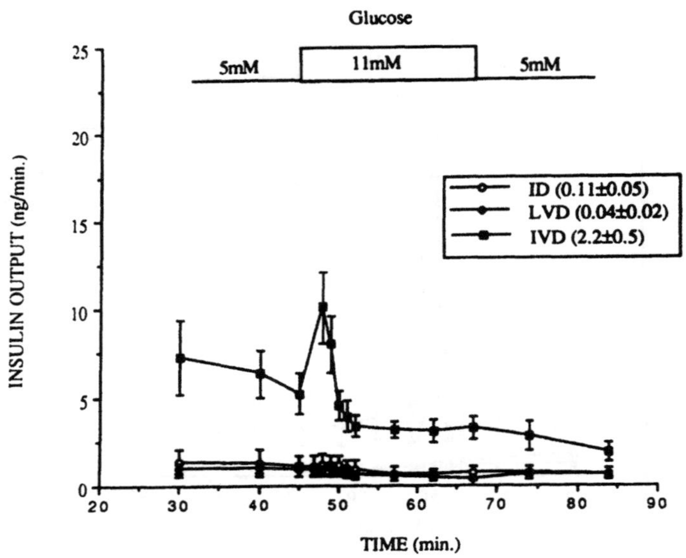

Fig. 3. Effect of treatments of 8-day diabetic rats with low dosage $VOSO_4$, insulin, or the association of low dosage $VOSO_4$ plus insulin on the insulin response of isolated pancreas. Pancreas from 8 day diabetic rats treated for 8 days with low dosage $VOSO_4$ (LVD, n = 5), insulin (ID, n = 5) or the association of low dosage $VOSO_4$ plus insulin (IVD, n = 7) were isolated and perfused according to Materials and methods two months after treatment, and insulin secretion measured under two different conditions of glucose concentration. Among LVD animals, uncorrected animals only (representing 80% of the entire group) were perfused. Values in the legend indicate plasma insulin levels (ng/ml) at sacrifice. Values are means ± S.E.M.

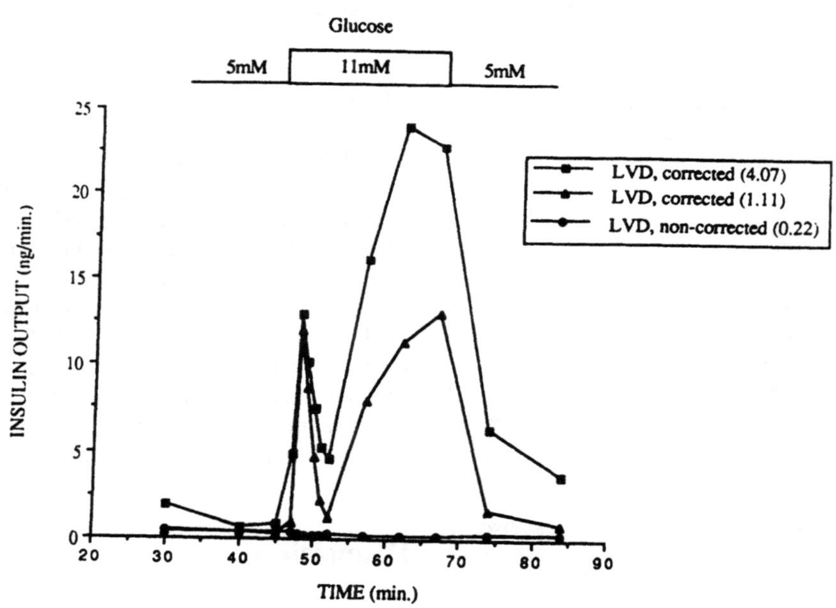

Fig. 4. Effect of treatment of 8-day diabetic animals with low-dosage $VOSO_4$ on individual insulin responses of isolated pancreas. Pancreas from 8-day diabetic rats treated for 8 days with low-dosage $VOSO_4$ (LVD) were isolated and perfused according to Material and methods four months after treatment, and insulin secretion measured under two different conditions of glucose concentration. Recordings of three animals which were euglycemic two months after treatment are shown. Two animals (■▲) were still euglycemic four months after treatment. Values in the legend indicate plasma insulin levels (ng/ml) at sacrifice.

insulin therapy alone (as compared to the association of insulin and $VOSO_4$ in similar conditions of glycemia) or food restriction of the animals.

The deleterious effect of hyperglycaemia on the evolution of pancreatic function during diabetes is well documented [21] and insulin therapy performed before or just at the onset of type I diabetes is able to preserve some of the insulin secreting capacities of the endocrine pancreas [22, 23]. It is therefore possible that partial or total restoration of pancreatic function could be obtained indirectly by way of glycaemic normalization. However, insulin treatment was not able to induce long-term diabetes correction. Although the mean level of glycemia obtained with insulin alone (≈8 mM) was higher than with high-dosage $VOSO_4$ treatment (≈3.5 mM), it was similar to the level obtained with the association of insulin and low-dosage $VOSO_4$, which induced long-term correction of diabetes. This result indicates that glycaemic normalization is not the main factor of the long-term activity of the metal. *In vivo* studies, including the present one, have shown a reduction of food intake induced by vanadium derivatives in diabetic animals and recently, Malabu *et al.* [24] postulated that the glucose-lowering effect of vanadate in STZ-induced diabetic rats could be entirely attributed to the suppression of feeding. Our results indicate that, if food restriction to the level of vanadium-treated animals was indeed able to induce normoglycemia in 50% of the animals, it was however totally ineffective in inducing long-term correction. It is possible that the long-term storage and release of vanadium in the tissues may have maintained some hypophagic activity contributing to its normoglycemic effect. However, our results obtained with insulin, vanadium (lower dosage) and their association indicate that reduction of feeding is probably not an essential factor in the long-term correction of the diabetic rats.

Our study also confirms the long-term tissue storage of vanadium previously described [25, 26]. In particular, amounts of vanadium similar to those obtained during oral $VOSO_4$ treatment (≈ 1 µg/g, 20) were still found in liver 2 months after cessation of treatment, raising the possibility of an insulin-like or an insulin-enhancing effect of long-term stored vanadium that may explain its long-term activity. In addition, in other studies on normal animals, we could show that pancreatic islet and liver vanadium contents were similar (unpublished data), suggesting that a direct and prolonged effect of vanadium on the endocrine pancreas is possible, at least for a period of two months following treatment.

Concerning insulin secretion, two original findings of our study were (a) an increased insulin secretion under basal conditions (i.e. 5 mM glucose), and (b) the occurence of a small but significant first phase of β-cell response to glucose up two months after treatment withdrawal. *In vitro* studies of the effect of vanadate on isolated islets have given controversial results: Fagin *et al.* [27] demonstrated that vanadate (0.5–1 mM) stimulated insulin release, and this result was confirmed by Zhang *et al.* [28] who showed the enhancement of glucose-induced stimulation of insulin secretion for similar concentrations. In contrast, Voss *et al.* [29] found an inhibition of insulin secretion and biosynthesis in normal rat islets by sodium orthovanadate (1–50 µM). It is therefore possible that vanadium could exert a dual effect on insulin secretion. In any case, our estimation of pancreas vanadium levels (≈10^{-4}–10^{-5} M) indicates that active concentrations could be attained *in vivo*. The direct effect of vanadyl on insulin secretion during therapeutic conditions, as well as its relevance in vanadyl and-diabetic properties remained to be assessed. Another striking result of our study was the recurrence of a biphasic secretory pattern in 2 cases four months after treatment, when significant amounts of vanadium could be detected in bones only. This result, as well as our histological studies showing complete recovery of the islet aspect argue for the possibility that vanadyl might also act through endocrine pancreas protection and/or regeneration. A similar structural recovery was described by Pederson *et al.*[17] 13 weeks after having treated 3-day diabetic rats for 3 weeks.

More recently, data obtained by Cam *et al.*[16] indicate that the immediate response to vanadyl treatment as well as its long-term effects after treatment withdrawal are directly correlated to the insulin levels at the beginning of treatment. Our own data also indicate that some insulin secreting capacities are still present 8 days after STZ injection, i.e. at initiation of treatment. As well, although Cam *et al.* [30] have obtained similar response to oral $VOSO_4$ 3, 10 or 17 days after 55 mg/kg STZ, our own experience with *i.p.* $VOSO_4$ treatment indicates that a significant glycemic response is difficult to obtain when treatment is delayed for one month after 60 mg/kg STZ (unpublished data). These lines of evidence are rather in favor of a *protective* effect of $VOSO_4$ against the evolution of STZ-induced alterations, allowing the residual insulin secretion to induce normoglycemia in the (enhancing) presence of vanadium [13]. This is probably true for the animals remaining corrected for some time after treatment, before returning to diabetes. On the other hand, the occurence of *very* long-term corrections associated with normal insulin secretion and histological aspect suggests that a β-cell *regeneration* is equally susceptible to occur. Such a possibility is worthy of further investigations, all the more since vanadium has previously been shown to enhance the proliferating effects of growth factors on cultured cells [31].

The possibility that euglycemia induced by the insulin treatment may favor the effect of $VOSO_4$, allowing reduction of dosage and thereby toxicity, led us to associate insulin to a non-toxic dose of $VOSO_4$. Our result indicate that, although insulin treatment induces no long-term correction, it enhances the activity of vanadyl, as 50% of the animals treated with the association were still corrected 2 months after treatment. These data suggest that, although insulin-induced euglycemia or insulin *per se* cannot produce long-term correction, they may favor the long-term correcting activity of $VOSO_4$. Surprisingly, high vanadium tissue levels were found in animals which have been concommitantly treated with insulin, and the better vanadium efficiency in inducing long-term normoglycemia might be explained by the higher vanadium tissue concentrations obtained in these conditions of treatment. Whether insulin-induced normoglycemia (through a decrease rate of vanadium urinary elimination?) or another mechanism is responsible for the increased vanadium accumulation is not known at the present time. This result indicates anyway that increased tissue levels of vanadium are not necessarily associated with increased toxicity, which might be important to consider for the future of vanadium therapy.

In summary, our data demonstrate that $VOSO_4$ treatment has a beneficial effect on pancreatic B-cell function in STZ animals. This effect cannot be reproduced by food restriction or insulin treatment, and is enhanced when $VOSO_4$ is associated with insulin. The protective or corrective role of $VOSO_4$ on diabetes-related pancreatic alterations, as well as the potential of the $VOSO_4$-insulin association should be further studied in view of the possible use of vanadium derivatives in the treatment of diabetes.

References

1. Shechter Y, Karlish SJD: Insulin-like stimulation of glucose oxydation in rat adipocytes by vanadyl (IV) ion. Nature 284: 556–558, 1980
2. Rodriguez-Gil JE, Gomez-Foix AM, Fillat C, Bosch F, Guinovart J-J: Activation by vanadate of glycolysis in hepatocytes from diabetic rats. Diabetes 40: 1355–1359, 1991
3. Aguis L, Vaartjes WJ: The effects of orthovanadate on fatty acid synthesis in isolated rat hepatocytes. Biochem. J. 202: 791–794, 1982
4. Castro J, Maquedano A, Olive M: Lipid synthesis in isolated rat hepatocytes: activation by insulin and vanadate and inhibition by ouabain. Biochem Inter 9: 413–420, 1984
5. Ramanadham S, Mongold JJ, Brownsey R, Cros GH, McNeill JH: Oral vanadyl sulfate in treatment o diabetes mellitus in rats. Am. J. Physiol. 257 (Heart Circ. Physiol. 26): H904–H911, 1989
6. Sakurai H, Tschiya K, Nukatsuka M, Sofue M, Kawada J: Insulin-like effect of vanadyl ion on streptozotocin-induced diabetic rats. J. Endocrinol 126: 451–459, 1990
7. Heyliger CE, Tahiliani AG, McNeill JH: Effect of vanadate on elevated blood glucose and depressed cardiac performance of diabetic rats. Science 227: 1474–1477,1985
8. Meyerovitch J, Farfel Z, Sack J, Shechter Y: Oral administration of vanadate normalizes blood glucose levels in streptozotocin-diabetic rats. Characterization and mode of action. J Biol Chem 262: 6658–6662, 1987
9. Shisheva A, Ikonomov O, Shechter Y: The protein phosphatase inhibitor, pervanadate, is a powerful antidiabetic agent in streptoztocin-treated rats. Endocrinlogy 134: 507–510, 1994
10. Meyerovitch J, Rothenberg P, Shechter T, Bonner-Weir S, Kahn CR: Vanadate normalizes hyperglycemia in two mouse models of non-insulin-dependent diabetes mellitus . J Clin Invest 87: 1286–1294, 1991
11. Pugazhenthi S, Angel JF, Khandelwal RL: Long-term effects of vanadate treatment on glycogen metabolizing and lipogenic enzymes of liver in genetically diabetic (db/db) mice. Metabolism 40: 941–946,1991
12. Battell ML, Yuen VG, McNeill JH: Treatment of BB rats with vanadyl sulfate. Pharmacol Commun 1: 291–301, 1992
13. Ramanadham S, Cros GH, Mongold JJ, Serrano JJ, McNeill JH: Enhanced *in vivo* sensitivity of vanadyl-treated diabetic rats to insulin. Can J Physiol Pharmacol 68: 486–491, 1990
14. Ramanadham S, Brownsey RW, Cros GH, Mongold J-J, McNeill JH: Sustained prevention of myocardial and metabolic abnormalities in diabetic rats following withdrawal from oral vanadyl treatment. Metabolism 38: 1022–1028,1989
15. Dai S, Thompson KH, McNeill JH: One-year treatment of Streptozotocin-induced diabetic rats with vanadyl sulfate. Pharmacology and Toxicology 74: 101–109,1994
16. Cam MC, Faun J, McNeill JH: Concentration-dependent glucose lowering effects of oral vanadyl are maintained following treatment withdrawal in streptozotocin-diabetic rats. Metab Clin Exp 44: 332–339, 1995
17. Pederson RA, Ramanadham S, Buchan AMJ, McNeill JH: Long-term effects of vanadyl treatment on streptozotocin-induced diabetes in rats. Diabetes 38: 1390–1395,1989
18. Loubatieres A, Mariani MM, Ribes G, De Malbosc H, Chapal J: Etude experimentale d'un noueau sulfamide hypoglycémiant particulierèment actif, le HB 419 ou glibenclamide. I. Action betacytotrope et insulinosécrétrice. Diabetologia 5: 1–10,1969
19. Herbert V, Lau KS, Gottlieb CW, Bleicher SJ: Coated charcoal immunoassay of insulin. J Clin Endocrinol Metab 25: 1375–1384, 1965
20. Mongold JJ, Cros GH, Vian L, Tep A, Ramanadham S, Siou G, Diaz J, McNeill JH, Serrano JJ: Toxicological aspects of vanadyl sulfate on diabetic rats: effects on vanadium levels and pancreatic B-cell morphology. Pharmacol Toxicol: 67: 192–198, 1990
21. Unger RH, Grundy: Hyperglycemia as an inducer as well as a consequence of impaired islet cell function and insulin resistance: implication for the management of diabetes. Diabetologia 28: 119–121, 1985
22. Guastamacchia E, Ciampolillo A, Lattanzi V, Lollino G, Rosco M, Lucivero G, Dell'Osso A, Giorgino R, Nardelli GM: In search of predictive markers of remission from insulin dependence in type 1 diabetes: a preliminary report. Diab Res Clin Pract 16: 145–149, 1992
23. Buschard K: The functional state of the beta-cells in the pathogenesis of insulin-dependent diabetes mellitus. Autoimmunity 10: 65–69,1991
24. Malabu UH, Dryden S, McCarthy DMC, Kilpatrick A, Williams G: Effects of chronic vanadate administration in the STZ-induced diabetic rat. The antihyperglycemic action of vanadate is attributable entirely to its suppression of feeding. Diabetes 43: 9–15, 1994
25. Parker RDR, Shama RP: Accumulation and depletion of vanadium in selected tissues of rats treated with vanadyl sulfate and sodium orthovanadate. J Environ Pathol Toxicol 2: 235–245, 1978
26. Al-Bayati M, Raabe OG, Giri SN, Knaak JB: Distribution of vanadate in the rat following subcutaneous an oral routes of administration. J Am Coll Toxicol 10: 233–241, 1991
27. Fagin JA, Ikejiri K, Levin SR: Insulinotropic effects of vanadate. Diabetes 36: 1448–1452, 1991
28. Zhang A, Gao Z-Y, Gilon P, Nenquin M, Drews G, Henquin JC: Vanadate stimulation of insulin release in normal mouse islets. J Biol Chem 266: 21649–21656, 1991
29. Voss C, Herrmann I, Hartmann K, Zuhlke H: *In vitro* effect of vanadate on content, secretion and biosynthesis of insulin in isolated islets of normal Wistar rats. Exp Clin Endocrinol 99: 159–163, 1992
30. Cam MC, Pederson RA, Brownsey RW, McNeill JH: Long-term effectiveness of oral vanadyl sulphate in streptozotocin-diabetic rats. Diabetologia 36: 218–224, 1993
31. Rijksen G, Voller MVW, Van Zoellen EJJ: Orthovanadate both mimics and antagonizes the transforming growth factor z action on normal rat kidney cells. J Cell Physiol 154: 393–401, 1993

Molecular and Cellular Biochemistry **153**: 205–209, 1995.

Antihypertensive effects of vanadium compounds in hyperinsulinemic, hypertensive rats

Sanjay Bhanot, Aspasia Michoulas and John H. McNeill
Division of Pharmacology and Toxicology, Faculty of Pharmaceutical Sciences, The University of British Columbia, Vancouver, B.C. V6T 1Z3, Canada

Abstract

Although considerable evidence lends credence to the association between insulin resistance, hyperinsulinemia and essential hypertension, the precise nature of this relationship remains unexplained. In the present investigation, we examined the proposition that these metabolic defects contribute causally to the development of high blood pressure. If these metabolic abnormalities were responsible for the development of hypertension, then drug interventions that improve these defects should also decrease high blood pressure. Since previous studies have demonstrated that vanadium compounds enhance insulin action and lower plasma insulin levels in nondiabetic rats, we examined the effects of these compounds on insulin sensitivity, plasma insulin concentration and blood pressure in two hyperinsulinemic models of experimental hypertension. The animal models studied were the genetically predisposed spontaneously hypertensive rat and the fructose-hypertensive rat, where hypertension is induced in normotensive rats by feeding them a high fructose diet. Vanadium compounds caused marked and sustained decreases in plasma insulin concentration and blood pressure in both the animal models studied. Furthermore, the effect of the drugs on blood pressure was reversed by restoring plasma insulin levels in the drug-treated rats to those observed in their untreated counterparts. These data suggest that either hyperinsulinemia contributes to the development of hypertension in both the spontaneously hypertensive and the fructose-hypertensive rats or that the underlying mechanism is closely related to the expression of both these disorders. (Mol Cell Biochem **153**: 205–209, 1995)

Key words: hyperinsulinemia; insulin resistance; hypertension; vanadium compounds

Overview

Considerable epidemiological, clinical and experimental data lend credence to the association between essential hypertension and abnormalities in carbohydrate and lipid metabolism [1, 2]. Of these metabolic defects, two that seem to be frequently associated with hypertension are insulin resistance (or resistance to the glucoregulatory effects of insulin) and hyperinsulinemia [1–5]. These defects in glucose metabolism are associated with a highly atherogenic risk profile and a good deal of evidence now suggests that they may play a central role in the development of hypertension, dyslipidemia and atherosclerosis [4–7]. The presence of insulin resistance and hyperinsulinemia in young, non obese subjects with untreated, uncomplicated hypertension strengthens the contention that there may exist an intrinsic link between these metabolic abnormalities and an increase in blood pressure [8]. However, an important limitation of studies linking insulin resistance and hyperinsulinemia to hypertension is that they do not establish causality. Although associations and correlations favor such a link, this issue requires further experimental evaluation. Essentially, if these metabolic abnormalities were responsible for the development of hypertension, then drug interventions that improve these defects may also decrease high blood pressure.

One such class of drugs that has been of particular interest to us includes the inorganic and organic derivatives of the trace element vanadium. Recently discovered, and of particular interest in diabetic research, are the insulin-mimetic effects of vanadium, which include increased glucose uptake in rat diaphragm, liver and fat cells [9], enhanced glucose transport and oxidation in rat adipocytes and skeletal muscle [10, 11,12] as well as inhibition of lipolysis [13] and activation of lipogenesis [14]. Recent work from our laboratory

Address for offprints: J.H. McNeill, Division of Pharmacology and Toxicology, Faculty of Pharmaceutical Sciences, The University of British Columbia, Vancouver, B.C. V6T 1Z3, Canada has shown the effectiveness of oral vanadyl sulfate in correcting various abnormalities in the heart and adipose tissue in streptozotocin-diabetic rats and in enhancing the effects of insulin *in vivo* [15, 16]. An early observation that surfaced from our studies was that vanadyl treatment not only decreased plasma glucose levels in diabetic rats (without an increase in plasma insulin) but that it also reduced insulin levels in control, non-diabetic rats [16, 17]. These findings suggested that vanadyl either potentiated or replaced the glucose-lowering effects of endogenous insulin, leading to a decreased requirement of insulin in non-diabetic rats.

We recently initiated a series of experiments in which we employed vanadium compounds as a tool to elucidate the relationship between insulin resistance, hyperinsulinemia and hypertension. In an effort to broaden the nature of our enquiry, we used both a genetic and an acquired model of experimental hypertension in our studies. These are the spontaneously hypertensive rat, which is thought to closely resemble human essential hypertension [18] and the fructose hypertensive model, where hypertension is induced in normotensive rats by feeding them a high fructose diet [19]. The intent of this review is to summarize and discuss the results obtained from these studies in context to the general hypothesis under study.

Specific research problem and research strategy

One of the major issues that requires resolution is to determine whether the insulin resistance and hyperinsulinemia associated with hypertension are a cause or an effect of the 'hypertensive syndrome.' It has been documented that epinephrine, acting primarily through the beta adrenergic receptor, markedly impairs hepatic as well as peripheral tissue sensitivity to increments in plasma insulin [20]. It could, therefore, be argued that a primary increase in sympathetic activity (increases in plasma catecholamines) may antagonize insulin action and lead to secondary insulin resistance. However, decreasing blood pressure with most antihypertensive drugs does not improve insulin sensitivity or decrease plasma insulin levels [21, 22], which suggests that these metabolic defects are not secondary to hypertension. Resolution of this issue requires more direct and specific experimental interventions, which were attempted in the experiments described below. We examined the hypothesis that insulin resistance and hyperinsulinemia are causally related to hypertension by attempting to improve the defects in glucose metabolism (with vanadium compounds) and studying the consequent changes in blood pressure. We reasoned that if insulin resistance and hyperinsulinemia played a role in the development of hypertension, then such metabolic improvements would lead to a decrease in high blood pressure. However, if these metabolic defects were not causally related to hypertension, or were secondary to it, then specific improvements in these abnormalities should not cause any resultant change in blood pressure.

Effects of vanadium compounds in spontaneously hypertensive rats

Spontaneously hypertensive rats (SHR) exhibit a genetic propensity for hypertension and have been shown to be hyperinsulinemic as compared to their controls, the Wistar Kyoto (WKY) strain [18, 23]. In addition, the SHR exhibit a decreased insulin clearance, which may also result in higher circulating insulin levels [24]. Recent studies have demonstrated that SHR exhibit postabsorptive hyperinsulinemia as compared to their Wistar Kyoto (WKY) controls [25]. The primary reason for postabsorptive hyperinsulinemia in the SHR appears to be 'an enhanced pancreatic beta-cell responsiveness', which results in hypersecretion of insulin in response to a glucose load [26]. It has been proposed that hyperinsulinemia may contribute causally towards the development of high blood pressure in the SHR.

(A) Studies with vanadyl sulfate

In the initial series of experiments, the effects of chronic vanadyl sulfate treatment in the SHR were tested in two stages: a prevention study, where the drug treatment was initiated when the SHR were only 6 weeks of age and a reversal study, where it was given after the SHR were fully hypertensive (11 weeks of age). Vanadyl sulfate, in doses of 0.4–0.6 mmol/kg/day markedly attenuated the increase in both plasma insulin and systolic BP in the SHR when it was given before the SHR were fully hypertensive (Table 1, [27]). When vanadyl was given after the SHR became fully hypertensive, it again caused significant and sustained decreases in plasma insulin and blood pressure. Vanadyl treatment did not affect plasma glucose levels in either the SHR or the WKY (Table 1). No changes in BP or plasma insulin were seen in the normotensive WKY. Since vanadyl sulfate decreased food/fluid consumption and body weight in the treated rats, a separate study was initiated in which one group of rats (SHR as well as WKY) was pair-fed with the corresponding vanadyl treated group, but was not given vanadyl. This was done to observe if a decrease in food and fluid intake per se caused a lowering of plasma insulin levels and thereby contributed towards the amelioration of hypertension in the vanadyl treated rats. We observed that the pair-fed SHR remained as hypertensive as the untreated rats, thereby excluding any contribution of weight loss towards the antihypertensive effects observed with vanadyl (Table 1). Ten weeks of vanadyl sulfate treatment had no effect on plasma GOT (glutamic oxaloacetic transaminase), GPT (glutamic pyruvic transaminase) or urea levels [27], indicating that vanadyl did not affect hepatic or renal function in the SHR and WKY.

Table 1. Various parameters from the experimental groups in the vanadyl sulfate studies

Rats	B.P. (mmHg)	Insulin (pmol/L)	Glucose (mmol/L)
SHR	189 ± 3	336 ± 12	6.7 ± 0.1
SHRV	158 ± 2†	252 ± 22†	6.8 ± 0.2
SHRP	193 ± 4	293 ± 18′†	6.9 ± 0.2
WKY	135 ± 1†	264 ± 22†	6.7 ± 0.1
WKYV	127 ± 12†	262 ± 12†	6.6 ± 0.3
WKYP*	138 ± 2†	270 ± 18†	7.0 ± 0.3

*SHR (untreated, n = 15); SHRV (vanadyl treated, n = 8); SHRP (pair-fed, n = 7); WKY (untreated, n = 12); WKYV (vanadyl treated, n = 6) and WKYP (pair-fed, n = 8). Values are mean ± SE.†p < 0.001, different from SHR. Calculated in part from: reference [27]; Bhanot, S and McNeill, J.H.: Hypertension, 1994.

(B) Studies with bis(maltolato)oxovanadium(IV)

Vanadyl sulfate is poorly absorbed from the gastrointestinal tract (absorption ranges from 1–10%) and also causes some gastrointestinal side effects [28, 29]. In an effort to increase the absorption of vanadyl from the gut and to decrease its gastrointestinal toxicity, an organic vanadium complex, bis(maltolato)oxovanadium(IV) (BMOV), was synthesized by complexing one molecule of vanadyl with two molecules of the common food additive maltol [30]. We speculated that an organic compound would be more lipophilic than its inorganic counterpart and may, therefore, be better absorbed from the GI tract. Preliminary studies indicated that BMOV also exhibited insulin enhancing effects and that besides lowering plasma glucose levels in diabetic rats, it also decreased insulin levels in non-diabetic rats [30]. In addition, BMOV also had a lesser negative effect on weight gain in rats as compared with vanadyl sulfate [30]. Furthermore, in the previous experiments, we had not examined the effect of vanadyl on insulin sensitivity in the SHR. Although euglycemic clamp studies conducted in anesthetized SHR demonstrate that they are insulin resistant, recent studies done in conscious SHR have challenged this notion. These investigators have proposed that the SHR are more responsive to stress (anesthesia or restraint) as compared to the WKY [25]. Consequently, there may be a greater release of endogenous catecholamines in the SHR, which may in turn antagonize insulin action and lead to secondary insulin resistance in the anesthetized SHR. Their studies done in conscious, minimally restrained rats showed no difference in insulin sensitivity in the SHR as compared to the WKY. However, the presence of hyperinsulinemia in the SHR was confirmed even in those studies [25, 26] and it was proposed that increased insulin levels may contribute to the development of high BP in the SHR.

Therefore, we initiated chronic BMOV treatment in the SHR with the following aims: (a) to assess insulin sensitivity in SHR and WKY by performing euglycemic hyperinsulinemic clamps in conscious, minimally restrained rats

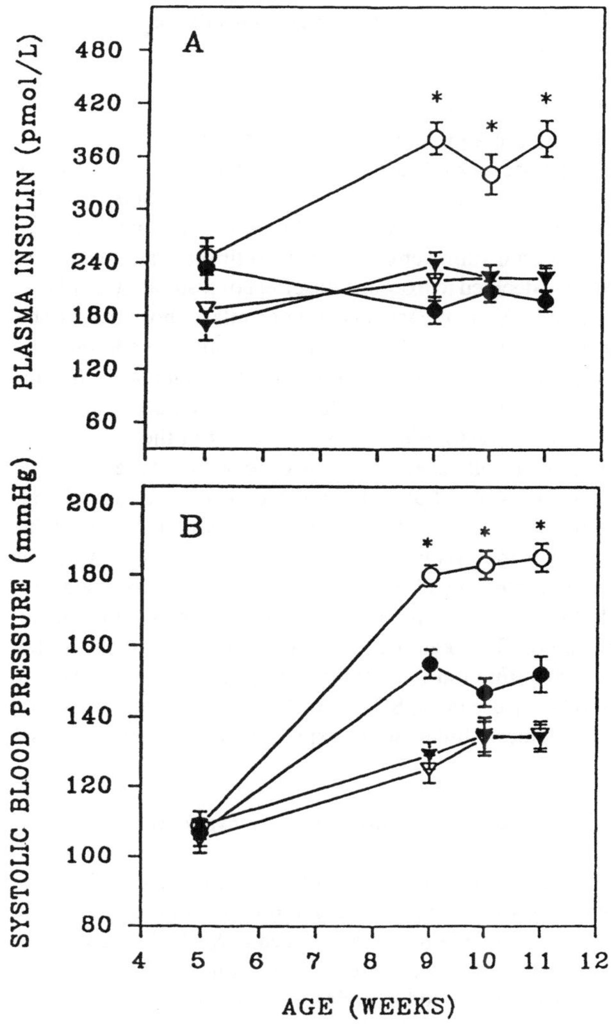

Fig. 1. (A) Plasma insulin levels and (B) Systolic blood pressure in the four groups: O SHR (untreated, n = 9), ■ SHR BMOV treated (n = 11), ∇ WKY (untreated, n = 11) and ▼ WKY BMOV treated, n = 9). Values are means ± SE. *p < 0.0001, SHR different from the other 3 groups.

and (b) to examine the effects of the drug on insulin sensitivity, plasma insulin levels and systolic blood pressure. BMOV (0.35–0.45 mmol/kg/day) caused a marked reduction in plasma insulin levels in the hyperinsulinemic SHR (Fig. 1A, [31]); the average percentage decrease ranged from 30–35% (5 h fasted values). BMOV also lead to a marked reduction in systolic BP (30–35 mmHg) in the SHR (Fig. 1B). No effect on insulin or BP was observed in the WKY rats. BMOV caused no reduction in food intake or body weight in either the SHR or the WKY until 11 weeks of age [31]. Euglycemic, hyperinsulinemic clamps conducted in conscious, minimally restrained rats indicated that the SHR were not insulin resistant as compared to the WKY and that

BMOV caused a further increase in insulin sensitivity in the SHR [31].

(C) Insulin implant studies

Vanadium compounds have been shown to affect other enzyme systems such as the Na-K-ATPase [32]. It may, therefore, be argued that vanadyl may also affect factors other than insulin; ie, it may lower blood pressure independent of its effects on insulin action. In order to further examine this issue, we decided to examine the effect of restoration of plasma insulin levels in the vanadyl-treated rats to those that existed in the untreated SHR. In a separate study, we administered exogenous insulin (14000 pmol/kg/day given as a subcutaneous insulin implant) to the vanadyl-treated rats and observed that restoration of insulin levels in the drug-treated group reversed the effects of the drug on blood pressure [27]. Reversal of blood pressure was independent of any change in plasma glucose or body weight [27], which indicated that whatever the precise mechanism of vanadyl might be, the anti-hypertensive effects of the drug could be reversed simply by raising plasma insulin levels to those seen in the untreated SHR. Taken together, these data suggest that either hyperinsulinemia may contribute towards the increase in blood pressure in the SHR or that the underlying mechanism is very closely linked to the expression of both the disorders.

Effects of vanadyl sulfate in fructose hypertensive rats

The fructose rat model represents an acquired form of systolic hypertension, where the rise in blood pressure is not genetically determined but is diet induced [19, 33]. Some very interesting findings have emerged from studies in which insulin resistance and hyperinsulinemia were induced in normotensive Sprague Dawley rats by giving them a fructose enriched diet [33]. Induction of these metabolic defects was associated with a concomitant increase in blood pressure in these rats [19, 33]. Furthermore, exercise training (which resulted in improved insulin sensitivity) and somatostatin administration (which decreased hyperinsulinemia) to the fructose fed rats attenuated the fructose induced increase in blood pressure in the animals [34, 35]. Although the precise mechanism of fructose-induced hypertension has not been elucidated, it has been suggested that the rise in blood pressure is secondary to the development of insulin resistance and hyperinsulinemia [34]. In order to determine if these metabolic defects were causally related to the fructose induced increase in blood pressure, we initiated chronic vanadyl sulfate treatment in the fructose-fed rats and examined the resultant changes in plasma insulin levels and blood pressure. Furthermore, we studied the effects of vanadyl treatment on insulin sensitivity by conducting euglycemic clamps in the rats. We observed that fructose feeding caused a marked impairment in insulin sensitivity (an 80% decrease as compared to control untreated rats). Vanadyl treatment completely prevented this decrease in insulin sensitivity in the fructose-fed rats [36]. In addition, vanadyl completely prevented the fructose induced increases in plasma insulin and blood pressure. Once again, restoration of plasma insulin levels in the fructose-vanadyl-treated rats to pre-treatment levels reversed the anti-hypertensive effects of vanadyl. Reversal of vanadyl's antihypertensive effects was independent of any change in plasma glucose or body weight. These results are consistent with the hypothesis that hyperinsulinemia may contribute causally towards the development of fructose induced increases in blood pressure.

Since our initial demonstration of the anti-diabetic effects of vanadium compounds *in vivo* [15], impressive advances have been made in understanding the glucose lowering properties and the mechanism of action of these compounds. We believe that our studies are the first to demonstrate that these compounds also exhibit antihypertensive effects in two well established models of experimental hypertension. Further studies are needed to examine the effects of these compounds on specific organ systems and to elucidate the mechanism/s of action underlying their anti-hypertensive properties.

Acknowledgements

The studies quoted in the text were partly supported by funds from the Heart and Stroke Foundation of B.C. and Yukon. S.B. is a Medical Research Council of Canada Fellow.

References

1. DeFronzo RA, Ferrannini E: Insulin resistance: a multifaceted syndrome responsible for NIDDM, obesity, hypertension, dyslipidemia and atherosclerotic cardiovascular disease. Diabetes Care 14: 173-194, 1991
2. Ferrannini E, Natali A: Essential hypertension, metabolic disorders and insulin resistance. Am Heart J 121: 1274–1282, 1991
3. Reaven GM: Role of insulin resistance in human disease. Diabetes 37: 1595–1607, 1988
4. Modan M, Halkin H, Almog S, Lusky A, Eskol A, Shefi M, Shitrit A, Fucho I: Hyperinsulinemia: a link between hypertension, obesity and glucose intolerance. J Clin Invest 75: 809–417, 1985
5. Bonora E, Zavaroni I, Pezarossa A, Alpi O, Brushi F, Dall'Aglio E, Guerra L, Coscelli C, Butturini U: Relationship between blood pressure and plasma insulin in nonobese and obese nondiabetic subjects. Diabetologia 30: 719–723, 1987
6. Pollare T, Lithell H, Berne C: Insulin resistance is a characteristic feature of primary hypertension independent of obesity. Metabolism 39: 167–174, 1990
7. Haffner SM, Fong D, Hazuda HP, Pugh JA, Patterson JK: Hyperinsulinemia, upper body adiposity and cardiovascular risk factors in non-diabetics. Metabolism 37: 338–345, 1988
8. Ferrannini E, Buzzigoli G, Bonadonna R, Giorico MA, Oleggini M,

Graziadei L, Pedrinelli R, Brandi L, Bevilacqua S: Insulin resistance in essential hypertension. N Eng J Med 317: 350–354, 1987

9. Shechter Y: Insulin-mimetic effects of vanadate: possible implications for future treatment of diabetes. Diabetes 39: 1–5, 1990
10. Duckworth WC, Solomon SS, Loiepnieks J, Hamel FG, Hand S, Peavy DE: Insulin like effects of vanadate in isolated rat adipocytes. Endocrinology 122: 2285–2289, 1988
11. Clark AS, Fagan JM, Mitch WE: Selectivity of the insulin-like action of vanadate on glucose and protein metabolism in skeletal muscle. Biochem J 232: 273–276, 1985
12. Dubyak GR, Kleinzeller A: The insulin-mimetic effects of vanadate in isolated rat adipocytes. Dissociation from effects of vanadate as (Na/K)ATPase inhibitor. J Biol Chem 255: 5306–5312, 1980
13. Degani H, Gochin M, Karlish SJD, Shechter Y: Electron paramagnetic studies and insulin-like effects of vanadium in rat adipocytes. Biochemistry 20: 5795–5799, 1981
14. Shechter Y, Ron A: Effect of depletion of phosphate and bicarbonate ions on insulin action in rat adipocytes. J Biol Chem 261: 1494–14950, 1986
15. Heyliger CE, Tahiliani AG, McNeill JH: Effect of vanadate on elevated blood glucose and depressed cardiac performance of diabetic rats. Science 227: 1474–1477, 1985
16. Ramanadham S, Mongold JJ, Brownsey RW, Cros GH, McNeill JH: Oral vanadyl sulphate in the treatment of diabetes mellitus in the rat. Am J Physiol 257: H904–H911, 1989
17. Ramanadham S, Cros GH, Mongold JJ, Serrano JJ, McNeill JH: Enhanced in vivo sensitivity of vanadyl treated diabetic rats to insulin. Can J Physiol Pharmacol 68: 486–491, 1990
18. Mondon CE, Reaven GM: Evidence of abnormalities of insulin metabolism in rats with spontaneous hypertension. Metabolism 37: 303–305, 1988
19. Hwang IS, Ho H, Hoffman BB, Reaven GM: Fructose induced insulin resistance and hypertension in rats. Hypertension 10: 512–516, 1987
20. Diebert DC, DeFronzo RA: Epinephrine induced insulin resistance in man: a beta receptor mediated phenomenon. J Clin Invest 65: 717–721, 1980
21. Swislocki ALM, Hoffman BB, Reaven GM: Insulin resistance, glucose intolerance and hyperinsulinemia in patients with hypertension. Am J Hypert 2: 419–423, 1989
22. Shen DC, Shieh SM, Fuh MMT, Wu DA, Chen YDI, Reaven GM: Resistance to insulin stimulated glucose uptake in patients with hypertension. J Clin Endocrinol Metab 66: 580–583, 1988
23. Hulman S, Falkner B, Chen YQ: Insulin resistance in the spontaneously hypertensive rat. Metabolism 40: 359–361, 1991
24. Mondon CE, Reaven GM, Azhar S, Lee CM, Rabkin R: Abnormal insulin metabolism by specific organs from rats with spontaneous hypertension. Am J Physiol 257: E491–E498, 1989
25. Buchanan TA, Sipos GF, Madrilejo N, Chaplin L, Campese VM: Hypertension without peripheral insulin resistance in spontaneously hypertensive rats. Am J Physiol 262: E14–E19, 1992
26. Buchanan TA, Youn JH, Campese VM, Sipos GM: Enhanced glucose tolerance in spontaneously hypertensive rats: pancreatic β cell hyperfunction with normal insulin sensitivity. Diabetes 41: 872–878, 1992
27. Bhanot S, McNeill JH: Vanadyl sulfate lowers plasma insulin levels and blood pressure in spontaneously hypertensive rats. Hypertension 24: 625–632, 1994
28. Conklin AW, Skinner TL, Fetin TL, Sanders CL: Clearance and distribution of intratracheally instilled vanadium-48 compounds in the rat. Toxicol Letter 11: 199–203, 1982
29. Underwood EJ: In 'Trace elements in human and animal nutrition" Academic Press, New York, pp 388–397, 1977
30. McNeill JH, Yuen VG, Hoveyda HR, Orvig C: Bis(maltolato)oxovanadium(IV) is a potent insulin mimic. J Med Chem 35: 1489–1491, 1992
31. Bhanot S, Bryer-Ash M, Cheung A, McNeill JH: Bis(maltolato)-oxovanadium(IV) attenuates hyperinsulinemia and hypertension in spontaneously hypertensive rats. Diabetes 43: 857–861, 1994
32. Cros GC, Mongold JJ, Serrano JJ, Ramanadham S, McNeill JH: Effects of vanadyl derivatives on animal models of diabetes. Mol Cell Biochem 109: 163–166, 1992
33. Hwang IS, Huang WC, Wu JN, Shian LR, Reaven GM: Effect of fructose induced hypertension on the renin-angiotensin-aldosterone system and atrial natriuretic factor. Am J Hypert 2: 424–427, 1989
34. Reaven GM, Ho H, Hoffman BB: Attenuation of fructose induced hypertension in rats by exercise training. Hypertension 12: 129–32, 1988
35. Ho H, Hoffman BB: Somatostatin inhibition of fructose induced hypertension. Hypertension 14: 117–120, 1989
36. Bhanot S, McNeill JH, Bryer-Ash M: Vanadyl sulfate prevents fructose-induced hyperinsulinemia and hypertension in rats. Hypertension 23: 308–312, 1994

Molecular and Cellular Biochemistry **153:** 211–215, 1995.

Vanadate induces normolipidemia and a reduction in the levels of hepatic lipogenic enzymes in obese Zucker rat

Subbiah Pugazhenthi, Atta Hussain, Bing Yu, Roger W. Brownsey[1], Joseph F. Angel and Ramji L. Khandelwal

Department of Biochemistry, University of Saskatchewan, 107 Wiggins road, Saskatoon, Saskatchewan, Canada, S7N 5E5 and [1]Department of Biochemistry and Molecular Biology, University of British Columbia, Vancouver, British Columbia, V6T 1Z3, Canada

Abstract

The effects of vanadate administration on the plasma lipids and hepatic lipogenic enzymes were investigated in Zucker (fa/fa) rat, a model for obesity and non insulin-dependent diabetes. These animals were administered sodium orthovanadate through drinking water for a period of four months. The plasma levels of insulin, triacylglycerols and total cholesterol were significantly ($p < 0.001$) elevated in untreated obese control rats as compared to the lean animals. In the livers of obese rats, the number of insulin receptors decreased by 60% and the activities of lipogenic enzymes acetyl-CoA carboxylase and ATP-citrate lyase increased by 4.7- and 5.6-folds, respectively . The messenger RNA for ATP-citrate lyase as measured by Northern blot analysis showed a parallel increase in obese control rats. Treatment of these rats with vanadate caused 56–77% decreases in the plasma levels of insulin, triacylglycerols and total cholesterol. The insulin receptor numbers in vanadate-treated obese rats increased (119%) compared to levels in untreated obese animals. The elevated activities of acetyl-CoA carboxylase and ATP-citrate lyase observed in livers of obese rats were significantly reduced by vanadate. The messenger RNA for ATP-citrate lyase also decreased in vanadate-treated obese rats back to the lean control levels. This study demonstrates that vanadate exerts potent actions on lipid metabolism in diabetic animals in addition to the recognized effects on glucose homeostasis. (Mol Cell Biochem **153:** 211–215, 1995)

Key words: vanadate, obese Zucker rat, acetyl-CoA carboxylase, ATP-citrate lyase, insulin receptor, hypertriglyceridemia, hypolipidemic agent

Introduction

Genetically obese Zucker (fa/fa) rats exhibit characteristic hyperinsulinemia, hyperlipidemia and normoglycemia [1]. Lipogenesis is significantly elevated in adipose tissue and liver of these rats, perhaps accounted for in part by hyperinsulinemia [2]. Though hyperphagia is an important defect in these animals, food restriction does not decrease the body fat content [3]. There is a preferential channeling of nutrients towards lipid synthesis resulting in the elevation of plasma lipids [3]. Long-term complications of diabetes such as coronary heart disease are known to be associated with hypertriglyceridemia [4]. Hence the Zucker rat is a suitable animal model for testing the effects of hypolipidemic agents.

Sodium orthovanadate, a potent inhibitor of phosphotyrosine phosphatase, has been shown to exert insulin-like effects *in vitro* and *in vivo* [5, 6]. Chronic administration of vanadate to streptozotocin-induced diabetic rats and db/db mice results in significant lowering of plasma glucose and normalization of many metabolic abnormalities [7, 8]. Long term treatment of obese Zucker rats with vanadate leads to significant improvement in glucose homeostasis and increases glucose utilization in muscle [9, 10]. Vanadate stimulates lipid synthesis and potentiates the insulin-induced

Address for offprints: R.L. Khandelwal, Department of Biochemistry, University of Saskatchewan, 107 Wiggins Road, Saskatoon, SK, S7N 5E5, Canada lipogenesis in adipocytes [11]. However, the enhanced hepatic lipogenesis observed in rats fed a high sucrose diet is corrected *in vivo* by this insulin-mimetic agent [12]. The aim of this study was to examine whether the abnormality of enhanced lipid storage in obese Zucker rat might be corrected by chronic treatment with vanadate.

Materials and methods

Chemicals

[γ-^{32}P]ATP, [α-^{32}P]dCTP and [^{125}I]insulin were obtained from New England Nuclear (Lachine, Canada). Sodium [^{14}C]-bicarbonate was obtained from Amersham International (Oakville, ON, Canada). Sodium orthovanadate and other fine chemicals were purchased from Sigma Chemical Co. (St. Louis, Missouri, USA).

Experimental animals

Five week-old Zucker rats were divided into three groups of 8 rats each: lean (Fa/fa) control, obese (fa/fa) control and obese (fa/fa)-vanadate treated. Control animals drank 0.5 g% sodium chloride solution whereas rats in the treated group drank 0.5 g% sodium chloride solution containing 0.8 mg/ml sodium orthovanadate. The concentration of vanadate was increased gradually from 0–0.8 mg/ml over a period of one week. The treatment was continued at this dose for a period of four months. At the end of treatment, animals were killed in fed condition in the morning. Blood samples collected by orbital sinus bleeding, were centrifuged at 3000 × g for 10 min and the plasma was stored at –80°C. The liver samples were frozen in liquid N_2 and stored at –80°C.

Plasma analyses, enzyme assays and insulin binding

The concentrations of glucose, insulin, triacylglycerols and cholesterol in plasma and the activity of ATP-citrate lyase were determined by the procedures described earlier [12]. Assay of hepatic acetyl-CoA carboxylase was carried out according to the method of Winz *et al* [13]. Partial purification of insulin receptor from liver and [^{125}I]insulin binding were carried out by standard procedures [14].

Northern blot hybridization

Total RNA was isolated from the liver samples by the method of Chomczynski and Sacchi [15]. RNA samples (50 µg) were separated on Agarose-formaldehyde gels and blotted to Nylon Hybond C (Amersham). Two hundred base pairs fragment from the 5′ prime end of ATP-citrate lyase cDNA clone [16] was labelled with [α-^{32}P]dCTP using Amersham random primer kit. The filter was hybridized with this probe overnight at 65°C, washed in 5 X Saline-Sodium phosphate-EDTA buffer (SSPE) for 15 min at 42°C (twice), in 1 X SSPE, 0.1% (w/v) Sodium dodecyl sulfate (SDS) for 30 min at 42°C and finally in 0.1 X SSPE, 0.1% (w/v) SDS for 15 min at room temperature. It was dried and exposed to Kodak X-OMAT x-ray film with an intensifying screen at –80°C. Quantitation of mRNA was carried out by the densitometric scanning of autoradiograms using an LKB 2202 Ultroscan laser densitometer.

Results

Body weight, the plasma levels of glucose, insulin, triacylglycerols and cholesterol, and insulin binding in different groups of animals are shown in Table 1. Obese rats had 57% higher body weight compared to the lean animals. The treatment of obese rats with vanadate led to a reduction of body weight to values which were comparable to the lean rats. The mean daily food consumption per rat was 42 g in the obese control group as compared to 28 g in the lean control group. Vanadate treatment of the obese rats decreased the food intake to the lean control level. The average water consumptions were 8.0, 5.5 and 8.7 ml/100 g body weight/day in the lean control, obese control and obese-vanadate treated rats respectively. Plasma glucose level in the vanadate-treated obese rats was slightly lower when compared to the untreated obese rats. In the untreated obese rats, plasma insulin level was 5-fold higher than in the lean control animals. After treatment with vanadate, the insulin level decreased by 77%, resulting in values similar to those found in lean control animals. Obese Zucker rats were characterized by severe lipemia. The plasma levels of triacylglycerols were 77.7 ± 16.8 and 2.0 ± 0.2 mM in obese and lean control rats, respectively. Vanadate treatment of the obese rats resulted in 67% decrease in triacylglycerols. Similarly, plasma cholesterol concentration in the obese rats was 8.7-fold higher than in the lean animals and this level was brought down by 56% after chronic administration of orthovanadate. The total insulin binding capacity in livers of the obese rats decreased by 60% and this was improved to 87% of lean control levels after treatment with vanadate. The affinity of the insulin receptor for its ligand (as expressed by Kd values) was very similar in the three groups of livers.

In order to determine whether the levels of plasma triacylglycerols in various groups of animals had any correlation with lipogenic enzyme activities in liver, the activities

Table 1. Effect of vanadate treatment on body weight, plasma levels of glucose, insulin, triacylglycerols and cholesterol, and insulin receptor properties in obese zucker rats

			Plasma			Insulin receptor	
Animals	Body weight (g)	Glucose (mM)	Insulin (pM)	Triacylglycerols (mM)	Cholesterol (mM)	Total insulin binding capacity (fmol/mg protein)	Kd (nM)
Lean contol	466 ± 6.3	8.4 ± 0.6	565 ± 47	2.0 ± 0.2	1.44 ± 0.1	253.3 ± 5.1	1.33 ± 0.04
Obese control	732 ± 26.3[d]	9.4 ± 0.8	2804 ± 449[d]	77.7 ± 16.8[d]	12.47 ± 2.3[d]	100.0 ± 4.9[d]	1.11 ± 0.03
Obese vanadate-treated	423 ± 33.9[c]	7.1 ± 0.4[a]	650 ± 148[c]	25.6 ± 6.4[b]	5.43 ± 1.4[a]	219.3 ± 9.3[c]	1.38 ± 0.15

Values are mean ± S.E.M. of 8 rats. Compansons were made by student's t test [a]$p < 0.05$, [b]$p < 0.01$ and [c]$p < 0.001$ vs obese control. [d]$p < 0.001$ vs lean control.

of ATP-citrate lyase and acetyl-CoA carboxylase in the livers of these animals were determined. As shown in Fig. 1A, the activity of ATP-citrate lyase in the obese untreated rats was 4.7-fold higher as compared to the lean animals and vanadate treatment resulted in a 61% decrease. Similarly, acetyl-CoA carboxylase was 5.6-fold higher in the livers of obese rats and this activity was decreased by 40% after vanadate treatment (Fig. 1B). The activities of both enzymes in different animal groups paralleled the plasma levels of insulin and triacylglycerol. The level of messenger RNA (mRNA) encoding ATP-citrate lyase was also measured in these animals by Northern blot analysis and results are illustrated by an autoradiogram (Fig. 1C). Results of quantitation by scanning densitometry of autoradiograms are summarized in Fig. 1D. These results confirm substantial increase in the mRNA levels for ATP-citrate lyase in obese rats and subsequent decrease after vanadate treatment.

Discussion

Previous *in vivo* studies with vanadate have examined essentially the anti-hyperglycemic effect of vanadate. The aberrations in lipid metabolism in a diabetic animal model like streptozotocin-treated rat are secondary to changes in glucose homeostasis. Our present investigation looks at the hypolipidemic effect of vanadate using a normoglycemic hyperlipidemic animal model, genetically obese Zucker rats. Findings of this study indicate that vanadate can induce normolipidemia and a reduction in the expression and activities of hepatic lipogenic enzymes in these animals.

The obese Zucker rats are characterized by increased food intake leading to an elevation of body fat content [1]. In the younger obese rats, enhanced lipogenesis is especially apparent in adipose tissue. Marked hyperresponsive insulin-mediated glucose transport is observed in adipose cells as a result of increased number of glucose transporters in the intracellular pool [17]. Insulin receptor kinase has been also shown to be hyperresponsive in adipocytes of young obese rats [18]. Hence hyperinsulinemia and increased responsiveness to insulin can lead to marked elevation of lipogenesis in adipose tissue. However, Godbole and York [19] demonstrated that in older obese rats lipogenesis is decreased in adipose tissue and increased in liver. Our study being in older rats, therefore, focussed on plasma triacylglycerol levels and hepatic lipogenic enzymes.

Hyperlipidemia of obese rats in our study was associated with significant increases in the activities of the lipogenic enzymes, ATP-citrate lyase and acetyl-CoA carboxylase. The enhanced activity of ATP-citrate lyase was accompanied by a parallel increase in the expression of this enzyme (Fig. 1). The *de novo* fatty acid synthesis has been found to be higher in hepatocytes from obese rats compared to the lean animals [20]. Enhanced hepatic lipogenesis in these animals could be due to hyperinsulinemia. The activities of hepatic lipogenic enzymes are elevated in other animal models for hyperinsulinemia such as db/db mice [8] and rats fed high sucrose diets [12].

Vanadate treatment led to significant improvement in the hyperinsulinemic status. The decrease in the insulin receptor numbers was also corrected by this insulin-mimetic agent. Hypolipidemic effects of vanadate in obese rats could, therefore, be due to decrease in the plasma insulin levels. We had earlier observed similar effects of vanadate on plasma insulin levels and hepatic lipogenic enzyme activities in db/db mice and high-sucrose fed rats [8, 12]. Vanadate seems to have a dual effect on lipogenic enzyme activities in the animal models for two types of diabetes. In insulin-deficient diabetic rats (streptozotocin-induced) vanadate increases the enzyme activities to the normal levels whereas in hyperinsulinemic obese animals vanadate decreases the activities to the lean control levels.

Vanadate exerted differential effects on various parameters

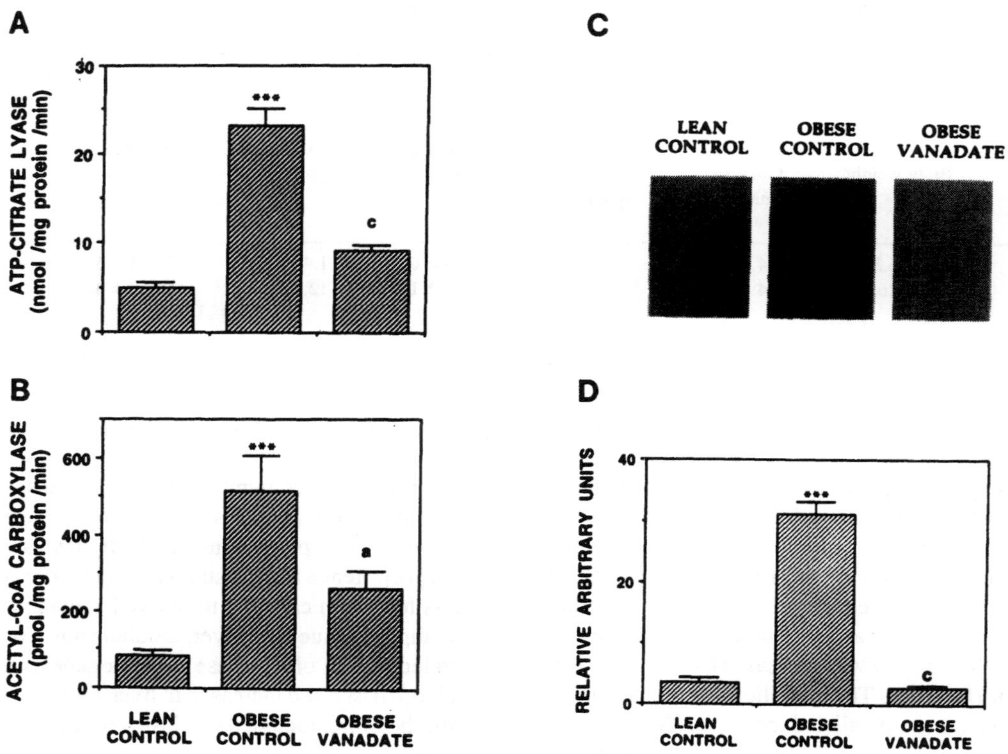

Fig. 1. The activities of hepatic ATP-citrate lyase (A) and acetyl-CoA carboxylase (B), and mRNA levels of ATP-citrate lyase (C and D) in lean control, obese control and obese-vanadate treated rats. The enzyme activities are mean ± S.E.M. of 8 samples and are expressed as nmol of NADP reduced to NADPH/mg protein/min (ATP-citrate lyase) and pmol of malonyl-CoA produced/mg protein/min (acetyl-CoA carboxylase). The mRNA levels for ATP-citrate lyase were determined by North blot hybridization and results are presented by an autoradiogram of a representative experiment with three RNA samples from each group (C) and quantitation (mean + S.E.M. of 6–8 samples) by scanning densitometry (D). Comparisons were made by Student's *t*-test. ***$p < 0.001$ vs. lean control; $^a p < 0.05$ and $^c p < 0.001$ vs. obese control.

measured in obese Zucker rats. The plasma insulin level in obese rats was completely normalized to the level in lean rats. However, in case of plasma levels of triacylglycerols and cholesterol and the activities of ATP-citrate lyase and acetyl-CoA carboxylase in liver, similar normalization was not observed. Though these parameters were significantly (40–67%) reduced as compared to the obese control rats, they were above the lean control levels. The probable reasons are (i) hyperinsulinemia may not be the only cause of enhanced lipogenesis in obese Zucker rats, and (ii) the plasma insulin level in treated obese rats, in presence of the insulin-mimetic agent, vanadate could have a more stimulatory effect on hepatic lipogenesis as compared to the lean rats. It has been shown earlier that vanadate decreases the plasma insulin level significantly in normal rats without causing any metabolic abnormality [7]. There seems to be a potentiation of insulin action in these animals.

In vanadate-treated diabetic rats there is a decrease in body weight gain. It has been indicated that the hypoglycemic action of vanadate in diabetic rats could be partially due to its body weight lowering effect . However, Brichard *et al* [9] showed by using a pair-fed control that the beneficial effects of vanadate in obese Zucker rats are not secondary to decrease in body weight gain. This agent also exerts anti-hyperglycemic effects in mild-diabetic rats at a low dose which does not affect the food consumption [21]. Further, it has been demonstrated that the elevated hepatic lipogenesis in obese Zucker rats can not be corrected by calorie restriction [3]. When these obese rats were pair-fed with lean rats to get similar body weight, the *in vivo* lipogenic rate and the activities of key lipogenic enzymes such as glucose-6-phosphate dehydrogenase and ATP-citrate lyase in adipocytes decreased to the lean control levels. However, the lipogenic rate and the activities of lipogenic enzymes in liver were significantly higher in obese rats inspite of pair feeding as compared to the lean ones indicating that hyperphagia is not the cause of increased lipogenesis in liver. There seems to be a shift in the utilization of nutrients when food intake is restricted in obese rats. Hence, the decreases in plasma triacylglycerol levels and the expression of ATP-citrate lyase in vanadate-treated obese rats as compared to the untreated ones are not due to decreased body weight gain. Since the non-insulin dependent type of diabetes mellitus is associated with obesity, the lipid lowering action of vanadate is an additional beneficial effect of this insulin-mimetic agent in diabetic animal models

Acknowledgement

This work was supported by grants from the Medical Research Council of Canada. The authors thank Gordon Dong for excellent technical assistance.

References

1. Bray GA: The Zucker-fatty rat: a review. Federation Proc 36: 148–153, 1977
2. Dugail I, Quignard-Boulange A, Le Liepvre X, Ardouin B, Lavau M: Gene expression of lipid storage related enzymes in adipose tissue of the genetically obese Zucker rat. Biochem J 281: 607–611, 1992
3. Martin RJ: In vivo lipogenesis and enzyme levels in adipose and liver tissues from pair-fed genetically obese and lean rats. Life Sciences 14: 1447–1453, 1974
4. Santen RJ, Willis PW, Fajans SS: Atherosclerosis in diabetes mellitus. Correlations with serum lipid levels, adiposity, and serum insulin level. Arch Intern Med 130: 833–843, 1972
5. Shechter Y: Insulin-mimetic effects of vanadate. Possible implications for future treatment of diabetes. Diabetes 39: 1–5, 1990
6. Khandelwal RL, Pugazhenthi S: *In vivo* effects of vanadate on hepatic glycogen metabolizing and lipogenic enzymes in insulin-dependent and insulin-resistant diabetic animals. Mol Cell Biochem 153: 87–94
7. Pugazhenthi S, Khandelwal RL: Insulin-like effects of vanadate on hepatic glycogen metabolism in nondiabetic and streptozocin-induced diabetic rats. Diabetes 39: 821–827, 1990
8. Pugazhenthi S, Angel JF, Khandelwal RL: Long-term effects of vanadate treatment on glycogen metabolizing and lipogenic enzymes of liver in genetically diabetic (db/db) mice. Metabolism 40: 941–946, 1991
9. Brichard SM, Pottier AM, Henquin JC: Long term improvement of glucose homeostasis by vanadate in obese hyperinsulinemic fa/fa rats. Endocrinology 125: 2510–2516, 1989
10. Brichard SM, Assimacopoulos-Jeannet F, Jeanrenaud B: Vanadate treatment markedly increases glucose utilization in muscle of insulin-resistant fa/fa rats without modifying glucose transporter expression. Endocrinology 131: 311–317, 1992
11. Fantus IG, Ahmad F, Deragon G: Vanadate augments insulin binding and prolongs insulin action in rat adipocytes. Endocrinology 127: 2716–2725, 1990
12. Pugazhenthi S, Angel JF, Khandelwal RL: Effects of vanadate administration on the high sucrose diet-induced aberrations in normal rats. Mol Cell Biochem 122, 69–75, 1993
13. Winz R, Hess D, Aebersold R, Brownsey RW: Unique structural features and differential phosphorylation of the 280-kDa component (Isozyme) of rat liver acetyl CoA carboxylase. J Biol Chem 269: 14438–14445, 1994
14. Pugazhenthi S, Khandelwal RL: Does the insulin-mimetic action of vanadate involve insulin receptor kinase? Mol cell Biochem 127/128, 211–218, 1993
15. Chomczyhnski P, Sacchi N: Single-step method of RNA isolation by acid guanidinium thiocyanate-phenol-chloroform extraction, Anal Biochem 162, 156–159, 1987
16. Elshourbagy NA, Near JC, Kmetz PJ, Sathe GM, Southan C, Strickler JE, Gross M, Young JF, Wells TNC, Groot PHE. Rat ATP Citrate-lyase. J Biol Chem 265: 1430–1435, 1990
17. Guerre-Millo M, Lavau M, Horne JS, Wardzala LJ: Proposed mechanism for increased insulin-mediated glucose transport in adipose cells from young, obese Zucker rats. J Biol Chem 260: 2197–2201, 1985
18 Debant A, Guerre-Millo M, Marchand-Brustel YL, Freychet P, Lavau M, Van Obberghen E: Insulin receptor kinase is hyperresponsive in adipocytes of young obese Zucker rats. Am J Physiol 252: E273–E278, 1987
19. Godbole V, York DA: Lipogenesis in situ in the genetically obese Zucker fatty rat (fa/fa): Role of hyperphagia and hyperinsulinemia. Diabetologia 14: 191–197, 1978
20. McCune SA, Durant PJ, Jenkins PA, Harris RA: Comparative studies on fatty acid synthesis, glycogen metabolism, and gluconeogenesis by hepatocytes isolated from lean and obese Zucker rats. Metabolism 30: 1170–1178, 1981
21. Henquin JC, Carton F, Ongemba LN, Becker DJ: Improvement of mild hypoinsulinemic diabetes in the rat by low non-toxic doses of vanadate. J Endocrinol 142: 555–561, 1994

Molecular and Cellular Biochemistry **153**: 217–231, 1995.

In vivo and *in vitro* studies of vanadate in human and rodent diabetes mellitus

Allison B. Goldfine, Donald C. Simonson, Franco Folli, Mary-Elizabeth Patti and C. Ronald Kahn

Research Division, Joslin Diabetes Center, Department of Medicine, Brigham and Women's Hospital, and Harvard Medical School, Boston, MA 02215, USA

Abstract

In vivo vanadate and vanadyl have been shown to mimic the action of insulin and to be effective treatment for animal models of both Type I and Type II diabetes. The molecular mechanism of action of the vanadium salts on insulin sensitivity remains uncertain, and several potential sites proposed for the insulin-like effects are reviewed. In human trials, insulin sensitivity improved in patients with NIDDM, as well as in some patients with IDDM after two weeks of treatment with sodium metavanadate. This increase in insulin sensitivity was primarily due to an increase in non-oxidative glucose disposal, whereas oxidative glucose disposal and both basal and insulin stimulated suppression of hepatic glucose output (HGP) were unchanged. Clinically, oral vanadate was associated with a small decrease in insulin requirements in IDDM subjects. Of additional benefit, there was a decrease in total cholesterol levels in both IDDM and NIDDM subjects. Furthermore, there was an increase in the basal activities of MAP and S6 kinases to levels similar to the insulin-stimulated levels in controls, but there was little or no further stimulation with insulin was seen. Further understanding of the mechanism of vanadium action may ultimately be useful in the design of drugs that improve glucose tolerance. (Mol Cell Biochem **153**: 217–231, 1995)

Key words: vanadate, vanadium, insulin dependent diabetes mellitus, noninsulin dependent diabetes mellitus, mechanism of action, phophatases

Introduction

In the late 1800's vanadium was proposed to have medicinal value and to be of benefit in nutrition, diabetes, atherosclerosis, anemia, metabolism of lipids, prevention of dental caries, and treatment of infection, especially tuberculosis and syphilis [1, 2]. Over the past 15 years, considerable evidence has accumulated to show that vanadium salts, specifically tetravalent vanadyl, usually found as the divalent cation VO^{2+}, and pentavalent vanadate, VO_3^-, will mimic insulin action in a number of isolated cell systems and produce dramatic glucose lowering effects when given orally to animal models of both Types I and II diabetes mellitus [1, 3]. *In vitro*, vanadate was first found to mimic insulin action on hexose uptake and glucose metabolism in rat adipocytes in about 1980 and has since been demonstrated to have insulin-like actions in many cell types [3, 4]. The insulinomimetic properties include stimulation of hexose transport in rat adipocytes [3, 4] and rat skeletal muscle [5], stimulation of lipogenesis [6], inhibition of lipolysis [7], stimulation of glucose oxidation [4], and stimulation of glycogen synthase in rat adipocytes [8]. In addition to the effects on glucose metabolism, these compounds appear to enhance K^+ uptake in cardiac muscle cells [9] and stimulate DNA synthesis in cultured cells [10–12] similar to insulin (reviewed in Shechter [13]).

Animal studies

In vivo vanadate and vanadyl have been shown to be effective treatment for animal models of both Type I and Type II diabetes [13–18]. In rodent models of Type I diabetes, oral vanadate has been found to improve blood glucoses in STZ-

Address for offprints: C. Ronald Kahn, Joslin Diabetes Center, One Joslin Place, Boston, MA 02215, USA

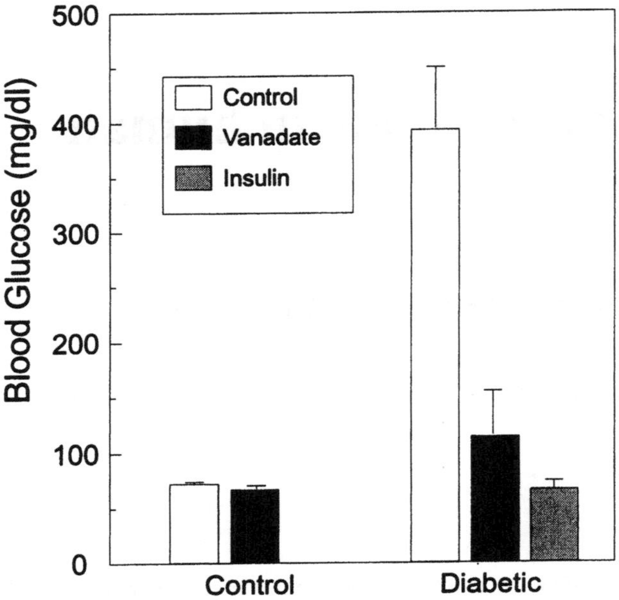

Fig. 1. The effect of vanadate on blood glucose in the STZ-diabetic rat.

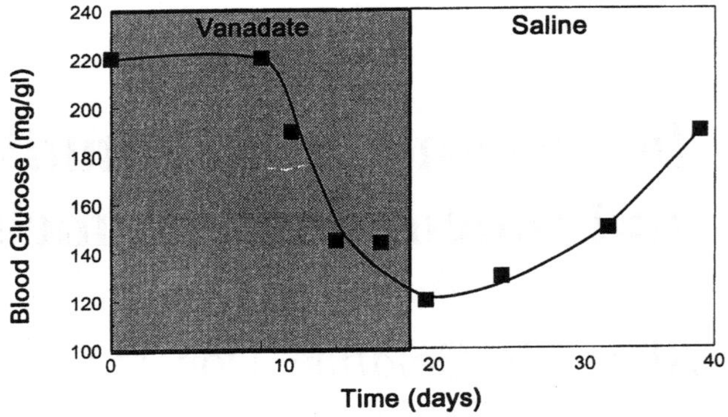

Fig. 2. The effect of vanadate on blood glucose in ob/ob mice.

diabetic rat. Treatment for 5 days with either insulin (14 U/rat/day) or with vanadate (0.6 mg/ml in the drinking water) produced a near normalization of blood glucose in the diabetic rats within 3–4 days, without any exogenously administered insulin (Fig. 1) [19]. Similar improvements in glucose were seen in the BB rat, in the presence of decreased, although continued insulin [20]. The catabolic state induced by the diabetes was reversed in both models by vanadate treatment, and the animals became anabolic. In addition to the improvement in glucose levels, 2 weeks of vanadate therapy restored to normal the decreased hepatic glycogen, fructose-2,6-bisphosphate concentrations and 6-phosphofructo-2-kinase activity present in the STZ-diabetic rat [21]. These beneficial effects were reversible following the removal of vanadate from the drinking water. Furthermore, abnormalities in isolated working heart function and glycerol output from adipose tissue are corrected by vanadyl treatment, and there is an apparent persistence of a therapeutic effect demonstrated up to 13 weeks following completion of a three week treatment period [22].

Vanadate administered in the drinking water has also been shown to lower blood glucose to near normal levels in three models of NIDDM, including ob/ob and db/db mice, and fa/fa rats [14–16]. Ob/ob and db/db mice are two commonly studied rodent models of type II diabetes mellitus, characterized by obesity, hyperglycemia, hyperinsulinemia and a blunted response to insulin at the receptor and postreceptor levels. To investigate the effects of vanadate on blood glucose levels rodent models of NIDDM, vanadate (0.25 mg/ml drinking water) was administered to ob/ob and db/db mice for three weeks [14]. Fasting blood glucose levels were reduced from 236–143 mg/dl in ob/ob mice and from 170–114 mg/dl in ob/+ mice (Fig. 2), and from 228–141 mg/dl in db/db mice and from 126–81 mg/dl in db/+ mice. Vanadate also lowered glucose levels in the fed state, improved oral glucose tolerance and restored early insulin secretion [16]. In both mouse models of Type II diabetes, some effect of vanadium was seen after 2–10 days of treatment, but the effect was not maximal until 25–50 days. This is in contrast to the rapid effects seen after 3–4 days in the STZ-diabetic rats noted above. There was no effect on body weight. The effect of vanadate of blood glucose levels was reversible and after withdrawal the blood glucose levels gradually returned to the original hyperglycemic level. There was no evidence of hepatotoxicity by electron microscopic examination after 47 days of treatment. Furthermore, disappearance rates for intravenous glucose was doubled in treated animals as compared to the controls, the insulin response to the glucose challenge reappeared, hepatic glycogen content doubled, and pancreatic insulin stores were increased. There was also an increase in basal glucose oxidation rates by muscle (hemidiaphragm) in the vanadate-treated animals, although insulin stimulation was similar to that of the control [16].

When given to obese, hyperinsulinemic fa/fa rats, vanadate also improved oral glucose tolerance without decreasing body weight or changing counter-regulatory hormones [15]. In euglycemic, hyperinsulinemic clamp studies, the glucose infusion rate required for stable glycemia was higher in the vanadate treated rats due to higher rates of peripheral glucose disposal. There was no effect of vanadate on inhibition of hepatic glucose production by insulin [23]. In obese mice or rats treated with vanadate, there was no change in the percent lean body mass, again suggesting that the effects of vanadate do not require a major decrease in food intake.

Mechanisms of vanadium's insulin-like actions

The exact mechanism of the vanadium salts effects remains uncertain. In both STZ-diabetic rats and db/db mice oral administration of vanadate improves blood glucoses without increasing serum insulin levels [18, 24], thus indicating that the primary site of action is on insulin target tissues. Insulin action at the cellular level is complex (reviewed in refs [25, 26] (Fig. 3). Insulin initiates its actions by binding to its tetrameric membrane receptor. This receptor is a member of the family of receptor tyrosine kinases and following insulin binding undergoes autophosphorylation on multiple tyrosine residues which in turn activates the receptor kinase toward other substrates. In most cells, the major substrate is a high molecular weight cytosolic protein termed IRS-1 (insulin receptor substrate-1) [27, 28]. IRS-1 possesses 22 potential sites of tyrosine phosphorylation, many in repetitive sequence motifs (YMXM and YXXM). Following phosphorylation IRS-1 serves as a 'docking protein' for other intracellular proteins, including enzymes like phosphatidylinositol 3-kinase (PI 3-kinase) and SHPTP2, and adaptor proteins such as Grb2 [29]. The latter links IRS-1 to a series of closely linked serine/threonine kinases and phosphatases such as the MAP kinases, S6 kinases and protein phosphatase-1A via the ras-GTPase system. These serine kinase act on enzymes like glycogen synthase, transcription factors, and other proteins to produce many of the final biological effects of the hormone. In adipose tissue and muscle, insulin stimulation also increases glucose uptake by promoting translocation of an intracellular pool of glucose transporters to the plasma membrane [30]. Exactly how this action is linked to the phosphorylation cascade is unknown, but several recent studies suggest that this is down-stream of PI 3-kinase and may involve a member of the Rab family of GTPases [31]. In the absence of continued insulin secretion, the cellular actions of the hormone are 'turned off' in two ways: first insulin ei-

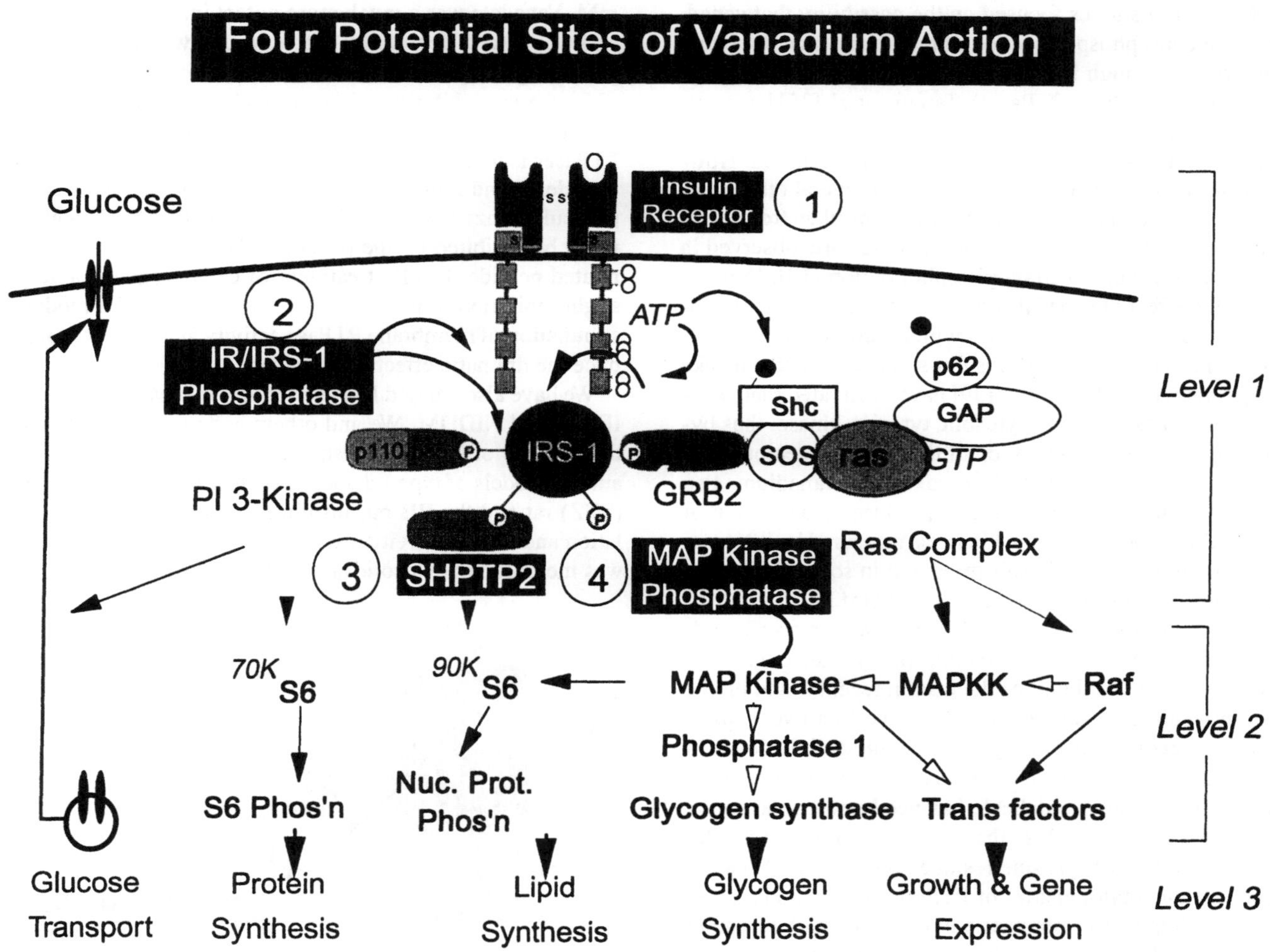

Fig. 3. The cellular mechanism of insulin action and possible sites of vanadium effects.

ther dissociates from its receptor or is internalized and degraded [32], and secondly the activated receptor kinase and IRS-1 molecule are dephosphorylated by the action on a group of enzymes called phosphotyrosine phosphatases [33].

Several potential sites have been proposed for the insulin-like effect of vanadium (Fig. 3). To determine possible mechanisms of vanadate action, serum insulin levels, insulin receptor function and some postreceptor sites of insulin action were analyzed. Plasma insulin levels in both ob/ob and db/db mice showed only small changes following vanadate therapy which could not account for the normalization of glucose. As previously reported [34], a decrease in insulin binding to partially purified insulin receptor is present in ob/ob mice as compared to control ob/+ mice. Vanadate treatment tended to down-regulate the level of insulin binding in the ob/+ treated animals, but had no effect on the already reduced level of insulin binding in ob/ob mice, consistent with a post-receptor mechanism of vanadate action [19].

Most early studies focused on the possibility that vanadate stimulates phosphorylation of the insulin receptor either directly or through its inhibitory effect on phosphotyrosyl protein phosphatase (PTPases). Tamura *et al.* [35] have presented evidence that vanadate might directly stimulate insulin receptor β-subunit tyrosine autophosphorylation. However this action has not been observed in several other studies [19, 36, 37], such that when tissues were taken from vanadate-treated diabetic animals, no increase wass observed in either receptor or substrate phosphorylation [19]. Furthermore, the effect of vanadate on hexose transport is not inhibited by quercitin, a compound which inhibits both insulin receptor tyrosine kinase and insulin-stimulated hexose uptake [38]. Shisheva and Shechter [39] have postulated that vanadate stimulates a soluble cytosolic tyrosine kinase, thus bypassing the need for activation of the insulin receptor itself. By contrast to vanadate itself, peroxides of vanadium, also called pervanadates, do appear to produce an activation of the insulin receptor tyrosine kinase as measured by ^{32}P incorporation into a synthetic substrate, and in some studies are even more potent than insulin in this regard [40].

The best studied site of action for vanadium compounds is at the level of the PTPases. As phosphorylation of tyrosine residues on the β-subunit of the insulin receptor and on the insulin receptor substrate IRS-1 form the first two steps in transduction of the insulin signal at the intracellular level, in normal tissues, the level of tyrosine phosphorylation of these and other proteins is a delicate balance between the action of tyrosine kinases such as the insulin receptor and the dephosphorylation of tyrosine residues by specific PTPases. There are two major classes of PTPases: one class is represented by large, transmembrane molecules which resemble receptors (although no ligands have been found to date) and the other are smaller cytoplasmic enzymes [33]. Genes for over 20 of these PTPases have been cloned, however, the role of many these enzymes in normal physiology has yet to be defined [41, 42]. Vanadate is a potent inhibitor of cellular PTPases, especially the cytosolic PTPases [41, 43]. Thus, one could imagine that vanadate would enhance insulin receptor and/or substrate phosphorylation indirectly by inhibiting the dephosphorylation of these proteins. In isolated cell systems, this effect of vanadate is easily demonstrable, albeit at somewhat higher concentrations than those achieved *in vivo* during treatment of diabetic animals, or humans (Fig. 4).

The effects of insulin and vanadate on *in vitro* PTPase activity in the well-differentiated Fao rat hepatoma cell line was studied using a ^{32}P-labeled peptide corresponding to the major site of insulin receptor autophosphorylation [44]. Of the PTPase activity in Fao cells, 14% was in the cytosolic fraction, whereas 86% was in the particulate fraction; this latter fraction also had a 4-fold higher specific activity against the insulin receptor phosphopeptide. PTPase activity in both fractions was inhibited by zinc at concentrations of 0.1–1.0 mM. Vanadate was a much more potent inhibitor than zinc, and was also differentially active in the two cell fractions producing complete inhibition of the cytosolic PTPase at 20 μM, whereas inhibition of the particulate enzymes required 200 μM for maximal effect (Fig. 4). The effect on cytosolic PTPases is in the range observed during oral administration to rodents and humans (see Section 4). The cytosolic and particulate enzymes could also be distinguished by their ability to be inhibited by the phosphorylated insulin receptor-related peptide. Insulin treatment of cells produced a very slight inhibition of cytosolic PTPase activity and a modest stimulation of membrane PTPase activity, but did not reproduce the dramatic effects of vanadate.

We have also studied PTPase activity in *in vivo* models of IDDM and NIDDM. We and others had previously shown that in spite of increased insulin receptor concentrations in animal models of type I diabetes, such as the streptozotocin (STZ) rat and the BB rat, insulin receptor autophosphorylation and kinase activity is decreased [45]. PTPase activity was measured in cytosolic and particulate fractions of liver

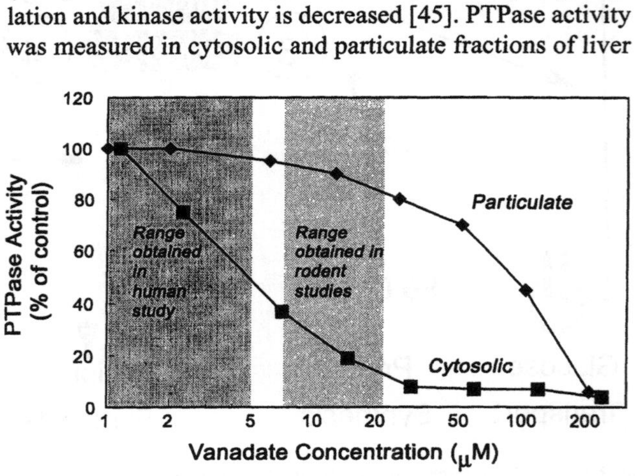

Fig. 4. The effect of vanadate on PTPase activity in Fao cells.

in STZ and BB diabetic rats before and after vanadate therapy using a phosphopeptide substrate with a sequence identical to the major insulin receptor autophosphorylation site [19]. In STZ- and BB-diabetic rats, cytosolic PTPase activity increased to 180% of the control value. In the particulate fraction, PTPase activity was also increased up to 80% at eight days of diabetes in the STZ rat, however this increase was not sustained at 30 days. Thus insulin deficient diabetes is associated with marked alterations in hepatic PTPase activity.

To investigate the effects vanadate on *in vivo* PTPase activity STZ-diabetic rats and BB rats were treated for 5 days with either insulin (14 U/rat/day) or vanadate (0.6 mg/ml in the drinking water) which produced a near normalization of blood glucose. Three days of treatment with either insulin or vanadate resulted in a reduction of PTPase activity in the particulate fraction (Fig. 5). In contrast, no change in cytosolic PTPase activity was seen with either insulin or vanadate. Thus, in two animal models ofType I diabetes there is an early increase in cytosolic and membrane PTPase activity, and treatment with either insulin or vanadate improves glucose homeostasis and reduces membrane PTPase activity.

Similar to changes in insulin receptor tyrosine kinase activity seen in IDDM, insulin receptor kinase activity is also reduced in hepatocytes [46], adiposcytes [47] and skeletal muscle tissue [48] isolated from obese NIDDM subjects. In animal models of NIDDM, cytosolic and membrane associated PTPase activity in ob/ob mice was 55% of that seen in ob/+ mice as assayed using a phosphorylated peptide substrate *in vitro*, whereas in the db/db mice, only the cytosolic PTPase was reduced [19]. Vanadate treatment did not alter the cytosolic or particulate PTPase activity in either ob/ob or db/db mice, in contrast to the findings in insulin deficient models of diabetes mellitus described above. Thus, under the conditions of these experiments, vanadate<apos>s therapeutic effect was observed without any measurable change in hepatic PTPase activity. Furthermore, vanadate treatment was not associated with increases in the low levels of insulin receptor β-subunit or insulin receptor substrate (IRS-1) phosphorylation. Also, no new phosphotyrosyl proteins were detected in the vanadate treated animals. These studies demonstrate that vanadate is an effective hypoglycemic agent in these two rodent models of NIDDM and suggest that vanadate acts distally to the insulin receptor and the insulin receptor substrate IRS-1 and do not involve the PTPase activity in NIDDM models.

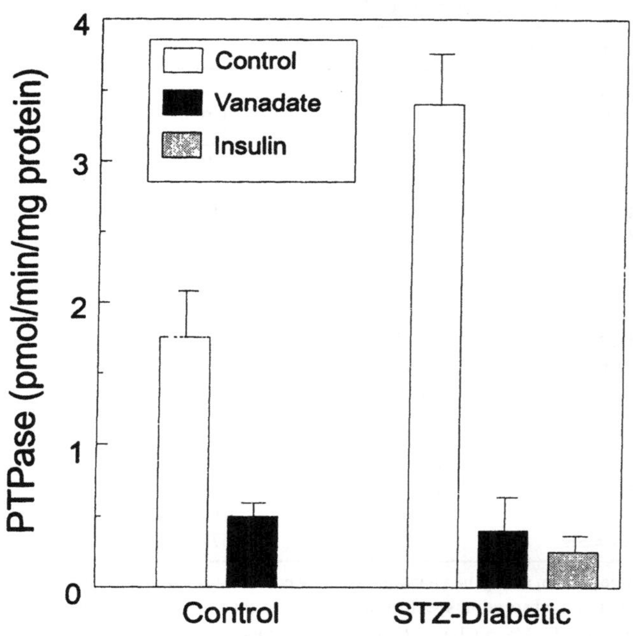

Fig. 5. Effect of vanadate on particulate PTPase activity in STZ-diabetic rats.

Three other post-receptor sites of vanadate action have been explored in a limited manner. In animals that have received vanadate in the drinking water, the uptake of 3-O-methyl glucose, a non-metabolized analogue of glucose, is doubled in muscle and liver tissues of both control and STZ-rats suggesting that the observed lowering of blood glucoses is due to enhanced tissue glucose uptake and metabolism [24]. It is worth noting, however, that insulin itself does not increase hepatic glucose uptake, hence the increased transport of 3-O-methylglucose observed in liver is not simply another insulin-like action. In isolated rat fat pads, vanadate increases the activity of the low K_m (Type IV) cAMP phosphodiesterase (PDE) in the microsomal fraction [49]. Preincubation of the fat pads with inhibitors of protein kinase C results in a partial inhibition of the vanadate effect, but has no effect on insulin-stimulated PDE. Vanadate/glutathione complexes also inhibit the membrane-bound PDE IV in eosinophils [50].

In an attempt to further characterize the possible mechanism by which vanadate improves blood glucose levels in ob/ob mice, several key enzymes and transport proteins involved in carbohydrate metabolism were analyzed at the mRNA level, including PEPCK and the Glut2 glucose transporter [37, 51–53]. Despite the increased gluconeogenesis present in these diabetic animals, ob/ob mice exhibited 90% reductions in hepatic levels of PEPCK mRNA, and 2–5 fold increases in glyceraldehyde-3-phosphate dehydrogenase (GAPDH) and β-cell/liver glucose transporter (GLUT2) mRNA levels, as compared to their lean ob/+ controls. Vanadate treatment resulted in an increase in PEPCK mRNA levels similar to that seen with 24–48 h of starvation [54]. Vanadate also stimulates PEPCK mRNA in hepatoma cells in culture. However, vanadate had no effect on PEPCK mRNA levels in ob/+ controls and did not alter mRNA levels of any of the other proteins analyzed. While vanadate does have effects to increase the expression of several of these

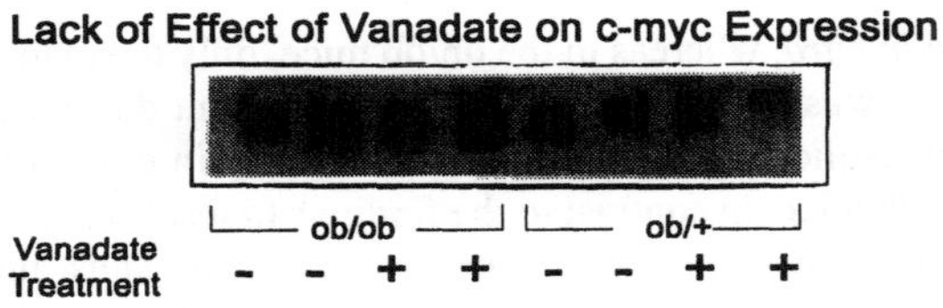

Fig. 6. mRNA levels of c-myc in ob/ob mouse hepatocytes.

proteins at the mRNA level, it is not at all clear if these account for any of the insulin-like effects. Indeed, the effect on PEPCK is exactly opposite that of insulin itself.

Since chronic vanadate treatment has been noted to produce a somewhat altered appearance of cultured fibroblasts suggestive of transformation, mRNA levels of the proto-oncogene c-myc were analyzed. Although 3–4 fold higher levels of c-myc mRNA were seen in the liver of ob/ob mice as compared to their lean controls, six weeks of vanadate administration did not alter the steady-state mRNA levels of this proto-oncogene (Fig. 6).

Human studies of the use of vanadium salts in diabetes mellitus

To investigate the efficacy and mechanism of action of vanadate as oral hypoglycemic agent, 5 subjects with IDDM (age 45 ± 6 yrs, BMI 24.0 ± 0.8 kg/m^2, HbA1c 11.1 ± 1.0%)and 5 with NIDDM (age 52 ± 3 yrs, BMI 28.7 ± 1.7 kg/m^2, HbA1c 10.9 ± 2.0%) were studied before and after oral sodium metavanadate ($NaVO_3$) 125 mg/day. Subjects were studied for 4 weeks including a 1 week baseline period, 2 weeks of treatment with vanadate, and a 1 week post-treatment period. Baseline laboratory testing included glycosylated hemoglobin and fructosamine levels, chest radiograph, electrocardiogram, CBC, chemistry profile, coagulation times, thyroid function tests, urine analysis, and urine microalbumin. Subjects were instructed to monitor blood glucoses four times a day (Lifescan, One Touch Meter) and keep records of insulin dosage and dietary intake for the first, baseline, week. At the end of the first study week, subjects were admitted to the General Clinical Research Center and a two step euglycemic, hyperinsulinemic clamp study was performed to quantitate insulin sensitivity. Deuterated glucose was used to measure endogenous hepatic glucose production and continuous indirect calorimetry was used to measure substrate oxidation and energy expenditure. In addition, subjects with NIDDM on oral hypoglycemic therapy and/or diet alone were evaluated with a 2 h hyperglycemic clamp to assess insulin secretion on the morning preceding the euglycemic clamp.

Subjects were then given sodium metavanadate 125 mg orally each day (50 mg with breakfast, 50 mg with lunch, and 25 mg with supper) for two weeks. Blood glucose monitoring, as well as insulin and dietary logs were continued. Physical examination, blood tests and urine profiles were repeated weekly throughout the study to evaluate for biochemical evidence of toxicity. At the completion of the third study week (the second week of therapy with vanadate), the hyperglycemic clamp (in NIDDM) and the euglycemic insulin clamp studies (in all subjects) were repeated, and the sodium metavanadate was discontinued. The subjects were then followed for an additional week to monitor for delayed adverse effects.

Pharmacology in humans

Although a few subjects experienced mild nausea when vanadate therapy was initiated, this rapidly dissipated. Only one study subject could not tolerate the full dose of sodium metavanadate due to nausea and vomiting. He became asymptomatic when the dose was reduced to 25 mg po tid, and this dose was continued throughout the study. One subject noted increased salivation. Serum vanadium levels were measured by electrothermal atomization atomic absorption spectroscopy at the Metals Laboratory of the Mayo Clinic. The assay has a limit of sensitivity of 1 ng/ml (< 20 nM). Human serum vanadium concentrations are consistently less than this in normal subjects by this assay (T. P. Moyer, personal communication). In addition, 24 h collections of urine were obtained for analysis of vanadium content at the end of the two week treatment period. Serum V levels varied from 19–261 ng/ml (Table 1). Mean vanadium levels were 187

Table 1. Serum and urine vanadium levels measured by atomic absorption spectroscopy at the conclusion of 2 weeks of $NaVO_3$ therapy.

Vanadium levels:		
IDDM		
Patient	Serum	Urine
		(ng/ml) (μg/l)
MV[1]	114	145
KG	139	982
BB	283	683
SH	238	2293
RR	162	792
Mean ±	187.2 ± 28.3	979.0 ± 319.0
NIDDM		
Patient	Serum	Urine
	(ng/ml)	(μg/l)
VS	136	372
RM	19	155
BF	224	418
MB	70	305
WT	261	1638
Mean ± SE	142.0 ± 40.6	577.6 ± 240.4

Source: Serum and urine vanadium levels were measured by graphite furnace atomic absorption spectroscopy. (Special thanks to Dr. Thomas P. Moyer, Ph. D. Mayo Clinic, Rochester, Minnosota)

Note: [1]Dose of vanadate reduced to 75 mg. po qd, divided.

and 142 ng/ml (~3 mM) for the IDDM and NIDDM groups, respectively. There was a 15 fold-range of serum vanadium levels in the NIDDM group which did not correlate with either apparent bioactivity or side effects. Urinary vanadium excretion was about 1–2 mg/day or slightly less than 1% of the administered dose.

Effect of vanadate on insulin sensitivity and insulin action
As a primary end-point of short term therapy, insulin sensitivity was measured using two-step euglycemic, hyperinsulinemic clamps at 0.5 mU insulin/kg/min and 1.0 mU insulin/kg/min [55]. Patients were studied in the postabsorptive state after blood glucose was normalized by administration of a low-dose insulin infusion overnight prior to the study. The clamp studies were performed with 6,6-[$^2H_2$]-glucose to quantitate endogenous glucose production [56]. The clamp studies were performed with continuous indirect calorimetry to assess oxidative versus non-oxidative glucose disposal [57, 58] at baseline and during the last 60 min of each step of the clamp.

Although two of the five subjects with IDDM showed an improved rate of glucose utilization after treatment with sodium metavanadate, on average there was no change in the rate of glucose utilization with either the low (3.6 ± 0.6 *vs* 3.2 ± 0.5 mg/kg/min) or high (8.0 ± 0.8 *vs* 9.0 ± 1.1 mg/kg/min) insulin dose. In contrast, insulin sensitivity improved in all subjects with NIDDM during therapy. Glucose disposal increased by 29% at the 0.5 mU insulin dose (1.7 ± 0.4 *vs* 2.2 ± 0.3 mg/kg/min) and by 39% at the 1.0 mU insulin dose (4.1 ± 1.0 *vs* 5.7 ± 1.3 mg/kg/min) (combined p=0.03) (Figs 7A and 7B). In this small group, there was no apparent correlation between the serum vanadium level achieved and the change in insulin sensitivity.

Oxidative and non-oxidative glucose disposal were calculated from indirect calorimetry data (Fig. 8). Basal oxidative and non-oxidative glucose disposal were similar in subjects with IDDM and NIDDM before and after treatment. The improved insulin sensitivity in the patients with NIDDM during vanadate therapy appeared to be accounted for entirely by increased non-oxidative glucose disposal (1.6 ± 1.0 *vs* 2.9 ± 0.8 mg/kg/min, at the higher insulin dose; $p < 0.03$), with no change in oxidative glucose utilization. Subjects with IDDM demonstrated no significant change in the proportion of oxidative and non-oxidative glucose disposal at either insulin dose.

In contrast to its effect on peripheral glucose utilization, vanadate therapy had no measurable effect on hepatic glucose production (HGP). Basal HGP was unchanged before and after vanadate in both IDDM and NIDDM (Fig. 9). Furthermore, suppression of HGP was similar pre- and post-vanadate in both study groups during both lower and higher dose insulin clamps. These data suggest that the improvement in glucose utilization in NIDDM patients was due to enhanced insulin sensitivity in peripheral tissues (i.e., muscle) without a change in hepatic insulin sensitivity.

Insulin secretion
To assess the effects of vanadate on insulin secretion in subjects with NIDDM not requiring insulin, hyperglycemic clamps [55] were performed before and at the end of treatment. Subjects were studied in the fasted state on the day prior to the euglycemic clamp with blood glucose normalized by administration of a low-dose insulin infusion during the night prior to the study. After the collection of baseline samples for hormones and substrates, the plasma glucose concentration was raised by 100 mg/dl above fasting, and maintained at this level for 2 h. First phase insulin and C-peptide secretion was calculated as the increment in mean plasma insulin or C-peptide level during the first 10 minutes of the hyperglycemic clamp, second phase secretion was calculated as the increment in mean levels during the 10–120 min time period (Fig. 10). C-peptide levels in the basal state (0.69 ± 0.07 *vs* 0.70 ± 0.02 pM/ml) and during the first 10 min of hyperglycemia (0.76 ± 0.07 *vs* 0.81 ± 0.04) and the last 110 min of hyperglycemia (1.16 ± 0.17 *vs* 1.14 ± 0.17) were unchanged after vanadate therapy. Similarly, there was no significant change in basal, first phase or second phase insulin secretion.

Diabetic control
After 2 weeks on vanadate individuals with IDDM showed a small (14%), but significant, decrease in mean daily insulin requirements (39.1 ± 6.6 to 33.8 ± 4.7 units insulin/day, $p < 0.05$), but no change in mean blood glucose (MBG) (157 ± 15 *vs* 152 ± 8 mg/dl). The decrease in insulin requirements did not correlate with the improvement in insulin sensitivity or with any change in caloric intake. There was no change in the dose of the oral hypoglycemic agent in any NIDDM subject. In view of the short duration of the study, it was not surprising that there was no significant change in glycohemoglobin, although values in both groups decreased by ~10% (11.1 ± 1.0 *vs* 10.2 ± 0.8, 10.9 ± 2.0 *vs* 10.2 ± 1.1% in IDDM and NIDDM, respectively). Fructosamine decreased slightly during the baseline week , but no further drop was seen after two weeks on vanadate. No significant changes in average blood glucose or weight were demonstrated in either cohort (Fig. 11).

Effect of vanadate treatment on cholesterol
Vanadate has previously been studied in humans for its potential in treatment of hypercholesterolemia [59]. Serum cholesterol decreased over the course of two week treatment with vanadate in both IDDM and NIDDM subjects (175.2 ± 6.3 *vs* 165.0 ± 8.7 mg/dl, p = 0.06, 267.6 ± 29.0 *vs* 204.2 ± 17.9, $p < 0.05$, respectively) (Fig. 12). This is similar in magnitude to the decrease seen in non-diabetic individuals [59]. Apolipoprotein A-1 and apolipoprotein B levels did not

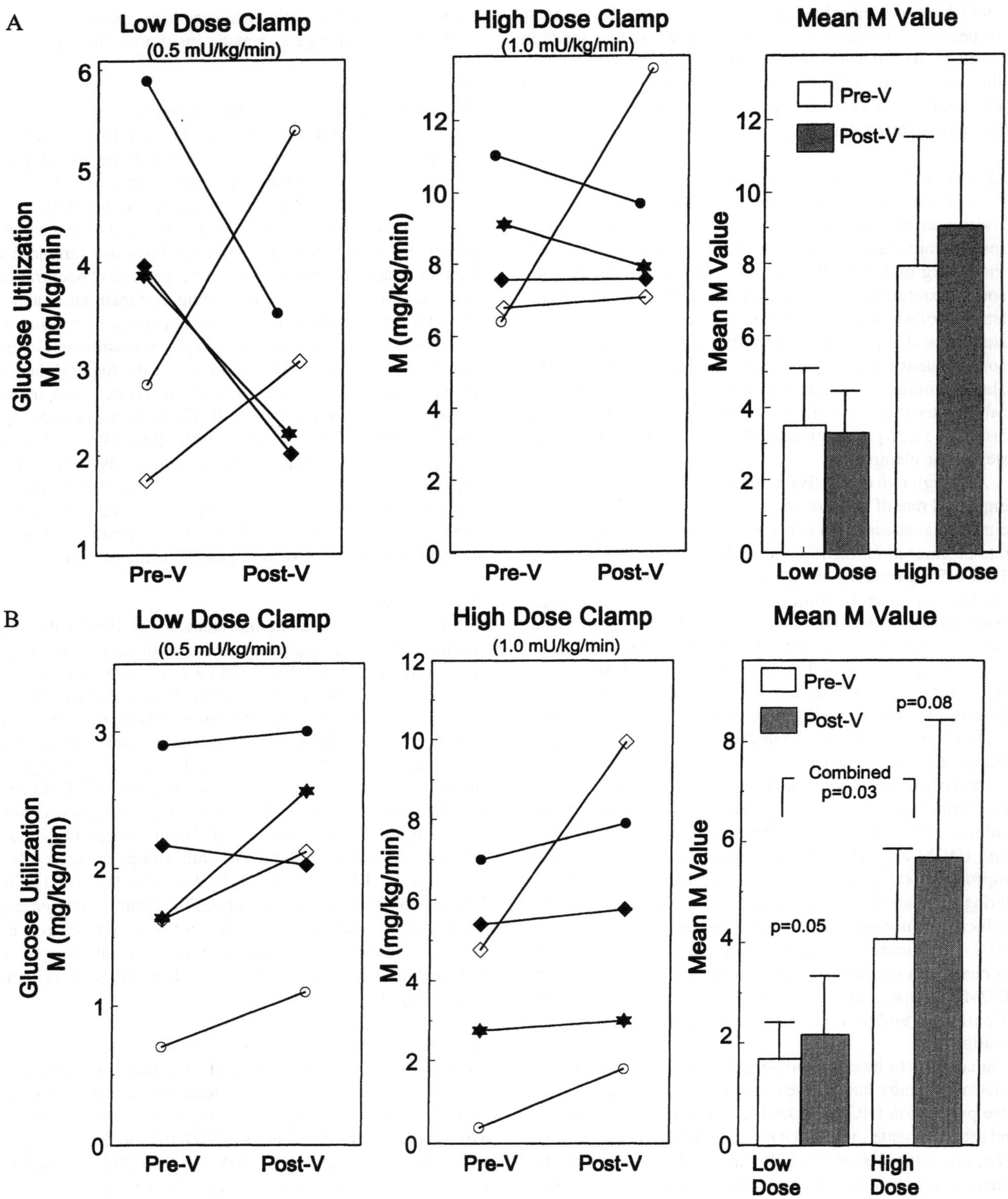

Fig. 7. Effect of vanadate on insulin sensitivity in patients with IDDM (A) and NIDDM (B). Figures on the left represent individual responses, and the figure on the right demonstrates the mean ± SE for the group. Statistical analysis was performed using paired, one tailed, *t*-tests.

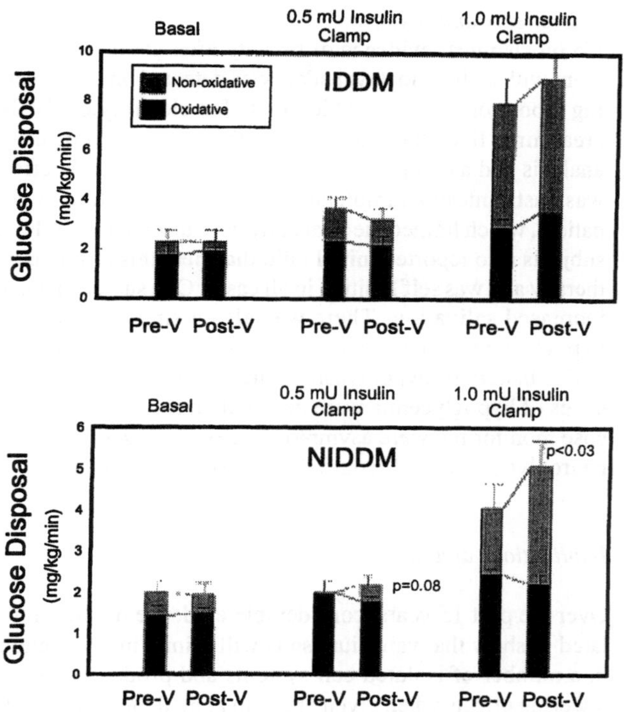

Fig. 8. Oxidative and non-oxidative glucose disposal in IDDM (top panel) and NIDDM (bottom panel) before and after two weeks on sodium metavanadate.

change significantly in either IDDM or NIDDM subjects or in the combined cohort.

Effects of vanadate treatment on map and s6 kinases

MAP and S6 kinases are insulin-stimulated enzymes which have been shown to play important roles in insulin signalling downstream of the insulin receptor [60–62]. To assess possible mechanisms of vanadate action on this post-receptor pathway, MAP and S6 kinase activity was measured in circulating mononuclear cells isolated from subjects pre- and post-vanadate therapy and stimulated with insulin *in vitro*. Each subject's sample was assayed in parallel with that of an age-matched healthy control who did not take vanadate.

In non-diabetic subjects, MAP and S6 kinase activity increased about 2-fold after *in vitro* stimulation with 100 nM insulin. In contrast, there was no significant insulin stimulation of MAP or S6 activity in leukocytes taken from untreated patients with IDDM or NIDDM. Following two weeks of oral therapy in diabetic subjects with sodium metavanadate, there was a 1.7–3.9-fold increase in *basal* MAP and S6 kinase activity in both IDDM and NIDDM to levels similar to those produced by *in vitro* stimulation with insulin in healthy controls. However, following vanadate, there was little further stimulation with insulin *in vitro* in cells from both IDDM and NIDDM subjects (Fig. 13).

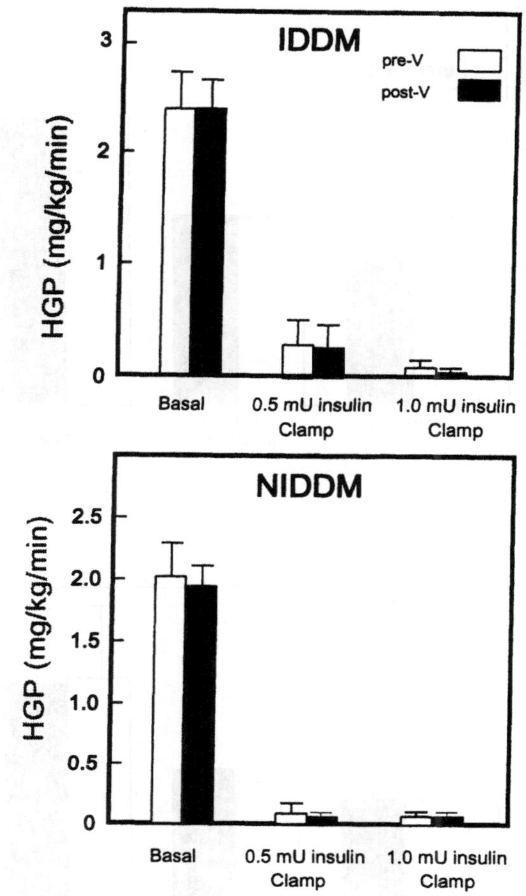

Fig. 9. Hepatic glucose production (HGP) in subjects with IDDM (top panel) and NIDDM (bottom panel) in the basal state and at the steady state of each step of the euglycemic clamp.

Na^+/Li^+ countertransport and na^+/H^+ exchange

Increased Na^+/Li^+ countertransport activity is seen in a subset of patients with essential hypertension and has been attributed to insulin resistance and hypertension [63]. Vanadyl sulfate has been shown to lower plasma insulin and systolic blood pressure without affecting plasma glucose in animal models of hypertension, including the SH rat [64] and the fructose-fed rat [65]. Insulin has been shown to increase the maximal transport rate [V_{max}] and substrate concentration for half maximal velocity [K_m] of Na^+/Li^+ exchange, as well as the V_{max} and K_m for Na^+, but not the K_m for intracellular H^+ [66]. These effects are mimicked by okadaic acid, an inhibitor of protein phosphatases. Thus, we measured the effect of vanadate and vanadyl on Na^+/Li^+ and Na^+/H^+ transport activity in a limited number of patients. Although the V_{max} of Na^+/Li^+ countertransport activity was not affected by 2 weeks of sodium metavanadate (0.36 ± 0.09 *vs* 0.38 ± 0.14 mmol/L cell/h), the K_m was significantly increased (40.8 ± 18.9 *vs* 70.8

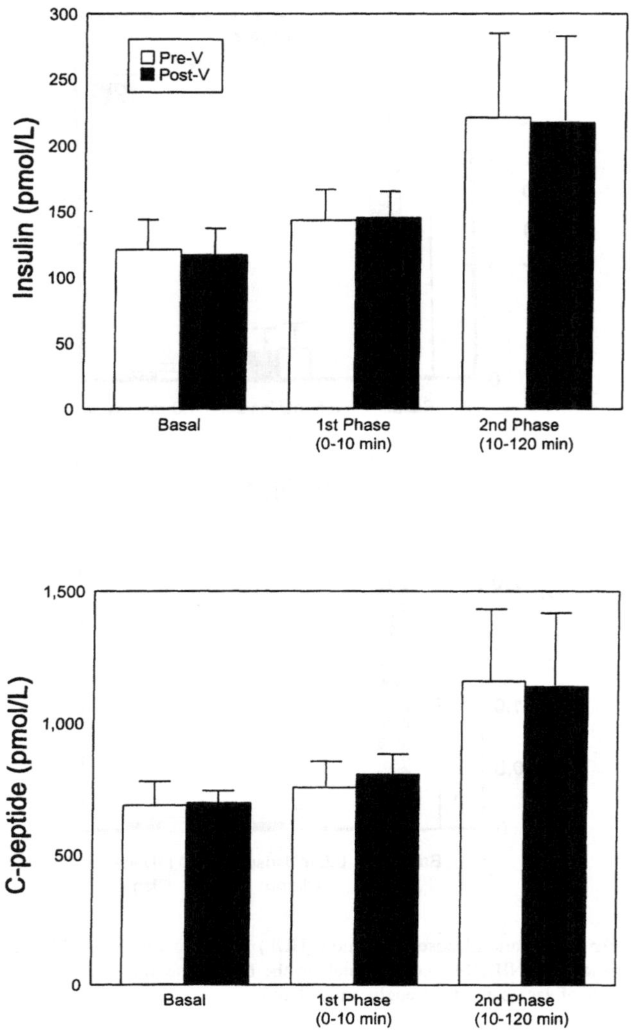

Fig. 10. Insulin and C-peptide levels during a + 5.5 mmol/L hyperglycemic clamp in individuals with NIDDM on oral agent and/or diet treatment.

± 21.2 mM, $p < 0.05$). (Fig. 14) Furthermore, the V_{max} of Na^+/H^+ exchange tended to increase (12.3 ± 6.3 *vs* 18.8 ± 5.5 mmol/L cell/h, $p = 0.08$). This is in contrast to the results of Simonson and Wolpert [67] who found that Na^+/Li^+ transporter activity did not change in obese individuals despite dietary intervention which improved insulin sensitivity. Since hypertensive subjects have an elevated V_{max} Na^+/Li^+ and Na^+/H^+ transport activity, and there is an association between hyperinsulinemia, insulin resistance and hypertension, one might expect vanadium treatment to improve insulin sensitivity and to lower the V_{max} of Na^+/H^+ exchange. The area of interaction of insulin, sodium-ion exchange and vanadium requires further investigation.

Side effects and toxicity

No biochemical evidence of toxicity after short term treatment with sodium metavanadate was detected on the screening laboratory profiles which included electrolytes, BUN, creatinine, liver function studies, thyroid functions, urine analysis and a complete blood count. The major side effect was gastrointestinal intolerance, including vomiting in one patient, which limited the total daily dose to 75 mg daily. Four subjects also reported initial mild diarrhea; this required no therapy and was self limiting in all cases. One subject noticed increased salivation. There was also a single episode of hypoglycemia requiring assistance in a subject with IDDM with a history of hypoglycemia unawareness. All other episodes of hypoglycemia were detected on routine blood glucose monitoring, were asymptomatic or not associated with neuroglycopenia and required no external assistance.

Implications and future directions

Over the past 15 years, considerable evidence has accumulated to show that vanadium salts will mimic insulin action in a number of isolated cell systems and produce glucose lowering effects when given orally to animal models of both IDDM and NIDDM [19]. The present investigation shows that two weeks of treatment with sodium metavanadate at a dose of 125 mg po daily appears to be relatively well tolerated, and improves insulin sensitivity in patients with NIDDM, as well as in some patients with IDDM. The increase in insulin sensitivity is primarily due to an increase in non-oxidative glucose disposal, whereas oxidative glucose disposal and both basal and insulin stimulated suppression of hepatic glucose output (HGP) were unchanged. Thus the improvement in insulin action in patients receiving vanadate is due primarily to increased peripheral (i.e., skeletal muscle) insulin sensitivity, although it is difficult to rule out an effect on hepatic insulin sensitivity, since HGP was maximally suppressed even by low dose insulin. Clinically, this is associated with a small decrease in insulin requirements and glycohemoglobin. Of additional benefit, there is a decrease in total cholesterol levels in both IDDM and NIDDM subjects. Furthermore, there is an increase in the basal activities of MAP and S6 kinases to levels similar to the insulin-stimulated levels in controls, but there is little or no further stimulation with insulin. Although this study focused on sodium metavanadate, in preliminary studies, we and others [68] have seen similar effects with oral vanadyl sulfate treatment.

The present data in humans are encouraging and consistent with previous studies in animal models of both human IDDM and NIDDM [4, 14–18, 21, 24, 69, 70]. The smaller effect observed in humans as compared to rodents may reflect differences in dose, attained blood levels, and time course of

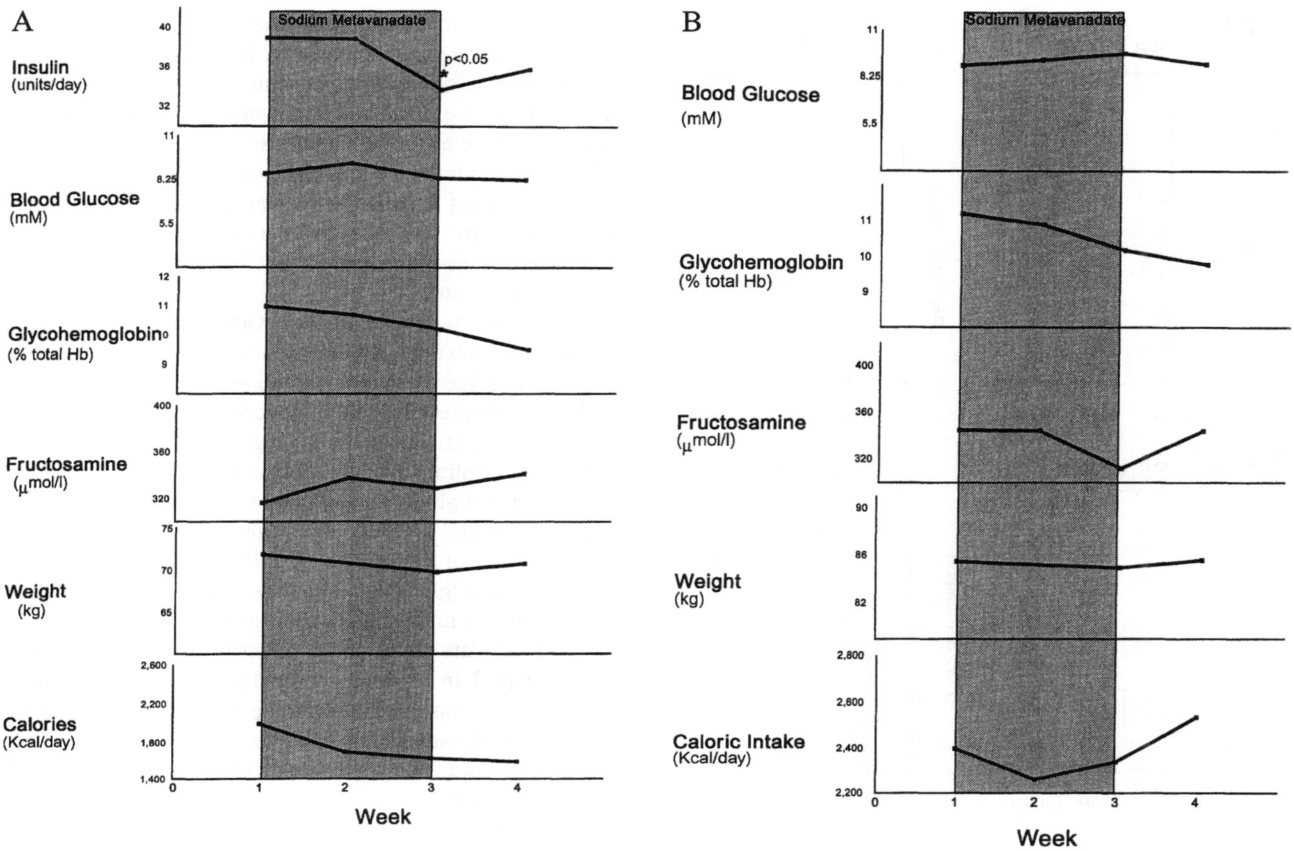

*Fig. 11.*Clinical parameters in IDDM (A) and NIDDM (B) patients receiving sodium metavanadate. Each panel shows the mean value for each parameter at the indicated time in the study.

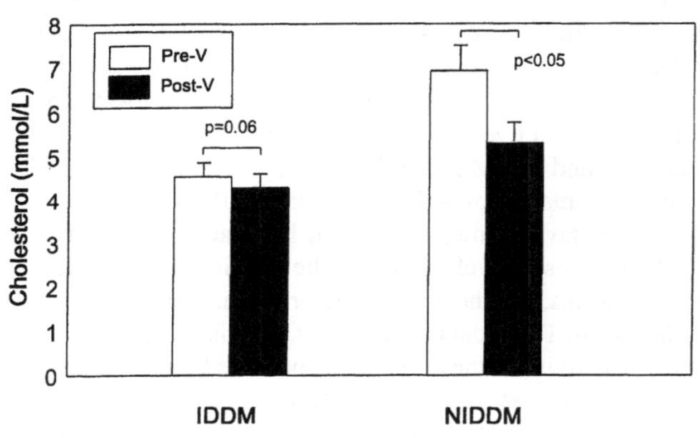

Fig. 12. Serum cholesterol levels before and after treatment with sodium metavanadate.

action. For example, in rodents, the dose of oral vanadate which has been found to improve blood glucoses is about 100 mg/kg/day, while the dose used in the current study was about 1.5 mg/kg/day. Likewise the blood level of vanadium achieved in rodent studies is between 10 and 20 μM, while the blood level in the present human study was between 1 and 5 μM. Finally, in humans the entire treatment period was only two weeks. In STZ-diabetic and BB rats (in the presence of continued, reduced insulin doses) vanadate added to the drinking water lowers glucose to near normal values within 3–4 days [18, 24, 69], however, the effect of vanadate to lower blood glucose to near normal levels in ob/ob and db/db mice takes 10–20 days [14, 16]. Thus, studies of longer duration in humans are warranted in order to fully evaluate the potential of this compound on glucose metabolism in humans.

It has recently been suggested that one factor which may contribute to the hypoglycemic action of vanadate is a decrease in food intake [71]. While there are some mild, transient gastrointestinal symptoms in humans receiving vanadate, we were unable to detect a significant difference in food intake as assessed by dietary history and food records or body weight during the vanadate treatment period in either NIDDM or IDDM, thus decreased intake does not appear

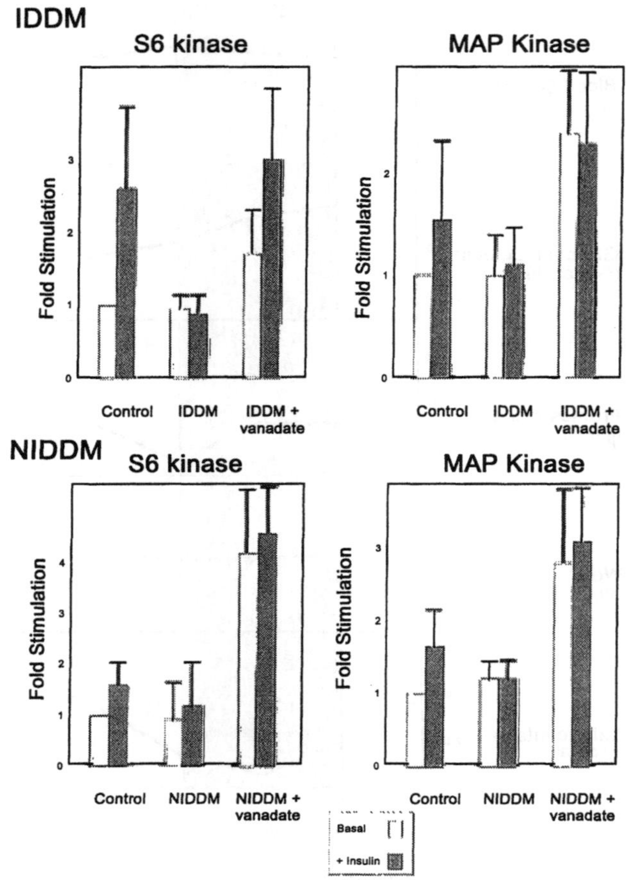

Fig. 13. MAP and S6 kinase activity in monocytes of control subjects, IDDM and NIDDM patients in the basal state and after two weeks on vanadate. (control individuals were not treated with $NaVO_3$).

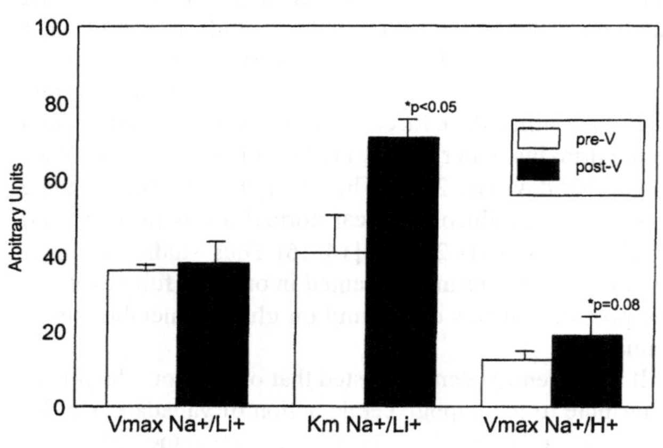

Fig. 14. Na^+/Li^+ and Na^+/H^+ transport activity in human erythrocytes before and after sodium metavanadate.

to be the mechanism of action in humans. These data are consistent with most previous animal data. In obese mice or rats treated with vanadate, there is no change in the percent lean body mass suggesting that the effects of vanadate do not require a major decrease in food intake [14, 16, 24]. When given to obese, hyperinsulinemic fa/fa rats, vanadate also improves oral glucose tolerance without decreasing body weight or changing counter-regulatory hormones [53, 70]. The differences in reported findings could be due to different responses in different animal models studied, different vanadium salts administered, different dosages employed, or the nutritional status (fed vs. fasted) of the rats at the time of study.

The exact tissue site of action of vanadium salts is remains unclear. In ob/ob mice, the disappearance rate for intravenous glucose is doubled in treated animals as compared to controls, and the insulin response to the glucose challenge reappears [14, 16]. In addition, hepatic glycogen content doubles, and pancreatic insulin stores are increased. There is also an increase in basal glucose oxidation by the hemidiaphragm in vanadate-treated animals, although insulin stimulation is similar to that of the control [70]. In vanadate treated STZ diabetic rats, euglycemic, hyperinsulinemic clamp studies indicate higher rates of peripheral glucose disposal, but no effect of vanadate on inhibition of hepatic glucose production by insulin [69]. In humans, we found no change in basal rates of oxidative or non-oxidative glucose disposal, and no change in oxidative glucose disposal at either low or high insulin levels, but a modest increase in non-oxidative glucose disposal in the NIDDM subjects. This improvement in glucose utilization in humans appears to be due to enhanced insulin sensitivity in muscle, since there was no change in basal or stimulated insulin secretion during the hyperglycemic clamp and no effect of vanadate on hepatic glucose production.

Improved insulin sensitivity, however, may be related to actions which increase basal activities of S6 and MAP kinases. Like insulin, sodium orthovanadate administration has been shown to increase both S6 phosphorylation and S6 activation in rat liver [72, 73]. In the present study we found that untreated IDDM and NIDDM subjects had a decreased insulin stimulation of activity of both MAP and S6 kinases. Sodium metavanadate increased the basal activities of MAP and S6 kinases to levels similar to the insulin-stimulated levels in controls, but there was little or no further stimulation with insulin. These data suggest that vanadate enhances the intrinsic activity of these enzymes, perhaps by inhibition of a phosphatase that would inhibit MAP kinase.

A potential additional clinical benefit of vanadate treatment was the reduction in serum cholesterol levels in the NIDDM patients. These results are consistent with the observations of Curran *et al.* who found that non-diabetics given oxytartratovanadate for eight weeks had significant reductions in total and free cholesterol levels [59]. The mechanism of this effect is most likely on cholesterol synthesis. Vanadium salts inhibit cholesterol synthesis by interference with formation and utilization of mevalonic acid [74]. Vanadium compounds may inhibit coenzyme A and thus interfere with the conversion of HMG to β-methyl crotonate [75]. Furthermore, early studies indicate that deposition of cholesterol is reduced and mobilization of predeposited cholesterol is increased in rabbits fed vanadium [76, 77].

Additional *in vivo* and *in vitro* studies with the vanadium compounds are warranted to better understand the mechanism of action of these interesting compounds.

Acknowledgements

The results of these studies in part were performed (Inpatient and Outpatient CRC, and Core Laboratory) supported by the National Institutes of Health grant NCRR GCRC MO1 RR02635; by an NIH grant to CRK (DK 47462), and the Diabetes and Endocrinology Research Center grant at Joslin Diabetes Center (DK 36836).

References

1. Nechay BR, Nanninga LB, Nechay PSE, Post RL, Grantham JJ, Macara IG, Kubena LF, Phillips TD, Nielsen FH: Role of vanadium in biology. Fed Proc 45: 123–132, 1986
2. Lyonett, Martz, Martin: L'emploithérapeutique de derivés du vanadium. La Presse Medicale: 191–192, 1899
3. Dubyak GR, Kleinzeller A: The insulin-mimetic effects of vanadate as a Na-K ATPase inhibitor. J Biol Chem 255: 5306–5312, 1980
4. Shechter Y, Karlish SJD: Insulin-like stimulation of glucose oxidation in rat adipocytes by vanadyl (IV) ions. Nature 284: 556–558, 1980
5. Dlouha H, Teisinger T, Vyskocil F: The effect of vanadate on the electrogenic Na^+/K pump, intracellular Na^+ concentration and electrophysiological characteristics of mouse skeletal muscle fiber. Physiol Bohemoslov 30: 1–10, 1981
6. Shechter Y, Ron A: Effect of depletion of bicarbonate or phosphate ions on insulin action in rat adipocytes. Further characterization of the receptor-effector system. J Biol Chem 261: 14951–14954, 1986
7. Degani H, Gochin M, Karlish SJD, Shechter Y: Electron paramagnetic studies and insulin-like effects of vanadium in rat adipocytes. Biochemistry 20: 5795–5799, 1981
8. Tamura S, Brown TA, Dubler RE, Larner J: Insulin-like effect of vanadate on adipocyte glycogen synthase and on phosphorylation of 95,000 dalton subunit of insulin receptor. Biochemical & Biophysical Research Communications 113: 80–86, 1983
9. Werdan K, Bauriedel G, Fisher B, Krawietz W, Erdmann E, Schmitz W, Scholz H: Stimulatory and inhibitory action of vanadate on potassium uptake and cellular sodium and potassium in heart cells in culture. Biochim Biophys Acta 23: 79–83, 1982
10. Hori C, Oka T: Vanadate enhances the stimulatory action of insulin on DNA synthesis in cultured mouse mammary gland. Biochim Biophys Acta 610: 235–240, 1980
11. Smith JB: Vanadium ions stimulate DNA synthesis in Swiss mouse 3T3 and 3T6 cells. Proc Natl Acad Sci (USA) 80: 6162–6166, 1983
12. Canalis E: Effect of sodium vanadate on deoxyribonucleic acid and protein synthesis on cultured rat clavaria. Endocrinology 116: 855–862, 1985
13. Shechter Y: Insulin mimetic effects of vanadate: possible implications for future treatment of diabetes. Diabetes 39: 1–5, 1990
14. Meyerovitch J, Rothenberg P, Shechter Y, Bonner-Weir S, Kahn CR: Vanadate normalizes hyperglycemia in two mouse models of non-insulin dependent diabetes mellitus. J Clin Invest 87: 1286–1294, 1991
15. Brichard SM, Baily CJ, Henquin JC: Long term improvement of glucose homeostatis by vanadate in obese hyperinsulin fa/fa rats. Endocrinology 125: 2510–2516, 1989
16. Brichard SM, Baily CJ, Henquin JC: Marked improvement of glucose homeostasis in diabetic ob/ ob mice given oral vanadate. Diabetes 39: 1326–1332, 1990
17. Ramanadham S, Mongold JJ, Brownsey RW, Cros GH, McNeill JH: Oral vanadyl sulfate in treatment of diabetes mellitus in rats. Am J Physiol 257: H904–911, 1989
18. Heyliger CE, Tahiliani AG, McNeill JH: Effect of vanadate on elevated blood glucose and depressed cardiac performance of diabetic rats. Science 227: 1474–1477, 1985
19. Meyerovitch J, Backer JM, Kahn CR: Hepatic phosphotyrosine phosphatase activity and its alterations in diabetic rats. J Clin Invest 84: 976–983, 1989
20. Battell ML, Yuen VG, McNeill JH: Treatment of BB rats with vanadyl sulphate. Pharma Commun 1(4): 291–301, 1992
21. Gil J, Miralpeix M, Carreras J, Bartrons R: Insulin-like effects of vanadate on glucokinase activity and fructose 2,6-busphosphate levels in the liver of diabetic rats. J Biol Chem 263: 1868–1871, 1988
22. Ramanadham S, Brownsey RW, Cros GH, Mongold JJ, McNeill JH: Sustained prevention of myocardial and metabolic abnormalities in diabetic rats following withdrawal from oral vanadyl treatment. Metabolism 38: 1022–1028, 1989
23. Brichard SM, Omgemba LN, Henquin JC: Oral vanadate decreases muscle insulin resistance in obese fa/fa rats. Diabetologia 35: 522–527, 1990
24. Meyerovitch J, Farfel Z, Sack J, Shechter Y: Oral administration of vanadate normalizes blood glucose levels in streptozotocin-treated rats. Characterization and mode of action. J Biol Chem 262: 6658–6662, 1987
25. Flier JS: Lilly Lecture: Syndromes of insulin resistance. From patient to gene and back again. Diabetes 41: 1207–1219, 1992
26. Kahn CR: Causes of insulin resistance. Nature 373: 384–385, 1995
27. Rothenberg PL, Lane WS, Karasik A, Backer J, White M, Kahn CR: Purification and partial sequence analysis of pp185, the major cellular substrate of the insulin receptor tyrosine kinase. J Biol Chem 266: 8302–8311, 1991
28. Sun X, Rothenberg P, Kahn CR, Backer JM, Araki E, Wilden PA, Cahill DA, Goldstein BJ, White MF: The structure of the insulin receptor substrate IRS-1 defines a unique signal transduction protein. Nature 352: 73–77, 1991
29. Koch CA, Anderson DJ, Moran MF, Ellis C, Pawson T: SH2 and SH3 domains: elements that control interactions of cytoplasmic signaling proteins. Science 252: 668–674, 1991
30. Gould GW, Derechin V, James DE, Tordjman K, Ahern S, Gibbs EM, Lienhard GE, Mueckler M: Insulin-stimulated translocation of the HepG2/erythrocyte-type glucose transporter expressed in 3T3-L1 adipocytes. J Biol Chem 264: 2180–2184, 1989
31. Cormont M, Tanti JF, Zahraoui A, Vanobberghen E, Tavitian A, LeMarchand-Brustel Y: Insulin and okadaic acid induce Rab4 redistribution in adipocytes. J Biol Chem 268(26): 19491–19497, 1993
32. Berman M, McGuire EA, Roth J, Zeleznik AJ: Kinetic modeling of insulin binding to receptors and degradation in vivo in rabbits. Diabetes 29: 50–59, 1980
33. Fischer EH, Charbonneau H, Tonks NK: Protein tyrosine phosphatases-a diverse family of intracellular and transmembrane enzymes. Science 253: 401–406, 1991

34. Papoulas O, Williams NG, Kingston RE: DNA binding activation of c-Myc purified from eukoryotic cells. J Biol Chem 267: 10470–10480, 1992
35. Tamura S, Brown TA, Whipple JH, Fujita-Yamaguchi Y, Dubler RE, Cheng K, Larner J: A novel mechanism for the insulin-like effect of vanadate on glycogen synthase in rat adipocytes. J Biol Chem 259: 6650–6658, 1984
36. Strout HV, Vicario PP, Saperstein R, Slater EE: The insulin-mimetic effect of vanadate is not correlated with insulin receptor tyrosine kinase activity nor phosphorylation in mouse diaphragm *in vivo*. Endocrinology 124: 1918–1924, 1989
37. Green A: The insulin-like effect of sodium vanadate on adipocyte glucose transport is mediated at a post-insulin-receptor level. Biochem J 238: 663–669, 1986
38. Shisheva A, Shechter Y: Quercetin selectively inhibits insulin receptor function in vitro and the bioresponses of insulin and insulinomimetic agents in rat adipocytes. Biochemistry 31: 8059–8063, 1992
39. Shisheva A, Shechter YA: Ccytosolic protein tyrsoine kinase in rat adipocytes. FEBS Lett 300: 93–96, 1982
40. Kadota S, Fantus GI, Deragon G, Guyda HJ, Hersh B, Posner BI: Peroxide(s) of vanadium: a novel and potent insulin-mimetic agent which activates the insulin receptor kinase. Biochem Biophys Res Commun 147: 259–266, 1987
41. Goldstein BJ: Protein-tyrosine phosphatase and the regulation of insulin action. J Cell Biochem 48: 33–42, 1992
42. Sale GJ: Anonymous Advances in Protein Phosphatases. 6th ed. Insulin receptor phosphotyrosyl protein phosphatases and the regulation of insulin receptor tyrosine kinase action. 1991, pp 159–186
43. Swarup G, Cohen S, Garbers DL: Inhibition of membrane phosphotyrosyl protein phosphatase activity by vanadate. Biochem Biophys Res Commun 107: 1104–1109, 1982
44. Meyerovitch J, Backer JM, Csermely P, Shoelson SE, Kahn CR: Insulin differentially regulates protein phosphotyrosine phosphatase activity in rat hepatoma cells. Biochemistry 31: 10338–10344, 1992
45. Burant CF, Treutelaar MK, Buse MG: Diabetes-induced functional and structural changes in insulin receptors from rat skeletal muscle. J Clin Invest 77: 260–270, 1986
46. Caro JF, Ittoop O, Pories WJ, Meelheim D, Flickinger EG, Thomas F, Jenquin M, Silverman JF, Khazanie PG, Sinha MK: Studies on the mechanism of insulin resistance in the liver from humans with noninsulin-dependent diabetes. J Clin Invest 78: 249–258, 1986
47. Freidenberg GR, Henry RR, Klein HH, Reichart DR, Olefsky JM: Decreased kinase activity of insulin receptors from adipocytes of non-insulin dependent diabetic subjects. J Clin Invest 79: 240–250, 1987
48. Arner P, Pollare T, Lithell H, Livingston JN: Defective insulin receptor tyrosine kinase in human skeletal muscle in obesity and Type II (noninsulin-dependent) diabetes mellitus. Diabetologia 30: 437–440, 1987
49. Ueki H, Okuhama R, Sera M, Inoue T, Tominaga N, Morita T: Stimulatory affect of vanadage on 3¢,5¢-cyclic guanosine monophosphate-inhibited low Michaelis-Menten constant 3¢,5¢-cyclic adenosine monophosphate phosphodiesterase activity in isolated rat fat pads. Endocrinology 131: 441–446, 1992
50. Souness JE, Maslen C, Scott LC: Effects of solubilization and vanadate\glutathione complex on inhibitor potencies agains eosinophil cyclic AMP-specific phosphodiesterase. FEBS Lett 302: 181–184, 1992
51. Soll AH, Kahn CR, Neville DM, Jr.: Insulin binding to liver plasma membranes in the obese hyperglycemic (ob/ob) mouse: Demonstration of a decreased number of functionally normal receptors. J Biol Chem 250: 7402–7407, 1975
52. Brichard SM, Assimacopoulos-Jeannet F, Jeanrenaud B: Vanadate treatment markedly increases glucose utilization in muscle of insulin-resistant fa/fa rats without modifying glucose transporter expression. Endocrinology 131: 311–317, 1992
53. Paquet MR, Romanek RJ, Sargeant RJ: Vanadate induces the recruitment of GLUT-4 glucose transporter to the plasma membrane of rat adipocytes. Mol Cell Biochem 109: 149–155, 1992
54. Ferber S, Meyerovitch J, Kriauciunas KM, Kahn CR: Vanadate normalizes hyperglycemia and PEPCK mRNA levels in ob/ob mice. (In Press) Metabolism 1994
55. DeFronzo RA, Robin JD, Andres R: Glucose clamp techniques: a method for quantifying insulin secreation and resistance. Am J Physiol 237: E214–223, 1979
56. Ferrannini E, DelPrato S, DeFronzo RA: Glucose kinetics and tracer methods. In: Clarke WL, Larner J and Pohl SL (eds). Methods in Diabetics Research. Vol II, Clinical Methods. Canada: Wiley Interscience, 1986
57. Ferrannini E: The theoretical basis of indirect calorimetry: A review. Metabolism 37: 287–301, 1988
58. Simonson DC, DeFronzo RA: Indirect calorimetry: methodological and interpretative problems. Am J Physiol 258: E399–412, 1990
59. Curran GL, Azarnoff DL, Bolinger RE: Effect of cholesterol synthesis inhibition in normocholesteremic young men. J Clin Invest : 1251–1261, 1959
60. Muggeo M, Bar RS, Roth J et al.: Insulin receptors: Biologically relevant regulation of concentration and affinity. In: Miescher PA, (ed). Menarini Series on Immunopathy: First Symposium on Organ Specific Autoimmunity. Schwabe and Company, Basel, 1978, pp 149–159
61. Maher DW, Davis I, Boyd AW, Morstyn G: Human interleukin-4: an immunomodulator with potential therapeutic applications. Progress in Growth Factor Research 3: 43–56, 1991
62. Tobe K, Kadowaki T, Hara K et al.: Sequential activation of MAP kinase activator, MAP kinases, and S6 peptide kinase in intact rat liver following insulin injection. J Biol Chem 267: 21089–21097, 1992
63. Doria A, Fioretto P, Avogaro A: Insulin resitance is associated with hig sodium-lithium countertransport in essential hypertension. Am J Physiol 261: E684–691, 1991
64. Bhanot S, McNeill JH: Vanadyl sulfate lowers plasma insulin and blood pressure in spontaneously hypertensive rats. Hypertension 24: 625–632, 1994
65. Bhanot S, McNeill JH, Bryer-Ash M: Vanadyl sulfate prevents fructose-induced hyperinsulinemia and hypertension in rats. Hypertension 23: 308–312, 1994
66. Pontremoli R, Zerbini G, Rivera A, Canessa M: Insulin activation of red blood cell Na^+/H^+ exchange decreases the affinity of sodium sites. Kidney International 46: 365–375, 1994
67. Wolpert HA, Steen SN, Istfan NW, Simonson DC: Disparite effects of weight loss on insulin sensitivity and erythrocyte sodium-lithium countertransport activity. Am J Hypertension 5: 754–757, 1992
68. Cohen N, Halberstam M, Shlimovich P, Chang CJ, Shamoon H, Rossetti L: Oral vanadyl sulfate improves hepatic and peripheral insulin sensitivity in patients with non-insulin-dependent diabetes mellitus. J Clin Invest 95: 2501–2509, 1995
69. Blondel O, Bailte D, Portha B: In vivo insulin resistance in streptozotocin-diabetic rats-evidence for reversal following oral vanadate treatment. Diabetologia 32: 185–190, 1989
70. Brichard SM, Ongemba LM, Henquin JC: Oral vanadate decreases muscle insulin resistance in obese fa/fa rats. Diabetologia 35: 522–527, 1992
71. Malabu UH, Dryden S, McCarthy HD, Kilpatrick A, Williams G: Effects of chronic vanadate administration in the STZ-induced diabetic rat. Diabetes 43: 9–15, 1994

72. Olefsky J: The insulin receptor: A multifunctional protein. Diabetes 39: 1008–1016, 1990
73. Kozma SC, Lane HA, Ferrari S, Luther H, Siegmann M, Thomas G: A stimulated S6 kinase from rat liver; identity with the mitogen activated S6 kinase of 3T3 cells. EMBO J 8: 4125–4132, 1989
74. Azarnoff DL: Site of vanadium inhibition of cholesterol biosynthesis. J Amer Chem Soc 79: 2968, 1957
75. Hudson TGF: The effects of vanadium on metabolism (continued). In: Anonymous Vanadium: Toxicology and Biological significance. Elselvier Publishing Co., New York, 1964 pp 30–43
76. Mountain JT, Stockell FR, Stokinger HE: Effect of ingested vanadium on cholesterol and phospholipid metabolism in the rabbit. Proc Soc Exp Biol Med 92: 582 1956
77. Curran GL, Costello RL: Reduction of excess cholesterol in the rabbit aorta by inhibition of endogenous cholesterol synthesis. J Exp Med 103: 49 1956

Molecular and Cellular Biochemistry **153:** 233–240, 1995.

Toxicology of vanadium compounds in diabetic rats: The action of chelating agents on vanadium accumulation

Jose L. Domingo[1], Mercedes Gomez[1], Domenec J. Sanchez[1], Juan M. Llobet[1] and Carl L. Keen[2]

[1]*Laboratory of Toxicology and Biochemistry, School of Medicine, 'Rovira i Virgili' University, Reus 43201, Spain and* [2]*Department of Nutrition, University of California, Davis, CA 95616, USA*

Abstract

The possible use of vanadium compounds in the treatment of diabetic patients is now being evaluated. However, previously to establish the optimal maximum dose for diabetes therapy, it should be taken into account that vanadium is a highly toxic element to man and animals. The toxic effects of vanadium are here reviewed. The tissue vanadium accumulation, which would mean an additional risk of toxicity following prolonged vanadium administration is also discussed. Recently, it has been shown that coadministration of vanadate and TIRON, an effective chelator in the treatment of vanadium intoxication, reduced the tissue accumulation of this element, decreasing the possibility of toxic side effects derived from chronic vanadium administration without diminishing the hypoglycemic effect of vanadium. However, previously to assess the effectiveness of this treatment in diabetic patients, a critical reevaluation of the antidiabetic action of vanadium and its potential toxicity is clearly needed. (Mol Cell Biochem **153:** 233–240, 1995)

Key words: vanadate, vanadyl, streptozotocin-induced diabetes, toxic effects, vanadium in tissues

Vanadium toxicity

In recent years, vanadium compounds have attracted the interest of a number of investigators due to their potential use in diabetes treatment. It has been shown that administration of vanadate (V^{+5}) causes a sustained decrease in blood glucose levels in insulin-deficient rats and substantially improves the glucose homeostasis of hyperinsulinemic insulin-resistant animals [1–4]. Vanadyl (V^{+4}) derivatives have also shown antidiabetic properties on various animal models of diabetes, characterized by potentiation of exogenous insulin, partial correction of pancreas alterations, and improvement of glucose homeostasis with long term effects after cessation of treatment [3, 5].

In spite of the concern that the biological effects of vanadium have raised, until recently the literature on the toxicology of this element has, however, been rather scarce since the industrial toxicology profile of vanadium was established early in 1950s–1960s [6–8]. Occupational toxicity of vanadium after inhalation of vanadium pentoxide dust is well documented. Inhalation of vanadium may cause rhinitis, pharyngitis, chronic productive cough, tracheobronchitis and bronchopneumonia [6, 8]. Studies on the acute oral toxicity of vanadium compounds in rats and mice showed that both vanadate and vanadyl are moderately toxic, with the severity of the toxic effects increasing as the valence increases [9].

In most of the recent studies on the effectiveness of vanadium in diabetic rats and mice, vanadate or vanadyl were given in the drinking water of these animals. In those studies, vanadium compounds were typically administered at concentrations ranging from 0.20 to 1.0 mg/ml, which might be considered as potentially toxic as it has been shown in nondiabetic rats [9–11]. Hematological and biochemical changes, reproductive and developmental toxicity, hepatotoxicity, as well as a number of miscellaneous toxic effects

Address for offprints: J.L. Domingo, Laboratory of Toxicology and Biochemistry, School of Medicine, San Lorenzo 21, Reus 43201, Spain have been reported to occur in healthy rats following vanadium exposure. Some of the vanadium toxic effects are here reviewed.

Hematological and biochemical changes

The hematological effects of vanadium on living organisms were recently reviewed by Zaporowska and Wasilewski [12]. It has been reported that vanadium reduces the deformability of erythrocytes and produces peroxidative changes in the erythrocyte membrane, this leading to hemolysis. Consequently, the depressed erythrocyte count observed in animals intoxicated with vanadium might be the consequence of both the hemolytic action of vanadium and the shortened time of survival of erythrocytes [12]. Following exposure to vanadium, some changes were also noted on white blood cell system. Although these changes might suggest the influence of vanadium on the resistance of organisms, the mechanism of action of vanadium still requires elucidation [12].

A decrease in the erythrocyte count, hemoglobin level and hematocrit index of healthy rats that received ammonium metavanadate in the drinking water at 0.01 and 0.05 mg V/ml for four weeks (equivalent to 0.023 and 0.115 mg/ml of ammonium metavanadate) was also observed by Zaporowska and co-workers [13]. The decrease in the erythrocyte count was associated with an increased percentage of reticulocytes in the peripheral blood of the animals drinking the solution with a higher vanadium content. Biochemical analyses demonstrated a decrease of ascorbic acid levels in the plasma and erythrocytes of animals drinking the metavanadate solutions. Among the enzymes examined in the erythrocytes (catalase, glucose-6-phosphate dehydrogenase, lactate dehydrogenase, and delta-aminolevulinic acid dehydratase [ALA-D]), only ALA-D activity was depressed.

Anemia, elevation in blood cholesterol concentrations, and a reduction in the levels of GPT, GOT, LDH and alkaline phosphatase, and systemic injuries in major organs (kidney, liver, spleen, heart, testes) were also found in rats which were injected s.c. with 0.9 mg V/kg/day for 16 days [14].

Reproductive and developmental toxicity of vanadium

The reproductive toxicity of vanadate has been evaluated in male mice. Animals were given sodium metavanadate in the drinking water at concentrations of 0.10, 0.20, 0.30 and 0.40 mg/ml for 64 days prior to mating with untreated females. Although metavanadate did not reduce fertility in male mice at 0.10 and 0.20 mg/ml, sperm count was significantly decreased at 0.20, 0.30 and 0.40 mg/ml but without affecting the sperm motility. The no observed adverse effect level (NOAEL) was 0.20 mg/ml [15].

Recent investigations have also shown that vanadium, administered as metavanadate or as orthovanadate, causes developmental toxicity in rats [16, 17], and mice [18–20], whereas vanadyl was embryotoxic and teratogenic in mice when given orally at 37.5, 75, and 150 mg/kg/day as vanadyl sulfate pentahydrate [21].

Miscellaneous toxic effects

An elevation in serum calcium and a reduction in phosphorus was observed in rats which were injected s.c. with 0.9 mg V/kg/day for 16 days [14]. This fact might be explained by the presence of renal damage as shown by the pathological and biochemical examination, and might also suggest an effect on bone [14]. Bone is likely to be susceptible to damage by vanadium because of its high uptake and retention [9, 14, 22]. The toxic effects of vanadium were cumulative, and the fibrosis observed at glomerular tuft, preglomeruli, pretubules, and interstitium being dose-dependent [23]. Lymphotoxicity of vanadate was also found when a single dose of 15.5 mg/kg of ammonium metavanadate was injected to mice [24]. The vanadium-treated animals showed neurological signs (ataxia, convulsions, dyspnoea, and paralysis of hind legs) between 5 min and 54 h post-treatment, but the concentrations of vanadium in the brain were very low during this period (< 5.2% of blood concentration).

On the other hand, chronic life-time treatment of rats with 20 ppm of vanadium as sodium metavanadate in the drinking water, affected collagen in their lungs. This was manifested by a decrease in total collagen and soluble collagen, and by an increase in insoluble collagen [25]. In a subsequent study, rats were treated in the period from conception up to the 105th day of age with 40 and 60 ppm of vanadium (as sodium metavanadate) in the drinking water. It was found that total collagen and soluble collagen within the lungs of vanadium-treated animals were significantly lower than in non-treated controls. However, the decrease in total collagen was not dependent on the dose of vanadium administered [26].

Immunotoxicology of vanadium compounds has also been reported. When mice were exposed to ammonium metavanadate (2.5, 5.0, or 10 mg/kg) by i.p. injection every 3 days for 3, 6, or 9 weeks, resistance to *Listeria monocytogenes* was significantly depressed in a time- and a dose-dependent manner, whereas the phagocytic activity of macrophages was also significantly diminished at 5.0 and 10 mg/kg [27].

Pro-oxidant effects and 'in vitro' *toxicity*

Vanadium also possess a pro-oxidant potential and it has been shown to oxidize a variety of biochemical substrates including lipids [28], glutathione [29], ascorbic acid [30] and

NAD(P)H [29]. Intraperitoneal administration of sodium metavanadate (3 mg/kg) to adult rats for 5 consecutive days resulted in major alterations in lipid profiles and protein concentrations in all the brain regions. Vanadium-induced stimulation of lipid peroxidation was characterized by a selective loss of brain polyunsaturated fatty acids [31].

Vanadate hepatotoxicity has been shown to occur both in isolated rat hepatocytes and in isolated perfused rat livers [32, 33]. When the toxic potential of sodium orthovanadate (2 mM) towards isolated perfused rat livers was investigated, the strong correlation noted between induction of lipid peroxidation and hepatotoxicity, and the inhibition of both processes in parallel by antioxidants, were suggestive of a causative role for lipid peroxidation in vanadate-induced hepatotoxicity [32]. The pro-oxidant effects of vanadate were found to be mainly responsible for its cytotoxic activity towards isolated perfused rat livers [33].

Toxic effects of vanadium in STZ-induced diabetic rats

In spite of the vanadium toxicity, most investigators involved in the evaluation of this element in diabetes treatment did not assess (or did not report) the toxic effects associated with prolonged vanadium administration. Thus, although Heyliger and associates [34] observed a severe diarrhoea in STZ-induced diabetic rats following administration of sodium orthovanadate in the drinking water at concentrations of 0.60 and 0.80 mg/ml, these investigators did not find toxic signs – including deaths – when orthovanadate was added to drinking water containing 0.5% NaCl and administered for four weeks after two weeks of diabetes. However, recent investigations showed that the addition of NaCl in the drinking water solutions did not improve fluid consumption in vanadate-treated rats [35], whereas Paulson and co-workers [36] noted that most of the diabetic rats did not drink the orthovanadate solutions containing 0.5% NaCl, which would probably be due to its high alkalinity. The majority of animals became dehydrated and died within four weeks. To avoid these effects, Paulson *et al.* [36] added 0.80 mg/ml of sodium orthovanadate to the drinking water and the pH was returned to neutrality by adding L-ascorbic acid.

Vanadium in drinking water

Because of the interest in the potential risks associated with vanadium therapy, the toxic effects derived from sodium metavanadate, sodium orthovanadate, and vanadyl sulfate administration to STZ-induced diabetic rats were assessed in our laboratory. Streptozotocin-induced diabetic rats drank aqueous solutions (NaCl, 80 mM) containing sodium metavanadate (0.15 mg/ml), sodium orthovanadate (0.23 mg/ml), or vanadyl sulfate pentahydrate (0.31 mg/ml) for 28 days. In a pilot study, we found that 0.15 mg/ml of sodium metavanadate was the lowest concentration of metavanadate effective in lowering blood glucose levels of STZ-treated rats. An aliquot (0.23 mg/ml) of sodium orthovanadate and 0.31 mg/ml of vanadyl sulfate pentahydrate provided an amount of vanadium equivalent to the 0.15 mg/ml of sodium metavanadate. Daily food and fluid intake were significantly decreased in the vanadium-treated animals relative to diabetic controls. Although vanadium treatment reduced the level of hyperglycemia in diabetic rats, serious signs of toxicity were also observed in all of the vanadium-treated animals as evidenced by some deaths, decreased weight gain, and increased serum concentrations of urea and creatinine [37]. In spite of these results, it was difficult to obtain conclusions about the comparative toxicity of the three vanadium compounds because no correction for the pH of the solutions was made. However, according to the above results and taking into account the vanadium intake, it seems that the toxicity ranking would be vanadyl < orthovanadate < metavanadate, which would be coincident with the effectiveness ranking.

Moreover, since toxic effects of vanadium are manifested primarily in the organs of accumulation, and vanadium can also be a potent mitogen in several types and capable of altering various enzyme activities [38–41], the extent of tissue accumulation of vanadium in the diabetic animals needed to be determined. It was found that vanadium accumulated significantly in bone, kidneys, spleen, liver, pancreas, heart, and muscle [37]. In a subsequent study, the effectiveness and toxicity of vanadyl sulfate treatment (at concentrations of 1.1 mg/ml in drinking water) was evaluated in STZ-induced diabetic rats, and compared with the effectiveness and toxicity derived from the administration of sodium metavanadate and sodium orthovanadate [42]. After two weeks of treatment, some signs of diabetes (hyperglycemia, hyperphagia, polydipsia) were again significantly ameliorated by vanadium administration, but negative side effects were also observed. Those effects included a high number of deaths (40–50%), decreased weight gain, and tissue vanadium accumulation.

While Pederson and associates [43] reported normoglycemia in STZ-induced diabetic rats 13 weeks after withdrawal of vanadyl treatment, other authors observed an increase in glycemia following cessation of treatment with metavanadate or orthovanadate [44–46]. In a recent study, Becker and co-workers [47] showed that although withdrawal of vanadium treatment was rapidly followed by a rise in the concentrations of glucose in plasma, those levels remained below those of untreated rats at least four weeks after cessation of treatment, independently of the vanadium salt used.

With regard to vanadium side effects in diabetes treatment, Dai *et al.* [48] did not find evidence of toxicity when STZ-induced diabetic rats were treated with vanadyl sulfate for one year, whereas Mongold and associates [49] did not report remarkable changes in a number of blood parameters or in the histology of various tissues after vanadyl sulfate treatment of diabetic rats for 39 days, although they found a significant accumulation of vanadium in bone, kidney and liver, which is in agreement with our results [37, 42]. Toxic effects of vanadium administration to STZ-treated rats were also reported by Okpere and associates [50]. Doses of 0.4 and 0.8 mg/ml of vanadate in drinking water caused diarrhoea, reduced physical activity, and (in most rats given 0.8 mg/ml) profound falls in food and water intake, leading to dehydration and death in many cases.

On the other hand, prolonged treatment (5 months) of STZ-diabetic rats with 0.75 mg/ml of vanadyl sulfate trihydrate did not cause any apparent hepatic toxicity as assessed histologically [51]. No correlation between vanadium levels in bone, kidney, liver and muscle, and the degree to which glucose was lowered in response to treatment in the diabetic animals could be demonstrated [51].

The effects of a new organic vanadium complex, bis-(maltolato)-oxovanadium (IV) (BMOV) were recently examined [52]. BMOV was administered in the drinking water for 25 weeks to control and STZ-induced diabetic rats. BMOV effectively produced the glucose-lowering effects at significantly lower doses than those previously used for inorganic vanadium salts, without any overt signs of vanadium toxicity, apart from hypoglycemia in a few animals [53].

Tissue oxidative damage induced by vanadium has also been reported. In a recent study, the effects of sodium metavanadate (1.2 mM) consumption on trace element metabolism, components of the antioxidant defense system and lipid oxidative damage were examined in STZ-induced diabetic rats [54]. The following parameters were similar among control and vanadate-treated groups: plasma zinc, copper and iron concentrations, plasma ceruloplasmin activity, liver zinc, copper, manganese and iron concentrations, kidney manganese and iron concentrations, liver non-selenium-dependent glutathione peroxidase, glutathione reductase and Mn-SOD activities, liver reduced glutathione, and oxidized glutathione concentrations and kidney non-selenium-dependent glutathione peroxidase activity . With regard to oxidative damage, tissue vanadium concentrations were positively correlated to TBARS (thiobarbituric acid reactive substances) production in both liver and kidney, which would indicate that vanadium therapy is associated with a marked deterioration in health of both control and diabetic rats [54].

The effect of vanadium treatment on TBARS production was similar to previous reports. Donaldson and LaBella [28] found that rat and mouse tissues incubated with sodium metavanadate exhibited high indices of lipid oxidative damage, as measured by tissue TBARS production compared to control tissues. Similarly, Elfant and Keen [55] reported that pregnant and lactating rats fed diets containing 75 µg $NaVO_3$/g diet (0.62 mM) for 43 days showed high liver whole homogenate and mitochondrial TBARS production compared to controls. These observations showed that vanadium could directly or indirectly induce oxidative stress to tissue. That this increased oxidative stress might be functionally significant would be suggested by the observation that ascorbic acid can alleviate signs of vanadium toxicity [56, 57]. In contrast, Thompson and McNeill [58] reported that chronic treatment with oral vanadyl sulfate (1.00–1.25 mg/ml) of diabetic rats for 12 weeks would appear not to represent a continuing pro-oxidant effect as demonstrated by cataract prevention, glutathione restoration and a lack of other toxic symptoms.

Taken into account the above reports and since vanadium – like insulin – would only be a palliative treatment of diabetes, the tissue accumulation of vanadium would mean an additional and serious risk of toxicity [37, 42] because vanadium compounds should be administered to diabetic patients throughout life-time.

Administration of vanadium by gavage

The effectiveness and toxicity of giving vanadium compounds by gavage, rather than via drinking water, to STZ-treated rats has been examined in our laboratory. The use of gavage to deliver the vanadium was based in two ideas. First, to assess if the reduction in food and fluid intake might be due in part to the bitter taste of vanadium, an effect that would be minimized by giving the vanadium via gavage. Second, giving the vanadium via gavage (2 doses per day) rather than as a dietary constituent would more closely reflect how the element could be used in the treatment of human diabetics.

In a first study, sodium metavanadate was given by gavage to STZ-induced diabetic rats for 28 days at doses of 0, 2.5, 5, 10 and 20 mg/kg/day, which were based on previous results [37, 42]. None of these doses showed noticeable beneficial effects on the diabetic status of the animals. In contrast, although not dose-related, vanadium treatment was associated with a higher mortality rate (50–62.5%) than that observed in the non vanadium treated diabetic group (12.5%) [59]. Vanadium accumulated in a dose-dependent manner in bone and kidneys, as well as in most of the remaining tissues analyzed.

In a subsequent experiment, vanadyl sulfate trihydrate was given by gavage to STZ-induced diabetic rats at 0, 25, 50, and 75 mg/kg/day, in two equally divided doses for a total of 21 days [60]. The choice of the dosage levels was again based on previous results [37]. In marked contrast to the reduction in plasma glucose observed in diabetic animals given vanadyl sulfate via drinking water, diabetic rats given vanadyl by gavage were not characterized by an alleviation in the signs of diabetes evaluated (hyperglycemia, hyperphagia, polydipsia). Notwithstanding, different signs of vanadium toxicity were seen including a significant number of deaths at 75 mg/kg/day [60]. In addition, vanadium was detected in all of the tissues analyzed.

Table 1 summarizes some results of a number of studies in which vanadium toxicity (including tissue accumulation of this element) was assessed.

Use of chelating agents in vanadium mobilization

Although it has been shown that vanadium compounds are effective to improve the glucose homeostasis of diabetic

Table 1. Vanadium administration in the streptozotocin-induced diabetic rat: a summary of side effects including tissue vanadium accumulation

Vanadium compound	Concentration/ dose	Period of administration	Route	Side effects/ Vanadium accumulation	Reference
Na_3VO_4	0.60 and 0.80 mg/ml	4 weeks	drinking water	Severe diarrhea which was avoided when vanadate was administered in 0.5% NaCl solution	Heyliger *et al.* [34]
$NaVO_3$	0.20 and 0.80 mg/ml	3 weeks	drinking water	Possible damage to kidneys at 0.80 mg/ml. Increases in serum GOT concentrations	Meyerovitch *et al.* [44]
Na_3VO_4	0.80 mg/ml	8 weeks	drinking water	Dehydration and death (prevented by addition of ascorbic acid and adjustment of pH to neutralization)	Paulson *et al.* [36]
Na_3VO_4	0.20 and 0.50 mg/ml	9 weeks	drinking water	Severe dehydration and deaths. Treatment did not impair kidney and liver function (urea, creatinine, GOT and GPT)	Brichard *et al.* [64]
Na_3VO_4	0.70 mg/ml	2 weeks	drinking water	Significant increase in the content of vanadium in liver	Gil *et al.* [65]
$NaVO_3$	0.20 up to 0.80 mg/ml	4 weeks	drinking water	Deaths at 0.80 mg/ml when the concentrations of $NaVO_3$ were not increased gradually	Bendayan and Gingras [45]
Na_3VO_4	0.10 up to 0.50 mg/ml	5 weeks	drinking water	Decreased body weight gain. Significant increases in plasma and liver vanadium levels	Pugazenthi and Khandelwal [61]
$VOSO_4.5H_2O$	0.25, 0.50, 0.75 and 1.0 mg/ml	39 days	drinking water	Vanadium accumulation in bone, kidney and liver.	Mongold *et al.* [49]
$NaVO_3$, Na_3VO_4 $VOSO_4.5H_2O$	0.15, 0.23 and 0.31 mg/ml	4 weeks	drinking water	Some deaths in the $NaVO_3$ and Na_3VO_4 groups. Decreased weight gain. Increases in serum urea and creatinine. Vanadium accumulation in bone, kidney, spleen, liver, pancreas, heart and muscle.	Domingo *et al.* [37]
$NaVO_3$, Na_3VO_4 $VOSO_4.5H_2O$	0.20, 0.50 and 1.10 mg/ml	2 weeks	drinking water	Mortality in all vanadium-treated groups (40–50%). Decreased weight gain. Vanadium accumulation in bone, kidney, spleen, liver, pancreas, heart and muscle.	Domingo *et al.* [42]
$NaVO_3$	2.5, 5, 10 and 20 mg/kg/day	4 weeks	gavage	Significant number of deaths in all the vanadium-treated groups (50–62.5%). Vanadium accumulation in bone, kidney and spleen at all doses	Domingo *et al.* [59]
$VOSO_4.3H_2O$	25, 50 and 75 mg/kg/day	3 weeks	gavage	Mortality (50%) at 75 mg/kg. Vanadium mg/kg/day accumulation in bone, kidney, spleen, liver, pancreas, heart, muscle and brain at 25, 50 and 75 mg/kg	Domingo *et al.* [60]
$VOSO_4.3H_2O$	0.75 mg/ml	5 months	drinking water	Vanadium accumulation in bone, kidney, liver and muscle	Cam *et al.* [51]
Bis(maltolato)-oxovanadium (IV)	0.50 up to 0.75 mg/ml	25 weeks	drinking water	No signs of vanadium toxicity, apart from hypoglycemia in a few animals	Yuen *et al.* [53]

animals [34, 36, 43–45, 61–67], the serious toxic side effects observed in the vanadium treated animals indicate that prolonged oral vanadium administration would argue against its possible use in human diabetes treatment [37, 42, 60]. In addition, the tissue accumulation of vanadium in diabetic animals underscores the possibility of vanadium toxicity derived from prolonged administration of this element [37, 42, 60, 68]. Consequently, it was thought to be of value if agents could be identified which when coadministered with vanadium compounds might eliminate or significantly reduce the vanadate or vanadyl toxic side effects (including tissue vanadium accumulation), while still leaving intact most of their insulin-like properties.

In recent years, the efficacy of a number of chelating agents on urinary excretion of vanadium and tissue distribution of this element following parenteral vanadate or vanadyl administration has been assessed in rats and mice [56, 69, 70]. Of the compounds examined, sodium 4,5-dihydroxybenzene-1,3-disulfonate (Tiron), desferrioxamine (DFOA) and 2-mercaptosuccinic acid were the most effective agents in increasing the urinary excretion of vanadium [56, 69, 70]. Those chelators were also effective in reducing the concentration of vanadium found in kidneys and bone, the main target tissues of vanadium accumulation.

However, until recently data about chelators to be used as potential mobilizing agents following prolonged oral vanadium exposure were not available. In view of the interest in the use of vanadium compounds as possible antidiabetic agents, in a recent investigation we evaluated the effectiveness of parenteral administration of Tiron, ascorbic acid, DFOA, and 2-mercaptosuccinic acid on the distribution and excretion of vanadium in rats. Chelation therapy followed six weeks of administration to sodium metavanadate or vanadyl sulfate in drinking water. Tiron was again the most effective agent for mobilizing vanadium in rats after oral administration of both vanadium compounds [71].

Coadministration of vanadium and Tiron in diabetes treatment

It seems quite reasonable that although vanadium compounds are effective in diminishing the expression of diabetes in the rat, because of the severe toxic side effects noted in the vanadium-treated animals, chronic oral administration of vanadium would not be an adequate treatment for human diabetes. However, before rejecting the hypothetic use of vanadate or vanadyl in the treatment of diabetes, we recently evaluated whether the toxic side effects of the vanadium could be eliminated while maintaining its insulin mimetic properties [68, 72].

It was found that the coadministration of i.p. Tiron and oral metavanadate to STZ-treated rats might be of value for treatment of diabetes. Intraperitoneal administration of Tiron (300 and 600 mg/kg) was initiated after 3 weeks of oral vanadate (0.20 mg/ml) administration and continued for 2 weeks. The significative reduction in plasma glucose levels caused by vanadate was not diminished by the administration of Tiron, but the accumulation of vanadium in kidney, bone, liver, heart, and muscle was significantly decreased in the Tiron-treated groups. Moreover, the diabetes associated increases in serum GOT, GPT, and cholesterol were diminished with Tiron treatment [68].

The previous results were recently corroborated through an additional investigation [72]. Streptozotocin-induced diabetic rats were given 0.20 mg/ml of sodium metavanadate in drinking water and 0, 125.6, 314 and 628 mg Tiron/kg/day by gavage for two weeks. Although most of the animals did not become normoglycemic, hyperglycemia was significantly alleviated by the treatment. Tiron administration did not diminish the beneficial effects of vanadate with respect to diabetes, but significantly decreased the levels of vanadium in target tissues [72]. These results showed that the insulin-like effects of vanadium are maintained in STZ-induced diabetic rats given Tiron, whereas the administration of the chelator resulted in a significant decrease in tissue vanadium accumulation reducing the risk of potential toxicity.

Conclusions

According to the above results, the coadministration of vanadate and Tiron might be of potential value for treatment of diabetes mellitus. Further studies should elucidate the possible toxic effects of prolonged administration of Tiron, the optimal pharmacological doses of Tiron and vanadium (as vanadate or vanadyl), as well as the frequency of Tiron administration. However, because recent investigations have shown that the antihyperglycemic action of vanadate would be attributable entirely to its suppression of feeding [73], or could occur at least in part, as an avoidance by the diabetic rat of food and fluid intake [74], before to pursue with the studies on the coadministration of vanadium and chelating agents, we would like to suggest that the antidiabetic actions of vanadium compounds are critically reevaluated. It should also include an assessment on the most suitable model of experimental diabetes: STZ-treated rats? [75].

Acknowledgements

This work was supported by grants SAL89-0853 and SAL91-0023 from the Interministerial Commission of Science and Technology (CICYT), Spain.

References

1. Shechter Y: Insulin mimetic effects of vanadate. Possible implications for future treatment of diabetes. Diabetes 39: 1–5, 1990
2. Brichard SM, Lederer J, Henquin JC: The insulin-like properties of vanadium: a curiosity or a perspective for the treatment of diabetes?. Diabet Metab 17: 435–440, 1991
3. Shechter Y, Shisheva A: Vanadium salts and the future treatment of diabetes. Endeavour 17: 27–31, 1993
4. Fantus IG, Deragon G, Posner BI: Insulin-mimetic effects of vanadate and pervanadate: potential therapeutic use in diabetes mellitus. Med Sci 7: 255–258, 1991
5. Cros GH, Mongold JJ, Serrano JJ, Ramanadham S, McNeill JH: Effects of vanadyl derivatives on animal models of diabetes. Mol Cell Biochem 109: 163–166, 1992
6. Roschin AV, Ordzhonikidze EK, Shalganova IV: Vanadium: toxicity, metabolism, carrier state. J Hyg Epidemiol Microbiol Immunol 24: 377–383, 1980
7. Kawai T, Seiji K, Watanabe T, Nakatsuka H, Ikeda M: Urinary vanadium as a biological indicator of exposure to vanadium. Int Arch Occup Environ Health 61: 283–287, 1989
8. Faulkner-Hudson TG: Vanadium Toxicity and Biological Significance. Elsevier, Amsterdam, 1964
9. Llobet JM, Domingo JL: Acute toxicity of vanadium compounds in rats and mice. Toxicol Lett 23: 227–231, 1984
10. Domingo JL, Llobet JM, Tomas JM, Corbella J: Short-term toxicity studies of vanadium in rats. J Appl Toxicol 5: 418–421, 1985
11. Zaporowska H, Wasilewski W: Some selected peripheral blood and haemopoietic system indices in Wistar rats with chronic vanadium intoxication. Comp Biochem Physiol 93C: 175–180, 1989
12. Zaporowska H, Wasilewski W: Hematological effects of vanadium on living organisms. Comp Biochem Physiol 102C: 223–231, 1992
13. Zaporowska H, Wasilewski W, Slotwinska M: Effect of chronic vanadium administration in drinking water to rats. Biometals 6: 3–10, 1993
14. Al-Bayati MA, Giri SN, Raabe OG: Time and dose-response study of the effects of vanadate in rats: changes in blood cells, serum enzymes, protein, cholesterol, glucose, calcium, and inorganic phosphate. J Environ Pathol Toxicol Oncol 10: 206–213, 1990.
15. Llobet JM, Colomina MT, Sirvent JJ, Domingo JL, Corbella J: Reproductive toxicity evaluation of vanadium in male mice. Toxicology 80: 199–206, 1993
16. Domingo JL, Paternain JL, Llobet JM, Corbella J: Effects of vanadium on reproduction, gestation, parturition and lactation in rats upon oral administration. Life Sci 39: 819–824, 1986
17. Paternain JL, Domingo JL, Llobet JM, Corbella J: Embryotoxic and teratogenic effects of sodium metavanadate administered to rats during organogenesis. Rev Esp Fisiol 43: 223–228, 1987
18. Sanchez DJ, Ortega A, Domingo JL, Corbella J: Developmental toxicity evaluation of orthovanadate in the mouse. Biol Trace Elem Res 30: 219–226, 1991
19. Gomez M, Sanchez DJ, Domingo JL, Corbella J: Embryotoxic and teratogenic effects of intraperitoneally administered metavanadate in mice. J Toxicol Environ Health 37: 47–56, 1992
20. Domingo JL, Bosque MA, Luna M, Corbella J: Prevention by Tiron (sodium 4,5-dihydroxybenzene-1,3-disulfonate) of vanadate-induced developmental toxicity in mice. Teratology 48: 133–138, 1993
21. Paternain JL, Domingo JL, Gomez M, Ortega A, Corbella J: Developmental toxicity of vanadium in mice after oral administration. J Appl Toxicol 10: 181–186, 1990
22. Parker RDR, Sharma RP: Accumulation and depletion of vanadium in selected tissues of rats treated with vanadyl sulfate and sodium orthovanadate. J Environ Pathol Toxicol 2: 235–245, 1978
23. Al-Bayati MA, Giri SN, Raabe OG, Rosenblatt LS, Shifrine M: Time and dose-response study of the effects of vanadate on rats: morphological and biochemical changes in organs. J Environ Pathol Toxicol Oncol 9: 435–455, 1989
24. Al-Bayati MA, Culberston MR, Schreider JP, Rosenblatt LS, Raabe OG: The lymphotoxic action of vanadate. J Environ Pathol Toxicol Oncol 11: 19–27, 1992
25. Kowalska M: The effect of vanadium on lung collagen content and composition in two succesive generations of rats. Toxicol Lett 41: 203–208, 1988
26. Kowalska M: Changes in rat lung collagen after life-time treatment with vanadium. Toxicol Lett 47: 185–190, 1989
27. Cohen MD, Wei CI, Tan H, Kao KJ: Effect of ammonium metavanadate on the murine immune response. J Toxicol Environ Health 19: 279–298, 1986
28. Donaldson J, LaBella F: Pro-oxidant properties of vanadate in vitro on catecholamines and on lipid peroxidation by mouse and rat tissues. J Toxicol Environ Health 12: 119–126, 1983
29. Bruech M, Quintanilla ME, Legrum W, Koch J, Netter KJ, Furhmann GF: Effects of vanadate on intracellular reduction equivalents in mouse liver and the fate of vanadium in plasma, erythrocites and liver. Toxicology 31: 283–295, 1984
30. Taqui Khan MM, Martell AE: Kinetics of metal ion and metal chelate catalyzed oxidation of ascorbic acid. III. Vanadyl ion catalyzed oxidation. J Am Chem Soc 90: 6011–6017, 1968
31. Sasi MM, Haider SS, El-Fakhri M, Ghwarsha KM: Microchromatographic analysis of lipids, protein, and occurrence of lipid peroxidation in various brain areas of vanadium exposed rats: A possible mechanism of vanadium neurotoxicity. Neurotoxicology 14: 57–64, 1993
32. Younes M, Strubelt O: Vanadate-induced toxicity towards isolated perfused rat livers: the role of lipid peroxidation. Toxicology 66: 63–74, 1991
33. Younes M, Kayser E, Strubelt O: Effect of antioxidants on vanadate-induced toxicity towards isolated perfused rat livers. Toxicology 70: 141–149, 1991
34. Heyliger CE, Tahiliani AG, McNeill JH: Effect of vanadate on elevated blood glucose and depressed cardiac performance of diabetic rats. Science 277: 1474–1477, 1985
35. Marfaing-Jallat P, Penicaud L: Effect of vanadium on regional brain glucose utilization in rats. Physiol Behav 54: 407–409, 1993
36. Paulson DJ, Kopp SJ, Tow JP, Peace DG: Effects of vanadate on in vivo myocardial reactivity to norepinephrine in diabetic rats. J Pharmacol Exp Ther 240: 529–534, 1987
37. Domingo JL, Gomez M, Llobet JM, Corbella J, Keen CL: Oral vanadium administration to streptozotocin-diabetic rats has marked negative side-effects which are independent of the form of vanadium used. Toxicology 66: 279–287, 1991
38. Jandhyala BS, Hom GJ: Physiological and pharmacological properties of vanadium. Life Sci 33: 1325–1340, 1983
39. Cantley LC, Josephson L, Warner M, Yanagisawa C, Lechene C, Guidotti G: Vanadate is a potent (NaK)ATPase inhibitor found in ATP derived from muscle. J Biol Chem 252: 7421–7423, 1977
40. Nechay BR: Mechanisms of action of vanadium. Annu Rev Pharmacol Toxicol 24: 501–524, 1984

41. Nechay BR, Saunders JP: Inhibition by vanadium of sodium and potassium dependent adenosinetriphosphatase derived from animal and human tissues. J Environ Pathol Toxicol 2: 247–262, 1978
42. Domingo JL, Gomez M, Llobet JM, Corbella J, Keen CL: Improvement of glucose homeostasis by oral vanadyl or vanadate treatment in diabetic rats is accompanied by negative side effects. Pharmacol Toxicol 68: 249–253, 1991
43. Pederson RA, Ramanadham S, Buchan AMJ, McNeill JH: Long-term effects of vanadyl treatment on streptozotocin-induced diabetes in rats. Diabetes 38: 1390–1395, 1989
44. Meyerovitch J, Farfel Z, Sack J, Shechter Y: Oral administration of vanadate normalizes blood glucose levels in streptozotocin-treated rats. J Biol Chem 262: 6658–6662, 1987
45. Bendayan M, Gingras D: Effect of vanadate administration on blood glucose and insulin levels as well as on the exocrine pancreatic function in streptozotocin-diabetic rats. Diabetologia 32: 561–567, 1989
46. Sekar N, Kanthasamy A, William S, Subramanian S, Govindasamy S: Insulinic actions of vanadate in diabetic rats. Pharmacol Res 22: 207–217, 1990
47. Becker DJ, Ongemba LN, Henquin JC: Comparison of the effects of various vanadium salts on glucose homeostasis in streptozotocin-diabetic rats. Eur J Pharmacol 260: 169–175, 1994
48. Dai S, Thompson KH, McNeill JH: One-year treatment of streptozotocin-induced diabetic rats with vanadyl sulphate. Pharmacol Toxicol 74: 101–109, 1994
49. Mongold JJ, Cros GH, Vian L, Tep A, Ramanadham S, Siou G, Diaz J, McNeill JH, Serrano JJ: Toxicological aspects of vanadyl sulphate on diabetic rats: effect of vanadium levels and pancreatic β-cell morphology. Pharmacol Toxicol 67: 192–198, 1990
50. Okpere BE, Williams G, Bloom SR: Failure of oral metavanadate to correct streptozotocin-induced diabetes in rats (Abstract). Diabetic Med 5 (S2): 8, 1988
51. Cam NC, Pederson RA, Brownsey RW, McNeill JH: Long-term effectiveness of oral vanadyl sulphate in streptozotocin-diabetic rats. Diabetologia 36: 218–224, 1993
52. McNeill JH, Yuen VG, Hoveyda HR, Orvig C: Bis(maltolato)-oxovanadium(IV) is a potent insulin mimic. J Med Chem 35: 1489–1491, 1992
53. Yuen VG, Orvig C, McNeill JH: Glucose-lowering effects of a new organic vanadium complex, bis(maltolato)oxovanadium(IV). Can J Physiol Pharmacol 71: 263–269, 1993
54. Oster MH, Llobet JM, Domingo JL, German JB, Keen CL: Vanadium treatment of diabetic Sprague-Dawley rats results in tissue vanadium accumulation and pro-oxidant effects. Toxicology 83: 115–130, 1993
55. Elfant M, Keen CL: Sodium vanadate toxicity in adult and developing rats. Role of peroxidative damage. Biol Trace Elem Res 14: 193–208, 1987
56. Domingo JL, Llobet JM, Tomas JM, Corbella J: Influence of chelating agents on the toxicity, distribution and excretion of vanadium in mice. J Appl Toxicol 6: 337–341, 1986
57. Zaporowska H: Effect of vanadium on L-ascorbic acid concentration in rat tissues. Gen Pharmac 25: 467–470, 1994
58. Thompson KH, McNeill JH: Effect of vanadyl sulfate feeding on susceptibility to peroxidative change in diabetic rats. Res Commun Chem Pathol Pharmacol 80: 187–200, 1993
59. Domingo JL, Ortega A, Llobet JM, Keen CL: No improvement of glucose homeostasis in diabetic rats by vanadate treatment when given by gavage. Trace Elem Med 8: 181–186, 1991
60. Domingo JL, Sanchez DJ, Gomez M, Llobet JM, Corbella J: Administration of vanadyl sulfate by gavage does not normalize blood glucose levels in streptozotocin-induced diabetic rats. Res Commun Chem Pathol Pharmacol 75: 369–372, 1992
61. Pugazhenthi S, Khandelwal RL: Insulinlike effects of vanadate on hepatic glycogen metabolism in nondiabetic and streptozotocin-induced diabetic rats. Diabetes 39: 821–827, 1990
62. Ramanadham S, Mongold JJ, Brownsey RW, Cros GH, McNeill JH: Oral vanadyl sulfate in treatment of diabetes mellitus in rats. Am J Physiol 257: H904–H911, 1989
63. Ramanadham S, Brownsey RW, Cros GH, Mongold JJ, McNeill JH: Sustained prevention of myocardial and metabolic abnormalities in diabetic rats following withdrawal from oral vanadyl treatment. Metabolism 38: 1022–1028, 1989
64. Brichard SM, Okitolonda W, Henquin JC: Long term improvement of glucose homeostasis by vanadate treatment in diabetic rats. Endocrinology 123: 2048–2053, 1988
65. Gil J, Miralpeix M, Carreras J, Bartrons R: Insulin-like effects of vanadate on glucokinase activity and fructose 2,6-bisphosphate levels in the liver of diabetic rats. J Biol Chem 263: 1868–1871, 1988
66. Blondel O, Bailbe D, Portha B: In vivo insulin resistance in streptozotocin-diabetic rats – evidence for reversal following oral vanadate treatment. Diabetologia 32: 185–190, 1989
67. Thompson KH, Leichter J, McNeill JH: Studies of vanadyl sulfate as a glucose-lowering agent in STZ-diabetic rats. Biochem Biophys Res Commun 197: 1549–1554, 1993
68. Domingo JL, Gomez M, Sanchez DJ, Llobet JM, Keen CL: Tiron administration minimizes the toxicity of vanadate but not its insulin mimetic properties in diabetic rats. Life Sci 50: 1311–1317, 1992
69. Domingo JL, Gomez M, Llobet JM, Corbella J: Chelating agents in the treatment of acute vanadyl sulphate intoxication in mice. Toxicology 62: 203–211, 1990
70. Gomez M, Domingo JL, Llobet JM, Corbella J: Evaluation of the efficacy of various chelating agents on urinary excretion and tissue distribution of vanadium in rats. Toxicol Lett 57: 227–234, 1991
71. Gomez M, Domingo JL, Llobet JM, Corbella J: Effectiveness of some chelating agents on distribution and excretion of vanadium in rats after prolonged oral administration. J Appl Toxicol 11: 195–198, 1991
72. Domingo JL, Sanchez DJ, Gomez M, Llobet JM, Corbella J: Oral vanadate and Tiron in treatment of diabetes mellitus in rats: improvement of glucose homeostasis and negative side-effects. Vet Hum Toxicol 35: 495–500, 1993
73. Malabu UH, Dryden S, McCarthy HD, Kilpatrick A, Williams G: Effects of chronic vanadate administration in the STZ-induced diabetic rat. The antihyperglycemic action of vanadate is attributable entirely to its suppression of feeding. Diabetes 43: 9–15, 1994
74. Domingo JL, Gomez M, Sanchez DJ, Llobet JM, Corbella J, Keen CL: Normalization of hyperglycemia by vanadate or vanadyl treatment in diabetic rats: pharmacological and toxicological aspects. Trace Elem Electr 11: 16–22, 1994
75. Sanchez DJ, Gomez M, Llobet JM, Domingo JL, Corbella J: Is the streptozotocin-treated rat a suitable model to assess the therapeutic potential of vanadium in diabetes treatment?. Can J Physiol Pharmacol 72(S3): 20, 1994

Molecular and Cellular Biochemistry **153**: 241–244, 1995.

Index to Volume 153

MIX
Papier aus verantwortungsvollen Quellen
Paper from responsible sources
FSC® C105338

If you have any concerns about our products,
you can contact us on
ProductSafety@springernature.com

In case Publisher is established outside the EU,
the EU authorized representative is:
Springer Nature Customer Service Center GmbH
Europaplatz 3, 69115 Heidelberg, Germany

Printed by Libri Plureos GmbH
in Hamburg, Germany